Annual Review of Materials Research

ANNUAL REVIEW OF MATERIALS RESEARCH

VOLUME 31, 2001

SAMUEL I. STUPP, *Guest Editor*
Northwestern University

DAVID R. CLARKE, *Co-Editor*
University of California, Santa Barbara

MANFRED RUHLE, *Co-Editor*
Max-Planck Institute, Stuttgart

JOHN C. BRAVMAN, *Associate Editor*
Stanford University

STEPHEN R. FORREST, *Associate Editor*
Princeton University

www.AnnualReviews.org science@AnnualReviews.org 650-493-4400

ANNUAL REVIEWS
4139 El Camino Way • P.O. BOX 10139 • Palo Alto, California 94303-0139

ANNUAL REVIEWS
Palo Alto, California, USA

International Standard Serial Number: 1531-7331
International Standard Book Number: 0-8243-1731-9
Library of Congress Catalog Card Number: 72-172108

∞ The paper used in this publication meets the minimum requirements of American National Standards for Information Sciences—Permanence of Paper for Printed Library Materials, ANSI Z39.48-1992.

TYPESET BY TECHBOOKS, FAIRFAX, VA
PRINTED AND BOUND IN THE UNITED STATES OF AMERICA

PREFACE

This volume constitutes the 31st of the Annual Reviews series devoted to materials, one of the 29 series published by Annual Reviews, a not-for-profit scientific publisher dedicated to promoting the advancement of the sciences. With this volume, we have changed the title to *Annual Review of Materials Research* from the previous title *Annual Review of Materials Science*. The aims and objectives of the series remain the same, however, namely to provide authoritative reviews of major advances in the field of materials. When the series began, it was relatively easy to define what constituted materials science. At the same time, circa 1970, materials science was being promoted as an independent academic discipline, and departments of metallurgy and ceramics were beginning to change their names to reflect the broader, interdisciplinary approach of a science of materials. So successful has interdisciplinary research in materials become and so rich a subject area that scientists from numerous academic and intellectual disciplines, ranging from physics, chemistry, engineering, and biology to marine science, now apply their skills and expertise to materials research. Thus to better reflect the broad appeal that materials research has for so many diverse groups of scientists and not simply those who identify themselves with the academic discipline of materials science, we have changed the name of our series to *Annual Review of Materials Research*. A small change, but one designed to mirror the extraordinary breadth and intellectual excitement in materials today.

This volume is the ninth of the series allocating a set of chapters to a particular topical area. The current theme—biomaterials—addresses the rapidly growing area of the design of materials for specific roles in biological systems. This much broader and more challenging area of materials contrasts with the previous, more traditional role of biomaterials as substitute structural materials (implants) in the body for repair or replacement purposes. It is clear that we are just near the beginning of a tremendously exciting era in the science and engineering of biomaterials and that in years to come we will have numerous occasions to pause and review the progress in this area of materials.

We thank the authors of all the chapters appearing in this volume for contributing their expertise and perspectives. We thank Dr. Samuel Stupp, who selected and invited authors to write chapters covering this year's Keynote Topic, Biomaterials, in his role as guest editor. The appearance of this volume gives us the opportunity to acknowledge the long, dedicated service to our series by Elton N. Kaufmann. Elton was editor of this series from 1994 until last year, having previously served as a member of the Editorial committee since 1990. In these capacities, he was instrumental in ensuring that the series continued to cover the broadest possible range of materials while also reflecting the newest, most exciting developments of materials and its enabling role in new technologies.

In addition to our authors' efforts, a volume such as this requires a great deal of support in the production process, becoming abundantly clear to an editor as a volume progresses from suggestions of possible authors to a final published volume. We therefore, as always, express our sincere thanks to series production editor Sandra Cooperman for her invaluable role in assembling and producing this volume.

DAVID R. CLARKE
For the Editors and the Editorial Committee

Annual Review of Materials Research
Volume 31, 2001

CONTENTS

ERRATA

An online log of corrections to *Annual Review of Materials Science* chapters (if any have yet been occasioned, 1997 to the present) may be found at http://MatSci.AnnualReviews.org/errata.shtm

Related Articles

From the ***Annual Review of Biomedical Engineering***, Volume 3 (2001)

Biomechanics of Trabecular Bone, Tony M. Keaveny, Elise F. Morgan, Glen L. Niebur, and Oscar C. Yeh

Biomechanics of Cardiovascular Development, Lawrence A. Taber

Fundamentals of Impact Biomechanics: Part 2—Biomechanics of the Abdomen, Pelvis, and Lower Extremities, Albert I. King

Micro- and Nanomechanics of the Cochlear Outer Hair Cell, W. E. Brownell, A. A. Spector, Robert Raphael, and Aleksander S. Popel

Soft Lithography in Biology and Biochemistry, George Whitesides, Emanuele Ostuni, Shuichi Takayama, Xingyu Jiang, and Donald E. Ingber

Vascular Tissue Engineering, Robert M. Nerem and Dror Seliktar

From the ***Annual Review of Physical Chemistry***, Volume 52 (2001)

Surface Biology of DNA by Atomic Force Microscopy, Helen G. Hansma

Mechanisms and Kinetics of Self-Assembled Monolayer Formation, Daniel K. Schwartz

Spectroscopy and Hot Electron Relaxation Dynamics in Semiconductor Quantum Wells and Quantum Dots, Arthur J. Nozik

Structure and Bonding of Molecules at Aqueous Surfaces, G. L. Richmond

Light-Emitting Electrochemical Processes, Neal R. Armstrong, R. Mark Wightman, and Erin M. Gross

Reactions and Thermochemistry of Small Transition Metal Cluster Ions, Peter M. Armentrout

Polymer Adsorption Driven Self-Assembly of Nanostructures, Arup K. Chakraborty and Aaron J. Golumbfskie

ANNUAL REVIEWS is a nonprofit scientific publisher established to promote the advancement of the sciences. Beginning in 1932 with the *Annual Review of Biochemistry*, the Company has pursued as its principal function the publication of high-quality, reasonably priced *Annual Review* volumes. The volumes are organized by Editors and Editorial Committees who invite qualified authors to contribute critical articles reviewing significant developments within each major discipline. The Editor-in-Chief invites those interested in serving as future Editorial Committee members to communicate directly with him. Annual Reviews is administered by a Board of Directors, whose members serve without compensation.

Annual Reviews of

Anthropology
Astronomy and Astrophysics
Biochemistry
Biomedical Engineering
Biophysics and Biomolecular Structure
Cell and Developmental Biology
Earth and Planetary Sciences
Ecology and Systematics
Energy and the Environment
Entomology
Fluid Mechanics
Genetics
Genomics and Human Genetics
Immunology
Materials Research
Medicine
Microbiology
Neuroscience
Nuclear and Particle Science
Nutrition
Pharmacology and Toxicology
Physical Chemistry
Physiology
Phytopathology
Plant Physiology and Plant Molecular Biology
Political Science
Psychology
Public Health
Sociology

SPECIAL PUBLICATIONS

Excitement and Fascination of Science, Vols. 1, 2, 3, and 4

Biomaterials Introduction

As we enter the 21st century, the field of biomaterials is becoming one of the most intellectually exciting areas of materials science and engineering. Even though activity in this field it not new, there is currently fresh energy in this subject brought about by the remarkable advances in molecular and cell biology of the past two decades. Early activities in this field (40 or 50 years ago) dealt with the selection and characterization of well-established synthetic materials to fabricate implants for use in medicine and as restorative materials in dentistry. Materials well established in technology and not originally developed as biomaterials were being used for human repair. For example, bone implants used by orthopedic surgeons were being fabricated from low-corrosion steels and other alloys or from ultrahigh-density polyethylene. Methacrylate polymers were being tested and used as bone cements and dental restoratives, and polyethylene terephthalate fibers were being woven into fabrics that could function as artificial blood vessels. During the early period of this field, cellulose-based membranes were being introduced in kidney dialysis machines. Therefore, biomaterials research was mostly centered on in vitro testing of these materials under conditions that simulated the biological environment (e.g., corrosion, fatigue, creep, wear). Extensive histological work ensued following the implantation of these materials into experimental animals in order to understand the biological response elicited by materials in various tissues, with scant research done to explore the use of alternative materials for similar applications. Therefore, the field focused on the application of materials science principles to materials in biomedical use. With only very few exceptions, biomaterials were synthetic and intended to be permanent implants that substituted for structure and/or function in diseased or traumatized tissues. Biomaterials research was mostly in the hands of engineers, physical scientists, pathologists, and clinicians who brought their skills to these new applications of technological materials.

The main objective in biomaterials research has become the design of materials that can interact specifically with the biological environment for a given purpose. This includes the so-called tissue regeneration approach: Biomaterials are considered as temporary "anchoring scaffolds" to which specific types of cells can colonize and use to elaborate three-dimensional tissues. The objective is to help biology regenerate the missing or dysfunctional body parts by providing a biodegradable substrate. Another more recent objective is to use materials as vehicles to deliver large and small molecules to specific tissues in order to restore normal physiological function. A good example in this genomic era is the use of biomaterials to deliver genes to cells to reverse disease, at least temporarily, or to alleviate the symptoms of disease. This is accomplished by transfecting cells with the missing normal genes encapsulated or dispersed in various structures. These structures play a key role in the success of such procedures, and therefore the correct biomaterials design for this purpose is of critical importance. In increasingly sophisticated ways, specially designed biomaterials are also being explored as delivery agents of therapeutic drugs or as encapsulants of non-autologous cells. In this last example, cells are exploited for the delivery of the biomacromolecules they produce but are hidden from the immune system by the biomaterials used to encapsulate them. The strategy of using materials as delivery agents provides many opportunities related to the design of micromaterials and, eventually, nanomaterials, which are miniaturized constructs designed to encapsulate, target, and deliver cargo to a specific tissue site. This strategy could contribute immensely to the eradication of cancer morbidity or the cure of other diseases. The general field of very smart micro- and nano-biomaterials will continue to expand with contributions to molecular design from materials scientists, chemists, and bioengineers.

The field of biomaterials has also grown and evolved in its capacity to study the molecular biology and cell biology of the implant-tissue interface. This work aims at understanding the biological response elicited by synthetic materials used in human repair, and great advances have occurred over the past few years that are useful in implant design. The surgical introduction of synthetic materials in living tissues elicits the expected inflammatory response, but depending on material composition, chemical degradation, surface texture, and other factors, a continuum of other responses can be observed. These include immune and foreign body responses often culminating in the fibrous encapsulation of implants with avascular tissue, which in turn contributes to a high risk of infection and limited implant durability. Over the past decade, biological response to materials has taken advantage of advances in molecular biology in an attempt to decipher the complex proteomics operating at these interfaces.

In this volume, experts in the field of biomaterials have written about the new areas and directions discussed above. One area covered is the biological response to synthetic materials, including specific mechanisms involved in osteoblast response to implant surfaces. This topic is of great interest in bone repair and is one of the most useful applications of biomaterials in medicine. Another article discusses the new field of biomaterials designed for specific biological or delivery functions. This area will certainly continue to grow with increasingly sophisticated molecular biomaterials designed to achieve targeted functions. Specific systems such as phospholipid strategies for biomineralization, photopolymerizable biomaterials, materials for gene delivery, and self-assembling polymer membranes and colloids are described by other reviews. Ongoing efforts in the field of designed biomaterials not specifically discussed in this volume include chemistry-intensive efforts to develop self-assembling biomaterials and also the biosynthesis of materials using genetically modified organisms. From the field of biomineralization other strategies are being explored that could lead to the development of biomimetic biomaterials for tissue repair. All these efforts, combined with emerging capabilities in nanoscale materials science, are defining an exciting new field of biomaterials science and engineering that not only targets human body repair but also science and technology in general.

SAMUEL I. STUPP
GUEST EDITOR

Annu. Rev. Mater. Res. 2001. 31:1–23

SYNTHESIS AND DESIGN OF SUPERHARD MATERIALS

J Haines, JM Léger, and G Bocquillon
Laboratoire de Physico-Chimie des Matériaux, Centre National de la Recherche Scientifique, 1 place Aristide Briand, 92190 Meudon, France;
e-mail: haines@cnrs-bellevue.fr; leger@cnrs-bellevue.fr

Key Words diamond, cubic boron nitride, carbon nitride, high pressure, stishovite

■ **Abstract** The synthesis of the two currently used superhard materials, diamond and cubic boron nitride, is briefly described with indications of the factors influencing the quality of the crystals obtained. The physics of hardness is discussed and the importance of covalent bonding and fixed atomic positions in the crystal structure, which determine high hardness values, is outlined. The materials investigated to date are described and new potentially superhard materials are presented. No material that is thermodynamically stable under ambient conditions and composed of light (small) atoms will have a hardness greater than that of diamond. Materials with hardness values similar to that of cubic boron nitride (cBN) can be obtained. However, increasing the capabilities of the high-pressure devices could lead to the production of better quality cBN compacts without binders.

INTRODUCTION

Diamond has always fascinated humans. It is the hardest substance known, based on its ability to scratch any other material. Its optical properties, with the highest refraction index known, have made it the most prized stone in jewelry. Furthermore, diamond exhibits high thermal conductivity, which is nearly five times that of the best metallic thermal conductors (copper or silver) at room temperature and, at the same time, is an excellent electrical insulator, even at high temperature. In industry, the hardness of diamond makes it an irreplaceable material for grinding tools, and diamond is used on a large scale for drilling rocks for oil wells, cutting concrete, polishing stones, machining, and honing. The diamonds used for industry are now mostly man-made because their cutting edges are much sharper than those of natural diamonds, which have been eroded by geological time. The synthesis of diamond has been a goal of science from Moissant at the end of the nineteenth century to the successful synthesis under high pressures in 1955 (1).

However, diamond has a major drawback in that it reacts with iron and cannot be used for machining steel. This has prompted the synthesis of a second superhard material, cubic boron nitride (cBN), whose structure is derived from that of diamond with half the carbon atoms being replaced by boron and the other half by nitrogen atoms. The resulting compound is half as hard as diamond, but it does not react with iron and can be used for machining steel. Cubic boron nitride does not exist in nature and is prepared under high-pressure high-temperature conditions, as is synthetic diamond. However, its synthesis is more difficult, and it has not been possible to prepare large crystals. Industry is thus looking for new superhard materials that will need to be much harder than present ceramics (Si_3N_4, Al_2O_3, TiC).

Hardness is a quality less well defined than many other physical properties. Hardness was first defined as the ability of one material to scratch another; this corresponds to the Mohs scale. This scale is highly nonlinear (talc = 1, diamond = 10); however, this definition of hardness is not reliable because materials of similar hardness can scratch each other and the resulting value depends on the specific details of the contact between the two materials. It is well known (2) that at room temperature copper can scratch magnesium oxide and at high temperatures cBN can scratch diamond (principle of soft indenter). Another, more accurate, way of defining and measuring hardness is by the indentation of the material by a hard indenter. According to the nature and shape of the indenter, different scales are used: Brinell, Rockwell, Vickers, and Knoop. The last two are the most frequently used. The indenter is made of a pyramidal-shaped diamond with a square base (Vickers), or elongated lozenge (Knoop). The hardness is deduced from the size of the indentation produced using a defined load; the unit is the same as that for pressure, the gigapascal (GPa). Superhard materials are defined as having a hardness of above 40 GPa. The hardness of diamond is 90 GPa; the second hardest material is cBN, with a hardness of 50 GPa.

The design of new materials with a hardness comparable to diamond is a great challenge to scientists. We first describe the current status of the two known superhard materials, diamond and cBN. We then describe the search for new bulk superhard materials, discuss the possibility of making materials harder than diamond, and comment on the new potentially superhard materials and their preparation.

DIAMOND AND CUBIC BORON NITRIDE

Diamond

The synthesis of diamond is performed under high pressure (5.5–6 GPa) and high temperature (1500–1900 K). Carbon, usually in the form of graphite, and a transition metal, e.g. iron, cobalt, nickel, or alloys of these metals [called solvent-catalyst (SC)], are treated under high-pressure high-temperature conditions; upon heating, graphite dissolves in the metal and if the pressure and temperature conditions are in the thermodynamic stability field of diamond, carbon can crystallize as diamond because the solubility of diamond in the molten metal is less than that of graphite. Some details about the synthesis and qualities of diamond obtained by this spontaneous nucleation method are given below, but we do not describe the growth of single-crystal diamond under high pressure, which is necessary in order to obtain large single crystals with dimensions greater than 1 mm. Crystals of this size are expensive and represent only a very minor proportion of the diamonds used for machining; they are principally used for their thermal properties.

It is well known that making diamonds is relatively straightforward, but controlling the quality of the diamonds produced is much more difficult. Improvements in the method of synthesis since 1955 have greatly extended the size range and the mechanical properties and purity of the synthetic diamond crystals. Depending on the exact pressure (P) and temperature (T) of synthesis, the form and the nature of the carbon, the metal solvent used, the time (t) of synthesis, and the pathways in P-T-t space, diamond crystals (3, 4) varying greatly in shape (5), size, and friability are produced. These three characteristics are used to classify diamonds; the required properties differ depending on the industrial application. Friability is related to impact strength. It is the most important mechanical property for the practical use of superhard materials, and low friability is required in order for tools to have a long lifetime.

In commercial literature, the various types of diamonds are classed as a function of their uses, which depend mainly on their friabilities, but the numerical values are not given, so it is difficult to compare the qualities of diamonds from various sources. The friability, which is defined by the percentage of diamonds destroyed in a specific grinding process, is obtained by subjecting a defined quantity of diamonds to repeated impacts by grinding in a ball-mill or by the action of a load falling on them. The friability values depend strongly on the experimental conditions used, and only values for crystals measured under the same conditions can be compared. The effect of various synthesis parameters on their quality can be evaluated by considering the total mass of diamond obtained in one experiment, the distribution size of these diamonds, and the friability of the diamonds of a defined size.

A first parameter is the source of the carbon. Most carbon-based substance can be used to make diamonds (6), but the nature of the carbon source has an effect on the quantity and the quality of synthetic diamonds. The best carbon source for diamond synthesis is graphite, and its characteristics are important. The effect of the density, gas permeability, and purity of graphite on the diamond yield have been investigated using cobalt as the SC (7). Variations of the density and gas permeability have no effect on the diamond yield, but carbon purity is important. The main impurity in synthetic graphite is CaO. If good quality diamonds are required, the calcium content should be kept below 1000 ppm in order to avoid excessive nucleation on the calcium oxide particles.

A second factor that alters the quality of diamonds is the nature of the SC. The friability and the size distribution are better with CoFe (alloy of cobalt with a small quantity of iron) than with invar, an iron-nickel alloy (Table 1: Ia, Ib; Figure 1*a*). Another parameter is how the mixture of carbon and SC is prepared. When fine or coarse powders of intimately mixed graphite and SC are used, a high yield of diamonds with high friabilities is obtained (8). These diamonds are very small,

TABLE 1 Friabilities of some diamonds as a function of the details of the synthesis process for a selected size of 200–250 μm

	SDA[a] 100+	MBS[a] 70	Ia	Ib	IIa	IIb	IIc	IIIa	IIIb
Synthesis detail			CoFe	Invar	1C-2SC stacking	1C-1SC stacking	2C-2SC stacking	Cycle A	Cycle B
Friability (%)	7	41	21	39	50	53	63	72	50

[a]The SDA 100+ is among De Beers best diamond with a very high strength that is recommended for sawing the hardest stones and cast refractories. MBS 70 is in the middle of General Electric's range of diamonds for sawing and drilling. Other diamonds were obtained in the laboratory using a belt–type apparatus with a working chamber of 40 mm diameter. (C, graphite; SC, solvent-catalyst.)

with metal inclusions, and they are linked together with numerous cavities filled with SC. A favorable geometry in order to obtain well-formed diamonds is to stack disks of graphite and SC. The effect of local concentration has been examined by changing the stacking of these disks (Table 1: IIa, IIb, IIc; Figure 1*b*). The method of stacking modifies the local oversaturation of dissolved carbon and thus the local spontaneous diamond germination. For the synthesis of diamond, the heating current goes directly through the graphite-SC mixture. Because the electrical resistivity of the graphite is much greater than that of the SC, the temperature of the graphite is raised by the Joule effect, whereas that of the SC increases mainly because of thermal conduction. Upon increasing the thickness of the SC disk, the local thermal gradient increases and the dissolved atoms of carbon cannot move as easily; the local carbon oversaturation then enhances the spontaneous diamond germination. This enables one to work at lower temperatures and pressures, which results in slower growth and therefore better quality diamonds.

Another important factor for the yield and the quality of the diamonds is the pathway followed in *P-T-t* space. The results of two cycles with the same final pressure and temperature are shown. In cycle A (Figure 1*d*), the graphite-SC mixture reaches the eutectic melting temperature while it is still far from the equilibrium line between diamond and graphite; as a result spontaneous nucleation is very high and the seeds grow very quickly. These two effects explain the high yield and the poor quality and small size and high friability of the diamonds compared with those obtained in cycle B (Figure 1*d*; see Table 1 IIIa and IIIb and Figure 1*c*). Large crystals (over 400 μm) of good quality are obtained when the degree of spontaneous nucleation is limited. The pathway in *P-T-t* space must then remain near the graphite-diamond phase boundary (Figure 1*d*), and the time of the treatment must be extended in the final *P-T-t* conditions.

Usually, friability increases with the size of the diamonds. Nucleation takes place at the beginning of the synthesis when the carbon oversaturation is important, and the carbon in solution is then absorbed by the existing nuclei, which grow larger.

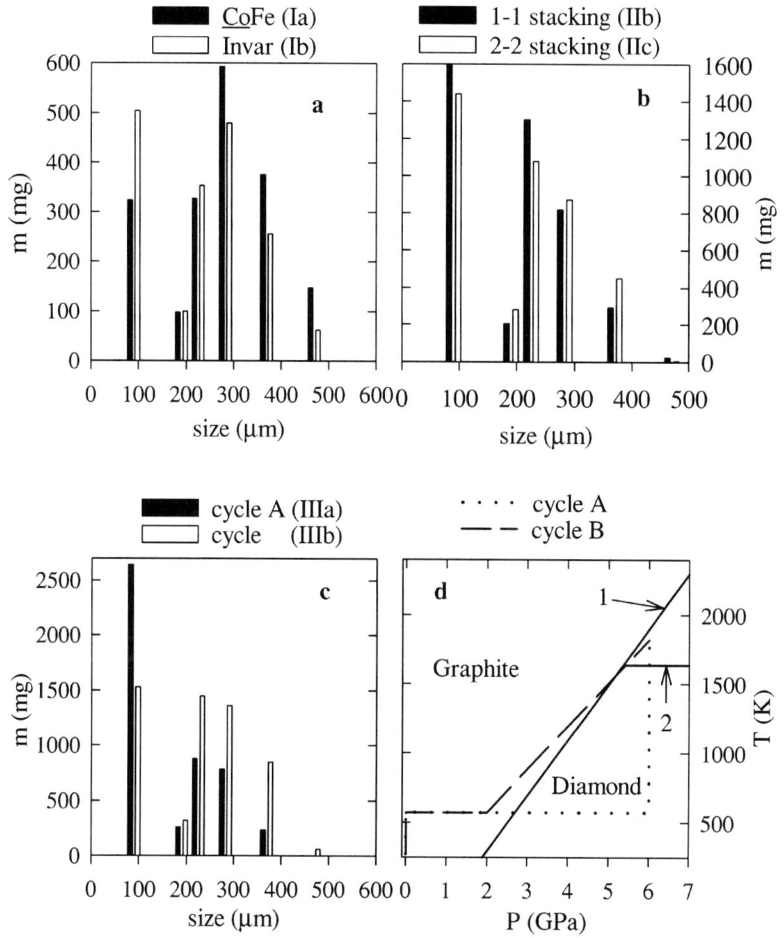

Figure 1 Size distribution of diamonds in one laboratory run for different synthesis processes; effects of (panel *a*) the nature of the metal, (panel *b*) the stacking of graphite and metal disks, (panel *c*) the P-T pathway. (Panel *d*) P-T pathways for synthesis. 1: graphite-diamond boundary and 2: melting temperature of the carbon-eutectic. The diamond synthesis occurs between the boundaries 1 and 2.

The growth time is about the same for all the crystals, thus those that can grow more quickly owing to a greater local thermal gradient become the largest. Owing to their rapid growth rate, they trap more impurities and have more defects, and therefore their friability is higher. Similarly, friability increases with the diamond yield. The diamonds produced by the spontaneous nucleation method range in size up to 800–1000 μm. The best conditions for diamond synthesis correspond to a compromise between the quantity and the quality of the diamonds.

Cubic Boron Nitride

Cubic boron nitride (cBN) is the second hardest material. The synthesis of cBN is performed in the same pressure range as that for diamond, but at higher temperatures, i.e. above 1950 K. The general process is the same; dissolution of hexagonal boron nitride (hBN) in a solvent-catalyst (SC), followed by spontaneous nucleation of cBN. However, the synthesis is much more complicated. The usual SCs are alkali or alkaline-earth metals and their nitrides (9): Mg, Ca, Li_3N, Mg_3N_2, and Ca_3N_2. All these SCs are hygroscopic, and water or oxygen are poisons for the synthesis. Thus great care must be taken, which requires dehydration of the materials and preparation in glove boxes, to avoid the presence of water in the high-pressure cell. Furthermore, the above compounds react first with hBN to form intermediate compounds, Li_3BN_2, $Mg_3B_2N_4$, or $Ca_3B_2N_4$, which become the true SC. These compounds and the hBN source are electrical insulators, thus an internal furnace must be used, which makes fabrication of the high pressure cell more complicated and reduces the available volume for the samples. In addition, the chemical reaction involved is complicated by this intermediate step, and in general the yield of cBN is lower than for diamond. Work is in progress to determine in situ which intermediate compounds are involved in the synthesis process. The crystals of cBN obtained from these processes are of lower quality (Figure 2) and size than for diamond. Depending on the exact conditions, orange-yellow or dark crystals are obtained; the color difference comes from a defect or an excess of boron (less than 1%); the dark crystals, which have an excess of born, are harder.

As in the synthesis of diamond, the initial forms of the SC source, hBN, play important roles, but the number of parameters is larger. For the source of BN, it is better to use pressed pellets of hBN powder rather than sintered hBN products, as the latter contain additives (oxides); a very fine powder yields a better reactivity. Doping of Li, Ca, or Mg nitrides with Al, B, Ti, or Si induces a change in the morphology and color of cBN crystals, which are dark instead of orange, are larger (500 μm), have better shapes and, in addition, gives a higher yield (10). Use of supercritical fluids enables cBN to be synthesized at lower pressures and temperatures (2 GPa, 800 K), but the resulting crystal size is small (11).

Diamond and cBN crystals are produced on a large scale, and the main problem is how to use them for making viable tools for industry. Different compacts of these materials are made (12) for various uses. Compacts of diamonds are made using cobalt as the SC. The mixture is treated under high-pressure high-temperature conditions, at which superficial graphitization of the diamonds takes place, and then under the *P-T-t* diamond synthesis conditions so as to transform the graphite into diamond and induce intergranular growth of diamonds. The diamond compacts produced in this way still contain some cobalt as a binder, but their hardness is close to that of single-crystal diamond. Compacts of cBN cannot be made in the same way because the SCs are compounds that decompose in air. Sintering without binders (13) is possible at higher pressures of about 7.5–8 GPa and temperatures higher than 2200 K, but these conditions are currently outside the range of those

Figure 2 SEM photographs of diamond (*top*) and cBN (*bottom*) crystals of different qualities depending on the synthesis conditions (the long vertical bar corresponds to a distance of 100 μm). *Top left*: good quality mid-sized diamonds of cubo-octahedral shape with well-defined faces and sharp edges; *top right*: lower quality diamonds; *bottom left*: orange cBN crystals; *bottom right*: very large black cBN crystals of better shapes.

used in industrial production. Compacts of cBN with TiC or TaN binders are of markedly lower hardness because there is no direct bonding between the superhard crystals, in contrast to diamond compacts. In addition, they are expensive, and this has motivated the search for other superhard materials.

SEARCH FOR NEW SUPERHARD MATERIALS

One approach for increasing hardness of known materials is to manipulate the nanostructure. For instance, the effect of particle size on the hardness of materials has been investigated. It is well known that high-purity metals have very low shear strengths; this arises from the low energy required for nucleation and motion of dislocations in metals. The introduction of barriers by the addition of impurities or grain size effects may thus enhance the hardness of the starting phase. In this case, intragranular and intergranular mechanisms are activated and compete with each other. As each mechanism has a different dependency on grain size, there can be a maximum in hardness as the function of the grain size. This effect of increasing the hardness with respect to the single-crystal value does not exist in the case of ceramic materials. In alumina, which has been thoroughly studied, the hardness (14) of fine-grained compacts is at most the hardness of the single crystal. When considering superhard materials, any hardness enhancement would have to come from the intergranular material, which would be by definition of lower hardness. In the case of thin films, it has been reported that it is possible to increase the hardness by repeating a layered structure of two materials with nanometer scale dimensions, which are deposited onto a surface (15). This effect arises from the repulsive barrier to the movement of dislocations across the interface between the two materials and is only valid in one direction for nanometer scale deformations. This could be suitable for coatings, but having bulk superhard materials would further enhance the unidirectional hardness of such coatings. In addition, hardness in these cases is determined from tests at a nanometer scale with very small loads, and results vary critically (up to a factor of three) with the nature of the substrate and the theoretical models necessary to estimate quantitatively the substrate's influence (16). We now discuss the search for bulk superhard materials.

Physics of Hardness

There is a direct relation between bulk modulus and hardness for nonmetallic materials (Figure 3) (17–24), and here we discuss the fundamental physical properties upon which hardness depends. Hardness is deduced from the size of the indentation after an indenter has deformed a material. This process infers that many phenomena are involved. Hardness is related to the elastic and plastic properties of a material. The size of the permanent deformation produced depends on the resistance to the volume compression produced by the pressure created by the indenter, the resistance to shear deformation, and the resistance to the creation and motion of dislocations. These various types of resistance to deformation indicate

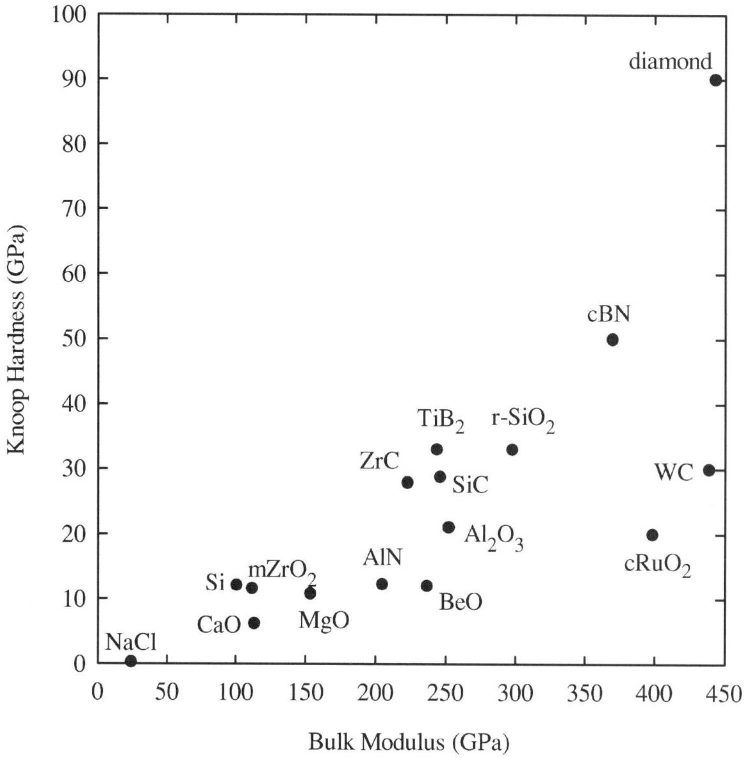

Figure 3 Hardness as a function of the bulk modulus for selected materials (r-, rutile-type; c, cubic; m, monoclinic). WC and RuO_2 do not fill all the requirements to be superhard (see text).

which properties a material must have to exhibit the smallest indentation possible and consequently the highest hardness. There are three conditions that must be met in order for a material to be hard: The material must support the volume decrease created by the applied pressure, therefore it must have a high bulk modulus; the material must not deform in a direction different from the applied load, it must have a high shear modulus; the material must not deform plastically, the creation and motion of the dislocations must be as small as possible. These conditions give indications of which materials may be superhard.

We first consider the two elastic properties, bulk modulus (B) and shear modulus (G), which are related by Poisson's ratio (ν). We consider only isotropic materials; a superhard material should preferably be isotropic, otherwise it would deform preferentially in a given direction (the crystal structure of diamond is isotropic, but the mechanical properties of a single crystal are not fully isotropic because cleavage may occur). In the case of isotropic materials, $G = (3/2)\ B\ (1 - 2\nu)/(1 + \nu)$; In order for G to be high, ν must be small, and the above expression reduces then to $G = (3/2)\,B\,(1 - 3\nu)$. The value of ν is small for covalent materials (typically $\nu = 0.1$), and there is little difference between G and B: $G = 1.1B$. A typical value of ν for ionic materials is 0.25 and $G = 0.6B$; for metallic materials ν is typically 0.33 and $G = 0.4B$; in the extreme case where ν is 0.5, G is zero.

The bulk and shear moduli can be obtained from elastic constants: $B = (c_{11} + 2c_{12})/3$, $G = (c_{11} - c_{12} + 3c_{44})/5$. Assuming isotropy $c_{11} - c_{12} = 2c_{44}$, it follows that $G = c_{44}$; Actually G is always close to c_{44}. In order to have high values of B and G, then c_{11} and c_{44} must be high with c_{12} low. This is the opposite of the central forces model in which $c_{12} = c_{44}$ (Cauchy relation).

The two conditions, that ν be small and that central forces be absent, indicate that bonding must be highly directional and that covalent bonding is the most suitable. This requirement for high bulk moduli and covalent or ionic bonding has been previously established (17–19, 21–24) and theoretical calculations (19, 25, 26) over the last two decades have aimed at finding materials with high values of B (Figure 3). The bulk modulus was used primarily for the reason that it is cheaper to calculate considering the efficient use of computer time, and an effort was made to identify hypothetical materials with bulk moduli exceeding 250–300 GPa. At the present time with the power of modern computers, elastic constants can be obtained theoretically and the shear modulus calculated (27).

The requirement for having directional bonds arises from the relationship between the shear modulus G and bond bending (28). Materials that exhibit limited bond bending are those with directional bonds in a high symmetry, three-dimensional lattice, with fixed atomic positions. Covalent materials are much better candidates for high hardness than ionic compounds because electrostatic interactions are omnidirectional and yield low bond-bending force constants, which result in low shear moduli. The ratio of bond-bending to bond-stretching force constants decreases linearly from about 0.3 for a covalent material to essentially zero for a purely ionic compound (29, 30). The result of this is that the bulk modulus has very little dependence on ionicity, whereas the shear modulus will exhibit a relative decrease by a factor of more than three owing entirely to the change in bond character. Thus for a given value of the bulk modulus, an ionic compound will have a lower shear modulus than a covalent material and consequently a lower hardness. There is an added enhancement in the case of first row atoms because *s-p* hybridization is much more complete than for heavier atoms. The electronic structure also plays an important role in the strength of the bonds. In transition metal carbonitrides, for example, which have the rock-salt structure, the hardness and c_{44} go through a maximum for a valence electron concentration of about 8.4 per unit cell (31). The exact nature of the crystal and electronic structures is thus important for determining the shear modulus, whereas the bulk modulus depends mainly on the molar volume and is less directly related to fine details of the structure. This difference is due to the fact that the bulk modulus is related to the stretching of bonds, which are governed by central forces. Materials with high bulk moduli will thus be based on densely packed three-dimensional networks, and examples can be found among covalent, ionic, and metallic materials. In ionic compounds, the overall structure is principally defined by the anion sublattice, with the cations occupying interstitial sites, and compounds with high bulk moduli will thus have dense anion packing with short anion-anion distances. The shear modulus, which is related to bond bending, depends on the nature of the bond and decreases dramatically as a function of ionicity. In order for the compound to have a high shear modulus and high hardness (Figure 4), directional (covalent) bonding and a rigid structural topology are necessary in addition to a high bulk modulus. A superhard material will have a high bulk modulus and a three-dimensional isotropic structure with fixed atomic positions and covalent or partially covalent ionic bonds.

Hardness also depends strongly on plastic deformation, which is related to the creation and motion of dislocations. This is not controlled by the shear modulus but by the shear strength τ, which varies as much as a factor of 10 for different materials with similar shear moduli. It has been theoretically shown that τ/G is of the order of 0.03–0.04 for a face-centered cubic metal, 0.02 for a layer structure such as graphite, 0.15 for an ionic compound such as sodium chloride, and 0.25 for a purely covalent material such as diamond (32). Detailed calculations must

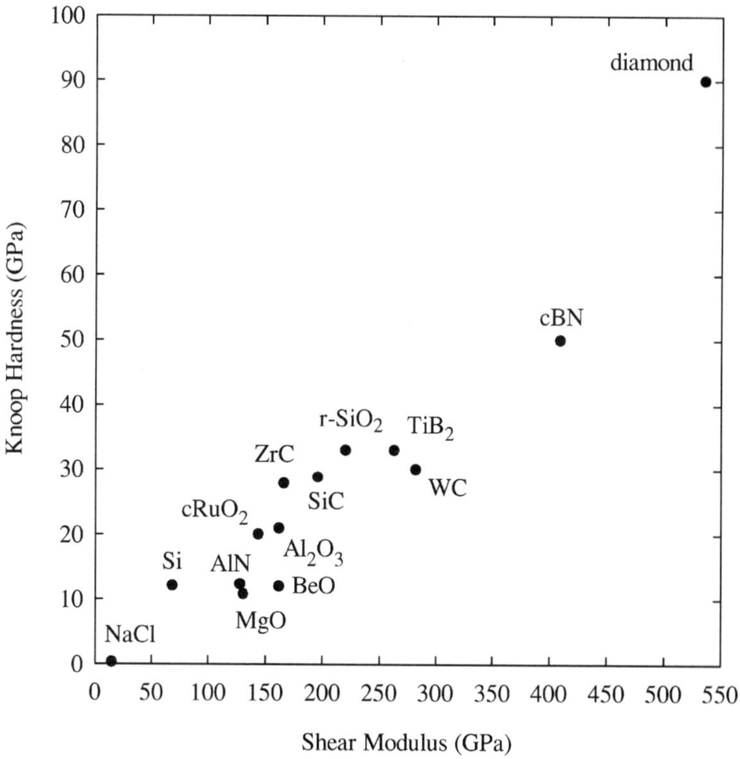

Figure 4 Hardness as a function of the shear modulus for selected materials (r-, rutile-type; c, cubic).

take into account the different slip systems in each material. The materials with the highest possible τ/G values are again covalently bonded solids and ionic materials with partially covalent bonds. The directional nature of the bond plays two roles; first, it yields a low Poisson's ratio, which increases the elastic shear modulus G; second, the nucleation and motion of dislocations are more difficult than in metallic materials, which increases the shear strength and further increases the hardness. If one considers a covalent, an ionic, and a metallic material, all having the same bulk modulus, the covalent material will be the hardest because of its high shear modulus and high shear strength. The ionic material will be less hard owing to its relatively low shear modulus, and the metal will have more than an order of magnitude lower hardness because of its low shear strength.

The bulk modulus of an ionic compound is (see Reference 33) $B \propto S^2 Z_a Z_c/V$, where S^2 is the ionicity (0.5 for oxides, 0.75 for halides, 0.20 for carbides), V the molar volume per ion pair, and Z_a and Z_c the anion and cation charges. This relationship indicates that a low molar volume and highly charged ions are required in order to have a high bulk modulus (Figure 5). A similar relationship (19) was proposed for semiconductors: $B = (N_c/4)\ (1972–220\lambda)d^{-3.5}$, where N_c is the coordination number, d is the distance between nearest neighbors, and λ is an empirical ionicity parameter ($\lambda = 0, 1, 2$, for IV, III-V, and II-VI semiconductors, respectively). In covalent solids, the bulk modulus (28) varies as a function of $V^{-5/3}$ or $1/d^5$ (Figure 5); $c_{11} - c_{12}$ is also proportional to $1/d^5$; assuming isotropy then c_{44}, or G, which is close to c_{44}, is proportional to $1/d^5$. In every case, the highest bulk or shear modulus is obtained for the lowest molar volume. Superhard materials are thus expected to be compounds with the shortest bond lengths; these are made of the light elements, B, C, O, N, Be, P, Si, and Al. Let us review typical bond lengths (Å) arranged in increasing order: C-O:1.43; C-N:1.47; B-O:1.48; P-O:1.52; C-C:1.54; B-N:1.57; B-C:1.57; P-N:1.60; Si-O:1.61; Be-O:1.65; B-B:1.71; Si-N:1.74; Ge-O:1.74; Al-O:1.74; and Si-C: 1.88.

Superhard and Potentially Superhard Materials

In considering potential and existing superhard materials it is useful to classify materials in terms of different types of chemical bonds.

C-C BOND The shortest covalent bond found in a three-dimensional network solid is the C-C bond in diamond, 1.54 Å (sp^3 carbon), which has the highest bulk modulus, shear modulus, and hardness thus far known: 442, 540, and 90 GPa, respectively. Much shorter bonds (sp^2 carbon) of 1.41 Å are found in graphite; however, these bonds lie within layers separated by much longer distances giving a low bulk modulus and a very low shear modulus. The discovery of fullerenes, such as C_{60}, with short C-C distances of 1.46 and 1.40 Å, renewed the hope of making new superhard materials from carbon compounds. These fullerenes are made of cage-like spheroids, consisting of a network of pentagons and hexagons; these balls lie on a cubic lattice and are linked by weak Van der Waals interactions. C_{60} is thus very compressible (34), and the balls approach each other rapidly under pres-

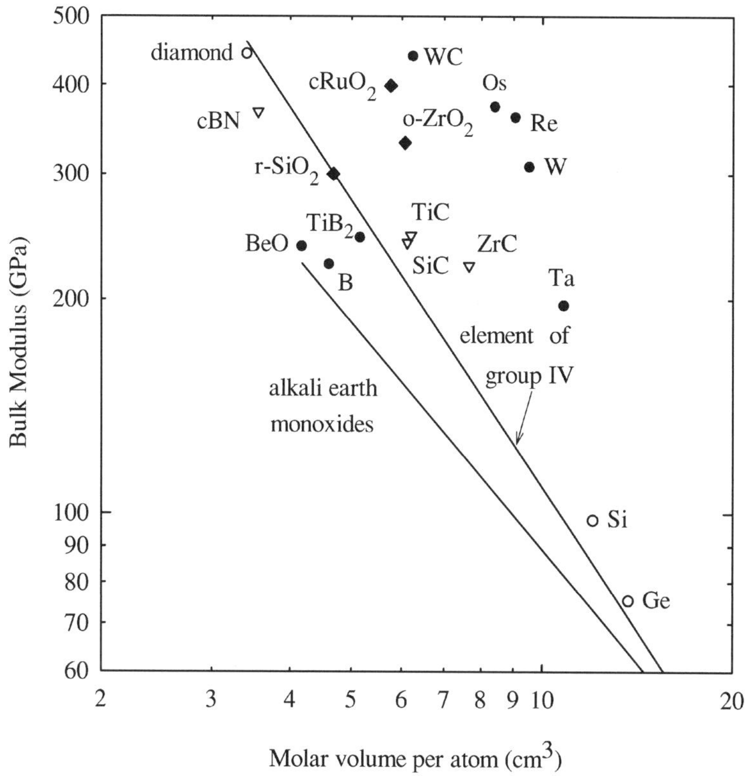

Figure 5 Bulk modulus as a function of the molar volume per atom (o-, orthorhombic cotunnite-type; r-, rutile-type; c, cubic).

sure. It was argued that these balls should be very stiff with a bulk modulus of 800–900 GPa (35), twice that of diamond, and that solid C_{60} would become rapidly less compressible than diamond. In fact, these balls are very stiff; however, when they approach one another, bonds between them are created and a three-dimensional network is produced giving an amorphous material above 20 GPa (34).

Reports that high-pressure forms of C_{60} could be harder than diamond were based on scratches produced on a diamond anvil by transformed fullerene under high pressure (36) in a diamond anvil cell. This only shows that a substance of comparable hardness (in the best case, or much lower, as found with a soft indenter; see above) was obtained. This is possible because any carbon-based substance will transform to diamond at high pressure and temperatures (such as those produced by friction). The only conclusion from this experiment is that a diamond-like compound may have been produced. Treatments of C_{60} at high pressures and high temperatures were reported to yield sizeable amounts of very hard substances that were recovered by quenching (37, 38). These materials consisted of three-dimensional polymerized amorphous and nanocrystalline diamond plus graphite composites and carbon phases. Maximum Vickers hardness values of 60 GPa were reported. This is certainly a method of making superhard materials; however, treating any carbon-based substance under these conditions will yield diamond-like materials and using fullerenes is certainly not the cheapest way. In addition, there is no hardness advantage over that of diamond produced by other methods, and the problem of chemical reactivity with iron remains unresolved.

B-B, B-N, B-O, AND B-C BONDS Apart from carbon, the other short homonuclear bond is that of boron; the B-B bond length is 1.71 Å, which is 11% larger than the C-C bond in diamond. Pure α-boron is made up of B_{12} icosahedra, as are boron-rich materials. Each boron atom has five nearest neighbors within its own icosahedron. If the bonding were of the conventional covalent type, then each boron atom would have to contribute five electrons to bonding within the icosahedron; however, boron only has three valence electrons. The result is that the bonds are electron deficient, and the bulk modulus, shear modulus, and hardness are consequently much lower than for diamond. The bulk modulus and hardness of elemental boron are only 200 and 25 GPa, respectively. Boron-rich materials, such as B_4C, B_6O, and B_6P, have bulk moduli of close to 200 GPa (39, 40) and are also made up of B_{12} icosahedra with a few additional intericosahedral chain atoms, and their hardness of about 30 GPa is similar to that of pure boron. The bond lengths B-O: 1.48; B-N: 1.57; B-C: 1.57 Å are close to the C-C bond length, but no compound of stoichiometry close to 1 is known with the exception of BN. BN is well known and is the second superhard material, H = 50 GPa. B_2O was claimed to have been obtained from high-pressure synthesis (41), and the measured Vickers hardness of the recovered material was found to be 33.5–40.5 GPa. The occurrence of a compound with this stoichiometry has not been corroborated in recent studies. Recently, it was reported that B_4C-B_6O solid solutions yield hardness values of up to 46 GPa (42). Compounds may exist in the B-C-O-N system (43), but their hardness, at best, would be equivalent to diamond-cubic boron nitride solid solutions. It should be noted that graphitic BC_2N, when treated under high-pressure high-temperature conditions, transforms to a mixture of diamond and cBN (44).

BE-O BOND The Be-O bond of 1.65 Å has the greatest ionic character among the known short bonds involving light atoms. BeO adopts the hexagonal wurtzite structure. The disadvantage of this structure is that the z-coordinate of oxygen is not fixed by symmetry. The bulk and shear moduli are 224 and 167 GPa, respectively (45), and the hardness is only 12.5 GPa (20). The value of ν (0.201) is already significantly greater than that of the covalent materials discussed above. The less than ideal structural topology and the ionic character of the bond result in only a moderately high shear modulus, and in addition τ/G is also reduced with respect to a covalent material. Theoretical calculations indicate that BeO should transform to a cubic NaCl-type phase at very high pressures of 137 GPa (46). The bulk modulus of this denser phase is calculated to be close to 270 GPa (46, 47), which is significantly lower than diamond and cBN. BeO, even with beryllium in sixfold coordination, will not be superhard.

SI-O BOND The Si-O bond of 1.61 Å is appreciably longer than the C-C bond in diamond (1.54 Å), and yet the polyhedral bulk modulus (33) of the SiO_4 tetrahedron, 500 GPa, is greater than the bulk modulus of diamond and also the polyhedral bulk modulus of the tetrahedral carbon unit, which is equal to the bulk modulus in the diamond structure. Many compounds with tetrahedral Si-O bonds are known, but their bulk moduli are low. This is due to their open crystal structures, which are built up of SiO_4 tetrahedra. In the case of the numerous polymorphs of silica, for example, these tetrahedra are linked only by their corners and can easily tilt. Under high pressure, it is possible to prepare a polymorph that has a much denser structure; stishovite (rutile-type SiO_2) is about 60% more compact than quartz. The arrangement of the oxygen atoms around silicon changes at the phase transition; there are no longer four oxygen atoms placed on the corners of a tetrahedron, but six oxygen atoms placed on the corners of an octahedron and the coordination number increases from four to six. This change is also accompanied by an increase in the length of the Si-O bond from 1.61 to 1.78 Å. At this transformation, the bulk modulus of the compound increases to 302 GPa, a value an order of magnitude higher than that of the initial quartz phase and significantly higher than that of corundum Al_2O_3 (252 GPa), whereas the polyhedral bulk modulus decreases from 500 GPa for the SiO_4 tetrahedron to 342 GPa for the SiO_6 octahedron (48). Pressure enables the preparation of this denser structure for which the bulk and polyhedral bulk moduli have similar values. In this rutile-type structure (tetragonal, $P4_2/mnm$, $Z = 2$) there is only one free positional parameter, the x coordinate for the oxygen atom, and the structure thus does not fulfill all the requirements needed to have a very high shear modulus. However, the compound is very hard, and indentation measurements indicate that stishovite (49) has the highest hardness (H = 33 GPa) of any well-characterized oxide, the hardness of corundum being only 21 GPa. This high value for the hardness of stishovite is in agreement with its bulk modulus of 302 GPa, which is a result of the low-molar volume and the high charge on the cation and its shear modulus of 220 GPa. Stishovite is only slightly harder than SiC, although its bulk modulus is much higher because of the shorter bond length, but there is one free atomic position in the rutile-type structure, whereas there are none for SiC, which adopts the same binary analog of the diamond structure as cBN. Stishovite is another example of the potential of high-pressure synthesis for making superhard materials. In each case the process is the same; high-pressure produces a phase that is markedly more dense than the low-pressure phase, by about 50% in diamond and cBN. Phase transformations induced by pressure typically increase the coordination number and, at the same time, increase the average bond length; the net effect is a large increase in the bulk modulus (Figure 6), shear modulus, and hardness. There is a major difference between stishovite and diamond or cBN; in contrast to these two classical superhard materials, stishovite is obtained directly by heating the lower pressure phase. It is

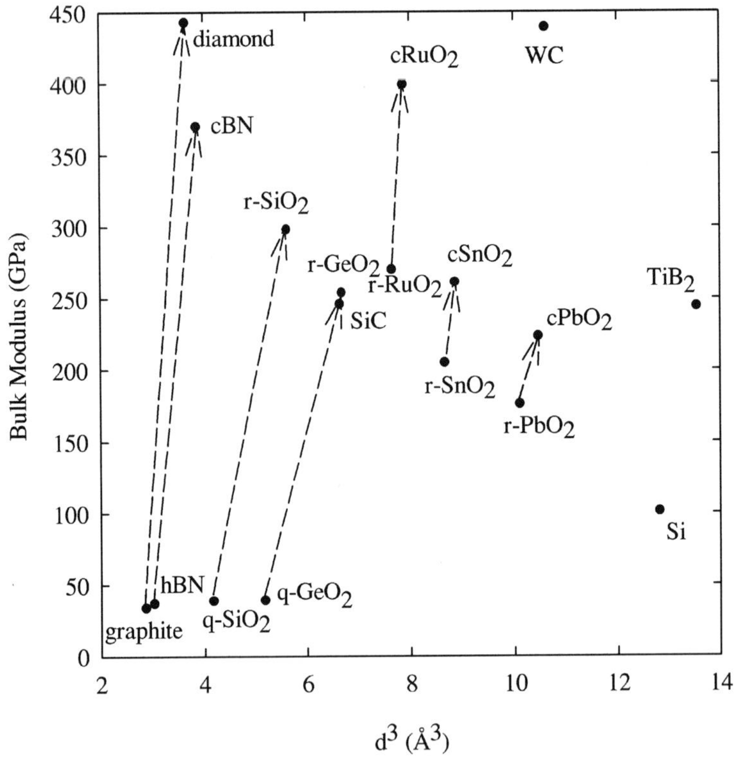

Figure 6 Bulk modulus as a function of the cube of the A-B bond length except for diamond and silicon; the dashed lines connect the normal pressure and the high pressure phases (c, cubic; r-, rutile-type; h, hexagonal; q-, quartz-type).

a phase transformation and not a catalytic synthesis involving complex chemistry. The resulting material is free of binder and impurities and may be of any shape.

SI- AND GE-N BONDS Silicon nitride has several polymorphs including a recently discovered dense cubic form (50). The bulk moduli of these materials are potentially high because of the short bond lengths and the high formal ionic charges. Silicon and nitrogen atoms are in fourfold and threefold coordination, respectively, in both the well-known α and β forms, with bulk moduli of 248 and 234 GPa and shear moduli of 144 and 123 GPa, respectively (51). The relatively low values of the shear moduli arise from the partially ionic character of the Si-N bond, the threefold coordination of nitrogen, which is far from optimal for the formation of a rigid three-dimensional network, the lower degree of *s-p* hybridization in the case of silicon, and the free atomic coordinates in these structures. The hardness of 21 GPa is equivalent to that of alumina. Dense, spinel-type γ-Si_3N_4 (50) and γ-Ge_3N_4 (52) have recently been synthesized under high-pressure high-temperature conditions, and the structure of the latter, in which two thirds of the cations are in sixfold coordination, was refined at ambient conditions. In this structure type, the cations are on fixed positions, and the anions lie on positions not far from a cubic close packing (ccp) (53) ($x = 0.375$); in Ge_3N_4, x is 0.3827, compared with 0.387 in spinel (Al_2MgO_4). This phase of germanium nitride is not very hard. With a smaller cation, silicon, the molar volume will be lower, and the x position may tend toward this value, which would be favorable for a high-shear modulus. However, from the calculated volume and bulk modulus of close to 300 GPa, spinel-type γ-Si_3N_4 should have at best a hardness close to that of stishovite.

RU-O The highest bulk modulus reported up to now, with the exception of diamond and metallic materials such as WC (54) and Co_6W_6C (55), is that of cubic RuO_2. In the common rutile form of this oxide, the ruthenium atom is sixfold coordinated to oxygen as is the case of silicon in stishovite, the high-pressure phase of silica discussed above. The bulk modulus is lower than that of stishovite as expected from the longer bond length of 1.97 Å. The density is 9% higher in the high-pressure form, with cubic $Pa\overline{3}$ symmetry (56, 57) in which the ruthenium atom is surrounded by eight oxygen atoms, six at 1.99 Å and two at 2.95 Å. An important increase in the bulk modulus from 270 GPa in the rutile form to 399 GPa in the cubic form is observed (58). Recent indentation measurements on a recovered specimen from high-pressure high-temperature treatments, however, yield hardness values of 20 GPa and the compound is thus not superhard. These measurements are consistent with the results obtained from Brillouin scattering, which indicate that $c_{44} = 144$ GPa (JM Léger, J Haines, F Ganot, AS Pereira & JAH da Jornada, unpublished data). The bulk modulus of this compound is only 10% lower than that of diamond; however, its hardness is less than a quarter of that of diamond. The value of the c_{44} elastic constant indicates that $\mu = 0.345$. The lower than expected shear modulus is due to the partial ionic character of the Ru-O bond and the crystal structure, which is a distortion of the fluorite-type structure. In both the $Pa\overline{3}$ and fluorite-type structures, the cations lie on 4a sites (0,0,0) and the anions lie on 8c sites (u,u,u). The difference between the two structures is in the oxygen sublattice. The oxygen coordinate u is fixed by symmetry to 0.25 in the fluorite structure, whereas it is 0.351 for the $Pa\overline{3}$ structure. The last structure is favorable for having a high bulk modulus because in the place of the empty cubic site in the true fluorite structure there is an oxygen-oxygen distance of only 2.5054 Å, which corresponds to a close anion-anion contact. Unfortunately, the same is not true for the shear modulus because the oxygen coordinate is not fixed by symmetry as in the fluorite-type structure. This produces an added degree of flexibility to the structure because the oxygen atom can move slightly, resulting in changes to the bond angles. The topological stiffness of the lattice is thus a major factor, largely governed by the site positions.

Potential Superhard Compounds

In addition to existing superhard materials, there are several compounds that have the potential to be superhard based on the bonds.

C-N BOND Much attention has been focused on the C-N bond, which is slightly shorter than the C-C bond in diamond. Over the past years, the search for C_3N_4 has been based on analogies to the homologue β-Si_3N_4 (26). The calculated bulk moduli of hypothetical carbon nitride phases including β-C_3N_4 are close to that of diamond and may exceed its value. This is not true for their shear moduli and hardness because their properties deviate from the optimum found for diamond in all respects. The C-N bond is not purely covalent, the connectivity is not ideal because nitrogen is in threefold coordination, and there are variable atomic coordinates, as in the case for Si_3N_4. Calculations indicate that the shear modulus will in fact be only 60% of that of diamond (26). There have been reports of the preparation of thin film containing β-C_3N_4 (59–61). Nanoindentation tests have been performed on non-stoichiometric carbon nitride films and carbon nitride/TiN composite coatings yielding hardness values of 8–35 GPa (62–65) and 45–55 GPa (66), respectively. Carbon nitrides would be expected to be less hard than cBN in view of their lower shear moduli, and the above hardness values probably represent the maximum of what can be obtained from these materials. It should also be noted that these materials are much more difficult to prepare than diamond and would pose the problem of carbide formation if they were to be used to machine ferrous metals. In a hypothetical spinel-type C_3N_4, the nitrogen x position may tend toward the value corresponding to a ccp lattice, which would be favorable for a high shear modulus. Formation of this phase would require that two thirds of the carbon atoms pass into sixfold coordination; at least several hundred GPa would probably be needed to prepare such a phase, which would likely not be metastable under ambient conditions.

C-O BOND The C-O bond is even shorter than the C-N bond. The bulk modulus of the hypothetical rutile-type (stishovite-like) phase of CO_2 can be estimated from the C-O distance; it is 1.43 Å for fourfold coordinated carbon and $\sim$1.58 Å for sixfold coordinated carbon. By comparison with the Si-O distance of 1.79 Å in the same structure, the bulk modulus would be about 440 GPa, as is that of diamond, but the covalent bond character would be much lower, and the shear modulus G and hardness H would probably be similar to those of cBN. The calculated pressure for obtaining this phase is as high as 400 GPa (67), and this phase will probably not be stable at normal pressure.

P-O AND P-N BONDS P-O and P-N bond lengths of 1.52 and 1.60 Å, respectively, are slightly shorter than Si-O and Si-N distances; however, no compound with a high bulk or shear modulus has been obtained. A phosphorus compound, PON, adopts structures that are similar to those found in silica compounds because phosphorus is also tetrahedrally coordinated. The structures are built of tetrahedra, which are much stiffer (68) ($B > 800$ GPa) than those in silica owing to the higher charges and shorter interatomic distances. However, in contrast to silica in which a phase transition occurs with an increase in the coordination number from four to six around 9 GPa, no similar increase in coordination number was found to occur for phosphorus up to at least 50 GPa, and no polymorph similar to stishovite was obtained for PON (JM Léger, J Haines, C Chateau, G Bocquillon, MW Schmidt, S Hull, F Gorelli, A Lesauze & R Marchand et al, submitted).

All compounds made up of light elements, such as carbon or boron, in which the atoms are in fourfold coordination, will have hardness values lower than that of diamond, for which low-molar volume, adequate structure, and optimum bonding are simultaneously found. Higher values may be obtained for high-pressure phases of these compounds in which these atoms are in sixfold coordination, but extremely high pressures (100 GPa – 2 TPa) will be necessary for their syntheses, and they will probably not be metastable at ambient pressure.

Is it not possible, however, to use another approach to design superhard compounds? The synthesis of the cubic high-pressure phase of RuO_2 is very encouraging; it shows that it is possible to simultaneously have a large-molar volume per atom and a high bulk modulus. The molar volume of RuO_2 is 75% higher than that of diamond, but the bulk modulus is only 10% lower. Unfortunately, the structure is not favorable from the point of view of high hardness because the atomic position of the oxygen atom is free in the cubic structure and the hardness is only 20 GPa, which is equivalent to that of corundum (Al_2O_3).

Could a higher-pressure phase among dioxides yield better mechanical properties? The dioxides adopt different structures (69) with increasing coordination numbers (CN) at high pressure: cubic $Pa\overline{3}$ (CN = 6 + 2), baddeleyite (CN = 7), orthorhombic-I ZrO_2 (CN = 7), and cotunnite (CN = 9). The high pressure forms of these oxides, which adopt the monoclinic baddeleyite-type structure or the orthorhombic-I-ZrO_2-type structure (these are both distortions of the fluorite structure), will have only moderate hardness values because the atomic positions of the anions and cations are not fixed. The dioxides in the cotunnite structure (CN = 9) may be expected to have still higher bulk moduli. For example, the bulk modulus of the high-pressure cotunnite-type phase of ZrO_2 is 332 GPa (70); in this structure, the z coordinates of the various atoms are fixed ($z = 0.25$ or 0.75), and this structure is more favorable than those discussed above for having high hardness values. However, this value will be far below that of diamond where all positions are fixed. All the crystal structures adopted at high pressure by the dioxides have free atomic positions, and thus their high-pressure phases are not good candidates for superhard materials.

Tungsten carbide, WC, has even a higher bulk modulus than cubic RuO_2. It has a simple hexagonal structure in which the positions of all the atoms are fixed. The bulk and shear moduli of this material, 439 and 282 GPa, respectively, are among the highest known (54). However, WC is not superhard, H = 30 GPa (20), and this can be attributed to the lack of a high degree of covalent bonding, as indicated by the value of ν (0.235), and the existence of a degree of metallic character that reduces the shear strength. Recently, a carbide with a bulk modulus higher than that of diamond, Co_6W_6C with B = 462 GPa (55), has been reported, but this material also has a degree of metallic character. Among the borides, TiB_2 has a Poisson's ratio of 0.104, a bulk modulus of 244 GPa, and a shear modulus of 263 GPa (20, 71). The measured hardness (20) of 33 GPa is close to what would be expected given this shear modulus. The simple hexagonal structure with fixed atomic coordinates is also favorable. One possible candidate compound, which may adopt this structure at high pressure in order to have a sufficiently low-molar volume, is beryllium diboride. It could potentially rival cBN in hardness, if it were to adopt this hexagonal structure at high pressure.

PERSPECTIVES

Diamond meets all the conditions needed to be the hardest material: lowest molar volume, highest electronic density, greatest degree of covalent bonding (homonuclear), a high degree of *s-p* hybridization, and a three-dimensional-bonding cubic (isotropic) structure with fixed atomic positions (topological rigidity). The resulting Poisson's ratio is small ($\nu = 0.07$), and its shear modulus is greater than its bulk modulus, which is already the highest known owing to its smallest volume. Diamond is a high-pressure form of carbon, in which the coordination number has increased from three in graphite to four. At present it is the hardest material that can be prepared and retained as a metastable phase at normal pressure. Furthermore, it is nearly stable at ambient pressure as demonstrated by the synthesis of diamond films by sputtering under vacuum.

Attempts to prepare compounds with shorter bond lengths (C-C, C-N) may result in materials with higher bulk moduli. However, their hardness will be lower than that observed for diamond, where all the properties are simultaneously at their optimum. Diamond is the easiest to make of all materials using tetrahedrally coordinated carbon; it is nearly stable at ordinary pressure and thus requires the lowest synthesis pressures. Other potentially hard high-pressure phases with short bond lengths and higher coordination will require higher pressures for their synthesis, and their stability at normal pressure is not guaranteed. The synthesis of C_3N_4 in the spinel structure, for example, would require extremely high pressures because carbon would be four- and sixfold coordinated.

At present, it has not been possible to make superhard compounds with bonds just a little shorter than those found in diamond, although stiffer bonds such as the C-C bond in the basal plane of graphite or the Si-O bond in the SiO_4 tetrahedron are well known. The detailed crystal structure and the crystal packing are important in order to have a high shear modulus and a high hardness value; thus far it has not been possible to identify a compound potentially harder than diamond containing these bonds.

Materials with bulk moduli close to that of diamond have been synthesized, and the existence of materials with bulk moduli higher than that of diamond has been reported, but no compound with a higher shear modulus or hardness has been predicted. These latter properties depend on the nature of the chemical bonds and fine structural details, which have not yet been taken into account. Theoretical calculations are now able to determine from first principles the elastic properties of high-pressure phases, but they are not capable of predicting which structures will be adopted because calculations are principally used to determine the properties of a given structure.

As stated above, diamond has the great disadvantage of reacting with iron. Is it possible, however, to imagine materials with a hardness about half that of diamond that would be useful for industry? Cubic BN, of course, fulfills the required properties, but it is difficult and expensive to prepare the compacts required for current applications. It is possible to make new compounds with a hardness lower than that of diamond but similar to that of cBN. Taking note from the above discussion, it is also possible to make compounds containing B-C-N, B-C-O, B-O-N, or B-C-O-N under high pressure, but apart from simple solid solutions, their synthesis will require a complex chemistry and in addition, their elastic properties will be inferior to that of diamond.

It will also be worthwhile to search for superhard materials that are not composed entirely of light elements. The preparation of the cubic phase of RuO_2 shows compounds with high bulk moduli can be obtained with heavy (large) atoms. The search of covalent or partially covalent bonds in simple structures, which would be obtained under high pressure, thus appears promising. Borides, carbides, and nitrides may be good candidates. One should look for dense and simple isotropic structures with fixed atomic positions. It is also possible that complex structures with atoms of different sizes with a high-packing index could be superhard materials. These directions are still largely unexplored and may be promising for new superhard materials. These materials will be most probably high-pressure phases because high pressure generally increases the packing efficiency.

Another approach would be to improve the currently known superhard materials, diamond and cBN. This improvement could come from an increase of the pressure capabilities of the devices used for their synthesis. At present, the capabilities for the production of diamond and cBN are restricted to pressures of about 6 GPa. This pressure limit comes from the general rule that the pressure available decreases when the volume increases. Increasing the pressure limit will permit the sintering of fine powders of diamond or boron nitride without any binder and may permit their direct synthesis from the low-pressure graphitic phases. This would be completely different from the present catalytic synthesis, which produces only small isolated crystals, and from the lengthy process required for growing large single crystals of diamond. Another way of producing these two materials is to lower the pressure required for their synthesis; the use of solvents under supercritical conditions for cBN has been outlined. These processes yield fine crystal sizes, and the powders obtained must be sintered. Again, new higher-pressure devices will produce better compacts, and this promises to be a way to produce superhard materials.

Visit the Annual Reviews home page at www.AnnualReviews.org

LITERATURE CITED

1. Bundy FP, Hall TH, Strong HM, Wentorf RH Jr. 1955. *Nature* 176:51–54
2. Field JE, ed. 1992. *The Properties of Natural and Synthetic Diamond*. London: Academic. 710 pp.
3. Strong HM. 1989. *Am. J. Phys.* 57:794–802
4. Bocquillon G. 1996. *Rev. Scient. Techn. Défense* 3:203–10
5. Giardini AA, Tydings JE. 1962. *Am. Min.* 47:1393–421
6. Wentorf RH Jr. 1965. *J. Phys. Chem.* 69: 3063–69
7. Bocquillon G, Loriers-Susse C, Loriers J. 1985. *J. Mater. Sci. Lett.* 4:141–43
8. Bakul V, Prikhna J. 1974. *J. Cryst. Growth* 26:127–28
9. Rapoport E. 1985. *Ann. Chim. Fr.* 10:607–38
10. Bocquillon G, Loriers-Susse C, Loriers J. 1993. *Mater. Sci.* 28:3547–56
11. Solozhenko VL. 1995. *Proc. Joint XV AIRAPT & XXXIII EHPRG Int. Conf., Warsaw, Poland*, ed. WA Trzeciakowski, pp. 338–42. Singapore: World Sci.
12. Hibbs IE Jr, Wentorf RH Jr. 1974. *High Temp. High Pres.* 6:409–13
13. Tanigushi T, Akaishi M, Yamaoka S. 1999. *J. Mater. Res.* 14:162–69
14. Rice RW, Wu CCm, Borchelt F. 1994. *J. Am. Ceram. Soc.* 10:2539–53
15. Shinn M, Hultman L, Barnett SA. 1992. *J. Mater. Res.* 7:901–11
16. Nowak R, Li CL, Maruno S. 1997. *J. Mater. Res.* 12:64–69
17. Léger JM, Haines J, Blanzat B. 1994. *J. Mater. Sci. Lett.* 13:1688–90
18. Léger JM, Haines J. 1997. *Endeavour* 21:121–24
19. Cohen ML. 1993. *Science* 261:307–8
20. Teter DM. 1998. *Mater. Res. Soc. Bull.* 23:22–27
21. Haines J, Léger JM. 1998. *Superhard Mater.* 2:3–8
22. Sung CM, Sung M. 1996. *Mater. Chem. Phys.* 43:1–18
23. Riedel R. 1994. *Adv. Mater.* 6:549–60
24. Goble RJ, Scott SD. 1985. *Can. Mineralogist* 23:273–85
25. Lowther JE, Dewhurst JK, Léger JM, Haines J. 1999. *Phys. Rev. B* 60:14485–88
26. Teter DM, Hemley RJ. 1996. *Science* 271:53–55
27. Tse JS, Klug DD, Uehara K, Li ZQ, Haines J, Léger JM. 2000. *Phys. Rev. B* 61:10029–34
28. Harrison WA. 1980. *Electronic Structure and the Properties of Solids*. San Francisco: Freeman. 582 pp.
29. Martin RM. 1970. *Phys. Rev. B* 1:4005–11
30. Phillips JC. 1973. In *Treatise on Solid State Chemistry, The Chemical Structure of Solids*, ed. NB Hannay, 1:34–39. New York: Plenum
31. Jhi SH, Ihm J, Louie SG, Cohen ML. 1999. *Nature* 399:132–34
32. Kelly A. 1973. In *Strong Solids*, ed. CEH Bawn, H. Fröhlich, PB Hirssch, NF Mott, pp. 4–51. Oxford, UK: Clarendon
33. Hazen RM, Finger LW. 1982. *Comparative Crystal Chemistry*. New York: Wiley & Sons. 231 pp.
34. Haines J, Léger JM. 1994. *Solid State Comm.* 90:361–63
35. Ruoff RS, Ruoff AL. 1991. *Nature* 350:663–64
36. Blank V, Popov M, Buga S, Davydov V, Denisov VN, et al. 1994. *Phys. Lett. A* 188: 281–86
37. Brazhkin VV, Lyapin AG, Popova SV, Klyuev Yu A, Naletov AM. 1998. *J. Appl. Phys.* 84:219–26
38. Levin VM, Blank VD, Prokhorov VM, Soifer Ja M, Kobelev NP. 2000. *J. Phys. Chem. Solids* 61:1017–24
39. Nelmes RJ, Loveday JS, Wilson RM,

Marshall WG, Besson JM, et al. 1995. *Phys. Rev. Lett.* 74:2268–71

40. Loveday JS, Nelmes RJ, Marshall WG, Besson JM, Klotz S, et al. 1996. *High Pres. Res.* 14:303–9
41. Endo T, Sato T, Shimada M. 1987. *J. Mater. Sci. Lett.* 6:683–85
42. Itoh H, Maekawa I, Iwahara H. 2000. *J. Mater. Sci.* 35:693–98
43. Knittle E, Kaner RB, Jeanloz R, Cohen ML. 1995. *Phys. Rev. B* 51:12149–56
44. Sasaki T, Akaishi M, Yamaoka S, Fujiki Y, Oikawa T. 1993. *Chem. Mater.* 5:695–99
45. Simmons G, Wang H. 1971. *Single Crystal Elastic Constants and Calculated Aggregate Properties*. Cambridge, MA: MIT Press. 370 pp.
46. Van Camp PE, Van Doren VE. 1996. *J. Phys. Condens. Matter* 8:3385–90
47. Boettger JC, Wills JM. 1996. *Phys. Rev. B* 54:8965–68
48. Ross NL, Shu JF, Hazen RM. 1990. *Am. Mineral.* 75:739–47
49. Léger JM, Haines J, Schmidt M, Petitet JP, Pereira AS, da Jornada JAH. 1996. *Nature* 383:401
50. Zerr A, Miehe G, Serghiou G, Schwarz M, Kroke E, et al. 1999. *Nature* 400:340–41
51. Yeheskel O, Gefen Y. 1985. *Mater. Sci. Eng.* 71:95–99
52. Leinenweber K, O'Keeffe M, Somayazulu M, Hubert H, McMillan PF, et al. 1999. *Chem. Eur. J.* 5:3076–78
53. Hyde BG, Andersson S. 1989. In *Inorganic Crystal Structures*. New York: Wiley & Sons. 430 pp.
54. Lee M, Gilmore RS. 1982. *J. Mater. Sci.* 17:2657–60
55. Dubrovinskaia NA, Dubrovinsky LS, Saxena SK, Selleby M, Sundman B. 1999. *J. Alloys Compounds* 285:242–45
56. Haines J, Léger JM, Schmidt MW, Petitet JP, Pereira AS, et al. 1998. *J. Phys. Chem. Solids* 59:239–43
57. Haines J, Léger JM, Schulte O. 1996. *Science* 271:629–31
58. Haines J, Léger JM. 1993. *Phys. Rev. B* 48: 13344–49
59. Niu C, Lu YZ, Lieber CM. 1993. *Science* 261:334–37
60. Yu KM, Cohen ML, Haller EE, Hansen WL, Liu AY, Wu IC. 1994. *Phys. Rev. B* 49: 5034–37
61. Chen Y, Guo L, Wang EG. 1997. *Phil. Mag. Lett.* 75:155–62
62. Li D, Chung Y-W, Wong M-S, Sproul WD. 1993. *J. Appl. Phys.* 74:219–23
63. Sjostrom H, Stafsstrom M, Sundgren JE. 1995. *Phys. Rev. Lett.* 75:1336–39
64. Sjostrom H, Stafsstrom M, Sundgren JE. 1996. *Phys. Rev. Lett.* 76:2205
65. Martin-Gil J, Martin-Gil FJ, Sarikaya M, Qian M, José-Yacaman M, Rubio A. 1997. *J. Appl. Phys.* 81:2555–59
66. Li D, Chu X, Cheng S-C, Li X-W, Dravid VP. 1995. *Appl. Phys. Lett.* 67:203–5
67. Holm B, Ahuja R, Belonoshko A, Johansson B. 2000. *Phys. Rev. Lett.* 85:1258–61
68. Léger JM, Haines J, de Oliveira LS, Chateau C, Lesauze A, Marchand R. 1996. *J. Phys. Condens. Matter* 8:L773–77
69. Léger JM, Haines J. 1997. *Eur. J. Solid State Inorg. Chem.* 34:785–94
70. Haines J, Léger JM, Atouf A. 1995. *J. Am. Ceram. Soc.* 78:445–48
71. Spoor PS, Maynard JD, Pan MJ, Green DJ, Hellmann JR, Tanaka T. 1997. *Appl. Phys. Lett.* 70:1959–61

Annu. Rev. Mater. Res. 2001. 31:25–46

MATERIALS FOR NON-VIRAL GENE DELIVERY

Tatiana Segura and Lonnie D Shea
Department of Chemical Engineering, Northwestern University, Evanston, Illinois 60208; e-mail: l-shea@northwestern

Key Words gene therapy, cationic lipids, cationic polymers, drug delivery, DNA

■ **Abstract** Novel therapeutic strategies can be envisioned based on altering the expression level of target genes involved in cellular processes and disease progression; however, our ability to efficiently manipulate gene expression is limited. Non-viral-based gene therapy provides a relatively safe approach to increase or decrease the expression of a specific gene using DNA or antisense sequences; however, synthetic systems are required to direct plasmids and oligonucleotides to a specific tissue and to enhance cellular uptake and intracellular trafficking. Numerous materials are being developed that interact with DNA to enhance its properties (e.g. stability, charge density) and thus direct its biodistribution and facilitate cellular interactions. The development of synthetic delivery systems to manipulate gene expression efficiently is a powerful tool that will ultimately lead to novel therapeutic strategies for the treatment of numerous disorders.

INTRODUCTION

Altering or manipulating genes or gene expression has tremendous therapeutic potential for the treatment of a variety of inherited or acquired disorders. Recent advances in molecular biology combined with the culmination of the Human Genome Project have provided a genetic understanding of cellular processes and disease pathogenesis (1). To date, numerous genes involved in disease and cellular processes have been identified as targets for therapeutic approaches. In addition, the development of high throughput screening techniques (e.g. cDNA microarrays, differential display, and database mining) may drastically increase the rate at which these targets are identified (2, 3). However, the development of novel therapeutic strategies using these targets is dependent on the ability to manipulate the expression of these target genes in the desired cell population.

Gene therapy aims to treat disease by delivering DNA, RNA, or antisense sequences that alter gene expression within a specific cell population thereby manipulating cellular processes and responses. Although it was originally devised for the treatment of inherited genetic disorders, such as cystic fibrosis, recent work has expanded the applications of gene therapy to develop strategies for HIV, cancer, and wound healing applications (4–6). For genetic disorders such as hemophilia, cancer, and cystic fibrosis, the replacement of a deficient gene could restore normal cellular processes. Alternatively, the altered expression of a gene could be used to stimulate desired cellular responses, such as immune responses or tissue regeneration. The delivery of genes encoding for therapeutic proteins is the traditional view of somatic gene therapy; however, we include in our definition antisense and antigene approaches, which aim to block expression at the level of transcription or translation (7, 8). Gene therapy can be performed by manipulating gene expression either ex vivo or in vivo. The ex vivo approach involves genetically modifying cells, which are isolated from either the patient or a donor, and subsequently implanting them into the patient (for review see 9). Ex vivo gene delivery provides a relatively high gene transfer efficiency; however, these procedures are costly and may not be appropriate for all situations (10).

The challenge of in vivo gene therapy is to develop safe and efficient gene-delivery systems (4). Viral vectors generally provide the most efficient gene transfer, which is a result of the viruses evolving to develop efficient methods to introduce DNA into host cells. The premise of viral gene delivery is to use viruses that have sections of their genome removed to make them replication deficient. Within the section that was removed, genes encoding for therapeutic proteins can be inserted. Viruses are produced using helper cell lines creating attenuated viruses that can efficiently deliver the therapeutic gene but are incapable of replicating in vivo. The limited space to insert genes in the viral genome, combined with issues associated with virus production and safety, have inspired the development of systems to deliver non-viral DNA. Non-viral approaches typically involve either plasmid DNA (i.e. circular loops of DNA) or oligonucleotides (ONs), short single-stranded pieces of DNA. Plasmid DNA and ONs are generally considered to be safe; however, the gene transfer efficiency is significantly less than viral vectors and must be improved for many therapeutic applications.

Barriers to Effective In Vivo Gene Delivery

Effective in vivo gene delivery of naked DNA requires that the plasmid or ON be delivered to the desired cell population, be efficiently internalized by the cell, and be transported to the appropriate cellular compartment. Although the path is known, many barriers exist that limit the efficiency of delivery. Endonucleases, present in the extracellular space, can degrade non-viral DNA within 30 min (11). Furthermore, intravenous delivery of DNA leads to rapid clearance from the circulation owing to extensive uptake by the liver via scavenger receptors, thus preventing extravasation to any organ or tissue (12). DNA injected directly into a tissue drains rapidly into the lymphatics (13, 14). The inability to cross biological membranes, such as the endothelium, the plasma membrane, and the nuclear membrane, is due to the size and charge density of naked DNA. Plasmid DNA is typically 10^3 to 10^4 base pairs in length, has a supercoiled tertiary structure in aqueous solutions, has a molecular weight of 10^6–10^7 and has an effective hydrodynamic diameter of greater than 100 nm (10). The surface charge density of naked DNA, which has zeta potentials ranging from −30 to −70 mV, creates repulsion between the DNA and the negatively charged cell surface. This large size and negative surface charge density most likely limit the uptake of DNA by cells.

Delivery systems are being developed to overcome the barriers of stability, size, charge density, and biodistribution. Numerous mechanical techniques are being investigated for naked DNA delivery such as the gene gun, hydrostatic pressure, electroporation, continuous infusion, and sonication (15–18). These techniques are capable of transfecting cells in vivo, possibly by compromising the integrity of the cell membrane thus allowing entry of DNA into the cell. However, the details of the mechanism by which DNA is internalized for the various systems are not well understood (19). An alternative approach involves the development of materials that stabilize DNA against degradation and simultaneously present it to cells in a manner that enhances uptake. This review article focuses on the materials that are being developed, their interaction with the biological systems, and the strategies underlying their design and development.

BIOMATERIALS

The various materials designed to increase gene transfer function to enhance the stability, to increase the efficiency of cellular uptake and intracellular trafficking, or to alter the biodistribution of DNA. The most common approach involves the complexation of naked DNA with cationic lipids or poly-cationic polymers that package the DNA for effective delivery. This approach typically alters the large negative surface charge on the DNA and facilitates internalization by the cell. These packages can also provide for protection against nuclease activity, targeting of a specific cell population, and efficient internalization and intracellular trafficking, all of which can significantly increase gene transfer efficiencies. A complementary approach to packaging entails the incorporation and release of DNA from polymeric systems. Polymeric delivery systems have been proven effective for various small-molecule drug and protein delivery and have recently been applied to the delivery of nucleic acids. In this section, we review the materials that have been developed, their physical properties and interaction with DNA, and their colloidal properties.

Cationic Lipids

The most investigated approach for condensing non-viral DNA for efficient gene transfer is the use of cationic lipids. Cationic lipids were used for the first time to deliver DNA by Felgner et al (20) and later by Bennett et al for oligonucleotides (21). The structures of several frequently used cationic lipids are shown in Figure 1. Although some cationic lipids are used individually to deliver non-viral DNA (e.g. DOTAP), many formulations of cationic lipids contain a zwitterionic or neutral colipid, such as DOPE (Figure 1) or cholesterol, to enhance transfection.

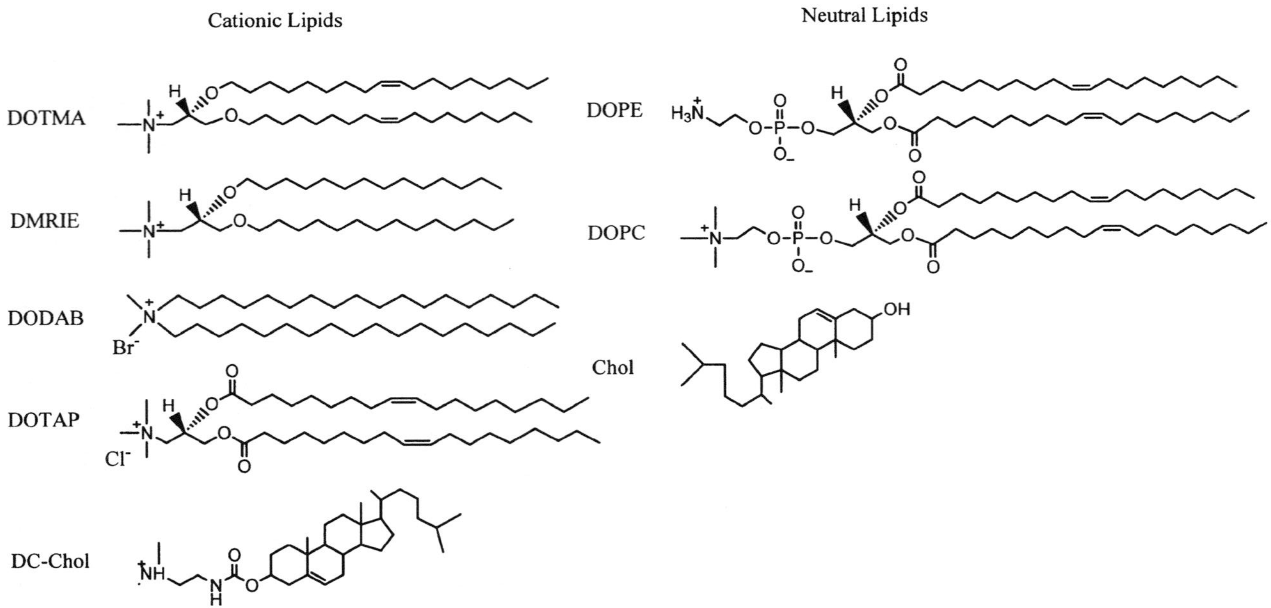

Figure 1 Structure of common lipids used as materials for gene therapy. Cationic lipids: DOTMA: [2,3-bis(oleoyl)propyl]-trimethyl] ammonium chloride; DMRIE: 1,2-dimyristyloxypropyl-3-dimethyl-hydroxyethyl ammonium bromide; DODAB: dioctadecyldimethyl ammonium bromide; DOTAP: 1,2-diacyl-3-trimethylammonium propane; DC-Chol: 3[*N*-(*N*′,*N*′-dimethylaminoethane)-carbamoyl] cholesterol. Neutral Lipids: DOPE: dioleoylphosphatidyl ethanolamine; DOPC: dioleoylphosphatidyl choline; Chol: cholesterol.

Formulations of cationic lipids have been widely applied for in vitro nucleic acid transfection, and more than 30 products are commercially available for this purpose, including Lipofectin® (a 1:1 mixture of DOTMA and DOPE), Transfectam®, LipofeACE™, LipofectAMINE™, and LipoTAXI™ (22).

The main components of a cationic lipid are a hydrophilic lipid anchor, a linker group, and a positively charged headgroup. The lipid anchor is typically either a fatty chain (e.g. derived from oleic or myristic acid) or a cholesterol group, which determines the physical properties of the lipid bilayer, such as flexibility and the rate of lipid exchange (23). The linker group is an important determinant of the chemical stability, biodegradability, and transfection efficiency of the cationic lipid. Biodegradable lipids are being developed that can be metabolized by various enzymes (e.g. esterases, peptidases) to minimize any toxicity (24, 25). The linker can also provide sites for the introduction of novel side chains to enhance targeting, uptake, and trafficking. The positively charged headgroup on the cationic lipid is responsible for interacting with the negatively charged DNA and is a critical determinant of the transfection and cytotoxicity of liposome formulations. The headgroups differ markedly in structure and may be single- or multiple-charged as primary, secondary, tertiary, and/or quaternary amines. The synthesis of novel cationic lipids from libraries of building blocks has provided insight into some structure-activity relationships (26–28). The hydrophobicity of the lipid moiety has a crucial effect on in vitro gene transfer. Multivalent headgroups, such as spermine, in a T-shape configuration tend to be more effective than their monovalent counterparts at facilitating gene transfer (23, 28). Generally, increases in the linker length correspond to increases in the gene-delivery activity (26). Continued progress toward a comprehensive relationship between lipid structure, its complexation with DNA, and the subsequent interaction with the biological system is necessary to facilitate the design of cationic lipids with optimal properties.

Mixing of DNA and cationic lipid results in the collapse of DNA to form a condensed structure—termed lipoplex—in which nucleic acids are buried within the lipid. The thermodynamic driving force for association of the DNA and lipid is the entropy increase from the release of counter ions and bound water associated with DNA and the lipid surface (29–31). Liposome association with DNA has resulted in tube-like bilayers (32, 33) and multilamellar complexes (29, 34, 35), as well as structures containing non-bilayer elements (36, 37). Multilamellar organization, in which DNA is intercalated between cationic lipid bilayers, has been reported for lipolexes resulting from the interaction of DNA with unilamellar vesicles of a cationic lipid, such as DOTAP (29) and DODAB (35), respectively, combined with DOPE and cholesterol (see Figure 1). Both of these studies used X-ray scattering to show that a periodicity of 6.5 nm occurs in the lamellar structure. The cationic lipid bilayer thickness was 3.9 nm, and the thickness of the water layer was 2.6 nm, which is sufficient to include a hydrated DNA double helix with a total diameter of 2.5 nm. A second periodicity was observed, which was attributed to DNA-DNA correlation, that ranged from 2.45 to 5.71 nm as the concentration of helper lipid was increased. These lamellar complexes can be converted into hexagonal complexes with DNA confined in inverted lipid micelles by changing the membrane spontaneous curvature or the membrane flexibility (37). This hexagonal arrangement has demonstrated increased transfection, which is attributed to the relative instability of the complexes through rapid fusion with anionic vesicles and subsequent DNA release. X-ray diffraction studies and microscopy have illustrated the high degree of variability in lipoplex structure and the need to understand and control the parameters that govern the organization of these structures.

The colloidal properties (e.g. size, stability) of the lipoplexes are principally determined by the cationic lipid/DNA charge ratio and not the composition of the lipid or the helper lipid (38). The charge ratio (+/−) is typically defined as the number of amines on the cationic lipid relative to the number of phosphate groups on the DNA. A neutral charge ratio (1:1 charge ratio for lipid:DNA) is usually avoided because it results in the formation of large aggregates (>1 μm) (38). Lipoplexes prepared at a positive charge ratio and a negative charge ratio likely represent structures with different lipid and DNA packaging (38). At a positive charge ratio, large multilamellar vesicles (LMV) (diameter 300–700 nm) transfected cells more efficiently than the small unilamellar vesicles (SUV) (diameter 50–200 nm) (23, 39, 40). These observations were consistent whether the structures were formed as LMV or as SUV that aggregated to form LMV (39). The order in which DNA and lipid are mixed is critical and significantly affects the lipid and DNA packing (30, 38). For the addition of DNA to lipid, a gradual increase in size is observed. When adding lipid to DNA, the particle size remains roughly constant until the amount of lipid positive charge exceeds the nucleic acid negative charge, whereupon the particles grow rapidly in size (30).

The net charge on the lipoplex affects its interactions with other components present in vivo and in vitro (e.g. media, serum, extracellular matrix glycoproteins, mucosal secretions), which can limit the transfection efficiency. A positive charge ratio, which facilitates interactions with the cell membrane, is frequently used for in vitro studies (3:1), whereas in vivo studies may require the charge ratio to be altered because of interactions with components of the physiological environment (41). The charge ratio of the complex determines the zeta potential, which ranges from −55 to +55 mV as the charge ratio is increased (42). Multivalent anions present in the serum or media can facilitate fusion of the lipids causing an increase in the size of the particle. Polyanions with adequate anionic charge density (e.g. heparin) released DNA from the complex by binding the cationic lipid (42). Serum can be a complicating factor for positively charged complexes, possibly causing premature release of the DNA from the complex and enhancing degradation by nucleases. For ON:lipid complexes, the various components of serum [e.g. bovine serum albumin (BSA), lipoproteins, and macroglobulin] interact with the complexes and alter the complex diameter and zeta potential, and interfere with cellular uptake and nuclear trafficking (43).

Aggregation of lipoplexes occurs rapidly, which can result in loss of activity in less than 24 h; thus strategies are being developed to stabilize the particles and to prolong their shelf-life stability. To improve lipoplex stability, polyethyleneglycol-phosphatidylethanolamine (PEG-PE) has been incorporated into the cationic liposome. PEG-containing liposomes are prevented from aggregating and interacting with serum components, which increases their stability (44–46). Alternatively, new preparation methods include a detergent solution containing cationic lipid and DNA (47). Removal of the detergent by dialysis allows the formation of uniform complexes. This process yielded a lipid/DNA suspension that was able to transfect tissue culture cells up to 90 days after formation with no loss in activity. Lyophilization, which is a common approach used for many pharmaceuticals, is also being applied to lipoplexes to increase their shelf life. Cryoprotectants (e.g. sucrose, trehalose) are typically added to prevent aggregation and fusion of plasmid/lipid complexes during lyophilization. Lyophilization of the complexes in the presence of 0.5 M sucrose or trehalose maintains their transfection rates and maintains the sizes of rehydrated complexes compared with those of non-lyophhilized controls (48).

Cationic Polymers

Cationic polymers have been used since the late 1980s (49) as a material for gene delivery. All cationic polymers contain high densities of primary amines, which are protonatable at neutral pH. This high density of positive charges allows the cationic polymers to form stable complexes with non-viral DNA. The cationic polymers self-assemble with DNA to generate condensed structures (40–100 nm in diameter) capable of entering the cell. Figure 2 shows the structure of some of frequently used cationic polymers, although many novel polymers are being developed including polyallylamine (50), peptoids (51), methacrylamide (52, 53), and cyclodextrin-containing polymers (54). These polymers vary widely in their structures, which range from linear to highly branched molecules and influence their complexation with nucleic acids and their transfection efficiency. In addition to providing positive charges for DNA complexation, the primary amines also serve as functional groups with which to chemically modify the polymers with ligands and peptides that can enhance one or more of the steps in the transfection process.

Poly-L-lysine (PLL) and poly-L-ornithine are the most commonly used cationic polymers for gene delivery. Differing in chemical structure by only the length of their side chain (four versus three carbons), they are capable of mediating gene delivery both in vitro and in vivo (55). In order to form uniform complexes, these polymers are generally synthesized on a solid support using a series of protecting/de-protecting synthetic steps [e.g. fluoren-9-ylmethoxy-carbonyl (fmoc) chemistry] to obtain mono-disperse peptides. PLL is typically used at charge ratios (+/−) ranging from 3:1 to 6:1. As increasing amounts of PLL are added to DNA, the structure changes from circular to thick, flattened to compact, and finally to toroids and rods at a charge ratio of 6:1 (56). The diameter and cross section of the toroids are approximately 140 and 44 nm, respectively. The ideal length of the

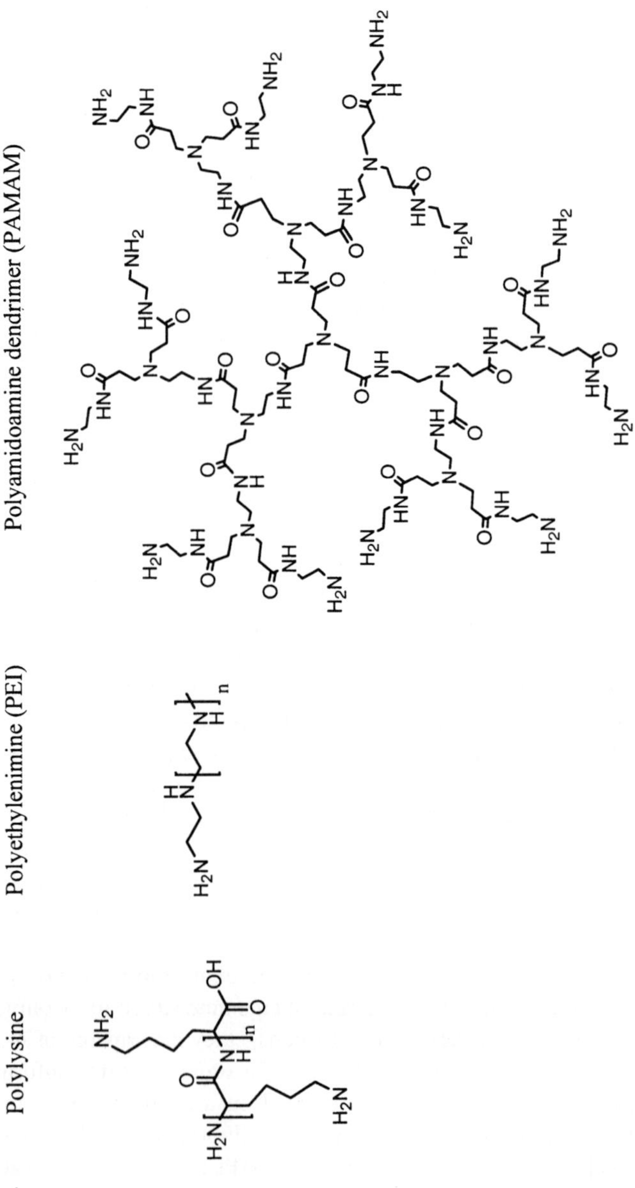

Figure 2 Structure of common cationic polymers used as materials for gene delivery.

PLL represents a balance between two competing effects: effective condensation and cytotoxicity. Relative to low-molecular–weight, the high-molecular–weight PLL forms tighter smaller condensates that are more resistant to the effects of salt concentration and sonication (57). However, cytotoxicity has been found to be inversely related to particle size (55), likely due to the high molecular weight PLL.

Poly(ethylenimine) (PEI) has been widely used to mediate gene delivery because of its high cationic charge density resulting from the protonatable amine on every third carbon (58). PEI can be synthesized by the acid-catalyzed ring opening polymerization of aziridine as either a linear or a branched structure (59). Similar to the PLL complexes, low-molecular–weight PEI is less cytotoxic than high-molecular–weight PEI (60). Furthermore, PEI/DNA complexes must also bear a net positive charge (6-13:1 nitrogen/phosphate ratio) in order to efficiently transfect cells (58, 60, 61). PEI/DNA complexes form spheres or toroids with mean diameters ranging from 30 to 100 nm. The presence of NaCl during complexation can result in particles with amorphous shapes and diameters greater than 1 μm.

Dendrimers are an attractive architecture for gene transfer because their well-defined structure and robust chemistry enable the synthesis of many generations of protonatable amines. Starburst dendrimers such as polyamidoamine dendrimers (PAMAM) have been shown capable of mediating gene delivery (62–64). PAMAM has been synthesized from either an ammonia or ethylenediamine core by successive addition of methyl acrylate and ethylenediamine (65). The surface charge and diameter of the dendrimer is determined by the number of synthetic steps (i.e. number of generations) (65). The generation of PAMAM used to complex DNA determines particle size and transfection efficiency. Complexation of DNA with fifth-generation PAMAM dendrimers produces monodisperse condensates with a radius below 200 nm (66). At charge ratios (+/−) above or equal to one, no free DNA is observed (62, 66). Higher levels of transfection were observed for PAMAM/DNA complexes formed from generation five and six dendrimers than with lower PAMAM generations (62). The success of the PAMAM dendrimers has also led to the development of a monodispersed PLL dendrimers capable of transfection (67).

Cationic polymers have also been used in combination with liposomes. DNA is initially complexed with PLL at low charge ratios and cationic lipids are subsequently added to completely condense the DNA. Alternatively, the PLL condensed DNA containing a net positive charge can subsequently be complexed with an anionic lipid (68). Precondensation with polylysine has been shown to reduce serum inhibition and also enhanced the transfection efficiency (69, 70).

Polymeric Delivery

Recently the strategies developed to deliver biologically active proteins have been adapted to deliver non-viral DNA. Polymer-based systems provide a number of potential advantages for the therapeutic delivery of drug (71). Drug encapsulation within the polymer can protect against degradation until release. Injection or implantation of the polymer into the body can be used to target a particular cell type or tissue. Drug release from the polymer and into the tissue can be designed to occur rapidly, as in a bolus delivery, or to occur over an extended period of time; thus a delivery system can be tailored to a particular application. For bolus delivery, drug levels quickly rise and decline as the drug is cleared or degraded. For sustained delivery, the drug concentration is maintained within a therapeutic window by adjusting the release rate (e.g. through the polymer choice). A result of sustained delivery is a possible reduction in the number of dosages or in the required cumulative dose (72). Sustained delivery can be particularly effective for drugs with unacceptable pharmacokinetics (e.g. short half-lives).

Non-viral plasmids and ONs have been incorporated and released from both synthetic and natural polymers fabricated into a variety of geometries and configurations (e.g. reservoirs, matrices, and microspheres). The structures of some of the synthetic materials in use for DNA delivery are provided in Figure 3. Release of DNA from polymeric vehicles has been observed for times ranging from hours to several weeks (73–75). Alternatively, the DNA may never be released and is internalized only following occupation of the polymer by the surrounding cells (76). Variations in the polymer composition and physical form, which affects the polymer degradation, can be used to control the time scale of release. In addition, the physical form of the polymer, which includes properties such as porosity, mass, and size, affects the diffusion of DNA from the vehicle and hence the release rate. Alternative mechanisms of release, such as chemical or enzymatic cleavage, are also being investigated for protein release and may eventually be applied to DNA (71).

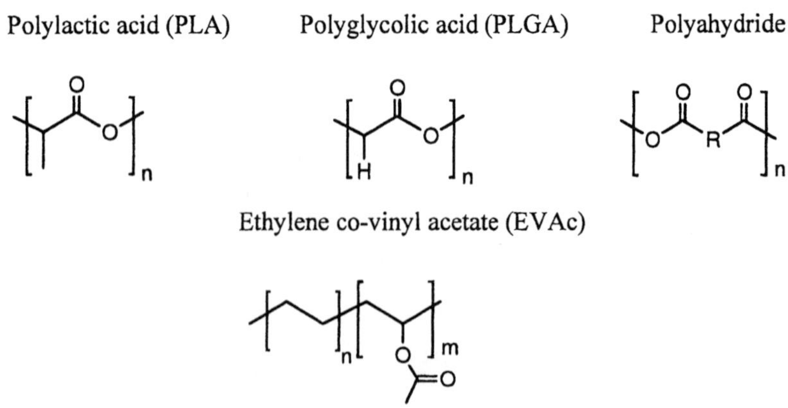

Figure 3 Structure of polymeric materials used for gene delivery.

Polymers composed of lactide and glycolide are perhaps the most widely used and recognized biodegradable synthetic polymers because of their use as materials for sutures (77). These polymers are FDA approved and are generally considered to be biocompatible. Hydrolytic degradation of the polymer produces lactic acid and glycolic acid, two substances naturally involved in metabolic pathways of the body. The degradation rate can be controlled through the composition of the polymer and molecular weight of the chains. These polymers exhibit relatively even polymer chain scission throughout their bulk after being placed in vivo. These properties have led to the use of these materials for drug delivery and as scaffolds for tissue engineering (78). Drug release from these systems typically occurs by a combination of polymer degradation and drug diffusion from the polymer.

Biodegradable polyanhydride polymers were originally developed as vehicles for drug delivery (71), which has subsequently led to their use in other biomaterial applications. These polymers are synthesized from hydrophobic monomers connected by water-labile anhydride bonds. Through appropriate choice of monomers, which controls the hydrophobicity, degradation of these materials can be designed to occur over periods ranging from one day to many months. However, the feature that makes these materials desirable for various applications is the degradation mechanism. These polymers exhibit chain scission mainly at the surface of the polymer matrix, and the matrix degrades and loses mass from the exterior to the interior (79), analogous to a bar of soap dissolving. In drug delivery applications, drug is released continuously as the matrix degrades and is particularly attractive because it prevents dose-dumping. The observation that the material maintains its structural integrity as it degrades has motivated its use in other applications, including structure replacement in a load-bearing environment (80).

Ethylene vinyl co-acetate (EVAc) has been used as a model system for the fabrication of non-degrading controlled release systems (78). Because the polymer is non-degradable, the release of high-molecular–weight drugs is dependent upon the presence of interconnecting network of channels leading to the external environment. The presence of interconnecting network of channels is a function of the drug loading. At low drug loading, complete entrapment of isolated drug particles within the polymer prevents release. Higher drug loadings or the addition of carrier proteins increases the possibility of an interconnecting network of drug particles with a path to the surroundings. As drug adjacent to the surface dissolves and diffuses out of the matrix, open channels remain that permit drug located deeper within the matrix to be solubilized and released. Modifications that affect the network of pores, such as drug loading, polymer molecular weight, and presence of co-dispersants, can be used to control the release rate of the protein.

Collagens are the major component of mammalian connective tissue and thus have been widely investigated as biomaterials for cell growth and drug delivery. Although there are at least 14 types of collagen, the most abundant is type I collagen. Type I collagen is found in high concentrations in tendon, skin, bone, and fascia, which serve as sources for its isolation. Following isolation, typically from bovine or porcine sources, collagen is preserved using techniques such as fixation with aldehydes or other chemical preservatives, β-irradiation, or lyophilization (81, 82). Collagen can be lyophilized to form a spongy product or can be formed into a hydrogel by thermal mechanisms or covalent crosslinking using aldehydes or carbodiimides (76, 83–85).

The current DNA delivery systems can generally be categorized into two different forms: microspheres or pellets, and polymer matrices. Microspheres or pellets loaded with non-viral DNA can be fabricated from non-degradable and degradable polymers in sizes ranging from 0.1 to 100 μm and have been used for applications such as DNA vaccines systemic protein delivery (78). One of the main advantages for delivery vehicles of this size is that they can be administered in a minimally invasive manner (e.g. direct injection, oral delivery). The use of biodegradable polymers provides the additional advantage of not having to retrieve the implant following DNA release. Unfortunately, these two qualities make removal of the devices difficult, should the therapy need to be terminated prematurely. The loading of DNA into the polymer ranges from 0.1 to 10 μg DNA per mg of polymer using various techniques based on either an emulsion or phase inversion process (73, 86–88). A double emulsion process has been used to incorporate aqueous solutions of non-viral DNA, both supercoiled and complexed with PLL, into polymer. Incorporation efficiencies ranging from 20 to 80% have been obtained, although the typical incorporation efficiency is approximately 30%. When supercoiled DNA is incorporated using this procedure, the incorporated DNA is structurally intact with 39% of the DNA in the supercoiled conformation. Protecting the DNA either by condensation (86) or by a cryogenic approach (88) can increase the DNA present in the supercoiled form to more than 80%.

Alternatively, DNA has also been delivered from polymer matrices for applications in tissue repair and wound healing (74, 76). These matrices are typically implanted at a specific anatomic location where they serve multiple roles (6). Initially, the matrix functions to create and maintain a space in vivo. However, the matrix also acts as a scaffold to support cell migration, proliferation, and differentiation of healthy cells from the surrounding tissue. As cells invade the matrix, they encounter DNA that is either released from or entrapped within the matrix. A novel technique was recently developed to fabricate scaffolds from PLG copolymers using a gas foaming/particulate leaching process (74, 89). The incorporation efficiency for this process was determined to be at least 50%. Release studies demonstrated a sustained release of DNA for more than 30 days.

CELLULAR EFFECTS

The ultimate applicability of the various biomaterials for gene delivery rests on their ability to effectively deliver the DNA to the nucleus of the target cell. An effective gene delivery system will protect the nucleic acid from degradation, target the appropriate cell population, be efficiently internalized by the cell and avoid the degradative pathway, and ultimately localize to the nucleus. The materials under development can specifically enhance one or more of these steps in the transfection process. The following sections describe how the materials interact with biological systems and how they can be designed to overcome the obstacles to effective gene transfer.

Tissue-Specific Delivery

In vivo production and secretion of therapeutic proteins by DNA delivery can be achieved through either systemic or local administration, each providing a unique opportunity for gene therapy. Systemic delivery allows noninvasive access to many target cells and tissue that are not accessible by direct administration. The major factors limiting the in vivo effectiveness for systemically delivered lipoplexes and polyplexes may be the poor availability or the colloidal instability. For intravenous administration, complexes are rapidly eliminated from the bloodstream (<30 min) and found mostly in the liver and to a lesser extent in organs with fine capillary beds like skin, muscle, and intestine (90). The interaction of biomolecules from the physiological fluid with the complexes may induce aggregation or dissociation of the complexes, which may limit bioavailability and internalization (10). Lipoplexes and polyplexes have demonstrated an increase in size (1–2 mm) within minutes after delivery, most likely the result of an interaction with serum components (91). The association of serum components (albumin, heparin, lipoprotein, or specific opsonins), which can target the complex for clearance by macrophages, is affected by the surface charge density and surface morphology (41). Cationic formulations (i.e. lipoplexes and polyplexes) can provide an enhanced stability against degradation and may allow for a more efficient interaction with the negatively charged endothelial cell; however, non-cationic or less cationic formulations may be required to avoid accumulation in the reticuloendothelial system. Variations in the charge ratio may be used as part of a passive targeting strategy. For lipoplexes delivered intravenously, different formulations (e.g. charge ratio) can be used to target the lung, liver, blood cells, or tumors (10, 41).

To avoid the problems of crossing the endothelium and of nonspecific uptake by the liver, DNA can be delivered directly to the target tissue by injection of polyplexes, lipoplexes, or microspheres with encapsulated DNA. Injection of lipoplexes and polyplexes into the desired tissue can be used to achieve in vivo gene transfer. However, transfection depends upon the physicochemical properties of the complexes and the rate of clearance from the tissue, which varies from tissue to tissue and depends on factors such as the lymph supply. For example, PEI/DNA complexes are sufficiently small and stable so as to diffuse throughout the brain ventricular spaces following injection at a single site (61). Sustained release systems are one approach to overcome the difficulties associated with clearance from the tissue (71). A continuous infusion of lipoplexes into the brain provides an increased gene transfer (17). Implantation of a PLG scaffold with a sustained release of DNA leads to significantly greater physiological responses compared with a scaffold implantation accompanied by a bolus delivery of the DNA on the scaffold (74). Sustained delivery has the potential to release larger quantities of DNA over a longer period of time than can be delivered by a single injection.

Polymeric systems have been hypothesized to increase transfection by overcoming the mass transfer limitations associated with transporting the DNA to the cell surface (92). PLG microspheres are being fabricated with diameters of ~5 μm to specifically target macrophages for internalization (93–96). These microspheres are too large to enter cells by endocytosis; however, they can be phagocytosed by macrophages, which are the antigen-presenting cells for the immune system. For preparations that are not internalized, many of the polymeric materials used for encapsulation also support cell adhesion (e.g. collagen, PLG); thus the cells are adhered to the material from which the DNA is being released, which creates a high concentration at the cell surface. Association of DNA with the extracellular matrix is a technique that viruses have used to enhance uptake (97). Implanted collagen scaffolds function similarly by maintaining the DNA in situ until endogenous fibroblasts arrive and internalize the DNA (76, 85). Injectable (84) and ingestible vehicles (98) achieve direct contact with the surrounding tissue and localized gene transfer is observed. Although gene transfer is observed, the mechanism by which internalization occurs is not well understood.

Cellular Internalization

The plasma membrane of the cell provides a protective shell that limits the transport of undesired molecules, such as DNA. However, cells must communicate with their environment and have thus developed mechanisms to transport high- and low-molecular weight macromolecules across the membrane. Non-clathrin-coated pit internalization can occur through smooth invaginations of 150–300 nm or via potocytosis, which involves the invagination of caveolae-rich 50–100 nm diameter vesicles from the cell surface. DNA complexes are thought to enter cells primarily through clathrin-coated pits. These pits, with diameters of ~150 nm, are internalized from the plasma membrane to form coated vesicles. Lipoplex and polyplexes with a net positive charge may interact nonspecifically with the negatively charged cell surface and be internalized. For lipoplexes, internalization is thought to occur primarily through endocytosis (99); however, alternative mechanisms have been proposed (100).

Receptor-mediated gene delivery offers a promising approach to create a specific interaction with the cell surface and to target a particular internalization pathway. Synthetic materials termed molecular conjugates are being developed and are composed of two domains: a DNA-binding domain and a receptor-binding domain. The DNA-binding domain is in charge of linking the DNA to the rest of the molecular conjugate molecule and is typically composed of a cationic polymer or a cationic lipid, which self-assembles with DNA. The receptor-binding domain is frequently attached to functional groups on the cationic polymer or lipid prior to complexation with DNA (26); however, complexes can be initially formed and subsequently coupled to a ligand (45, 101). The function of the receptor-binding domain is to direct the DNA/molecular conjugate complex to a receptor and guide the complex through the internalization pathway that leads to endosomal entrapment. The intracellular fate of DNA complexes can depend on the type of endocytic process involved in its internalization (102). The diversity and abundance of membrane bound receptors and ligands and the high efficiency of internalization and turnover ratio makes receptor-mediated endocytosis a powerful tool for controlling uptake. Targets for which the attached ligand bind include the receptors for asialoglycoprotein (49, 103, 104), transferrin (105), folate (106–108), lectins (109–111), integrins (112), and epidermal growth factor (EGF) (113). The attachment of a targeting ligand to the lipoplex has been shown to increase transfection by 1000-fold compared with that of lipid with no targeting ligand (114). In principle, any monoclonal antibody, antigen-binding fragment of a monoclonal antibody, peptide, peptide mimetic, proteins, or peptide fragment can be added to the molecular conjugate using either a covalent or ionic attachment.

Maximal gene delivery through the receptor-mediated endocytosis pathway requires that the design parameters be optimized, which includes properties such as the ligand-binding affinity, the length of the linker between the ligand and the complex, the complex charge ratio and structure, and the number of ligands per complex (113, 115). Ligand-receptor interactions can be very specific and may be negatively affected when covalently bound to a molecular conjugate. Knowing if the receptor-binding domain is interacting with its receptor with the same efficiency as the free ligand is important to achieve high gene delivery levels. In a model using an EGF-polylysine conjugate, it was found that a longer spacer arm, between the two domains, resulted in more native-like EGF-receptor binding and higher transfection efficiency than shorter spacer arms (113). The presence of the ligand can alter the structure of the nucleic acid complex. Incorporation of asialoorosomucoid into PLL/DNA complexes was more efficient at condensing DNA than PLL alone, probably because of a large aggregation of PLL around the negatively charged protein (56). At low-charge ratios, specific (ligand-mediated) and nonspecific endocytosis had insignificant levels of transfection. As the charge ratio was increased to four, the transfection efficiency by specific binding enhanced transfection by a factor of three more than by nonspecific endocytosis (113). Finally, the number of ligands per complex can affect binding of the complexes to cells. An excessive number of ligands may inhibit the binding of complexes (113). Optimization of the parameters that affect the surface binding of the molecular conjugates improves the specificity and gene transfer efficiency for gene delivery.

Intracellular Trafficking

The steps following internalization of DNA or DNA complexes–endosomal escape and nuclear localization–are thought to be rate limiting for the transfection of many cell types. Internalization of the plasmid does not necessarily correlate to transfection (116). Much of the DNA that is internalized into the endosome is either retained or degraded within the endosome. The DNA that does escape the endosome and enter the cytoplasm must subsequently avoid degradation and be transported to the nucleus for successful gene transfer (117). In vitro studies have demonstrated that although >95% of the cells were positive for plasmid (>100, 000 copies per cells), <50% of the cells were expressing the transgene (118, 119).

ENDOSOMAL ESCAPE For lipoplex-mediated delivery, the interaction of the lipids with the endosomal membrane is thought to facilitate escape of the DNA to the cytoplasm prior to its degradation in the lysosome. Following internalization, the coated vesicle transforms into an early endosome, which is accompanied by acidification of the vesicular lumen that continues into the late endosomal and lysosomal compartments, reaching a final pH in the perinuclear lysosome of approximately 4.5. Although it is not well understood, some lipoplexes preferentially fuse with the early endosome whereas others fuse with the late endosome (23). pH-sensitive liposomes can take advantage of the acidification process to facilitate the release of plasmids into the cytoplasm prior to lysosomal degradation (120). Release of the DNA from the complex may occur at the wall of the endosome (42, 121). Xu & Szoka proposed a model in which destabilization of the endosomal membrane causes a flip-flop of anionic lipids from the cytoplasmic-facing monolayer, which diffuse laterally into the complex and form a charge neutral ion pair with the cationic lipid (42). This pairing results in displacement of the DNA from the cationic lipid and release of the DNA into the cytoplasm. Displacement of the cationic lipid from the DNA prior to its entering the nucleus is critical for the ultimate expression of the gene.

Materials containing secondary and tertiary amines that are protonated at acidic pH, and not at neutral pH like primary amines, can act as proton sponges that buffer the decrease in pH and ultimately cause DNA release into the cytoplasm. Imidazoles have been incorporated into polylysine/DNA complexes effectively buffering the endosomal space and releasing PLL/DNA complexes into the cytosol (122, 123). The buffering capacity of the spermine head group on a cationic lipid may facilitate endosome escape (114). This proton sponge effect is thought to result from the protonation of amino moieties on the polymer as the pH decreases inside the endocytic vesicle. The influx of counter ions, which maintains electroneutrality, induces osmotic swelling and rupture of the vesicle membrane (124, 125). As an alternative approach, branched structures of PAMAM and PEI, which contain secondary and tertiary amines, unfold as they are protonated, which causes the endosome to rupture (58, 126).

Endosomal disrupting peptides, also termed fusogenic peptides, can be covalently incorporated into DNA/lipid (82, 127) and DNA/cationic polymer (128, 129) complexes to enhance escape from the endosome. Viruses and bacteria have evolved sophisticated endosomal release pathways that release viral particles into the cytosol (130). Based on the viral and bacterial pathways, peptides for gene delivery have been identified from the viral fusion proteins and have been used successfully to enhance endosomal escape. An extensive list of fusogenic peptides can be found in Reference 130. The peptides range in length from 15 to 30 amino acids and form stable amphipathic α-helices primarily because of alternating hydrophilic and hydrophobic amino acids. Most of the peptides are amphipathic pH sensitive, and active only at low pH (130). There are two mechanisms by which fusion peptides disrupt endosomal membranes. First, the peptide causes the rearrangement of the lipid packaging, thereby changing the membrane integrity and causing release of the endosomal contents. Second, the peptide causes pore formation within the endosomal membrane, without affecting the membrane integrity (130). Most peptides induce endosomal release by membrane disruption rather than membrane fusion (130). For example, GALA, which was one of the first synthetic peptides to be designed and synthesized, undergoes a conformational change to form an α-helix at acidic pHs. The α-helix formed induces endosomal leakage by disrupting the endosomal membrane. A similar peptide, KALA, also functions to disrupt endosomal membranes but is positively charged and is capable of condensing DNA and mediating gene delivery (131).

NUCLEAR LOCALIZATION The transport of DNA from the cytoplasm to the nucleus may be the most significant limitation to successful gene transfer. Plasmids injected far from the nuclei (60–90 μm) had less protein expression than plasmids injected near the nuclei (132). In addition to cytoplasmic transport limitations, the size of DNA is problematic for crossing into the nucleus. The nuclear pores allow free diffusion entry of only small particles $\leq$70 kDa (133). Entry of the DNA can be facilitated by the breakdown of the nuclear membrane, which occurs during cell division (134). However, for delivery to non-dividing cells, nuclear localization sequences (NLS) can be incorporated into the DNA complexes in order to direct the plasmid into the nucleus.

NLS are short peptide sequences (5–25 amino acids) necessary and sufficient for nuclear localization of their respective proteins (133). These sequences can be incorporated into complexes with cationic lipids (135), cationic polymers (136, 137), or can be directly linked to the plasmid (138, 139). Some nuclear localization signals have stretches of positive charge, which has led to speculation that cationic polymers function as nuclear localization signals (69). The cationic polymer PEI rapidly accumulates in the nucleus of cells and to a greater extent than does PLL (140). Nevertheless, the attachment of NLS does significantly enhance nuclear accumulation (137). For the transfection of non-dividing endothelial cells, lipoplex transfection resulted in 5% of the cells testing positive for transfection; whereas, the incorporation of a NLS resulted in more than 80% of the cells testing positive (135).

SUMMARY AND FUTURE DIRECTIONS

Non-viral vectors for efficient gene transfer in vivo are not currently available, although the prospects for development of materials to enhance gene transfer are promising. The successful development of a system for the safe, efficient, and controlled delivery of plasmids and ONs will require a thorough understanding of the interaction between nucleic acids, the material, and the biological environment. Microscopy and X-ray diffraction studies have illustrated the high degree of variability in the structure of DNA complexes (spheroidal, toroidal, and longitudinal structures) and the need to understand and control the parameters that govern the organization of these structures. Although some properties have emerged from the available studies, a definitive relationship between complex structure and function has yet to be elucidated. A complete set of structure-function data should allow the formidable task of a rational design of self-assembled complexes for enhanced gene delivery from the ground up. In addition to the complexation strategies, systems will be needed to control the biodistribution of the delivery. Future studies will undoubtedly develop targeting strategies and novel polymeric delivery systems (e.g. nanospheres) that can tightly control the exposure of the gene to the target cells. The expression level produced by these vehicles ultimately depends on its interaction with the biological environment and the ability to overcome the barriers to gene transfer. The ideal vehicle will stabilize the DNA against degradation, target the DNA to the desired tissue, enhance its transport across biological membranes, and dissociate at the appropriate time to allow for expression. More investigation is required to understand the mechanisms underlying these processes in order to optimize the design of the non-viral vector. The combined understanding of the material properties and the biological system will serve to identify vehicles that are optimized to alter gene expression for the treatment of a multitude of disorders.

LITERATURE CITED

1. Fannon MR. 1996. *Trends Biotechnol.* 14: 294–98
2. Schena M, Heller R, Theriault T, Konrad K, Lachenmeier E, Davis R. 1998. *Trends Biotechnol.* 16:301–6
3. Silverman L, Campbell R, Broach JR. 1998. *Curr. Opin. Chem. Biol.* 2:397–403
4. Anderson WF. 1998. *Nature* 392:25–30
5. Andree C, Swain WF, Page CP, Macklin MD, Slama J, et al. 1994. *Proc. Natl. Acad. Sci. USA* 91:12188–92
6. Bonadio J, Goldstein SA, Levy RJ. 1998. *Adv. Drug Del. Rev.* 33:53–69
7. Bennett CF. 1998. *Biochem. Pharmacol.* 55:9–19
8. Chan PP, Glazer PM. 1997. *J. Mol. Med.* 75:267–82
9. Chang PL, Bowie KM. 1998. *Adv. Drug Del. Rev.* 33:31–43
10. Ledley FD. 1996. *Pharmacol. Res.* 13: 1595–614
11. Kawabata K, Takakura Y, Hashida M. 1995. *Pharmacol. Res.* 12:825–30
12. Takakura Y, Mahato RI, Hashida M. 1998. *Adv. Drug Del. Rev.* 34:93–108
13. Choate KA, Khavari PA. 1997. *Hum. Gene Ther.* 8:1659–65
14. Levy MY, Barron LG, Meyer KB, Szoka FC Jr. 1996. *Gene Ther.* 3:201–11
15. Wells DJ. 1995. *Mol. Cell Biol. Hum. Dis. Ser.* 5:83–103
16. Budker V, Zhang G, Danko I, Williams P, Wolff J. 1998. *Gene Ther.* 5:272–76
17. Imaoka T, Date I, Ohmoto T, Yasuda T,

Tsuda M. 1998. *Brain Res.* 780:119–28
18. Wyber JA, Andrews J, D'Emanuele A. 1997. *Pharmacol. Res.* 14:750–56
19. Budker V, Budker T, Zhang G, Subbotin V, Loomis A, Wolff JA. 2000. *J. Gene Med.* 2:76–88
20. Felgner PL, Gadek TR, Holm M, Roman R, Chan HW, et al. 1987. *Proc. Natl. Acad. Sci. USA* 84:7413–17
21. Bennett CF, Mirejovsky D, Crooke RM, Tsai YJ, Felgner J, et al. 1998. *J. Drug Target* 5:149–62
22. Ferrari ME, Nguyen CM, Zelphati O, Tsai Y, Felgner PL. 1998. *Hum. Gene Ther.* 9:341–51
23. Felgner JH, Kumar R, Sridhar CN, Wheeler CJ, Tsai YJ, et al. 1994. *J. Biol. Chem.* 269:2550–61
24. Wang J, Guo X, Xu Y, Barron L, Szoka FC Jr. 1998. *J. Med. Chem.* 41:2207–15
25. MacDonald RC, Rakhmanova VA, Choi KL, Rosenzweig HS, Lahiri MK. 1999. *J. Pharmacol. Sci.* 88:896–904
26. Byk G, Dubertret C, Escriou V, Frederic M, Jaslin G, et al. 1998. *J. Med. Chem.* 41:229–35
27. Byk G, Soto J, Mattler C, Frederic M, Scherman D. 1998. *Biotechnol. Bioeng.* 61:81–87
28. Lee ER, Marshall J, Siegel CS, Jiang C, Yew NS, et al. 1996. *Hum. Gene Ther.* 7:1701–17
29. Radler JO, Koltover I, Salditt T, Safinya CR. 1997. *Science* 275:810–14
30. Kennedy MT, Pozharski EV, Rakhmanova VA, MacDonald RC. 2000. *Biophys. J.* 78:1620–33
31. Bruinsma R. 1998. *Eur. J. Phys. B* 4:75–88
32. Sternberg B, Sorgi FL, Huang L. 1994. *FEBS Lett.* 356:361–66
33. Gershon H, Ghirlando R, Guttman SB, Minsky A. 1993. *Biochemistry* 32:7143–51
34. Gustafsson J, Arvidson G, Karlsson G, Almgren M. 1995. *Biochim. Biophys. Acta* 1235:305–12
35. Lasic DD, Strey H, Stuart MCA, Pogdornik R, Frederik PM. 1997. *J. Am. Chem. Soc.* 119:832–33
36. Mok KW, Cullis PR. 1997. *Biophys. J.* 73:2534–45
37. Koltover I, Salditt T, Radler JO, Safinya CR. 1998. *Science* 281:78–81
38. Xu Y, Hui SW, Frederik P, Szoka FC Jr. 1999. *Biophys. J.* 77:341–53
39. Turek J, Dubertret C, Jaslin G, Antonakis K, Scherman D, Pitard B. 2000. *J. Gene Med.* 2:32–40
40. Ross PC, Hui SW. 1999. *Gene Ther.* 6:651–59
41. Mahato RI, Smith LC, Rolland A. 1999. *Adv. Genet.* 41:95–156
42. Xu Y, Szoka FC Jr. 1996. *Biochemistry* 35:5616–23
43. Zelphati O, Uyechi LS, Barron LG, Szoka FC Jr. 1998. *Biochim. Biophys. Acta* 1390:119–33
44. Hong K, Zheng W, Baker A, Papahadjopoulos D. 1997. *FEBS Lett.* 400:233–37
45. Meyer O, Kirpotin D, Hong K, Sternberg B, Park JW, et al. 1998. *J. Biol. Chem.* 273:15621–27
46. Mok KW, Lam AM, Cullis PR. 1999. *Biochim. Biophys. Acta* 1419:137–50
47. Hofland HE, Shephard L, Sullivan SM. 1996. *Proc. Natl. Acad. Sci. USA* 93:7305–9
48. Anchordoquy TJ, Carpenter JF, Kroll DJ. 1997. *Arch Biochem. Biophys.* 348:199–206
49. Wu GY, Wu CH. 1987. *J. Biol. Chem.* 262:4429–32
50. Boussif O, Delair T, Brua C, Veron L, Pavirani A, Kolbe H. V. 1999. *Bioconjugate Chem.* 10:877–83
51. Murphy JE, Uno T, Hamer JD, Cohen FE, Dwarki V, Zuckermann RN. 1998. *Proc. Natl. Acad. Sci. USA* 95:1517–22
52. van de Wetering P, Cherng JY, Talsma H, Crommelin DJ, Hennink WE. 1998. *J. Control. Release* 53:145–53
53. van de Wetering P, Moret EE, Schuurmans-Nieuwenbroek NM, van Steenbergen MJ,

Hennink WE. 1999. *Bioconjugate Chem.* 10:589–97
54. Gonzales H, Hwang SJ, Davis ME. 1999. *Bioconjugate Chem.* 10:1068–74
55. Duguid JG, Li C, Shi M, Logan MJ, Alila H, et al. 1998. *Biophys. J.* 74:2802–14
56. Golan R, Pietrasanta LI, Hsieh W, Hansma HG. 1999. *Biochemistry* 38:14069–76
57. Adami RC, Collard WT, Gupta SA, Kwok KY, Bonadio J, Rice KG. 1998. *J. Pharmacol. Sci.* 87:678–83
58. Boussif O, Lezoualc'h F, Zanta MA, Mergny MD, Scherman D, et al. 1995. *Proc. Natl. Acad. Sci. USA* 92:7297–301
59. Dick RC, Ham GE. 1997. *J. Macromol. Sci. A* 4:1301–14
60. Fischer D, Bieber T, Li Y, Elsasser HP, Kissel T. 1999. *Pharmacol. Res.* 16:1273–79
61. Goula D, Remy JS, Erbacher P, Wasowicz M, Levi G, et al. 1998. *Gene Ther.* 5:712–17
62. Haensler J, Szoka FC Jr. 1993. *Bioconjugate Chem.* 4:372–79
63. Bielinska AU, Kukowska-Latallo JF, Baker JR Jr. 1997. *Biochim. Biophys. Acta* 1353:180–90
64. Qin L, Pahud DR, Ding Y, Bielinska AU, Kukowska-Latallo JF, et al. 1998. *Hum. Gene Ther.* 9:553–60
65. Tomalia DA, Naylor AM, Goddard WA. 1990. *Angew Chem. Int. Ed. Engl.* 29:138–75
66. Tomlinson E, Rolland AP. 1996. *J. Control. Release* 39:357–72
67. Choi JS, Lee EJ Choi YH, Jeong YJ, Park JS. 1999. *Bioconjugate Chem.* 10:62–65
68. Lee RJ, Huang L. 1996. *J. Biol. Chem.* 271:8481–87
69. Vitiello L, Chonn A, Wasserman JD, Duff C, Worton RG. 1996. *Gene Ther.* 3:396–404
70. Gao X, Huang L. 1996. *Biochemistry* 35:1027–36
71. Langer R. 1998. *Nature* 392:5–10
72. Sanders LM, Kell BA, McRae GI, Whitehead GW. 1986. *J. Pharmacol. Sci.* 75:356–60
73. Jong YS, Jacob JS, Yip K, Gardner G, Seitelman E, et al. 1997. *J. Cont. Rel.* 47:123–34
74. Shea LD, Smiley E, Bonadio J, Mooney DJ. 1999. *Nat. Biotechnol.* 17:551–54
75. Cleek RL, Rege AA, Denner LA, Eskin SG, Mikos AG. 1997. *J. Biomed. Mater. Res.* 35:525–30
76. Bonadio J, Smiley E, Patil P, Goldstein S. 1999. *Nat. Med.* 5:753–59
77. Wong WH, Mooney DJ. 1997. In *Synthesis and Properties of Biodegradable Polymers Used as Synthetic Matrices for Tissue Engineering*, ed. A Atala, DJ Mooney, pp. 49–80. Boston: Birkhauser
78. Baldwin SP, Saltzman WM. 1998. *Adv. Drug Del. Rev.* 33:71–86
79. Peppas NA, Langer R. 1994. *Science* 263:1715–20
80. Muggli DS, Burkoth AK, Anseth KS. 1999. *J. Biomed. Mater. Res.* 46:271–78
81. Pachence JM. 1996. *J. Biomed. Mater. Res.* 33:35–40
82. Schoen FJ, Levy RJ. 1999. *J. Biomed. Mater. Res.* 47:439–65
83. Pieper JS, Oosterhof A, Dijkstra PJ, Veerkamp JH, van Kuppevelt TH. 1999. *Biomaterials* 20:847–58
84. Ochiya T, Takahama Y, Nagahara S, Sumita Y, Hisada A, et al. 1999. *Nat. Med.* 5:707–10
85. Fang J, Zhu YY, Smiley E, Bonadio J, Rouleau JP, et al. 1996. *Proc. Natl. Acad. Sci. USA* 93:5753–58
86. Capan Y, Woo BH, Gebrekidan S, Ahmed S, DeLuca PP. 1999. *Pharmacol. Res.* 16:509–13
87. Capan Y, Woo BH, Gebrekidan S, Ahmed S, DeLuca PP. 1999. *J. Control. Release* 60:279–86
88. Ando S, Putnam D, Pack DW, Langer R 1999. *J. Pharmacol. Sci.* 88:126–30
89. Harris LD, Kim BS, Mooney DJ. 1998. *J. Biomed. Mater. Res.* 42:396–402

90. Dash PR, Read ML, Barrett LB, Wolfert MA, Seymour LW. 1999. *Gene Ther.* 6: 643–50
91. Barron LG, Gagne L, Szoka FC Jr. 1999. *Hum. Gene Ther.* 10:1683–94
92. Luo D, Saltzman WM. 2000. *Nat. Biotechnol.* 18:893–95
93. Hedley ML, Curley J, Urban R. 1998. *Nat. Med.* 4:365–68
94. Wang D, Robinson DR, Kwon GS, Samuel J. 1999. *J. Control. Release* 57:9–18
95. Jones DH, Corris S, McDonald S, Clegg JC, Farrar GH. 1997. *Vaccine* 15:814–17
96. Luo D, Woodrow-Mumford K, Belcheva N, Saltzman WM. 1999. *Pharmacol. Res.* 16:1300–8
97. Williams DA. 1999. *Ann. NY Acad. Sci.* 872:109–13; discussion 113–24
98. Mathiowitz E, Jacob JS, Jong YS, Carino GP, Chickering DE, et al. 1997. *Nature* 386:410–14
99. Friend DS, Papahadjopoulos D, Debs RJ. 1996. *Biochim. Biophys. Acta* 1278:41–50
100. Monkkonen J, Urtti A. 1998. *Adv. Drug Del. Rev.* 34:37–49
101. Templeton NS, Lasic DD. 1999. *Mol. Biotechnol.* 11:175–80
102. Schuele RK, Cheng SH. 1996. *Gene Transfer into Mammalian Cells Using Synthetic Cationic Lipids.* Washington, DC: Am. Chem. Soc.
103. Plank C, Zatloukal K, Cotten M, Mechtler K, Wagner E. 1992. *Bioconjugate Chem.* 3:533–39
104. Baatz JE, Bruno MD, Ciraolo PJ, Glasser SW, Stripp BR, et al. 1994. *Proc. Natl. Acad. Sci. USA* 91:2547–51
105. Wagner E, Zenke M, Cotten M, Beug H, Birnstiel ML. 1990. *Proc. Natl. Acad. Sci. USA* 87:3410–14
106. Gottschalk S, Cristiano RJ, Smith LC, Woo SL. 1994. *Gene Ther.* 1:185–91
107. Leamon CP, Low PS. 1991. *Proc. Natl. Acad. Sci. USA* 88:5572–76
108. Weitman SD, Weinberg AG, Coney LR, Zurawski VR, Jennings DS, Kamen BA. 1992. *Cancer Res.* 52:6708–11
109. Midoux P, Mendes C, Legrand A, Raimond J, Mayer R, et al. 1993. *Nucleic Acids Res.* 21:871–78
110. Erbacher P, Bousser MT, Raimond J, Monsigny M, Midoux P, Roche AC. 1996. *Hum. Gene Ther.* 7:721–29
111. Stewart AJ, Pichon C, Meunier L, Midoux P, Monsigny M, Roche AC. 1996. *Mol. Pharmacol.* 50:1487–94
112. Harbottle RP, Cooper RG, Hart SL, Ladhoff A, McKay T, et al. 1998. *Hum. Gene Ther.* 9:1037–47
113. Schaffer DV, Lauffenburger DA. 1998. *J. Biol. Chem.* 273:28004–9
114. Remy JS, Kichler A, Mordvinov V, Schuber F, Behr JP. 1995. *Proc. Natl. Acad. Sci. USA* 92:1744–48
115. Ziady AG, Ferkol T, Dawson DV, Perlmutter DH, Davis PB. 1999. *J. Biol. Chem.* 274:4908–16
116. Reimer DL, Kong S, Bally MB. 1997. *J. Biol. Chem.* 272:19480–87
117. Lechardeur D, Sohn KJ, Haardt M, Joshi PB, Monck M, et al. 1999. *Gene Ther.* 6:482–97
118. Zabner J, Fasbender AJ, Moninger T, Poellinger KA, Welsh MJ. 1995. *J. Biol. Chem.* 270:18997–9007
119. Tseng WC, Haselton FR, Giorgio TD. 1997. *J. Biol. Chem.* 272:25641–47
120. Budker V, Gurevich V, Hagstrom JE, Bortzov F, Wolff JA. 1996. *Nat. Biotechnol.* 14:760–64
121. Zelphati O, Szoka FC Jr. 1996. *Pharmacol. Res.* 13:1367–72
122. Pack DW, Putnam D, Langer R. 2000. *Biotechnol. Bioeng.* 67:217–23
123. Midoux P, Monsigny M. 1999. *Bioconjugate Chem.* 10:406–11
124. Behr JP. 1997. *Chimia* 51:34–36
125. Pouton CW, Seymour LW. 1998. *Adv. Drug Deliv. Rev.* 34:3–19
126. Haensler J, Szoka FC Jr. 1993. *Bioconjugate Chem.* 4:85–93
127. Pedroso MC, Simoes S, Pires P, Gaspar

R, Slepushkin V, Duzgunes N. 1999. *Mol. Membr. Biol.* 16:103–9
128. Curiel DT, Agarwal S, Wagner E, Cotten M. 1991. *Proc. Natl. Acad. Sci. USA* 88:8850–54
129. Cristiano RJ, Smith LC, Woo SL. 1993. *Proc. Natl. Acad. Sci. USA* 90:2122–26
130. Plank C, Zauner W, Wagner E. 1998. *Adv. Drug Deliv. Rev.* 34:21–35
131. Wyman TB, Nicol F, Zelphati O, Scaria PV, Plank C, Szoka FC Jr. 1997. *Biochemistry* 36:3008–17
132. Dowty ME, Williams P, Zhang G, Hagstrom JE, Wolff JA. 1995. *Proc. Natl. Acad. Sci. USA* 92:4572–76
133. Jans DA, Chan CK, Huebner S. 1998. *Med. Res. Rev.* 18:189–223
134. Tseng WC, Haselton FR, Giorgio TD. 1999. *Biochim. Biophys. Acta* 1445:53–64
135. Subramanian A, Ranganathan P, Diamond SL. 1999. *Nat. Biotechnol.* 17:873–77
136. Singh D, Bisland SK, Kawamura K, Gariepy J. 1999. *Bioconjugate Chem.* 10:745–54
137. Chan CK, Jans DA. 1999. *Hum. Gene Ther.* 10:1695–702
138. Sebestyen MG, Ludtke JJ, Bassik MC, Zhang G, Budker V, et al. 1998. *Nat. Biotechnol.* 16:80–85
139. Zanta MA, Belguise-Valladier P, Behr JP. 1999. *Proc. Natl. Acad. Sci. USA* 96:91–96
140. Godbey WT, Wu KK, Mikos AG. 1999. *Proc. Natl. Acad. Sci. USA* 96:5177–81

Annu. Rev. Mater. Res. 2001. 31:47–79

DEVELOPMENT IN UNDERSTANDING AND CONTROLLING THE STAEBLER-WRONSKI EFFECT IN a-Si:H

Hellmut Fritzsche
Energy Conversion Devices, Inc., Troy, Michigan 48084; e-mail: hellmutf@aol.com

Key Words hydrogen diffusion, photo-induced defects, photo-induced expansion, solar cells, solar cell efficiency, electronic states, hydrogen microstructure

■ **Abstract** Hydrogenated amorphous silicon (a-Si:H) exhibits a metastable light-induced degradation of its optoelectronic properties that is called the Staebler-Wronski effect, after its discoverers. This degradation effect is associated with the relatively high diffusion coefficient of hydrogen and the changes in local bonding coordination promoted by hydrogen. Reviewed are the fundamental aspects of the interplay between hydrogen and electronic energy states that form the basis of competing microscopic models for explaining the degradation effect. These models are tested against the latest experimental observations, and material and preparation parameters that reduce the Staebler-Wronski effect are discussed.

INTRODUCTION

Hydrogenated amorphous silicon (a-Si:H) is the most commonly used material for large-area electronic and optoelectronic devices because it can be deposited as thin layers, which absorb light much more efficiently than crystalline silicon, in which crystal momentum conservation restricts optical transitions, and because it can be doped both *n*-type and *p*-type, allowing the manufacture of junction devices and thin-film transistors (1). Hydrogenation is essential for obtaining these properties and for reducing the native defect concentration to about 10^{15} cm^{-3}. These defects are dangling bonds on threefold coordinated Si atoms; they produce amphoteric electronic states near midgap, which act as efficient recombination centers (2).

Extended illumination produces an additional 10^{17} cm^{-3} of these defects, appreciably degrading the optoelectronic properties of this material (3). This light-induced degradation is called the Staebler-Wronski effect (SWE). Even though these defects anneal away between 150 and 200°C, the SWE seriously limits the use of a-Si:H in solar photovoltaic applications.

The search for understanding the SWE has revealed fascinating scientific aspects and made a-Si:H one of the most thoroughly studied materials. Reversible photostructural changes and photodegradation are commonly observed in low-coordination glasses and polymers but are unexpected in rigid fourfold coordinated structures. In crystalline Si, atomic motions that could produce coordination defects occur only near 1000°C. The low-anneal temperature of the light-induced coordination defects (dangling bonds) in a-Si:H is related to the diffusion of bonded hydrogen. Above $T_E \sim 200°C$, hydrogen diffusion enables the equilibration of solid-state chemical reactions, which determine the concentration of coordination defects of silicon as well as of dopant atoms (4–8).

The doping efficiency thus changes with the equilibration conditions near and above T_E because only fourfold coordinated P and B act as donors and acceptors, respectively, whereas most P and B in a-Si:H fulfill their valency requirement and are threefold coordinated. Hydrogen can be considered a subsystem, a hydrogen glass, within the rigid Si structure (4, 5 and references therein). As the material cools below T_E, the glass transition temperature, equilibration proceeds exponentially more slowly.

Present theories of the SWE combine our understanding of the electronic and hydrogen energy states and of hydrogen diffusion. About nine years ago, an authoritative review (9) actually concluded that the SWE is understood, except for a minor disagreement about the range of H motion involved. Since that time, new experimental results have eroded some foundations of these theories: These new findings include (*a*) a lack of spatial correlation between the defects and hydrogen; (*b*) the observation that the SWE occurs at 4.2 K with essentially the same efficiency as at 300 K; (*c*) the realization that the effectiveness of light-induced defects as recombination centers depends on the light exposure conditions; and (*d*) the discovery that not only defects are produced by extended light exposure but also larger structural changes in the material involving the Si network.

This review highlights the recent developments in our understanding as well as some successes in reducing the SWE in present devices.

Hydrogen in a-Si:H

The concentration of hydrogen, about 10 at.%, depends remarkably little on the conditions of preparing a-Si:H by plasma-enhanced chemical-vapor deposition at RF frequencies from Si H_4 (or Si H_4 typically diluted 20:1 with H_2) at substrate temperatures between 200 and 300°C. NMR studies revealed (10, 11) that about 4 at.% H is in a dilute phase, atomically dispersed and bonded as monohydride (Si-H). About 1 at.% is found as H_2 interstitially and in voids (12). The rest forms clusters of about 6–8 Si-H, presumably at internal surfaces of di- or tri-vacancies (13). Near 350°C, hydrogen begins to effuse from a-Si:H without significantly changing the dangling bond defect concentration N (14, 15). This means that the broken Si bonds left behind must be close pairs and reconstruct to a Si-Si bond, albeit weak and stretched. Taking one H from these pair sites leaves behind a dangling bond, which costs energy. Taking two H costs less because a Si-Si bond is formed. This feature leads to a negative effective correlation energy (-U) of

H pair states in clusters (16). Hydrogen can be reintroduced into a-Si:H from a H plasma, essentially without a change in N, which proves the presence of empty H pair states, i.e. presumably stretched Si-Si bonds (17). At higher temperatures, near 500°C, H from isolated Si-H bonds effuses, leaving behind dangling bonds. The density of states of H bonds can be deduced from H effusion experiments (17).

Great progress has been made in calculating the energies of various H-bonding states in c-Si and a-S:H and in associating these with H diffusion (18, 19). Their relation with, and immediate significance for, the SWE is more tenuous because of the following: the 10^{17} cm^{-3} SWE defects must come from breaking Si-Si or Si-H bonds. However, $N = 10^{17}$ cm^{-3} is a small fraction of 5×10^{21} cm^{-3} Si-H bonds and, say, 10^{23} cm^{-3} Si-Si bonds. Moreover, it takes about 10^7 photons to produce one dangling bond. Hence, typical Si-Si or Si-H bond energy calculations probably fail to describe the very few that are important for the SWE.

Diffusion of H in a-Si:H probably occurs by thermal dissociation of H from an isolated Si-H or 2H from a (-U) pair state into a Si-H-Si center bond state (5, 18–20). In a-Si:H, the H is somewhat off the centerline. If shifted closer to one Si, the Si-H-Si structure resembles a Si-Hdb pair (21). The Si-H-Si center bond has been observed by muon repolarization measurements (21a). The Si-Hdb is mobile with an estimated $\Delta E \sim 0.3$ eV saddle point energy by insertion of H into a neighboring Si-Si and reforming the broken bond left behind. These reactions are

$$\text{Si-H} + \text{Si-Si} \rightarrow \text{db} + \text{Si-Hdb} \quad \text{dissociation,} \qquad 1.$$

$$\text{Si-HH-Si} + 2(\text{Si-Si}) \rightarrow \text{Si-Si} + 2(\text{Si-Hdb}) \quad \text{dissociation from (-U) pair state,} \qquad 2.$$

$$\text{Si-Hdb} + \text{Si-Si} \rightarrow \text{Si-Si} + \text{Si-Hdb} \qquad \text{mobile.} \qquad 3.$$

Two points should be emphasized. The Si-H bonds have nearly the same energy because H is unconstrained. Evidence for this is the common Si-H stretching mode at 2000 cm^{-1}. The effect of different energies of the three Si back bonds is relatively small. The largest energy for the dissociation Reactions 1 and 2 arises from the creation of the two dangling bonds appearing on the right-hand side. Because dangling bonds result in electronic midgap states, which can capture either an electron or a hole from the Fermi energy E_F thereby lowering the defect formation energy (9, 22), the energy spectrum of hydrogen states depends on the spectrum of electronic states, as well as on E_F and the presence of photo-excited carriers. To understand the H density of states, we, therefore, have to know the electronic density of states.

Figure 1 is a sketch of the electronic density of states and, on the left-hand side, the assignment of these states to bonding (VB = valence band), nonbonding (D° = neutral and D^- = charged dangling bonds), and antibonding states (CB = conduction band). The dashed lines at 0 and 1.8e V indicate the mobility edges; i.e. the demarcation energies between extended states and localized states that tail from E_V and E_C into the gap. The ^{29}Si hyperfine structure of the electron-spin resonance (ESR) signal from electrons trapped in a deep CB tail state

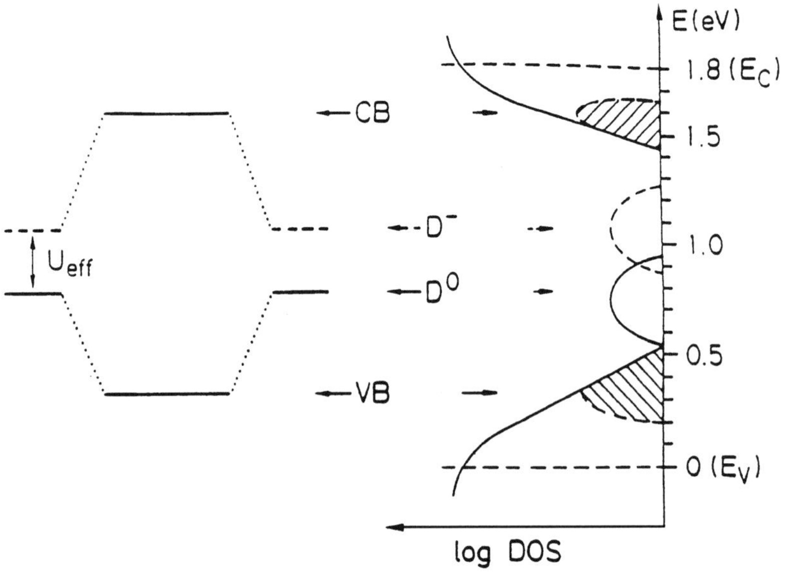

Figure 1 Electronic density of states as a function of energy. E_V and E_c are the hole and electron mobility edges. Localized valence band tail states (VB) are associated with the bonding levels, the conduction band tail states (CB) with the antibonding levels shown on the *left-hand side*. The nonbonding levels give rise to the dangling bond defect states in the gap center (after Reference 9).

show the wave function spread over only two Si atoms, as one would expect for an electron in an antibonding state of a weak Si-Si bond (23). In contrast, hopping transport in these tail states yields localization radii between 0.5 and 1.0 nm, which suggests that the wave functions are not confined to a weak bond, but rather are spread over strained regions (24). These two experiments may measure different parts of the wave function, the central region in one case and the exponential tail in the other. In the following we make the simplifying assumption that tail states are associated with one weak Si-Si bond, or antibond. The density of states at E_V and E_C are about 10^{21} eV^{-1} cm^{-3}; the exponential tail slopes are $E_O = 0.05$ eV for the valence band tail and 0.025 eV for the conduction band tail (2). Hence, there are only about 10^{19} cm^{-3} so-called weak bonds, out of a total of 10^{23} cm^{-3} bonds. According to the energy scale of Figure 1, it costs 2×0.5 eV to break a weak bond and thereby form two D° because two VB tail electrons are lifted from 0.3 to 0.8 eV—the location of D°. Local relaxations and charge transfer to D° lower this energy. The majority of bonds are stronger and the defect-formation energy higher.

Figure 2 shows the H density of states derived from these considerations (5, 20). According to Reactions 1 and 2, two dangling bonds are created by lifting one H into Si-Hdb states. H becomes mobile when it is among the majority of states

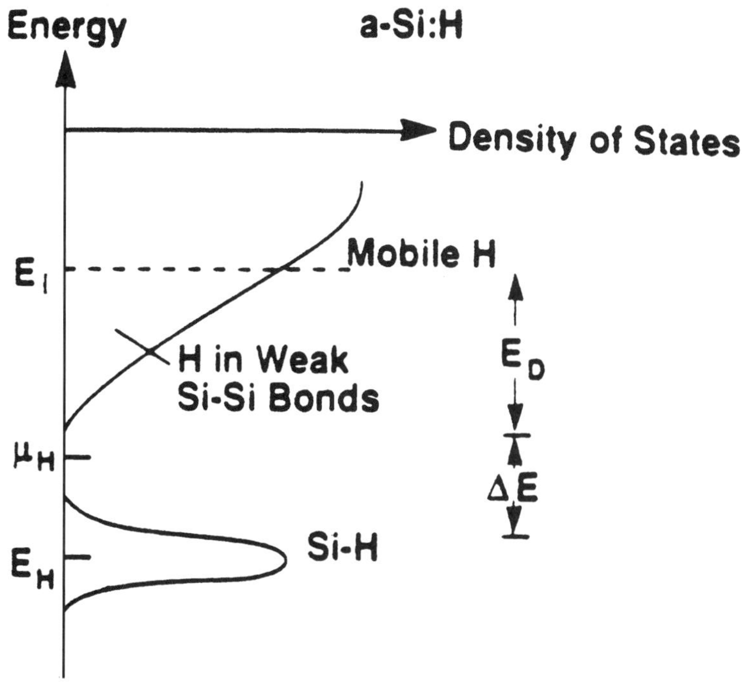

Figure 2 Hydrogen density of states as a function of energy. The mobile hydrogen at energy E_t are Si-Hdbs of Reaction 3 (after Reference 20).

near E_t and no longer restricted to the $\sim 10^{19}$ cm^{-3} weak bond states, which are considerably lower than E_t and, hence, not mobile without further activation.

In equilibrium above T_E, a hydrogen chemical potential μ_H is established, and the hydrogen diffusion constant D_H is activated by $E_D = (E_t - \mu_H) \approx 1.5$ eV (5, 20). Calculations agree well with the experimental value. The (-U) H pair states mentioned above are very important (16, 18, 25) but have not been included in Figure 2. In a manner of speaking, they are below and above μ_H and pin μ_H. However, pair-states statistics require the use of the partition function instead of the Fermi function and should not be plotted on a scale of one-H density of states. Si-H is about 0.3 eV below μ_H (14), as deduced from the temperature dependence of N in equilibrium above T_E (5, 20).

Despite the near-equal distribution of hydrogen in the quite different dilute and clustered phases, hydrogen appears to be diffusing with the same D_H and involving all H (5, 20). One aspect of H diffusion should be discussed here because it is the linchpin of the H glass model and proof of an intimate link between H diffusion and the kinetics of creation and relaxation of the SWE. This is the observation of a time dependence of the diffusion coefficient

$$D_H = D_o\, t^{\beta-1}, \qquad 4.$$

with $\beta \approx 0.8$, measured at 250°C in 1 at.% B-doped a-Si:H (5, 20), which likely results from multiple trapping of the diffusing H in the distribution of H states below E_t in Figure 2, the distribution having a logarithmic width kT_o, which would yield $\beta = T/T_o$ (4–6, 20). This interpretation of Equation 4 is, however, in conflict with the concept of the H chemical potential μ_H, which requires that all energy levels above μ_H reach thermodynamic equilibrium during the diffusion experiment. Moreover, subsequent experiments (26, 27) have yielded β values that do not fit a multiple-trapping model. They show that at least a major fraction of the observed time dependence of D_H can be caused by structural network relaxations at the diffusion temperatures. A more convincing link between H diffusion and T_E, the equilibration temperature of the H glass model, is the observation that the product of the H concentration with the value of D_H at T_E remains nearly the same as T_E is changing from 175 to 350°C in samples having different H concentrations (28).

The H glass model yields the thermodynamic equilibrium concentration of dangling bond defects as a function of temperature and Fermi level position, as well as the temperature dependence of the doping efficiency and the density of defects as a function of doping (7).

The concentrations of defects and dopants equilibrated at $T > T_E$ can be measured by quenching to lower temperatures where they are metastable (2, 5, 14). They relax by annealing to their equilibrium concentrations. Hydrogen diffusion enters the calculations only for explaining the kinetics of the defect and dopant reactions. Even if our understanding of the microscopic steps of diffusion will change in the future, perhaps forcing us to replace Reactions 1 to 3, the explanation of the kinetics will remain the same as it rests more on Equation 4 and measured diffusion activation energies.

As an example, Figure 3 shows the kinetics of metastable defect creation by abruptly changing the gate voltage V and, hence, the accumulated charge in a field effect transistor and the subsequent relaxation as V is removed (29). The kinetics can be described by a stretched exponential

$$N(t) = N_S - (N_S - N_O) \exp[-(t/\tau)^{\beta}]. \qquad 5.$$

Here, N(t) is to be understood as generic, representing the turn-on voltage of the FET in this case. The subscripts s and o refer to saturation and initial, respectively. Defect and dopant creation and relaxation follow Equation 5, where the value of β was found to be the same as in Equation 4 of H diffusion. The activation energy of τ is about 1 eV, which agrees with that of diffusion for diffusion over a constant distance (29). This refinement is necessitated by the dispersive nature, Equation 4, of diffusion. This link between creation and relaxation and H diffusion is further evidence for the validity of the H glass model and the involvement of H in the metastability reactions. However, one has to reexamine these issues again because structural relaxations can have a significant effect on the value of β (26, 27).

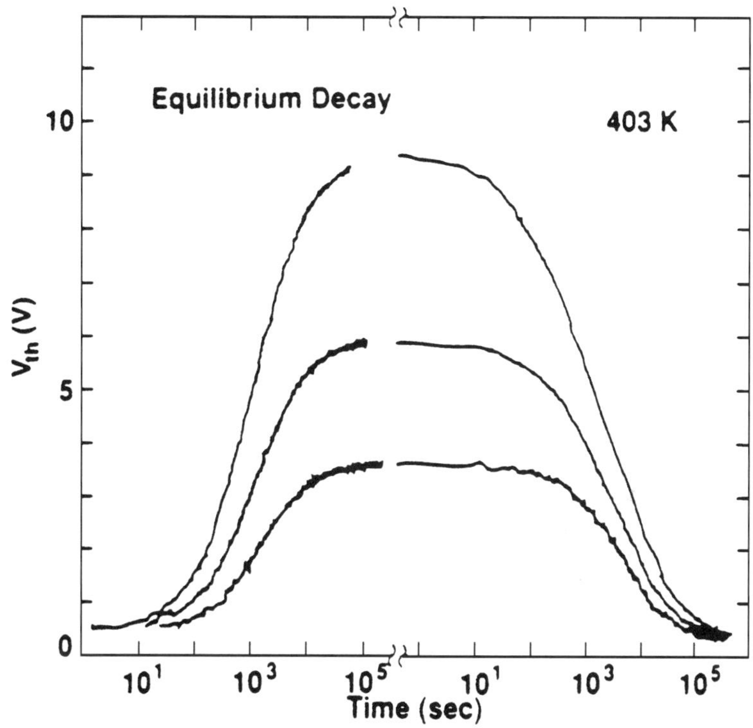

Figure 3 Threshold-voltage shift of a-Si:H thin-film transistors as a function of time for applied gate voltages V (*left*) and as a function of relaxation time after removal of the gate voltage (*right*) (after Reference 29).

Manifestations of the SWE in a-Si:H

Staebler & Wronski discovered their effect through observing a decrease of the photoconductivity and dark conductivity after light exposure (3). It quickly became clear that dangling bond defects were produced (30, 31), which, acting as recombination centers, decreased the photocarrier lifetime and, owing to their location near midgap shift E_F, thereby decreased the dark conductivity. The defect concentration can be measured by ESR, or by measuring the optical absorption coefficient α between $h\nu = 1$ and 1.3 eV—the range of optical transitions between the defects and the extended states at E_V and E_C in Figure 1. The increase of this defect absorption as a result of extended light exposure is shown in Figure 4. The exponential rise of α between $h\nu = 1.4$ and 1.9 eV is called the Urbach edge. It reflects quantitatively the exponential slope of the VB tail states E_O because the absorption is due to transitions from these states to E_C (2). The defect absorption near 1.2 eV is small but can be measured by photodeflection spectroscopy (PDS)

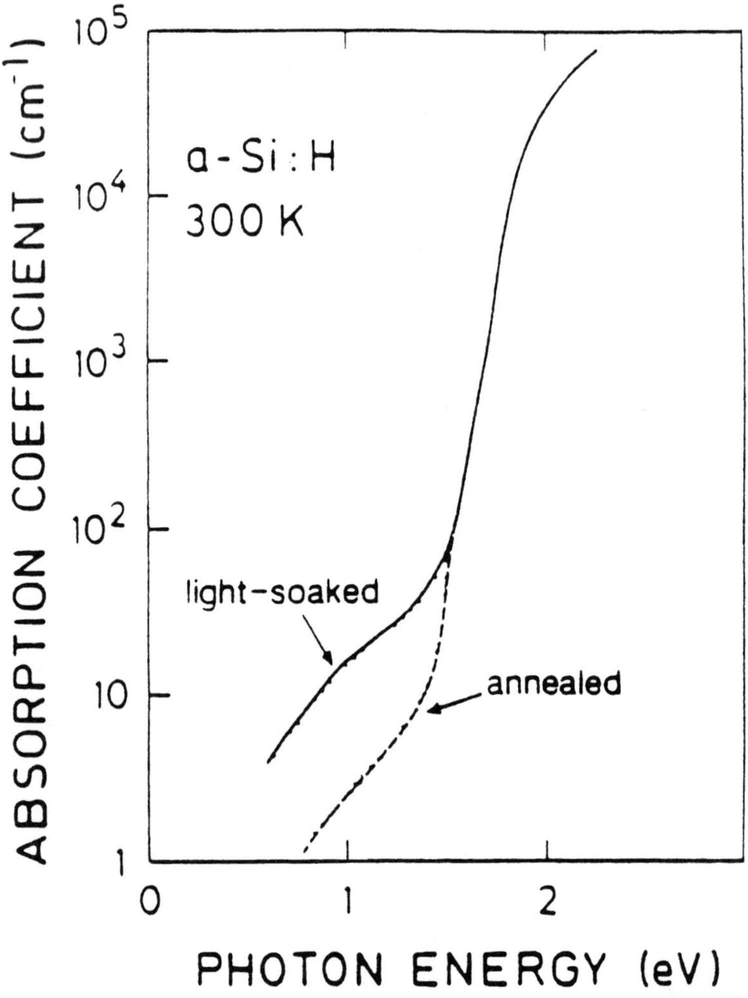

Figure 4 Absorption coefficient α of a-Si:H as a function of photon energy. The defect absorption band below 1.4 eV is used to measure the increase of N with light exposure. The exponential rise of α with $h\nu$ between 1.5 and 1.8 eV is a measure of the valence band tail width.

or by comparing the photon flux at $h\nu = 2$ eV, with that in the defect absorption region near 1.2 eV, when both yield the same photoconductivity value (2). This is the constant photocurrent method (CPM). Only neutral defects D^O are detected by ESR, and both ESR and PDS include surface states.

The photocarrier lifetime τ_n (electrons dominate the photoconductivity of undoped a-Si:H because of their larger mobility and the deeper trapping of holes in the wide VB tail) decreases with increasing N. The functional form is complicated by other recombination channels and by the fact that the capture cross section for electrons depends on the charge state D^O or D^+, and that D^-s cannot capture electrons. Whatever the form of the relation between μ_n and N, one expects this function to be single valued if all defects have the same properties. Figure 5 illustrates that this is not the case (32). The mobility lifetime product ($\mu_n\tau_n$) is plotted against the light-induced increase in absorption α of the defect band. There is no unique relation between the number of defects and ($\mu_n\tau_n$): The first defects created and the first to anneal change the photoconductivity much more than the majority of light-induced defects. This observation has been confirmed by several investigators (33–36). Other studies have shown that the same number of light-induced

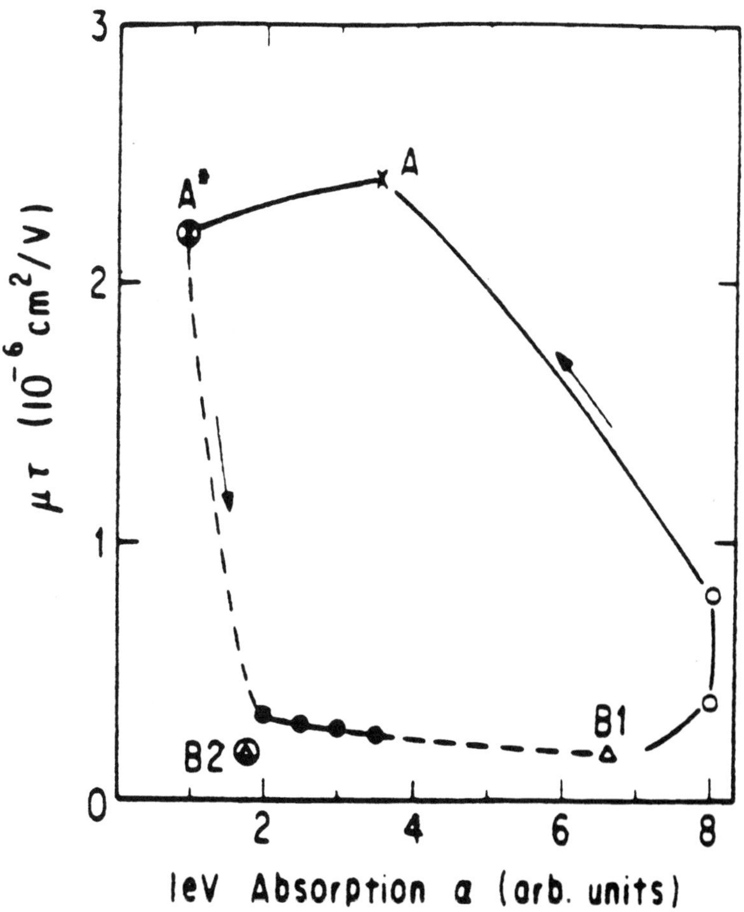

Figure 5 Mobility-lifetime product of electrons as a function of the change in defect absorption α at hν = 1 eV during light exposure (dashed curve) and thermal annealing (solid curve) (after Reference 32).

defects created by short light pulses (37), or created at low temperatures (34–38), decreases the photoconductivity at a given T more than those created by CW light at 300 K. In addition, the photoconductivity is less affected by metastable dangling bond defects produced by rapid quenching from temperatures above T_E than by light-induced dangling bond defects. Those defects that reduce τ_n most usually have a smaller annealing barrier and thus anneal more easily than the other dangling bond defects.

We therefore can conclude that photodegradation of the physical properties of a-Si:H is not defined by the concentration of dangling bond defects alone. Metastable dangling bonds either come with different properties and energies or light causes some additional changes (discussed in more detail below). Another case in which the question of metastable dangling bonds with different characteristics arises is as follows. Midgap defect states can be created by doubly injecting forward currents into PIN diodes (39). Their concentration can be detected by the magnitude of the reverse bias saturation current, which is proportional to N. The creation process saturates after time, and the saturation value N_s increases with larger forward current values. As shown in Figure 6, the N_s value corresponding to a given current can be reached either by increasing the current value from zero (curve D) or by setting this current value after a larger current has created a much higher N_s (curve B). Such low-current annealing of a larger number of defects fails to work when the same number of defects is created by light exposure of the PIN diode. Therefore, the defects produced by the two methods appear to be different.

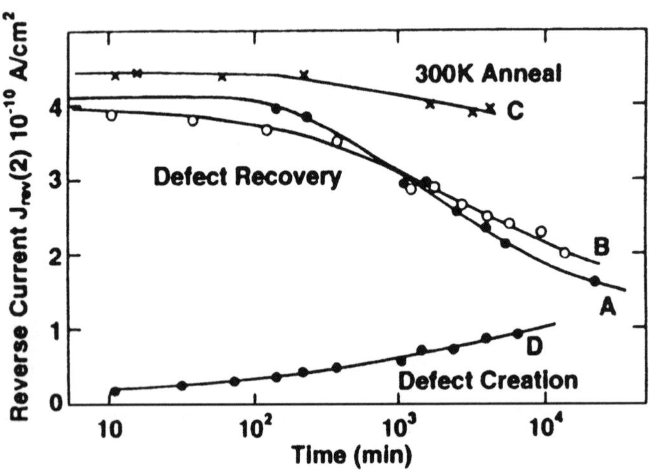

Figure 6 Current-induced defect recovery with current of 5 mA/cm^2 after inducing 2×10^{17} cm^{-3} defects with a current of 1 A/cm^2 for 1 h in a PIN junction (curves A and B) at 300 K. Curve D shows the creation of defects with 5 mA/cm^2. The reverse current is a measure of defect density. About 2×10^{17} cm^{-3} were created by light exposure but could not be removed by the 5 mA/cm^2 current (after Reference 39).

An alternative explanation may be that recombination and defect creation owing to double injection is confined to a limited region in the *i*-layer, whereas light induces defects uniformly.

Extended light exposure degrades the PIN solar-cell efficiency predominantly by decreasing the short circuit current I_{SC} and the fill factor FF. The thickness of the intrinsic layer is chosen by a compromise. If it is smaller than the drift length of the holes (the slower carriers), then most photocarriers are collected by the *n*- and *p*-layers before recombination and degradation is minimized. However, the thickness is actually made larger than this in order to absorb more photons and, thus, have a larger I_{SC}. As a consequence, recombination in the *i*-layer results in SWE defects and a decrease in efficiency by 10–15% (40). This percentage is much less than the order of magnitude changes in N and τ_n. The reasons are that most photocarriers do not recombine in the cell and, if so, not uniformly, but predominantly near the *p*-layer, and that the drift length of holes is already small and changes little.

These details are mentioned because it has been a frustrating experience that the quality of a-Si:H material as measured by dangling bonds and $\mu_n\tau_n$ rarely translates into quality of solar cells. There are other factors involved that need to be identified. Figure 5 illustrates the lack of a single-valued correlation between $\mu_n\tau_n$ and N. In Figure 7, we show an example in which solar cell degradation and light-induced recovery properties are well correlated with $\mu_n\tau_n$, but not with N. This figure is discussed further below.

Creation and Annealing

Metastable defects can be created by rapid quenching, charge accumulation, and double-injection currents, as mentioned above, as well as by keV electron bombardment (41, 42) and light exposure. The following describes those details about the SWE produced by light exposure that are most relevant for evaluating various models explaining the SWE.

The SWE occurs with nearly constant efficiency for photon energies as small as $h\nu = 1.2$ eV (43). Normally one uses $h\nu \approx 1.9$ eV bandgap light, which exposes a 1 μm-thick film uniformly. The electron-hole (e-h) recombination is an essential step in the SWE, as shown by the weakness of SWE in reverse biased PIN junctions in which most carriers are swept out before they can recombine (44). With a constant e-h pair generation rate G, the dangling bond concentration rises as (43)

$$N \propto t^{1/3} G^{2/3}, \qquad 6.$$

after an initial slower rise and with some variations in the values of the exponents. The sublinear time dependence results because the increasing density of dangling bonds provides an alternative path for e-h recombination. Evidence for this is an observed linear t-dependence when G is increased during exposure such that the carrier concentration $n = G\tau_n$ remains constant (45).

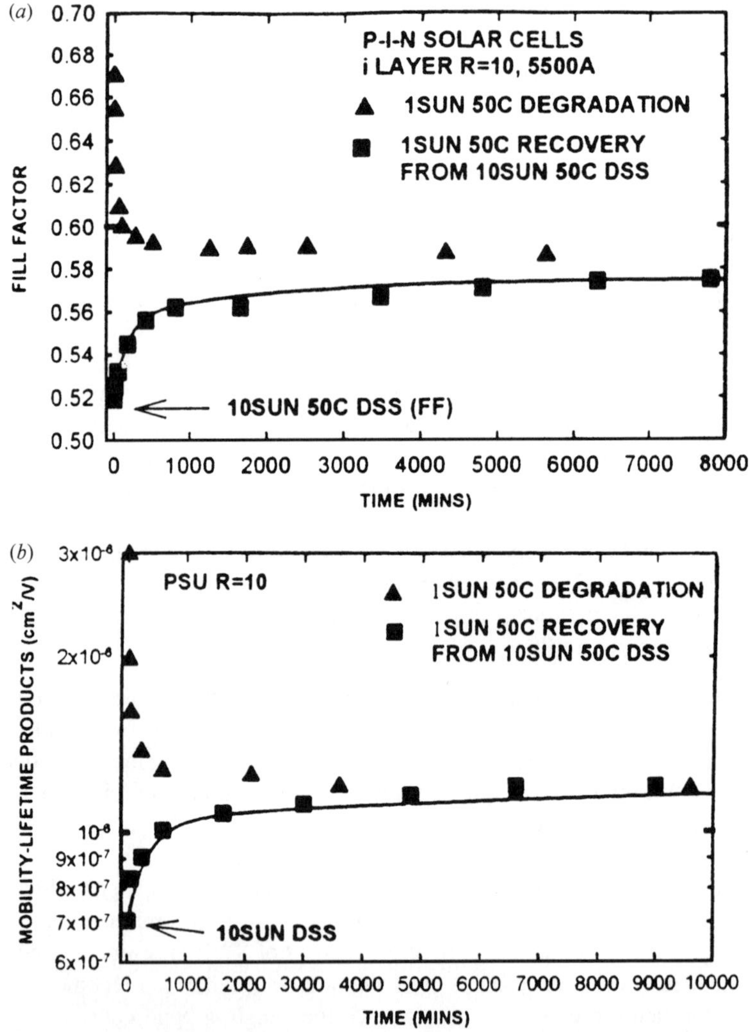

Figure 7 Upper panel (*a*) shows the decrease of the fill factor FF during exposure to a light intensity of 1 sun and the light-induced recovery with 1 sun light of a severe degradation of FF by 10 sun light. Both curves approach the 1 sun saturation value. The lower panel (*b*) shows that the mobility-lifetime product of electrons is well correlated with FF during these light exposures (after CR Wronski, private communication).

Alternatively, the kinetics can be fitted to a stretched exponential (46) that saturates at a defect density N_s between 5×10^{16} cm^{-3} and 2×10^{17} cm^{-3}. Saturation values near $N_s = 10^{18}$ cm^{-3} can be obtained by keV electron bombardment (41, 42), as well as by exposures to short light pulses (47), which indicate that saturation near 10^{17} cm^{-3} is not caused by exhaustion of a limited pool of potential defects. At $T > 80°C$, saturation is affected by thermal annealing, and at lower T by light-induced recovery (48, 49). These competing anneal processes cause N_s to increase slightly with increasing G and decreasing T.

Light-induced degradation and light-induced recovery of the fill factor of a PIN solar cell are shown in Figure 7*a* (CR Wronski, unpublished data). Exposure to 1 sun (~0.1 W/cm^2) degrades FF from 0.67 to 0.58. The same exposure causes FF, degraded more strongly by a previous 10-sun exposure, to recover from 0.52 to 0.58, the 1-sun saturation value. Figure 7*b* shows that the corresponding degradation and light-induced recovery of the $\mu_n\tau_n$ product are well correlated with those of FF. A measurement of N (not shown), in contrast, showed no such correlation.

The efficiency of dangling bond creation is difficult to quantify because of the sublinear dependencies of Equation 6. Roughly, 10^{24} cm^{-3} recombination events are needed to reach $N_s = 10^{17}$ cm^{-3} in undoped a-Si:H. The degradation rate is about three times larger in *n*-type a-Si:H, and 10 times larger in *p*-type a-Si:H, doped with 100 ppm P or B dopants, respectively. This trend follows the increase of the H diffusion coefficient with doping (2, 5, 6). Compensated a-Si:H containing 1000 ppm of both P and B, on the other hand, shows hardly any degradation even though the native dangling bond concentration $N_o \sim 5 \times 10^{15}$ cm^{-3} is as low as that of intrinsic material (50, 51). One hypothesis suggests that the large concentration of charged donor and acceptor states provides an alternative recombination path in compensated a-Si:H that does not create defects (50). Another hypothesis (51) holds that the internal fields, the result of the charged dopants, separate e-h pairs whose presence might be needed locally to initiate the SWE. Experiments with high-electric fields during light exposure at 4.2 K indeed show that a field of 10^5 V/cm reduces the SWE by 30% (52), as shown in Figure 8. This raises the question about the T-dependence of the SWE.

SWE degradation is observed down to the lowest temperatures (38). Figure 9 shows the relative rise of the defect band absorption after a 2-h exposure to $G = 3.7 \times 10^{21}$ cm^{-3} s^{-1} as a function of exposure temperature T_e (52). The degradation rate, but not the saturation value, first drops with decreasing T_e. However, at 20 and 4.2 K, the rate is essentially the same as at 300 K. Not only the magnitude but also the kinetics of the SWE is the same at 300 K and low temperatures, as shown in Figure 10 (52). The slopes exhibit the familiar $t^{1/3}$ dependence of Equation 6, whereas the G-dependence has an exponent between 0.4 and 0.5. The 300 K exposure curve in Figure 10 shows the beginning of saturation; not so the 20 K exposure curve. Even the lower efficiency 80 K exposure has been observed to cross the saturating 300 K curve, indicating much higher N_s values at low temperatures. In order to raise some appreciation for the complex relationship between $\mu_n\tau_n$ mentioned earlier, it should be noted that the increase of N by a factor 10, shown by the $T_e = 20$ K curve in Figure 10, decreases $\mu_n\tau_n$ (measured with

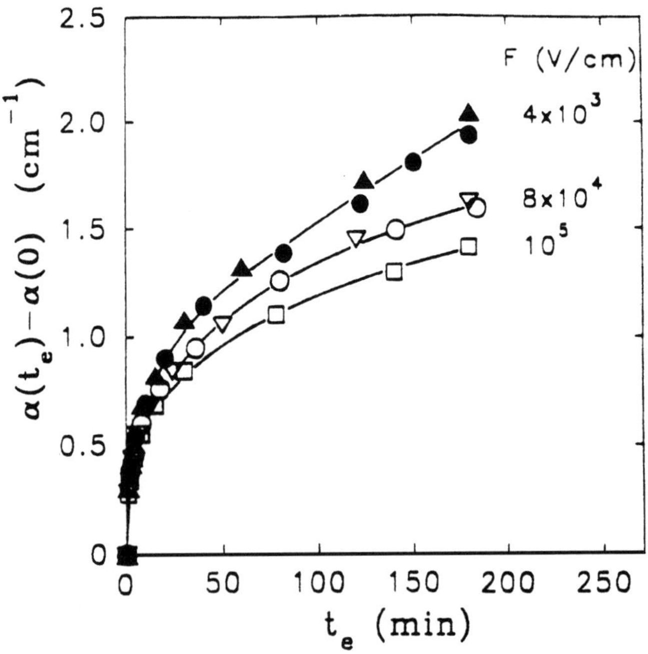

Figure 8 Increase of the defect absorption at $h\upsilon = 1.25$ eV during light exposure at T = 4.2 K at three different electric fields. The solid lines are fits to t_e^m power law (after Reference 52).

$G = 6 \times 10^{17}\ cm^{-3}\ s^{-1}$) by a factor 7000 at T = 250 K (34, 38). We contrast this with the factor 12 difference in $\mu_n\tau_n$ when two samples are compared with native defects $N_o = 5 \times 10^{15}$ and $5 \times 10^{16}\ cm^{-3}$, respectively.

Light-induced recovery of the SWE has not been observed at low temperatures, but thermal annealing already begins at 78 K for low T degradation. Fitting the initial anneal rate of the defects by

$$dN/dt = -N/\tau, \tag{7}$$

and using $\tau(s) = 10^{-10} \exp(E_A/kT)$, the distribution of anneal energies E_A shown in Figure 11 has been found for samples exposed at 77 K and, subsequently, kept at 77, 300, and 393 K for various lengths of time, as marked in the figure (36). There is a wide distribution of anneal barriers E_A, suggesting a wide stability spectrum of dangling bonds. Connecting this with the hysteresis-like creation and anneal pattern of Figure 5, which is also observed for low T defects, suggests that the less stable defects cause more severe degradation than the more stable ones. The less stable defects anneal out by light and temperature during 300 K SWE light exposures so that only those having higher anneal energies $E_A \approx$ 1.1 eV are observable. The distribution of dangling bonds created by very short,

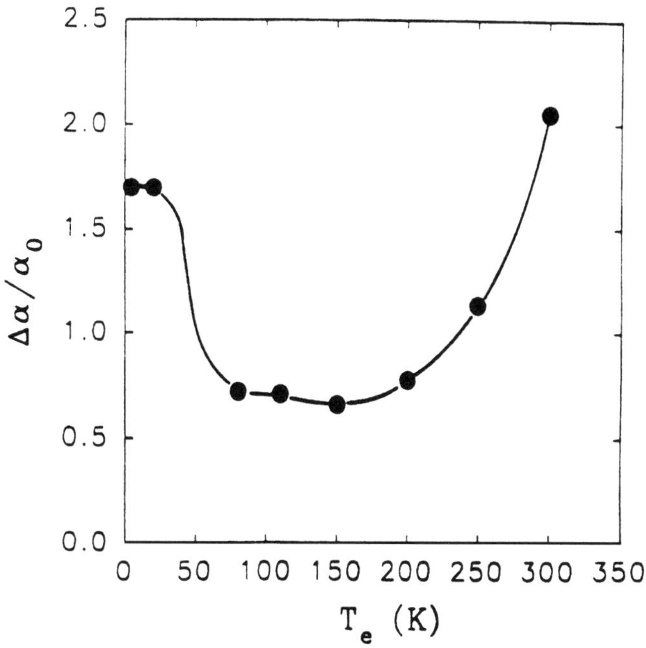

Figure 9 Relative change of the defect absorption after a 2-h exposure as a function of exposure temperature T_e. The line is a guide to the eye (after Reference 52).

high-intensity pulses is expected to be different again because different kinetics hold for both creation and opposing recovery processes. Degradation with intense pulses shows that less stable low E_A defects are observed, not only at low T but also at high light intensities (53).

Structural Environment of Dangling Bond Defects

The ESR signature of the dominant defect, $g = 2.0055$ with a line width of ≈10 G in the X-band, has now been firmly established (54 and references therein) to originate from a threefold coordinated Si atom with a missing nearest neighbor (a dangling bond) and not, as suggested (55), from a fivefold coordinated Si atom with an extra nearest neighbor (floating bond). According to ESR studies, the structure and the nearest neighbor environment of native dangling bonds and of those created by rapid quenching from $T > T_E$ or by the SWE are the same. Both native and SWE dangling bonds are separated from one another by at least 10 nm.

None of the common impurities—N, O, and C—are involved in the SWE. Samples enriched with ^{17}O and ^{13}C exhibited no hyperfine structure of light-induced dangling bonds related to these isotopes (56). Moreover, $N_s = 5 \times 10^{17}$ cm^{-3} SWE defects have been produced in highly pure a-Si:H that contain as few as 2×10^{15} cm^{-3} oxygen, 10^{16} cm^{-3} carbon, and 5×10^{14} cm^{-3} nitrogen (57).

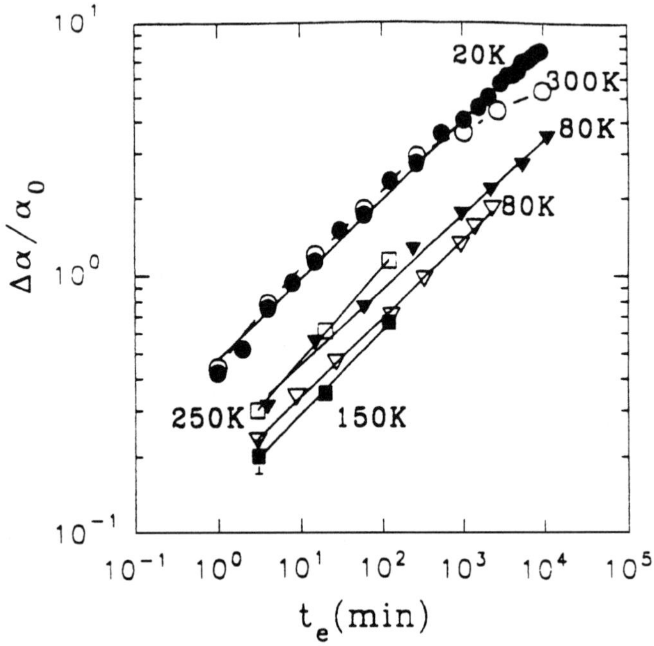

Figure 10 The relative increase in defect absorption measured at 4.2 K by CPM as a function of exposure time t_e at the different exposure temperatures marked in the figure. Solid lines are power law fits (after Reference 52).

The local environment of hydrogen or deuterium around dangling bonds was studied by electron-spin-echo envelope modulation (ESEEM) of pulsed ESR (58, 59). With this technique, small hyperfine and nuclear quadruple interactions can be extracted from within the ESR line broadened by random orientations and structural fluctuations. No hydrogen or deuterium was found closer than 0.4 nm to dangling bonds; the H density around dangling bonds actually appears to be less than average in the films. This is a most disturbing result because the majority of theories that explain the kinetics of equilibration of defects and dopants, as well as the creation and annealing of metastable defects, utilize Reactions 1 or 2, which predict that at least half of the defects should be spatially correlated with H.

Stutzmann et al (54) pointed out, however, that these ESEEM measurements, as well as their own very sensitive low-field studies of electrical detection of magnetic resonance, might have missed intimate H dangling bond pairs ($0.05 \leq r \leq 0.2$ nm). The hyperfine interaction of such close pairs contributes a background signal that is broadened by structural disorder over a spectral range of about 50 mTesla, making them very difficult to detect. As a consequence, we cannot dismiss the possibility that H dangling bond pairs exist, but we must also explore alternative

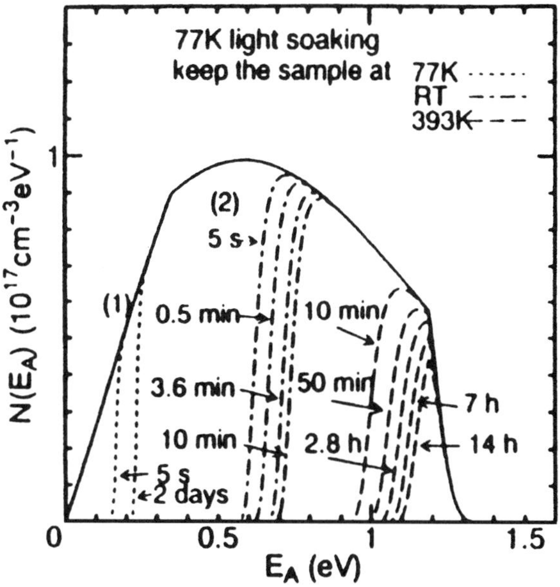

Figure 11 Distribution of anneal activation energies E_A of defects created by light exposure at 77 K. The sample was annealed at three temperatures for different lengths of time (after Reference 36).

explanations for the SWE, which are in agreement with the possibility that all metastable dangling bonds are spatially isolated from H.

EVIDENCE FOR PHOTO-INDUCED STRUCTURAL CHANGES

Recent experiments have shown that the SWE is accompanied by metastable structural changes of the underlying amorphous network (60). With X-ray photoemission spectroscopy (XPS), one observes a simultaneous shift of the Si-2s and the Si-2p peaks by 0.1 eV near SWE saturation toward lower binding energy (61). This shift was observed both with X-ray irradiation and 24-h light exposures of 0.1 W/cm^2 intensity. A solid peak shift indicates that essentially all Si atoms are involved. After a 7-h anneal at 110°C, the shifts are reversed with an anneal energy $E_A = 0.6$ eV, which is significantly lower than 1.1 eV for SWE defects produced at 300 K. The sampling depth of the XPS method is only about 0.6 nm; therefore, other evidence is needed to prove that one is dealing with a bulk effect.

This evidence may be supplied by a reversible volume change $\Delta V/V$ observed with a sensitive beam-bending method (62, 63). Figure 12 shows that the rise of $\Delta V/V$ with time of light exposure (0.3 W/cm^2) coincides with that of the

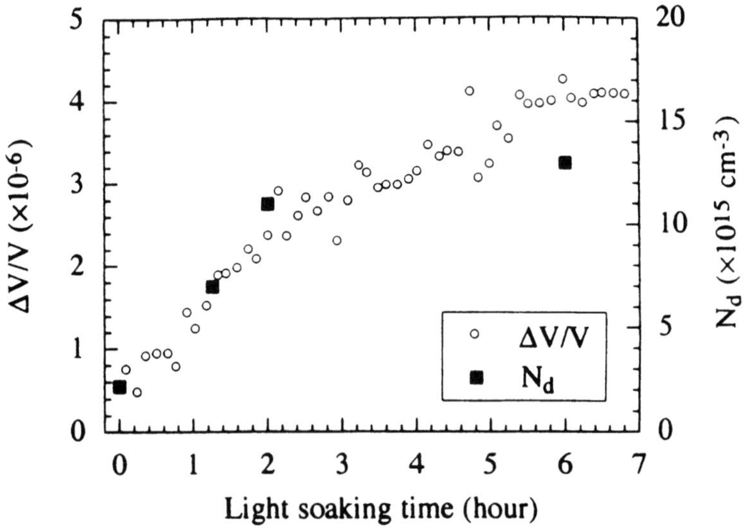

Figure 12 Photo-induced volume change and defect density as a function of light exposure time at 300 K (after Reference 62).

light-induced defects. One calculates a relative volume change per new defect of 3×10^{-22}, which is many times larger than a molecular Si unit in the lattice. This means either that each dangling bond involves a structural expansion over a 20-times larger volume or that $\Delta V/V$ should not be considered associated with dangling bonds. This is also borne out by the observation that $\Delta V/V$ is largest for high-quality, low N_o samples. Measuring $\Delta V/V$ at much higher light intensities between 1 and 10 W/cm^2, Stratakis et al (64) found $\Delta V/V$ reaching values as large as 10^{-4}, as shown in Figure 13, and continuing to grow after the dangling bond concentration had saturated. This suggests a wide-range global effect that is not limited to the environments of dangling bonds. These experiments disagree with findings of Shimizu et al (65, 66), who observed that light-induced changes in internal stress, volume, or polarized electro-absorption already saturate in less than 10 min upon exposure to a relatively weak 0.05 W/cm^2 light intensity.

Changes involving bonded hydrogen in a-Si:H were detected by Zhao et al (67) and Kong et al (68). They developed a sensitive differential method for detecting small changes of the 2000 cm^{-1} stretching mode vibration of Si-H. One of their results is shown in Figure 14. The white light intensity was 0.4 W/cm^2. The rate of change of $\Delta\alpha/\alpha$ follows a stretched exponential with the same dispersive parameter β and time τ as that of the rise in defect concentration $\Delta N/N$. That is a puzzling result because the number of defects $N = 10^{17}$ cm^{-3} is much less than $N_H = 5 \times 10^{21}$ cm^{-3}—the number Si-H bonds. It is more likely that $\Delta\alpha/\alpha$ is associated with $\Delta V/V$ owing to a change in effective charge and oscillator

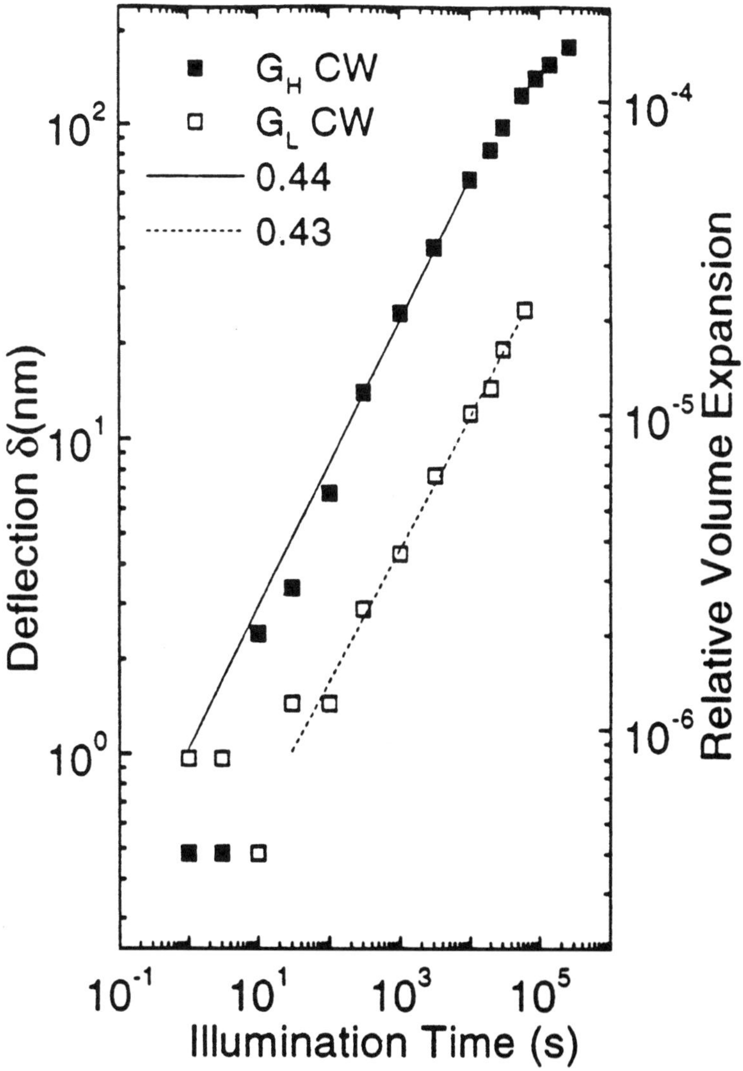

Figure 13 Micro-cantilever beam deflection and relative volume expansion of a-Si:H as a function of exposure time to photon fluxes $G_H = 3.2 \times 10^{23}$ cm^{-3} s^{-1} and $G_L = 2.5 \times 10^{22}$ cm^{-3} s^{-1}. The straight lines are power law fits with slope 0.44 and 0.43, respectively (after Reference 64).

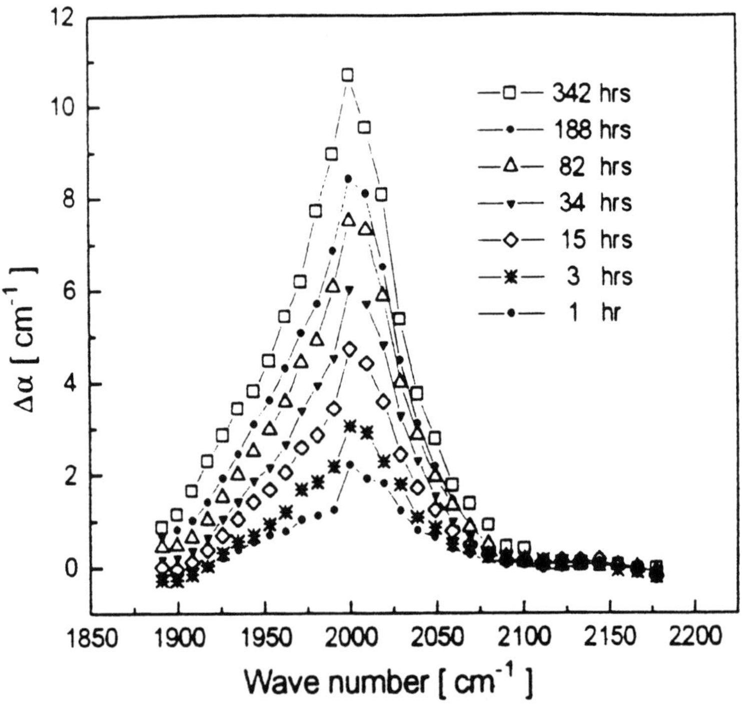

Figure 14 Increase in absorption $\Delta\alpha$ of the Si-H stretching mode at 2000 cm^{-1} for different exposure times to 0.4 W/cm^2 white light. $\alpha_o = 800\ cm^{-1}$ (after References 67, 68).

strength of the Si-H vibration. Because $\Delta V/V$ rises as $t^{0.4}$ according to Figure 13, the kinetics of $\Delta\alpha/\alpha$ is similar to that of $\Delta V/V$. Thus the result of Figure 14 might be from the global change in volume.

An ingenious experiment measuring the internal friction and the sound velocity of an a-Si:H film at low temperatures with a highly sensitive double-paddle oscillator technique has provided important new results (69–71). High-quality a-Si:H, in particular films containing 1–2 at.% H, prepared with the hot wire (HW) technique have an internal friction Q^{-1}, which is several orders of magnitude smaller than that of amorphous and vitreous films of other materials. The Q^{-1} value is a measure of two-level tunneling states (TLS), which give rise to the low-temperature anomalies of disordered materials. Q^{-1} might be a measure of a specific kind of disorder. The new experiments show that Q^{-1} rises with light exposure to the same value 6×10^{-6} for HW and for normal PECVD samples. The photo-induced Q^{-1} anneals out with an estimated activation energy $E_A = 0.8$ eV. This Q^{-1} value is not associated with dangling bonds, because when all H is driven out by heating and $N \approx 10^{19}\ cm^{-3}$, the Q^{-1} value is only 2×10^{-6}. In addition, the experiments revealed the presence of bulk H_2 in the films, or at the film-substrate interface, through the observation that the liquid-solid phase transition at the triple point was $T = 13.8$ K for H_2 and $T = 18.7$ K for D_2.

Material Parameters that Reduce the SWE

A significant reduction in photodegradation of solar cells was achieved by using hydrogen dilution of silane-processing gas during film growth (40). Beyond a certain dilution threshold value, the silicon films become microcrystalline. Shortly before this threshold, one observes evidence of greater structural order or substantial regions of intermediate range order (72) in the amorphous film (73, 74). Transmission electron micrographs show regions that have a crystalline-like appearance. These produce no microcrystalline X-ray diffraction peaks, but a narrowing of the amorphous Si (111) peak. In addition to the transverse optical (TO) vibration band near 475 cm^{-1}, one finds in the Raman spectra a TO band at 490 cm^{-1} that grows with the fraction of ordered regions.

Despite dilution ratios R of 20:1 and larger, the total H concentrations in the RF plasma-deposited films remains near 8–10 at.%. However, the distribution of H becomes more heterogeneous with the ordered regions containing less H and the surrounding amorphous material proportionately more. The shift of the Si-H wagging mode from 635–640 cm^{-1} to about 620 cm^{-1} observed in the infrared spectra of near-threshold material (74) suggests a change toward a surface-type Si-H bonding environment. Presumably, this gives rise to the sharp peak near 400°C in the hydrogen effusion spectrum (75) (Figure 15). These amorphous films already crystallize near 550°C, a temperature significantly below the 650°C crystallization temperature of films grown without H dilution.

The amorphous-microcrystalline phase boundary depends not only on the dilution ratio R but also on the film thickness d as shown in Figure 16 because the ordered regions grow with the growing film (76). Other parameters affecting the evolutionary phase boundary are substrate material, deposition rate and temperature, and the frequency (13.6 or 80 MHz, or microwave) of the plasma-deposition process. A sudden decrease in the open circuit voltage is a useful signal that the i-layer thickness has exceeded the phase boundary when the film is incorporated in a solar cell (77). As the fraction of amorphous-ordered regions increases with film thickness before the microcrystalline borderline, the dangling bond concentration decreases (78), as shown in Figure 17. This spatial variation suggests that optimized a-Si:H films might be obtained by varying (decreasing) the H dilution during growth.

Eight to ten at.% H is far more than needed to saturate potential dangling bonds. Because H diffusion decreases with H concentration, one also expects the SWE to diminish. The H concentration cannot be decreased with a plasma-deposition process without sacrificing material quality, but a-Si:H with H concentration as low as 0.5 at.% can be prepared by a HW deposition process (79, 80). Films containing about 1 at.% H have a reduced SWE and look promising. At present, they are still outperformed by a-Si:H near the threshold of crystallinity. The reason is the

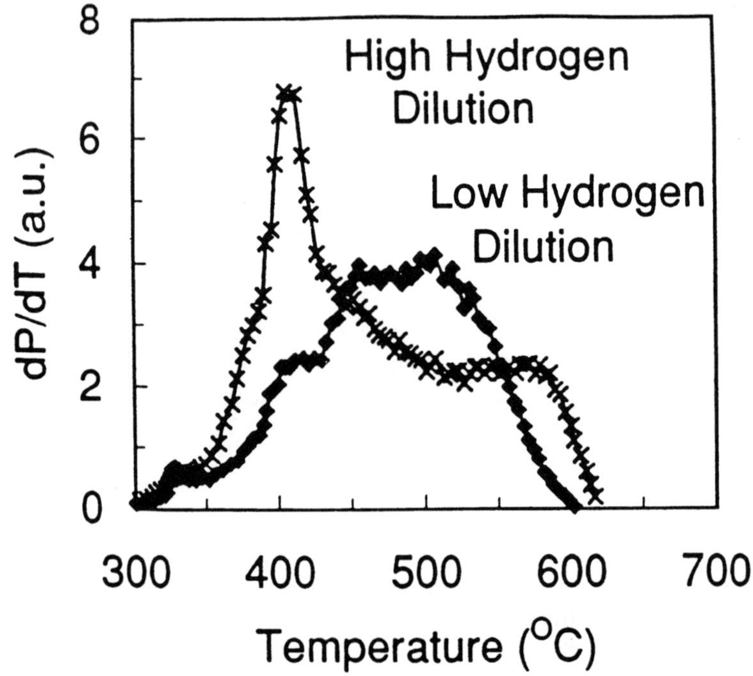

Figure 15 Hydrogen evolution spectra of a-Si:H prepared with a high and a low hydrogen dilution ratio (after Reference 75).

presence of a defect-rich layer during the first 50–100 nm of growth owing to H effusion at the high substrate temperature needed for the HW deposition (81).

Replacing hydrogen with deuterium has yielded material with reduced SWE (82, 83). Because of the expense of deuterium, only a few such experiments have been performed. Despite the positive outcome, one cannot rule out the possibility that the isotope effect changes the plasma conditions in a way to produce a film with a more favorable microstructure. However, the beneficial effect of deuterium is supported theoretically because the bending mode of Si-D is greatly damped as its frequency 460 cm^{-1} overlaps with the TO mode of the Si network (84, 85); it is this bending mode that assists the motion of H from Si-H to the center of a Si-Si bond of the same Si atom, forming the Si-H dangling bond mobile H state of Reaction 1.

THEORIES OF SWE DEFECT CREATION: Successes and Difficulties

In this section the essential features of the main theoretical contenders are briefly explained and their successes and difficulties in explaining some crucial aspects of the SWE are discussed. The question of a spatial correlation between H and

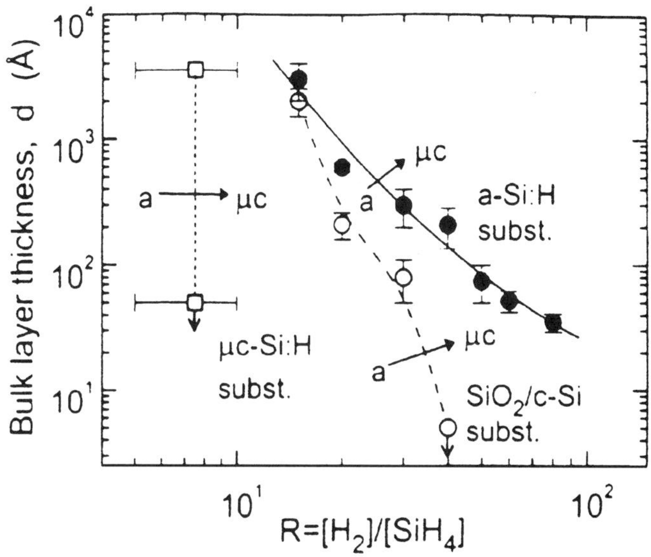

Figure 16 Evolutionary phase boundaries separating the deposition of amorphous and microcrystalline a-Si:H as a function of layer thickness *d* and dilution ratio R at 200°C on different substrates: a-Si:H, oxidized c-Si wafer, and *p*-type microcrystalline Si (after Reference 76).

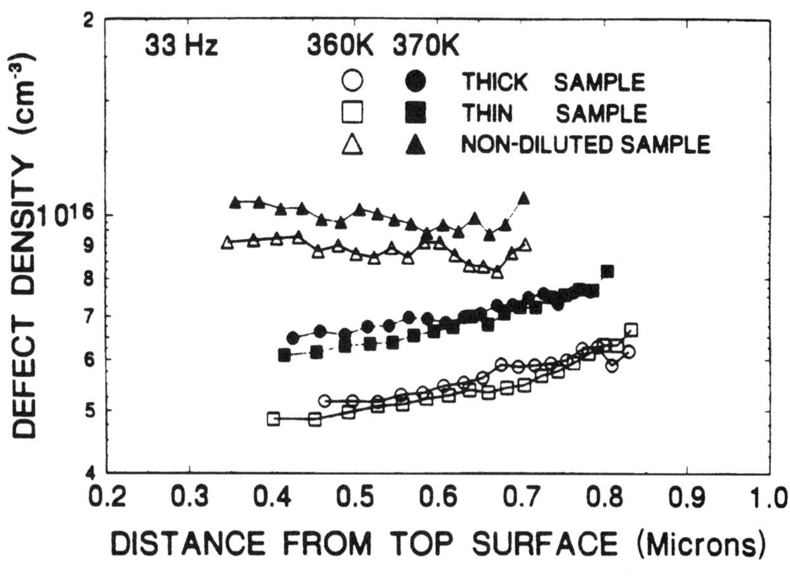

Figure 17 Defect density as a function of depth of a-Si:H deposited with and without hydrogen dilution. The defect density decreases in diluted samples as the microcrystalline phase boundary is approached (after Reference 78).

dangling bonds is an important issue but is not used to disqualify any theory as previously stated in this review.

Redfield & Bube (86) introduced a general rate equation with terms describing light-induced creation and annealing, as well as thermal creation and annealing. By assuming the rate parameters to be time dependent, which would arise from a distribution of formation energies for defects with the same dispersive factor as in Equation 4 for hydrogen diffusion, they obtain, as a solution, the stretched exponential of Equation 5, which fits the kinetics of dangling bond creation and relaxation near and above room temperature. The central straight region of Equation 5 can be made to match the sublinear relations of Equation 6 reasonably well. Aside from predicting that light-induced annealing should occur (87), which was verified experimentally soon afterward, the rate equation by itself does not reveal the microscopic mechanism of the SWE.

Building on the success in explaining defect and dopant equilibration, the equilibrium models mentioned in the section on hydrogen in a-Si:H have been extended to explain the SWE (88, 89). An important ingredient of these theories is the close link between electronic and hydrogen energy states and the dependence of the defect formation energy by transferring electrons between the Fermi energy and the defect states. The logical extension of these models holds that illumination increases the dangling bond concentration because the excess electrons and holes described by the splitting of E_F into quasi-Fermi levels reduce the energy barriers for defect formation and the defect formation energies. With plausible assumptions, these theories produce kinetics similar to Equations 5 and 6 and agreements with the observed magnitude and temperature dependence of the saturation value N_s above 250 K.

The weak bond-breaking model of Stutzmann et al (43) addresses the microscopic aspects of the SWE. The dangling bonds are created by the bimolecular recombination of electrons and holes at weak bonds. An exciton formed by a hole trapped in a valence tail–bonding state and an electron in the corresponding conduction tail–antibonding state break such a weak bond. The associated e-h recombination energy enables an H atom bonded to one of the Si to slip into the broken bond preventing the bond to reform. The probability for such bimolecular recombinations is $n_t p_t$, the product of electrons and holes in tail states (shown by the shaded regions in Figure 1), which is proportional to np, the product of free carriers. These in turn are given by $G\tau_e$ and $G\tau_p$, respectively. If the photocarrier lifetimes are limited by monomolecular recombination through dangling bonds of density N, then

$$dN/dt \propto np \propto \text{const. } G^2/N^2. \qquad 8.$$

Equation 6 is the solution of Equation 8 for $N^3 \gg N_o^3$. According to Reaction 3, the Si-Hdb can separate from its dangling bond partner by further insertion hops to satisfy the spatial separation of ~10 nm between dangling bonds demanded by the lack of exchange narrowing of the ESR line (90). Near the threshold photon energy for observing the SWE of $h\nu = 1.2$ eV, two photon absorption processes are apparently needed to excite from midgap defect states an electron and a hole (43). The excitonic precursor for bond breaking satisfies the demand of the Franck-Condon principle that the lifetime of the electronically excited state be longer than the inverse vibrational frequency of the atoms coupling to this electronic state, i.e. long enough to allow atomic motions. Evidence for the excitonic states comes from spin-dependent photoconductivity experiments (91), which show a broad spin resonance owing to e-h pairs within a distance of about 0.5 nm localized in the tail states.

Godet (92) has suggested a metastable H trapping model (MSH): A (-U) pair states exists, designated Si-HH-Si. An e-h pair recombination event initiates a dissociation according to Reaction 2. When the mobile Si-Hdb species produced thereby meet a dangling bond, they annihilate. If either one meets an isolated Si-H, on the other hand, an isolated dangling bond is produced as well as a new (-U) pair state according to reaction

$$\text{Si-Hdb} + \text{Si-H} \rightarrow \text{db} + \text{Si-HH-Si}. \qquad 9.$$

The models discussed so far involve H motion and yield defects in close spatial proximity with H, the Si-Hdb state of Reaction 1 or, alternatively, the dangling bond close to Si-HH-Si in Reaction 9. Two models that do not require hydrogen and thus yield isolated dangling bonds are the charged defect precursor model of Adler (93) and the floating bond model of Pantelides (94 and references therein). They are not discussed further because they have been ruled out by convincing counter-evidence, i.e. no large concentration of charged defects has been found (95), and the structure of floating bonds does not agree with spin resonance measurements, as mentioned above. The apparent lack of spatial correlation between H and dangling bonds has stimulated the H collision model of Branz (96, 97). This model suggests that the dissociation Reaction 1 occurs during a recombination process at a rate proportional to the recombination or the e-h generation rate G, as in a geminate recombination process. The mobile Si-Hdb, which is also considered to be the mobile species in H diffusion theories, moves through the material. When it meets a dangling bond, it is annihilated via the reverse reaction 1. Hence, its lifetime is proportional to $1/N$. Occasionally, one Si-Hdb meets another Si-Hdb. This is called an H collision, which results in a metastable Si-HH-Si* state. The asterisk means that the way of writing HH between two Si does not reflect the true structure of this metastable state. Because the probability of two unrelated Si-Hdbs colliding, each having a lifetime $\sim 1/N$, is proportional to G^2/N^2, one obtains the rate Equation 8, which yields the desired kinetics of Equation 6. Thermal- and light-induced annealing occurs by breaking up the metastable Si-HH-Si* state into two mobile Si-Hdb species. These, in turn, meet dangling bonds by diffusion and annihilate. The observed metastable dangling bonds are spatially isolated from H as well as from each other because the concentration of mobile Si-Hdbs always remains small compared with N. It is important to make a distinction between the metastable state Si-HH-Si* and the large concentration of (-U) pair states mentioned above in order to explain that both light-induced annealing as well as thermal annealing have E_A smaller than the activation energy of diffusion. If the two were identical and easily dissociated by a recombination process, one would be able to remove all N by light-induced annealing, and there would be no SWE (96).

We now compare these models with experimental results. The equilibrium models, as well as Godet's MSH model and Branz' H collision model, require diffusion of mobile Si-Hdb species; the first two require diffusion over a distance of about 1 nm, the average separation of Si-H and Si-HH-Si, whereas the H collision model deals with diffusion distances of 20–50 nm, the average separation of metastable dangling bonds. We should note that the diffusing Si-Hdb species have already been created by a recombination process and only have to overcome the barriers associated with Reaction 3. These saddle-point barriers have been estimated to be 0.2–0.4 eV. Even this relatively small barrier energy is in conflict with the nearly temperature-independent SWE creation efficiency shown in Figures 9 and 10. In order to solve this problem, it has been suggested that the large number of photo-excited carriers might reduce the saddle-point barriers or that the numerous recombination events provide sufficient local energy during illumination to help H overcome these barriers. However, similar distances of 20–50 nm have to be overcome by diffusing Si-Hdb species without such help in the dark during thermal annealing between 250 and 450 K. Note the broad distribution of anneal energies shown in Figure 11. This may be possible if the thermal dissociation energy of the metastable Si-HH-Si*, when created at low T, is 0.5 eV or less. With the help of light-induced diffusion of mobile Si-Hdb and favorable energies for Si-HH-Si*, it is thus possible to obtain isolated dangling bonds for defects created at 4.2 K.

This, however, might not be necessary because the evidence for isolated dangling bonds, if valid, was obtained only on material exposed to light at 300 K. We do not know the structural environment of defects created at low T. They behave differently. A large fraction anneal between 200 and 300 K, and many appear to have a larger recombination cross section. This behavior suggests close dangling bond Si-Hdb pairs (98) that are stable only at low T and would not be observable after light exposures at 300 K and above. This could be produced by the weak bond breaking process of Reaction 1 but presumably not by the H collision process, because the latter yields the kinetics of Equation 6 only when the mobile Si-Hdb species are given the chance to encounter many dangling bonds and thus to acquire a lifetime proportional to $1/N$, which they need in order to satisfy Equation 8.

This raises the question why Equation 6 is so robust, while the parent Equation 8 is based on a number of assumptions for electrons and holes that are not generally valid. The sublinear time and generation rate dependence of Equation 6 is observed at all temperatures, as shown in Figure 10, and at G values where monomolecular recombination through defects is not the dominant process determining the lifetimes of electrons and holes but, rather, recombination through a large concentration of photocarriers trapped in tail states. Moreover, the $t^{1/3}$ behavior is observed long before the $N^3 \gg N_o^3$ condition is satisfied, as can be seen in Figure 10. Is it possible that Equation 8 is not the parent of Equation 6 but that Equation 6 results essentially from a broad distribution of defect formation energies and barriers that lead to a dispersive nature of the rate coefficients, as argued by Bube & Redfield? These coefficients must then remain quite independent of temperature, which is most unusual. Moreover, the dispersive rate equation of these authors (86) has difficulties explaining the linear t-dependence of N observed during light exposures, which keep $G\tau_n$ constant (45).

When a-Si:H is exposed to short laser pulses of energy E, the dangling bond defect density increases as $N \propto EN_P^{0.5}$. The sublinear dependence on the number of pulses N_P was attributed to an increase in recombinations through defects that shorten the decay time of photocarriers after the pulse (99). This explanation is invalid because, at the pulse energies used, the recombination process is bimolecular and not defect controlled (100, 101). Moreover, most carriers recombine during the pulse, and the small residual carrier concentration between pulses does not contribute to defect creation (101). The H collision model, however, can account for the $EN_P^{0.5}$ dependence by assuming that the mobile Si-Hdb species created during the pulse keep moving and decay slowly during the dark time between pulses (96). A clever experiment that explores the time domain of defect creation is given in the following (101). By splitting the laser pulses of energy E into a pulse pair, one increases N_P by a factor 2 and diminishes the energy by the same factor. Because of the $EN_P^{0.5}$ dependence, the defect creation rate drops by 30% if the pulses are independent. This, indeed, is observed for pulses separated by more than 10 μs. For separations of 30 to 500 ns, however, the drop in defect creation rate is only between 10 and 20%, and the pulse pair does not act as two independent pulses. This suggests the presence of a precursor for defect formation with a lifetime of the order of microseconds at 300 K. During this time, a mobile Si-Hdb can diffuse about 1–2 nm.

One major difference between the H collision model, on the one hand, and the bond-breaking model, on the other, is the nature of metastability. In the latter, the metastable state is the Si-Hdb whose energy distribution forms the tail of energies below the transport energy E_t in Figure 2. The large anneal energy $E_A \approx 1$ eV of dangling bonds created at $T \geq 300$ K indicates that the mobile Si-Hdb species have found some deep traps; i.e. the low-lying Si-Hdb states associated with weak bonds. Many dangling bonds created at low T have a much smaller E_A, as shown in Figure 11. These Si-Hdb species appear to occupy the shallow traps or H energy states not far below the transport energy E_t in Figure 2. They have not moved far from their dangling bond partner and, hence, do not have a chance to get more deeply trapped in a weak bond state. In the H collision model, in contrast, the metastable state is Si-HHSi*, whereas the Si-Hdb species are very mobile and able to diffuse over 20–50-nm distances. This high mobility is the largest conceptual difficulty of this model. Light-induced H diffusion is known to exist (102), and a recent experiment succeeded in measuring D_H at 135°C under illumination with $G = 10^{23}\ cm^{-3}\ s^{-1}$. At this temperature, D_H with light is still thermally activated with 0.9 ± 0.1 eV. A temperature-independent regime of light-induced D_H has not yet been found that would explain light-induced H emission and diffusion at low temperatures. This problem has been discussed by Branz (103) who concludes that the question is still open. It seems that an attempt to measure D_H by using very long exposures to strong light at 300 K or below might find a value for D_H that is either too small or acceptable for the H collision model to work. Evidence for long-range H diffusion under 15 W/cm^2 illumination at 300 K from a-Si:H into an a-WO_3 overlayer has recently been discovered (104).

The larger scale structural changes produced by extended light exposures provide a new angle to view the SWE. Unfortunately, none of the structural changes appears to be a precursor of the SWE, as they seem to occur in parallel. There are roughly 10^7 recombination events happening for each defect produced. Because of the localization of wave functions in disordered materials, the recombinations are confined to about 0.6 nm regions. Aside from depositing energy in this small volume, the recombinations change the charge state of tail states and defects, which also can result in larger scale changes in the Si network as shown by molecular-dynamics simulations (105, 106). These can easily yield a higher energy state of increased volume and strain. Moreover, H diffusion of Reaction 3 probably leaves a wake of disturbed networks (96). It would be revealing if one could observe a reduction of the light-induced volume change similar to the reduction of the SWE in materials deposited near the threshold of crystallinity. Preliminary results unfortunately fail so far to show such a correlation.

CONCLUSION AND UNRESOLVED PROBLEMS

The lack of a spatial correlation between dangling bonds and hydrogen, which excludes Si-Hdb species as the metastable state, is the major difficulty of understanding the SWE. If it is true, as suggested by Stutzmann et al (54), that close H dangling bond pairs exist but have escaped detection, we are left with a number of lesser problems.

Increasingly, we become aware that the microstructure of a-Si:H is highly heterogeneous. There are more ordered regions containing less hydrogen and surrounding amorphous tissue with more hydrogen, in addition to clustered and atomically dispersed bonded hydrogen. We do not know exactly where the light-induced defects occur. There is evidence of different locations of deposition-related and light-induced dangling bonds (107). Recent experiments (108) with hydrogenated Si-Ge alloys show an unexpected interrelation between Si dangling bonds and Ge dangling bonds. The total dangling bond concentration, but not the individual ones, is proportional to $t^{1/3}$ during exposure.

Even though all dangling bonds in a-Si:H have the same ESR signature, their effect on the photocarrier lifetime can be profoundly different. It is important to understand why this is so because the degradation of the optoelectronic properties is not defined by the dangling bond density alone.

Light-induced degradation appears to be caused largely by one recombination process among several others. We have not clearly identified which is the most harmful. The reduction of the SWE in high-electric fields suggests that geminate recombinations play a role, whereas the rate Equation 8 is based on bimolecular recombination of non-geminate carriers. Geminate processes are certainly excluded in current-induced defect creation in PIN junctions (shown in Figure 6). An excess concentration of only electrons or holes also produces defects (shown in Figure 3) as well as by charge injection in PIP junctions (109). The distinction between these mechanisms and the SWE is not clear.

The general trend of improving the stability of a-Si:H to that of the present amorphous material near the threshold of microcrystallinity suggests that it is most important to decrease the presence of weak bonds (72). These give rise to deep traps for mobile hydrogen species in the density of H states. Without these deep traps, the anneal energy of the metastable SWE would be reduced and with it the magnitude of the SWE at and above room temperature.

ACKNOWLEDGMENTS

I am most grateful for the generous help and criticism of HM Branz, S Guha, J Kakalios, SR Ovshinsky, EA Schiff, and M Stutzmann.

LITERATURE CITED

1. Street RA, ed. 1999. *Technology and Applications of Amorphous Silicon*. Berlin: Springer
2. Street RA. 1991. *Hydrogenated Amorphous Silicon*. Cambridge, UK: Cambridge Univ. Press
3. Staebler DL, Wronski CR. 1977. *Appl. Phys. Lett.* 31:292–94
4. Street RA, Kakalios J, Tsai CC, Hayes TM. 1987. *Phys. Rev. B* 35:1316–33
5. Kakalios J. 1991. Hydrogen in semiconductors. In *Semiconductors and Semimetals*, ed. JI Pankove, NM Johnson, 34:381–445. New York: Academic
6. Street RA, Tsai CC, Kakalios J, Jackson WB. 1987. *Philos. Mag. B* 56:305–20
7. Street RA, Hack M, Jackson WB. 1988. *Phys. Rev. B* 37:4209–23
8. Street RA, Winer K. 1989. *Phys. Rev. B* 40:6236–49
9. Stutzmann M. 1992. Metastability in amorphous and microcrystalline semiconductors. In *Amorphous and Microcrystalline Semiconductor Devices: Materials and Device Physics*, ed. J Kanicki, pp. 129–87. Norwood, MA: Artech House
10. Reimer JA, Knights JC, Vaughan R. 1981. *Phys. Rev. B* 24:3360–68
11. Taylor PC. 1984. Magnetic resonance measurements in a-Si:H. In *Semiconductors and Semimetals*, ed. JI Pankove, 21C:99–154. New York: Academic
12. Su T, Taylor PC, Chen S, Crandall RS, Mahan AH. 2000. *J. Non-Cryst. Solids* 266–69:195–200
13. Gleason KK, Petrich MA, Reimer JA. 1987. *Phys. Rev. B* 36:3259

14. Zafar S, Schiff EA. 1991. *Phys. Rev. Lett.* 66:1493–96
15. Zellama K, Germain P, Squelard S, Bourdon B, Fontenille J, Danielou R. 1981. *Phys. Rev. B* 23:6648–67
16. Zafar S, Schiff EA. 1989. *Phys. Rev. B* 40:5235–38
17. Beyer W. 1999. Hydrogen phenomena in hydrogenated amorphous silicon. In *Semiconductors and Semimetals*. 61:165–239. New York: Academic
18. Tuttle B, Adams JB. 1998. *Phys. Rev. B* 57:12859–68
19. Van de Walle CG. 1999. Hydrogenated interactions with polycrystalline and amorphous silicon–theory. In *Semiconductors and Semimetals*, 61:241–81. New York: Academic
20. Street RA. 1991. *Physica B* 170:69–81
21. Biswas R, Li QM, Pan BC, Yoon Y. 1998. *Phys. Rev. B* 57:2253–56
21a. Davis EA. 1996. *J. Non-Cryst. Solids* 198–200:1–10
22. Stutzmann M. 1987. *Philos. Mag. B* 56: 63–70
23. Yamasaki S, Umeda T, Isoya J, Zhou JH, Tanaka K. 1998. *J. Non-Cryst. Solids* 227–30:332–37
24. Fritzsche H, Yoon BG, Chi DZ, Tran MQ. 1992. *J. Non-Cryst. Solids* 141:123–32
25. Zhang SB, Jackson WB, Chadi DJ. 1990. *Phys. Rev. Lett.* 65:2575–78
26. Asano A. 1991. *Physica B* 170:277–80
27. Shinar J, Shinar R, Wu XL, Mitra S, Girwan RF. 1991. *Phys. Rev. B* 43:1631–36
28. Branz HM, Iwaniczko E. 1993. *Phys. Rev. B* 48:17114–20
29. Jackson WB. 1990. *Phys. Rev. B* 41:1059–75
30. Hirabayashi I, Morigaki K, Nitta S. 1980. *Jpn. J. Appl. Phys.* 19:L357
31. Dersch H, Stuke J, Beichler J. 1981. *Appl. Phys. Lett.* 38:456–58
32. Han D, Fritzsche H. 1983. *J. Non-Cryst. Solids* 59–60:397–401
33. Shepard K, Smith ZE, Aljishi S, Wagner S. 1988. *Appl. Phys. Lett.* 53:1644–46
34. Stradins P, Tran MQ, Fritzsche H. 1993. *J. Non-Cryst. Solids* 164–66:175–78
35. Shimizu T. 1993. *J. Non-Cryst. Solids* 164–66:163–68
36. Zhang Q, Takashima H, Zhou JH, Kumeda M, Shimizu T. 1994. Recovery process for light-soaked a-Si:H. In *Amorphous Silicon Technology–1994*, ed. EA Schiff, M Hack, A Madan, M Powell, A Matsuda, 336:269–74. Pittsburgh: Mater. Res. Soc.
37. Tzanetakis P, Kopidakis N, Androulidaki M, Kalpouzos C, Stradins P, Fritzsche H. 1996. *J. Non-Cryst. Solids* 198–200:458–61
38. Stradins P, Fritzsche H. 1994. *Philos. Mag. B* 69:121–39
39. Street RA, Hack M. 1991. *J. Non-Cryst. Solids* 137–38:263–66
40. Guha S. 2000. Multijunction solar cells and modules. In *Technology and Applications of Amorphous Silicon*, ed. RA Street, pp. 252–305. Berlin: Springer
41. Schneider U, Schröder B, Finger F. 1989. *J. Non-Cryst. Solids* 114:633–35
42. Schneider U, Schröder B. 1988. Metastable defects in hydrogenated amorphous silicon (a-Si:H) produced by electron irradiation. In *Amorphous Silicon and Related Materials*, ed. H Fritzsche. pp. 687–717. Singapore: World Scientific
43. Stutzmann M, Jackson WB, Tsai CC. 1985. *Phys. Rev. B* 32:23–47
44. Guha S, Yang J, Czubatyj W, Hudgens SJ, Hack M. 1983. *Appl. Phys. Lett.* 42:588–89
45. Brandt MS, Stutzmann M. 1994. *J. Appl. Phys.* 75:2507–15
46. Kakalios J, Street RA, Jackson WB. 1987. *Phys. Rev. Lett.* 59:1037–40
47. Rossi MC, Brandt MS, Stutzmann M. 1992. *Appl. Phys. Lett.* 60:1709–11
48. Meaudre R, Meaudre M. 1992. *Phys. Rev. B* 45:12134–36
49. Gleskova H, Morin PA, Wagner S. 1993. *Appl. Phys. Lett.* 62:2063–65
50. Stutzmann M. 1990. *Appl. Phys. Lett.* 56: 2313–15
51. Tzanetakis P, Fritzsche H, Tran MQ,

Androulidaki M, Rynes E. 1993. *J. Non-Cryst. Solids* 164–66:607–10
52. Stradins P, Fritzsche H. 1996. *J. Non-Cryst. Solids* 198–200:432–35
53. Hata N, Ikeda T, Matsuda A. 1996. Nonlinear dependence of $\mu_e\tau_e$ on subgap absorption in pulse and CW light-soaked a-Si:H. In *Amorphous Silicon Technology–1996*. ed. H Hack, EA Schiff, S Wagner, R Schropp, A Matsuda, 420:575–80 Pittsburgh: Mater. Res. Soc.
54. Stutzmann M, Brandt MS, Bayerl MW. 2000. *J. Non-Cryst. Solids* 266–69:1–22
55. Pantelides ST. 1986. *Phys. Rev. Lett.* 57: 2979–82
56. Yamasaki S, Umeda T, Isoya J, Matsuda A, Tanaka K. 1996. Pulsed ESR study of light-induced metastable defects in ^{17}O and ^{13}C incorporated a-Si:H. In *Amorphous Silicon Technology–1996*, ed. H Hack, EA Schiff, S Wagner, R Schropp, A Matsuda, 420:587–92 Pittsburgh: Mater. Res. Soc.
57. Kamei T, Hata N, Matsuda A, Uchiyama T, Yoshioka Y, et al. 1996. *Appl. Phys. Lett.* 68:2380–82
58. Isoya J, Yamasaki S, Matsuda A, Tanaka K. 1994. *Philos. Mag. B* 69:263–68
59. Isoya J, Yamasaki S, Okushi H, Matsuda A, Tanaka K. 1993. *Phys. Rev. B* 47:7013–24
60. Fritzsche H. 1995. *Solid State Commun.* 94:953–55
61. Sheng S, Sacher E, Yelon A. 2000. X-ray photoemission spectroscopic study of structural changes in amorphous silicon. In *Amorphous and Heterogeneous Silicon Thin Films–2000*, ed. HM Branz, RW Collins, S Guha, H Okamoto, M Stutzmann, 609:In press. Warrendale, PA: Mater. Res. Soc.
62. Nonomura S, Gotoh T, Nishio M, Sakamoto T, Kondo M, et al. 1999. Photoinduced expansion in hydrogenated amorphous silicon. In *Amorphous and Heterogeneous Silicon Thin Films: Fundamentals to Devices–1999*, ed. HM Branz, RW Collins, H Okamoto, S Guha, R Schropp, 557:337–45. Warrendale, PA: Mater. Res. Soc.
63. Nonomura S, Yoshida N, Gotoh T, Sakamoto T, Kondo M, et al. 2000. *J. Non-Cryst. Solids* 266–69:474–80
64. Stratakis E, Spanakis E, Fritzsche H, Tzanetakis P. 2000. *J. Non-Cryst. Solids* 266–69:506–10
65. Shimizu K, Tabuchi T, Hattori K, Kida H, Okamoto H. 1998. Photoinduced structural changes in hydrogenated amorphous silicon. In *Amorphous and Microcrystalline Silicon Technology–1998*, ed. R Schropp, HM Branz, M Hack, I Shimizu, S Wagner, 507:735–46. Warrendale, PA: Mater. Res. Soc.
66. Shimizu K. 2000. *J. Non-Cryst. Solids* 266–69:486–90
67. Zhao Y, Zhang DL, Kong GL, Pan G, Liao XB. 1995. *Phys. Rev. Lett.* 74:558–61
68. Kong GL, Zhang DL, Yue GZ, Wang YQ, Liao XB. 1998. Light-excited structural instability of a-Si:H. In *Amorphous and Microcrystalline Silicon Technology–1998*, ed. R Schropp, HM Branz, M Hack, I Shimizu, S Wagner, 507:697–707. Warrendale, PA: Mater. Res. Soc.
69. Liu X, Pohl RO, Crandall RS. 1999. Structural origin of bulk molecular hydrogen in hydrogenated amorphous silicon. In *Amorphous and Heterogeneous Silicon Thin Films: Fundamentals to Devices–1999*, ed. HM Branz, RW Collins, H Okamoto, S Guha, R Schropp, 557:323–28. Warrendale, PA: Mater. Res. Soc.
70. Liu X, Pohl RO, Crandall RS. 1999. Light-induced increase in two-level tunneling states in hydrogenated amorphous silicon. In *Amorphous and Heterogeneous Silicon Thin Films: Fundamentals to Devices–1999*, ed. HM Branz, RW Collins, H Okamoto, S Guha, R Schropp, 557:389–94. Warrendale, PA: Mater. Res. Soc.
71. Liu X, Spiel CL, Pohl RO, Iwaniczko E, Crandall RS. 2000. *J. Non-Cryst. Solids* 266–69:501–5
72. Ovshinsky SR. 1994. *Solar Energy Mater. Solar Cells* 32:443–49
73. Tsu DV, Chao BS, Ovshinsky SR, Guha S,

Yang J. 1997. *Appl. Phys. Lett.*. 71:1317–19
74. Mahan AH, Yang J, Guha S, Williamson DL. 2000. *Phys. Rev. B* 61:1677–80
75. Yang J, Guha S. 1999. Amorphous silicon alloy materials and solar cells near the threshold of microcrystallinity. In *Amorphous and Heterogeneous Silicon Thin Films: Fundamentals to Devices–1999*, ed. HM Branz, RW Collins, H Okamoto, S Guha, R Schropp, 557:239–50. Warrendale, PA: Mater. Res. Soc.
76. Koh J, Ferulato AS, Rovira PI, Wronski CR, Collins RW. 1999. *Appl. Phys. Lett.* 75:2286–88
77. Yang J, Lord K, Guha S, Ovshinsky SR. 2000. Amorphous silicon alloy solar cells near the threshold of amorphous to microcrytalline transition. In *Amorphous and Heterogeneous Silicon Thin Films–2000*, ed. HM Branz, RW Collins, S Guha, H Okamoto, M Stutzmann, 609: In press. Warrendale, PA: Mater. Res. Soc.
78. Guha S, Yang J, Williamson DL, Lubianiker Y, Cohen JD, Mahan AH. 1999. *Appl. Phys. Lett.* 74:1860–62
79. Mahan AH, Carapella J, Nelson BP, Crandall RS, Balberg I. 1991. *J. Appl. Phys.* 69:6728–30
80. Papadopulos P, Scholz A, Bauer S, Schröder B, Oechsner H. 1993. *J. Non-Cryst. Solids* 164–66:87–90
81. Mahan AH, Reedy RC Jr, Iwaniczko E, Wang Q, Nelson BP, et al. 1998. H outdiffusion and device performance in n-i-p solar cells utilizing high temperature hot wire a-Si:H I-layers. In *Amorphous and Microcrystalline Silicon Technology–1998*, ed. R Schropp, HM Branz, M Hack, I Shimizu, S Wagner, 507:119–26. Warrendale, PA: Mater. Res. Soc.
82. Ganguly G, Yamasaki S, Matsuda A. 1991. *Philos. Mag. B* 63:281–92
83. Sugiyama S, Yang J, Guha S. 1997. *Appl. Phys. Lett.* 70:378–80
84. Van de Walle CG, Jackson WB. 1996. *Appl. Phys. Lett.* 69:2441
85. Biswas R, Li YP, Pan BC. 2000. *J. Non-Cryst. Solids* 266–69:178–79
86. Redfield D, Bube RH. 1989. *Appl. Phys. Lett.* 54:1037–39
87. Redfield D. 1988. *Appl. Phys. Lett.* 52:492–94
88. Santos PV, Jackson WB, Street RA. 1991. An alternative model for the kinetics of light-induced defect in a-Si:H. In *Amorphous Silicon Technology–1991*, ed. A Madan, Y Hamakawa, MJ Thompson, PC Taylor, PG LeComber, 219:15–20. Pittsburgh, PA: Mater. Res. Soc.
89. Hata N, Wagner S. 1992. *J. Appl. Phys.* 72:2857–72
90. Morigaki M. 1988. *Jpn. J. Appl. Phys.* 27:163–68
91. Brandt MS, Stutzmann M. 1991. *Appl. Phys. Lett.* 58:1620–22
92. Godet C. 1998. *J. Non-Cryst. Solids* 227–30:272–75
93. Adler D. 1983. *Solar Cells* 9:133–48
94. Pantelides ST. 1992. *Solid State Commun.* 84:221–25
95. Brandt MS, Asano A, Stutzmann M. 1993. Are there charged dangling bonds in device quality amorphous silicon? In *Amorphous Silicon Technology–1993*, ed. EA Schiff, MJ Thompson, A Madan, K Tanaka, PG LeComber, 297:201–6. Pittsburgh: Mater. Res. Soc.
96. Branz HM. 1999. *Phys. Rev. B* 59:5498–12
97. Branz HM. 2000. *J. Non-Cryst. Solids* 266–69:391–96
98. Fritzsche H. 1997. Search for explaining the Staebler-Wronski effect. In *Amorphous and Microcrystalline Silicon Technology–1997*, ed. S Wagner, H Hack, EA Schiff, R Schropp, I Shimizu, 467:19–30. Pittsburgh, PA: Mater. Res. Soc.
99. Stutzmann M, Rossi MC, Brandt MS. 1994. *Phys. Rev. B* 50:11592–605
100. Tzanetakis P, Kopidakis N, Fritzsche H. 1996. *J. Non-Cryst. Solids* 198–200:276–79

101. Stradins P, Kondo M, Matsuda A. 2000. A study of the time-scales of processes responsible for the light-induced degradation of a-Si:H by pulse illumination. In *Amorphous and Heterogeneous Silicon Thin Films–2000*, ed. HM Branz, RW Collins, S Guha, H Okamoto, M. Stutzmann, 609:In press. Warrendale, PA: Mater. Res. Soc.
102. Branz HM, Asher S, Gleskova H, Wagner S. 1999. *Phys. Rev. B* 59:5513–20
103. Branz HM. 1999. *Phys. Rev. B* 60:7725–27
104. Cheong HM, Lee SH, Nelson BP, Mascarenhas A, Deb SK. 2000. *Appl. Phys. Lett.* 77:2686–88
105. Fedders PA, Fu Y, Drabold DA. 1992. *Phys. Rev. Lett.* 68:1888–91
106. Biswas R, Li YP. 1999. *Phys. Rev. Lett.* 82:2512–15
107. Hata N, Matsuda A. 1993. *Appl. Phys. Lett.* 63:1948–50
108. Palinginis KC, Cohen JD, Yang JC, Guha S. 2000. *J. Non-Cryst. Solids* 266–69:665–69
109. den Boer W, Geerts MJ, Ondris M, Wentinck HM. 1984. *J. Non-Cryst. Solids* 66:363–68

Annu. Rev. Mater. Res. 2001. 31:81–110

Biological Responses to Materials

James M Anderson
Institute of Pathology, 2085 Adelbert Road, Case Western Reserve University, Cleveland, Ohio 44106

Key Words biocompatibility, inflammation, foreign body reaction, in vivo studies, toxicity

■ **Abstract** All materials intended for application in humans as biomaterials, medical devices, or prostheses undergo tissue responses when implanted into living tissue. This review first describes fundamental aspects of tissue responses to materials, which are commonly described as the tissue response continuum. These actions involve fundamental aspects of tissue responses including injury, inflammatory and wound healing responses, foreign body reactions, and fibrous encapsulation of the biomaterial, medical device, or prosthesis. The second part of this review describes the in vivo evaluation of tissue responses to biomaterials, medical devices, and prostheses to determine intended performance characteristics and safety or biocompatibility considerations. While fundamental aspects of tissue responses to materials are important from research and development perspectives, the in vivo evaluation of tissue responses to these materials is important for performance, safety, and regulatory reasons.

INTRODUCTION

The goal of this review is to provide material scientists and engineers with an appreciation of the fundamental aspects of tissue responses to materials, as well as the in vivo evaluation of tissue responses to materials. Fundamental aspects of tissue responses to materials include the tissue response continuum, which is initiated when a material (biomaterial), medical device, or prosthesis is implanted in living tissue. The tissue response continuum is the series of responses that are initiated by the implantation procedure, as well as by the presence of the biomaterial, medical device, or prosthesis. The fundamental aspects of the tissue response continuum are viewed from the classical medical perspective of the pathologist. It includes our current understanding of inflammatory and wound healing responses, foreign body reactions, and ultimately fibrous encapsulation (scar formation) of the biomaterial, medical device, or prosthesis.

The second part of this review addresses the in vivo evaluation of tissue responses to materials. From a practical perspective, i.e. manufacturing, clinical, and regulatory, the in vivo evaluation of prostheses and medical devices, i.e. biomaterials in their ready-to-use form, is necessary to determine their biocompatibility.

Biocompatibility is generally defined as the ability of a biomaterial, prosthesis, or medical device to perform with an appropriate host response in a specific application, and biocompatibility assessment, i.e. evaluation of biological responses, is a measure of the magnitude and duration of the adverse alterations in homeostatic mechanisms that determine the host response. Practically speaking, the evaluation of biological responses to a medical device is carried out to determine that the medical device performs as intended and presents no significant harm to the patient or user. Thus the goal of biological response evaluation is to predict whether a biomaterial, medical device, or prosthesis presents potential harm to the patient or user by evaluating conditions that simulate clinical use.

FUNDAMENTAL ASPECTS OF TISSUE RESPONSES TO MATERIALS

Injury

The process of implantation of a biomaterial, prosthesis, or medical device results in injury to tissues or organs (1, 2). It is this injury and the subsequent perturbation of homeostatic mechanisms that lead to the cellular cascades of wound healing. The response to injury is dependent on multiple factors including the extent of injury, the loss of basement membrane structures, blood-material interactions, provisional matrix formation, the extent or degree of cellular necrosis, and the extent of the inflammatory response. These events, in turn, may affect the extent or degree of granulation tissue formation, foreign body reaction, and fibrosis or fibrous capsule development. These events are summarized in Table 1. The host reactions are considered to be tissue, organ, and species dependent. In addition, it is important to recognize that these reactions occur very early, i.e. within 2 to 3 weeks of the time of implantation.

TABLE 1 Sequence of host reactions following implantation of medical devices

Injury
Blood-material interactions
Provisional matrix formation
Acute inflammation
Chronic inflammation
Granulation tissue
Foreign body reaction
Fibrosis/fibrous capsule development

In considering these early host reactions following injury, it is important to consider whether tissue resolution or organization occurs within the injured tissue or organ. In situations where injury has occurred and exudative inflammation is present, but no cellular necrosis or loss of basement membrane structures has occurred, the process of resolution occurs. Resolution is the restitution of the pre-existing architecture of the tissue or organ. On the other hand, with necrosis, granulation tissue grows into the inflammatory exudate and the process of organization with development of fibrous tissue occurs. With implants, the process of organization with development of fibrous tissue leads to the well-known fibrous capsule formation at the tissue/material interface. The proliferative capacity of cells within the tissue or organ also plays a role in determining whether resolution or organization occurs. In general, the process of implantation in vascularized tissues leads to organization with fibrous tissue development and fibrous encapsulation.

Blood-Material Interactions and Initiation of the Inflammatory Response

Blood-material interactions and the inflammatory response are intimately linked and, in fact, early responses to injury involve mainly blood and the vasculature (1–4). Regardless of the tissue or organ into which a biomaterial is implanted, the initial inflammatory response is activated by injury to vascularized connective tissue (Table 2). Because blood and its components are involved in the initial inflammatory responses, thrombi and/or blood clots also form. Thrombus formation involves activation of the extrinsic and intrinsic coagulation systems, the complement system, the fibrinolytic system, the kinin-generating system, and platelets. Thrombus or blood clot formation on the surface of a biomaterial is related to the well-known Vroman effect of protein adsorption. From a wound-healing perspective, blood protein deposition on a biomaterial surface is described as provisional matrix formation.

Immediately following injury, changes occur in vascular flow, caliber, and permeability. Fluid, proteins, and blood cells escape from the vascular system into the injured tissue in a process called exudation. Following changes in the vascular system, which also include changes induced in blood and its components, cellular events occur and characterize the inflammatory response (3–6). The effect of the injury and/or biomaterial in situ on plasma or cells can produce chemical factors that mediate many of the vascular and cellular responses of inflammation. Although injury initiates the inflammatory response, released chemicals from plasma, cells, and injured tissue mediate the response. Important classes of chemical mediators of inflammation are presented in Table 3. Several important points must be noted in order to understand the inflammatory response and how it relates to biomaterials. First, although chemical mediators are classified on a structural or functional basis, different mediator systems interact and provide a system of checks and balances regarding their respective activities and functions. Second, chemical mediators

TABLE 2 Cells and components of vascularized connective tissue

Intravascular (blood) cells
Erythryocytes (RBC)
Neutrophils
Monocytes
Eosinophils
Lymphocytes
Basophils
Platelets
Connective tissue cells
Mast cells
Fibroblasts
Macrophages
Lymphocytes
Extracellular matrix components
Collagens
Elastin
Proteoglycans
Fibronectin
Laminin

are quickly inactivated or destroyed, suggesting that their action is predominantly local (i.e. at the implant site). Third, generally the lysosomal proteases and oxygen-derived free radicals produce the most significant damage or injury. These chemical mediators are also important in the degradation of biomaterials.

The predominant cell type present in the inflammatory response varies with the age of the injury. In general, neutrophils predominate during the first several days following injury and then are replaced by monocytes as the predominant cell type. Three factors account for this change in cell type: (*a*) Neutrophils are short-lived and disintegrate and disappear after 24 to 48 h; neutrophil emigration is of short duration because chemotactic factors for neutrophil migration are activated early in the inflammatory response. (*b*) Following emigration from the vasculature, monocytes differentiate into macrophages, and these cells are very long-lived (up to months). (*c*) Monocyte emigration may continue for days to weeks, depending on the injury and implanted biomaterial, and chemotactic factors for monocytes are activated over longer periods of time.

Provisional Matrix Formation

Injury to vascularized tissue in the implantation procedure leads to immediate development of the provisional matrix at the implant site. This provisional

TABLE 3 Important chemical mediators of inflammation derived from plasma, cells or injured tissue

Mediators	Examples
Vasoactive agents	Histamines, serotonin, adenosine, endothelial-derived relaxing factor (EDRF), prostacyclin, endothelin, thromboxane a_2
Plasma proteases	
Kinin system	Bradykinin, kallikrein
Complement system	C3a, C5a, C3b, C5b-C9
Coagulation/fibrinolytic system	Fibrin degradation products, activated Hageman factor (FXIIA), tissue plasminogen activator (tPA)
Leukotrienes	Leukotriene B_4 (LTB_4), hydroxyeicosa-tetraenoic acid (HETE)
Lysosomal proteases	Collagenase, elastase
Oxygen-derived free radicals	H_2O_2, superoxide anion
Platelet activating factors	Cell membrane lipids
Cytokines	Interleukin 1 (IL-1), tumor necrosis factor (TNF)
Growth factors	Platelet derived growth factor (PDGF), fibroblast growth factor (FGF), transforming growth factor (TGF-α or TGF-β), epithelial growth factor (EGF)

matrix consists of fibrin, produced by activation of the coagulative and thrombosis systems, and inflammatory products, released by the complement system, activated platelets, inflammatory cells, and endothelial cells (7–9). These events occur early, within minutes to hours following implantation of a medical device. Components within or released from the provisional matrix, i.e. fibrin network (thrombosis or clot), initiate the resolution, reorganization, and repair processes such as inflammatory cell and fibroblast recruitment. Platelets, activated during the fibrin network formation, release platelet factor 4, platelet-derived growth factor (PDGF), and transforming growth factor β (TGF-β), which contribute to fibroblast recruitment (10, 11). Upon activation, monocytes and lymphocytes generate additional chemotactic factors, including LTB_4, PDGF, and TGF-β, to recruit fibroblasts.

Fibrin, the major component of the provisional matrix, has been shown to play a key role in the development of neovascularization, i.e. angiogenesis. Implanted porous surfaces filled with fibrin exhibit new vessel growth within four days. The intensity of this angiogenic response is enhanced when zymosan-activated serum or PDGF is incorporated in the fibrin matrix (12).

The provisional matrix is composed of adhesive molecules such as fibronectin and thrombospondin bound to fibrin, as well as platelet granule components released during platelet aggregation. Platelet granule components include thrombospondin, released from the platelet α-granule, and cytokines, including TGF-α, TGF-β, PDGF, platelet factor 4, and platelet-derived endothelial cell growth factor. The provisional matrix is stabilized by the cross-linking of fibrin by factor XIIIa.

The provisional matrix appears to furnish both structural and biochemical components to the process of wound healing. The complex three-dimensional structure of the fibrin network with attached adhesive proteins provides a substrate for cell adhesion and migration. The presence of mitogens, chemoattractants, cytokines, and growth factors within the provisional matrix supplies a rich milieu of activating and inhibiting substances for various cellular proliferative and synthetic processes.

The provisional matrix may be viewed as a naturally derived, biodegradable, sustained release system in which mitogens, chemoattractants, cytokines, and growth factors are released to control subsequent wound healing processes (13–18). In spite of the rapid increase in our knowledge of the provisional matrix and its capabilities, our knowledge of the control of the formation of the provisional matrix and its effect on subsequent wound healing events is poor. In part this lack is due to the fact that much of our knowledge regarding the provisional matrix has been derived from in vitro studies, and there is a paucity of in vivo studies that provide for a more complex perspective. Little is known regarding the provisional matrix that forms at biomaterial and medical device interfaces in vivo. Attractive hypotheses have been presented regarding the presumed ability of materials and protein adsorbed materials to modulate cellular interactions through their interactions with adhesive molecules and cells.

Temporal Sequence of Inflammation and Wound Healing

Inflammation is generally defined as the reaction of vascularized living tissue to local injury. Inflammation serves to contain, neutralize, dilute, or wall off the injurious agent or process. In addition, it sets into motion a series of events that may heal and reconstitute the implant site through replacement of the injured tissue by regeneration of native parenchymal cells, formation of fibroblastic scar tissue, or a combination of these two processes (3, 4).

The sequence of events following implantation of a biomaterial is illustrated in Figure 1. The size, shape, and chemical and physical properties of the biomaterial and the physical dimensions and properties of the prosthesis or device may be responsible for variations in the intensity and time duration of the inflammatory and wound healing processes. Thus intensity and/or time duration of inflammatory reaction may characterize the biocompatibility of a biomaterial, prosthesis, or device.

In general, the biocompatibility of a material with tissue has been described in terms of the acute and chronic inflammatory responses and of the fibrous capsule formation that is seen over various time periods following implantation (19, 20).

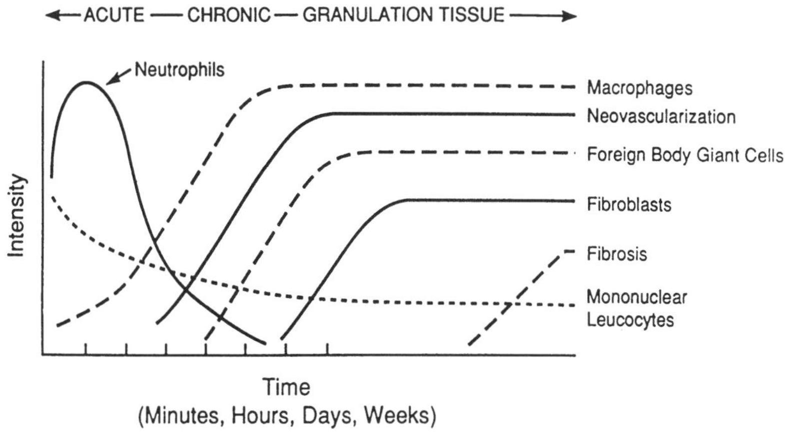

Figure 1 The temporal variation in the acute inflammatory response, chronic inflammatory response, granulation tissue development, and foreign body reaction to implanted biomaterials. The intensity and time variables are dependent upon the extent of injury created in the implantation and the size, shape, topography, and chemical and physical properties of the biomaterial.

Histological evaluation of tissue adjacent to implanted materials as a function of implant time has been the most commonly used method of evaluating the biocompatibility. Classically, the biocompatibility of an implanted material has been described in terms of the morphological appearance of the inflammatory reaction to the material; however, the inflammatory response is a series of complex reactions involving various types of cells, the densities, activities, and functions of which are controlled by various endogenous and autocoid mediators. The simplistic view of the acute inflammatory response progressing to the chronic inflammatory response may be misleading with respect to biocompatibility studies and the inflammatory response to implants. Studies using the cage implant system show that monocytes and macrophages are present in highest concentrations when neutrophils are also at their highest concentrations, i.e. the acute inflammatory response (21, 22). Neutrophils have short lifetimes—hours to days—and disappear from the exudate more rapidly than do macrophages, which have lifetimes of days to weeks to months. Eventually macrophages become the predominant cell type in the exudate, resulting in a chronic inflammatory response. Monocytes rapidly differentiate into macrophages, the cells principally responsible for normal wound healing in the foreign body reaction. Classically, the development of granulation tissue has been considered to be a part of chronic inflammation, but because of unique tissue-material interactions, it is preferable to differentiate the foreign body reaction—with its varying degree of granulation tissue development, including macrophages, fibroblasts, and capillary formation–from chronic inflammation.

Acute Inflammation

Acute inflammation is of relatively short duration, lasting from minutes to days, depending on the extent of injury. The main characteristics of acute inflammation are the exudation of fluid and plasma proteins (edema) and the emigration of leukocytes (predominantly neutrophils). Neutrophils and other motile white cells emigrate or move from the blood vessels to the perivascular tissues and the injury (implant) site (23–25).

The accumulation of leukocytes, in particular neutrophils and monocytes, is the most important feature of the inflammatory reaction. Leukocytes accumulate through a series of processes including margination, adhesion, emigration, phagocytosis, and extracellular release of leukocyte products (26). Increased leukocytic adhesion in inflammation involves specific interactions between complementary adhesion molecules present on the leukocyte and endothelial surfaces (27, 28). The surface expression of these adhesion molecules is modulated by inflammatory agents; mechanisms of interaction include stimulation of leukocyte adhesion molecules (C5a, LTB_4), stimulation of endothelial adhesion molecules (IL-1), or both effects, i.e. tumor necrosis factor (TNF). Integrins make up a family of transmembrane glycoproteins that modulate cell-matrix and cell-cell relationships by acting as receptors to extracellular protein ligands and also as direct adhesion molecules (29). An important group of integrins (adhesion molecules) on leukocytes include the CD11/CD18 family of adhesion molecules. These integrins have identical beta (CD18) subunits but different alpha (CD11a, b, c) subunits. Inflammatory mediators, i.e. cytokines, stimulate a rapid increase in these adhesion molecules on the leukocyte surface, as well as increased leukocyte adhesion to endothelium. Leukocyte-endothelial cell interactions are also controlled by endothelial-leukocyte adhesion molecules (ELAMs, E-selectins) or intracellular adhesion molecules (ICAM-1, ICAM-2, and VCAMs) on endothelial cells (30).

White cell emigration is controlled in part by chemotaxis, which is the unidirectional migration of cells along a chemical gradient. A wide variety of exogenous and endogenous substances have been identified as chemotactic agents (5, 23–34). Important to the emigration or movement of leukocytes is the presence of specific receptors for chemotactic agents on the cell membranes of leukocytes. These and other receptors may also play a role in the activation of leukocytes. Following localization of leukocytes at the injury (implant) site, phagocytosis and the release of enzymes occur following activation of neutrophils and macrophages. The major role of the neutrophils in acute inflammation is to phagocytose microorganisms and foreign materials. Phagocytosis is seen as a three-step process in which the injurious agent undergoes recognition and neutrophil attachment, engulfment, and killing or degradation. With regard to biomaterials, engulfment and degradation may or may not occur, depending on the properties of the biomaterial.

Although biomaterials are not generally phagocytosed by neutrophils or macrophages because of the size disparity (i.e. the surface of the biomaterial is greater than the size of the cell), certain events in phagocytosis may occur. The process of recognition and attachment is expedited when the injurious agent is coated by naturally occurring serum factors called opsonins. The two major opsonins are IgG and the complement-activated fragment C3b. Both of these plasma-derived proteins are known to adsorb to biomaterials, and neutrophils and macrophages have corresponding cell membrane receptors for these opsonization proteins. These receptors may also play a role in the activation of the attached neutrophil or macrophage. Because of the size disparity between the biomaterial surface and the attached cell, frustrated phagocytosis may occur (31, 32). This process does not involve engulfment of the biomaterial but does cause the extracellular release of leukocyte products in an attempt to degrade the biomaterial. Neutrophils adherent to complement-coated and immunoglobulin-coated non-phagocytosable surfaces may release enzymes by direct extrusion or exocytosis from the cell (31, 32). The amount of enzyme released during this process depends on the size of the polymer particle, with larger particles inducing greater amounts of enzyme release. This suggests that the specific mode of cell activation in the inflammatory response in tissue is dependent upon the size of the implant and that a material in a phagocytosable form (e.g. powder or particulate) may provoke a degree of inflammatory response different from that of the same material in a non-phagocytosable form (e.g. film).

Chronic Inflammation

Chronic inflammation is less uniform histologically than is acute inflammation. In general, chronic inflammation is characterized by the presence of macrophages, monocytes, and lymphocytes, with the proliferation of blood vessels and connective tissue (3, 4, 35, 36). It must be noted that many factors modify the course and histological appearance of chronic inflammation.

Persistent inflammatory stimuli lead to chronic inflammation. Although the chemical and physical properties of the biomaterial may lead to chronic inflammation, motion in the implant site by the biomaterial may also produce chronic inflammation. The chronic inflammatory response to biomaterials is confined to the implant site. Inflammation with the presence of mononuclear cells, including lymphocytes and plasma cells, is given the designation chronic inflammation, whereas the foreign body reaction with granulation tissue development is considered the normal wound healing response to implanted biomaterials (i.e. the normal foreign body reaction).

Lymphocytes and plasma cells are involved principally in immune reactions and are key mediators of antibody production and delayed hypersensitive responses. Their roles in non-immunologic injuries and inflammation are largely unknown. Little is known regarding humoral immune responses and cell-mediated immunity to synthetic biomaterials. The role of macrophages must be considered in the possible development of immune responses to synthetic biomaterials. Macrophages process and present the antigen to immunocompetent cells and thus are key mediators in the development of immune reactions.

The macrophage is probably the most important cell in chronic inflammation because of the great number of biologically active products it produces (35). Important classes of products produced and secreted by macrophages include neutral proteases, chemotactic factors, arachidonic acid metabolites, reactive oxygen metabolites, complement components, coagulation factors, growth-promoting factors, and cytokines.

Growth factors such as PDGF, FGF, TFG-β, TGF-α/EGF, and IL-1 or TNF are important to the growth of fibroblasts and blood vessels and the regeneration of epithelial cells. Growth factors, released by activated cells, stimulate production of a wide variety of cells; they initiate cell migration, differentiation, and tissue remodeling and may be involved in various stages of wound healing (37–42). It is clear that there is a lack of information regarding interaction and synergy among various cytokines and growth factors and their abilities to exhibit chemotactic, mitogenic, and angiogenic properties.

Granulation Tissue

Within one day following implantation of a biomaterial (i.e. injury), the healing response is initiated by the action of monocytes and macrophages, followed by proliferation of fibroblasts and vascular endothelial cells at the implant site, leading to the formation of granulation tissue, the hallmark of healing inflammation. Granulation tissue derives its name from the pink, soft granular appearance on the surface of healing wounds, and its characteristic histological features include the proliferation of new small blood vessels and fibroblasts. Depending on the extent of injury, granulation tissue may be seen as early as three to five days following implantation of a biomaterial.

The new, small blood vessels are formed by budding or sprouting of preexisting vessels in a process known as neovascularization or angiogenesis (43–45). This process involves proliferation, maturation, and organization of endothelial cells into capillary tubes. Fibroblasts also proliferate in developing granulation tissue and are active in synthesizing collagen and proteoglycans. In the early stages of granulation tissue development, proteoglycans predominate; later, however, collagen, especially type I collagen, predominates and forms the fibrous capsule. Some fibroblasts in developing granulation tissue may have features of smooth muscle cells. These cells are called myofibroblasts and are considered to be responsible for the wound contraction seen during the development of granulation tissue.

The wound healing response is generally dependent on the extent or degree of injury or defect created by the implantation procedure. Wound healing by primary union (or first intention) is the healing of clean, surgical incisions in which the wound edges have been approximated by surgical sutures. Healing under these conditions occurs without significant bacterial contamination and with a minimal loss of tissue. Wound healing by secondary union (or second intention) occurs when there is a large tissue defect that must be filled or where there is extensive loss of cells and tissue. In wound healing by second intention, regeneration of parenchymal cells cannot completely reconstitute the original architecture, and much more granulation tissue is formed, resulting in larger areas of fibrosis or scar formation.

Granulation tissue is distinctly different from granulomas, which are small collections of modified macrophages called epithelioid cells. Foreign body giant cells may surround non-phagocytosable particulate materials in granulomas. Foreign body giant cells are formed by the fusion of monocytes/macrophages in an attempt to phagocytose the material.

Foreign Body Reaction

The foreign body reaction is composed of foreign body giant cells and the components of granulation tissue, which consist of macrophages, fibroblasts, and capillaries in varying amounts, depending upon the form and topography of the implanted material. Relatively flat and smooth surfaces, such as those found on breast prostheses, have a foreign body reaction that is composed of a layer of macrophages one to two cells in thickness. Relatively rough surfaces, such as those found on the outer surfaces of expanded poly(tetrafluoroethylene) (ePTFE) vascular prostheses, have a foreign body reaction composed of macrophages and foreign body giant cells at the surface. Fabric materials generally have a surface response composed of macrophages and foreign body giant cells with varying degrees of granulation tissue subjacent to the surface response.

As previously discussed, the form and topography of the surface of the biomaterial determines the composition of the foreign body reaction. With biocompatible materials, the composition of the foreign body reaction in the implant site may be controlled by the surface properties of the biomaterial, the form of the implant, and the relationship between the surface area of the biomaterial and the volume of the implant. For example, high surface-to-volume implants such as fabrics or porous materials will have higher ratios of macrophages and foreign body giant cells in the implant site than will smooth-surface implants, which will have fibrosis as a significant component of the implant site.

The foreign body reaction, consisting mainly of macrophages and/or foreign body giant cells, may persist at the tissue-implant interface for the lifetime of the implant (1, 2, 46–48). Generally, fibrosis (i.e. fibrous encapsulation) surrounds the biomaterial or implant with its interfacial foreign body reaction, isolating the implant and foreign body reaction from the local tissue environment. Early in the inflammatory and wound healing response, the macrophages are activated upon adherence to the material surface. Although it is generally considered that the chemical and physical properties of the biomaterial are responsible for macrophage activation, the nature of the subsequent events regarding the activity of macrophages at the surface is not clear. Tissue macrophages, derived from circulating blood monocytes, may coalesce to form multinucleated foreign body giant cells. Very large foreign body giant cells containing large numbers of nuclei are typically present on the surface of biomaterials. Although these foreign body giant cells may persist for the lifetime of the implant, it is not known if they remain activated, releasing their lysosomal constituents, or become quiescent.

Efforts in our laboratory have focused on differential lymphokine regulation of macrophage fusion, which leads to morphological variants of multinucleated giant cells, and the role played by the surface chemistry and other properties of the foreign material in facilitating monocyte adhesion, macrophage development, and giant cell formation. Foreign body giant cells are observed at the tissue/material interface of medical devices implanted in soft and hard tissue and remain at the implant/tissue interface for the lifetime of the device in vivo, which in some cases may extend beyond 20 years. In addition, foreign body giant cells have been implicated in the biodegradation of polymeric medical devices. Foreign body giant cells and macrophages constituting the foreign body reaction at the tissue/device interface are surface-area dependent. Fabrics utilized as vascular grafts show high densities of foreign body giant cells, whereas flat surfaces such as those found on breast implants exhibit only a one- to two-cell layer of macrophages and foreign body giant cells at the tissue/material interface. For these and other reasons, we have sought to identify the mechanism of induction of foreign body giant cells on biomaterials and the physiological and material bases for their formation.

Early studies utilizing lymphokines in the induction of foreign body giant cell formation employed a wide variety of experimental conditions that resulted in both positive and negative modulation of these cells' formation. A number of these studies utilized conditioned media or supernatants. To provide a clearer identification of cell-derived agents that produce foreign body giant cells, we used recombinant human lymphokines with freshly isolated human monocytes in our culture systems. We believe that these conditions provide greater insight into foreign body giant cell formation and obviate unidentified problems that may result from the use of transformed cell lines and conditioned media and supernatants.

In our studies, human interleukin-4 (IL-4) induced the formation of foreign body giant cells from human monocyte-derived macrophages, an effect that was optimized with either GM-CSF or IL-3, dependent on the concentration of IL-4, and specifically prevented by anti-IL-4 (49, 50). Very large foreign body giant cells with randomly arranged nuclei and extensive cytoplasmic spreading (285 $\pm$ 121 nuclei and 1.151 $\pm$ 0.303 mm^2 per cell) were consistently obtained. Rates of macrophage fusion in this system were high, 72 $\pm$ 5%.

Figure 2 demonstrates the progression from circulating blood monocyte to tissue macrophage to foreign body giant cell development that is most commonly observed. Indicated in the figure are important biological responses considered to play an important role in foreign body giant cell development. Möst and colleagues have shown that the fusion rates of monocytes/macrophages decrease with advancing differentiation, and almost no giant cell formation was observed with 8-day-old macrophages that were derived from freshly isolated monocytes stimulated with cytokine-containing supernatants (51). A distinct difference in differentiation was seen in our studies when IL-4 was added to freshly adherent (2 h) monocytes. IL-4 under these conditions resulted in a detachment of adherent cells and an

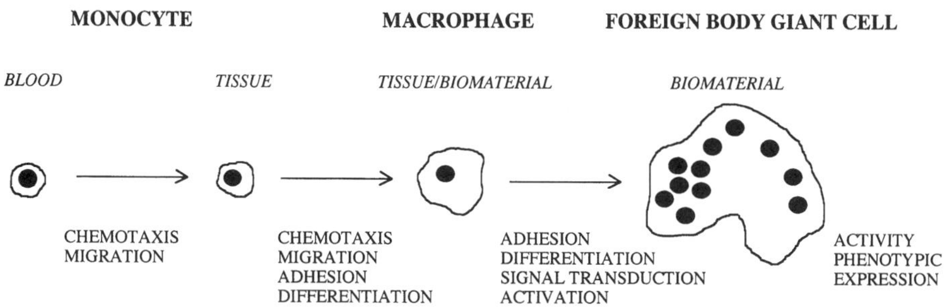

Figure 2 In vivo transition from blood-borne monocyte to biomaterial adherent monocyte/macrophage to foreign body giant cell at the tissue/biomaterial interface. Little is known regarding the indicated biological responses that are considered to play important roles in the transition to foreign body giant cell development.

inhibition of initial monocyte adhesion by IL-4. To accelerate the development of macrophage morphology, we added GM-CSF initially and at 3 days added IL-4 to induce macrophage fusion and foreign body giant cell formation. Although positive effects may result from the use of conditioned media or inflammatory cell-derived supernatants, it is also possible that negative autocrine or paracrine effects with down-regulation of biological interactions important to macrophage differentiation and foreign body giant cell development may occur. It is obvious that the utilization of human cells, together with appropriate recombinant human cytokines and antibodies, provides for cleaner and more relevant systems in mechanistic studies of macrophage differentiation and fusion with foreign body giant cell formation.

Figure 3 demonstrates the sequence of events involved in inflammation and wound healing when medical devices are implanted. In general, the polymorphonuclear neutrophil predominant acute inflammatory response and the lymphocyte/monocyte predominant chronic inflammatory response resolve quickly, i.e. within 2 weeks, depending on the type and location of implant. Studies utilizing IL-4 by ourselves and others demonstrate the role for Th2 helper lymphocytes in the development of the foreign body reaction at the tissue/material interface. Th2 helper lymphocytes have been described as anti-inflammatory based on their cytokine profile of which IL-4 is a significant component. Th2 helper lymphocytes also produce IL-13, and we have utilized this to demonstrate its similar effect to IL-4 on foreign body giant cell formation (52). In this regard, it is noteworthy that anti-IL-4 does not inhibit IL-13-induced foreign body giant cell formation nor does anti-IL-13 inhibit IL-4-induced foreign body giant cell formation. In our IL-4 and IL-13 foreign body giant cell culture systems, the macrophage mannose receptor (MMR) has been identified as critical to the fusion of macrophages in the formation of foreign body giant cells (52, 53). This formation can be prevented by competitive inhibitors of MMR activity, i.e. α-mannan, or inhibitors of glycoprotein processing that restrict MMR surface expression.

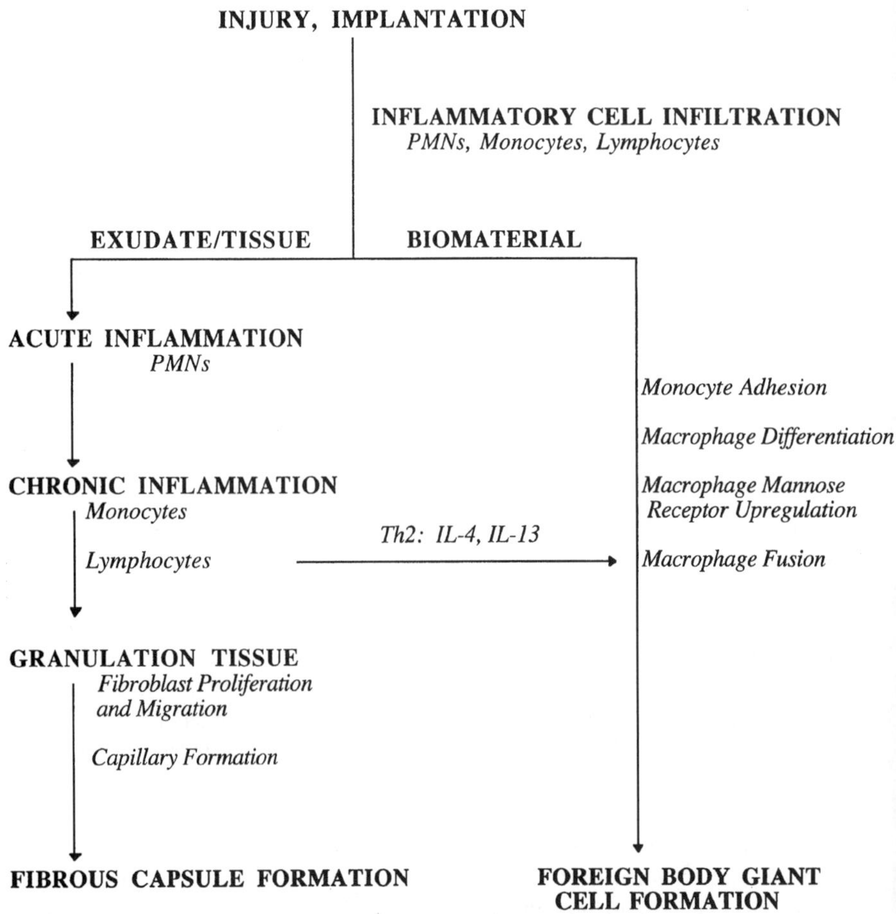

Figure 3 Sequence of events involved in inflammatory and wound healing responses leading to foreign body giant cell formation. This shows the importance of Th2 lymphocytes in the transient chronic inflammatory phase with the production of IL-4 and IL-13, which can induce monocyte/macrophage fusion to form foreign body giant cells.

Two factors that may play a role in multinucleated giant cell studies are the surface chemistry of the substrate onto which the cells adhere and the protein adsorption that occurs before cell adhesion. These two factors have been hypothesized to have significant roles in the inflammatory and wound healing responses to biomaterials and medical devices in vivo.

We have extensively investigated the effect of substrate surface chemistry on monocyte/macrophage adhesion, macrophage fusion, and foreign body giant cell development (54–58). The overall goal of these studies is to identify surfaces that do not permit monocyte/macrophage adhesion and/or macrophage fusion to form foreign body giant cells. Long-chain hydrocarbon groups on glass surfaces markedly reduce macrophage adhesion and nearly eliminate IL-4-induced foreign body giant cells (58). In contrast, polyethylene oxide (PEO) chains on glass surfaces do permit macrophage adhesion, but the level of IL-4-induced foreign body giant cell formation is markedly reduced (56). In the case of clean glass surfaces, adherent macrophage densities are high enough to allow maximal levels of foreign body giant cell formation; however, negligible formation is observed. A comparison of these three different types of surfaces supports the hypothesis that the composition and conformation of proteins adsorbed on surfaces provide signals or ligands for the adhesion of monocytes/macrophages as well as the macrophage fusion process itself. Thus long-term macrophage adhesion and IL-4- or IL-13-induced foreign body giant cell formation are surface-dependent phenomena.

Cytoskeletal and adhesive structure studies of in vitro FBGC formation have demonstrated that podosomal structures, and not focal contacts, are the major adhesive structures present within macrophages and foreign body giant cells on surfaces (59, 60). The podosomal structures found at the ventral periphery of the foreign body giant cells contain vinculin, talin, and paxillin in a ring-like structure surrounding an F-actin core. These podosomal adhesion structures are similar to those identified for osteoclast adhesion, and their presence at the ventral and peripheral surface implies a functional polarization and suggests frustrated phagocytosis via the formation of a closed compartment between the foreign body giant cells and the underlying substrate where degradative enzymes, reactive oxygen intermediates, and/or other products are secreted.

The lifetime of foreign body giant cells at tissue/material interfaces is still unknown. Early publications had suggested that they were relatively short-lived, lasting for several days. This is probably not true as clinical specimens show the presence of foreign body giant cells for years and, in some cases, decades. Honma & Hamasaki have reported on the ultrastructure of multinucleated giant cell apoptosis in a collagen sponge granuloma (61). They noted the disappearance of giant cells coincident with the resorption of the collagen sponge, which is most probably accurate because once the inciting agent for giant cell formation is no longer present, the presence of giant cells is no longer necessary.

The osteoclast, the multinucleated giant cell responsible for bone resorption, is the most widely studied of all types of giant cells. Unlike other types of giant cells, which are found with pathological conditions, the osteoclast is found at bone surfaces where it participates in the constant process of bone remodeling. Excessive osteoclast activity in bone resorption has been implicated in pathological processes such as the advanced stages of multiple myeloma, with lytic lesions in bone, and post-menopausal osteoporosis. The majority of studies suggest that the CFU-GM, the granulocyte-macrophage progenitor, a cell in the monocyte-macrophage lineage, is the earliest osteoclast precursor. While the osteoclast, like the Langhans giant cell and the foreign body giant cell, may have a hematopoietic precursor, molecular and cell biology studies have shown that the osteoclast has distinctly different functional and phenotypic characteristics (62, 63).

The calcitonin receptor is the best marker for distinguishing mammalian osteoclasts because this receptor is not expressed on monocyte/macrophage-derived giant cells. A wide variety of factors that influence osteoclast formation and function include systemic hormones, cytokines, and growth factors. It is noteworthy that neither IL-4 (FBGC formation) nor γ-interferon (Langhans giant cell formation) is described as a significant factor in the formation or activation of osteoclasts. These findings suggest that although the CFU-GM progenitor is of monocytic lineage, its differentiation does not include expression of IL-4 or IFN-γ receptors or, perhaps, even a common signal transduction pathway. This is somewhat surprising as both foreign body giant cells and osteoclasts adhere to substrates through podosomal structures.

Recent studies demonstrate the ability of IL-1 and TNF-α to induce both osteoclast formation and bone-resorbing activity (64–66). These studies suggest that activated macrophages may facilitate bone resorption by participating in osteoclast formation and activation. The role of TNF-α in regulating osteoclastic bone resorption continues to be elucidated with studies demonstrating that osteoblasts/stromal cells express a new member of the TNF-ligand family—osteoclast differentiation factor (ODF)/osteoprotegerin (OPGL)/TNF-related activation-induced cytokine (TRANCE)/receptor activation of NF-κB ligand (RANKL)—as a membrane associated factor (66–68).

Fibrosis and Fibrous Encapsulation

The end-stage healing response to biomaterials is generally fibrosis or fibrous encapsulation. However, there may be exceptions to this general statement (e.g. porous materials inoculated with parenchymal cells or porous materials implanted into bone).

Repair of implant sites involves two distinct processes: regeneration, which is the replacement of injured tissue by parenchymal cells of the same type, or replacement by connective tissue that constitutes the fibrous capsule (3, 69, 70). These processes are generally controlled by either (*a*) the proliferative capacity of the cells in the tissue or organ receiving the implant and the extent of injury as it relates to the destruction or (*b*) persistence of the tissue framework of the implant site. The regenerative capacity of cells permits classification into three groups: labile, stable (or expanding), and permanent (or static) cells. Labile cells continue to proliferate throughout life, stable cells retain this capacity but do not normally replicate, and permanent cells cannot reproduce themselves after birth. Perfect repair with restitution of normal structure theoretically occurs only in tissues consisting of stable and labile cells, whereas all injuries to tissues composed of permanent cells may give rise to fibrosis and fibrous capsule formation, with very little restitution of the normal tissue or organ structure. Tissues composed of permanent cells (e.g. nerve cells, skeletal muscle cells, and cardiac muscle cells) most commonly undergo an organization of the inflammatory exudate, leading to fibrosis. Tissues composed of stable cells (e.g. parenchymal cells of the liver, kidney, and pancreas), mesenchymal cells (e.g. fibroblasts, smooth muscle cells, osteoblasts, and chrondroblasts), and vascular endothelial and labile cells (e.g. epithelial cells and lymphoid and hematopoietic cells) may also follow this pathway to fibrosis or may undergo resolution of the inflammatory exudate, leading to restitution of the normal tissue structure. The condition of the underlying framework or supporting stroma of the parenchymal cells following an injury plays an important role in the restoration of normal tissue structure. Retention of the framework may lead to restitution of the normal tissue structure, whereas destruction of the framework most commonly leads to fibrosis. It is important to consider the species-dependent nature of the regenerative capacity of cells. For example, cells from the same organ or tissue but from different species may exhibit different regenerative capacities and/or connective tissue repair.

Following injury, cells may undergo adaptations of growth and differentiation. Important cellular adaptations are atrophy (decrease in cell size or function), hypertrophy (increase in cell size), hyperplasia (increase in cell number), and metaplasia (change in cell type). Other adaptations include a change in which cells stop producing one family of proteins and start producing another (phenotypic change) or begin a marked overproduction of protein. This may be the case in cells producing various types of collagens and extracellular matrix proteins in chronic inflammation and fibrosis. Causes of atrophy may include decreased workload (e.g. stress-shielding by implants), as well as diminished blood supply and inadequate nutrition (e.g. fibrous capsules surrounding implants).

Local and systemic factors may play a role in the wound healing response to biomaterials or implants. Local factors include the site (tissue or organ) of implantation, the adequacy of blood supply, and the potential for infection. Systemic factors may include nutrition, hematological and immunological derangements, glucocortical steroids, and preexisting diseases such as atherosclerosis, diabetes, and infection.

IN VIVO EVALUATION OF TISSUE RESPONSES TO MATERIALS

From a practical perspective, the in vivo assessment of tissue compatibility of medical devices is carried out to determine that the device performs as intended and presents no significant harm to the patient or user. Thus, the goal of the in vivo assessment of tissue compatibility of medical devices is to determine and predict whether such devices present potential harm to the patient or user by evaluations under conditions simulating clinical use.

Recently, extensive efforts have been made by government agencies, i.e. FDA and regulatory bodies, i.e. ASTM, ISO, USP, to provide procedures, protocols, guidelines, and standards that may be used in the in vivo assessment of tissue compatibility of medical devices (71–76). This chapter draws heavily on the ISO 10,993 standard, Biological Evaluation of Medical Devices, in presenting a systematic approach to such assessments (71).

TABLE 4 Biomaterials and components relevant to in vivo assessment of tissue compatibility

The material(s) of manufacture
Intended additives, process contaminants and residues
Leachable substances
Degradation products
Other components and their interactions in the final product
The properties and characteristics of the final product

In the selection of biomaterials to be used in device design and manufacture, the first consideration should be fitness for purpose with regard to characteristics and properties of the biomaterial(s), which include chemical, toxicological, physical, electrical, morphological, and mechanical properties. Relevant to the overall in vivo assessment of tissue compatibility of a biomaterial or device is a knowledge of the chemical composition of the materials, including the conditions of tissue exposure, as well as the nature, degree, frequency, and duration of exposure of the device and its constituents to the intended tissues into which it will be utilized. Table 4 presents a list of biomaterial components and characteristics that may impact the overall biological responses of the medical device. Therefore, knowledge of these components in the medical device, i.e. final product, is necessary. The range of potential biological hazards is broad and may include short-term effects, long-term effects, or specific toxic effects, which should be considered for every material and medical device. However, this does not imply that testing for all potential hazards is necessary or practical.

Selection of In Vivo Tests According to Intended Use

In vivo tests for assessment of tissue compatibility are chosen to simulate end-use applications. To facilitate the selection of appropriate tests, medical devices with their component biomaterials can be categorized by the nature of body contact of the medical device and by the duration of contact of the medical device. Table 5 presents medical device categorization by body contact and contact duration. The tissue contact categories and subcategories, as well as the contact duration categories, have been derived from standards, protocols, and guidelines utilized in the past for safety evaluation of medical devices. Certain devices may fall into more than one category, in which case testing appropriate to each category should be considered.

The ISO 10,993 standard and the FDA guidance document present a structured program for biocompatibility evaluation in which matrices are presented that indicate required tests according to specific types of tissue contact and contact duration. These matrices are not presented here but the in vivo tests are indicated in Table 6

TABLE 5 Medical device categorization by tissue contact and contact duration

Tissue contact	
Surface devices	Skin Mucosal membranes Breached or compromised surfaces
External communicating devices	Blood path, indirect Tissue/bone/dentin communicating Circulating blood
Implant devices	Tissue/bone Blood
Contact duration	Limited, $\leq$24 h Prolonged, >24 h and <30 days Permanent, >30 days

Significant Issues in In Vivo Testing

Two perspectives may be considered in the in vivo assessment of tissue compatibility of biomaterials and medical devices. The first perspective involves the utilization of in vivo tests to determine the general biocompatibility of newly developed biomaterials for which some knowledge of the tissue compatibility is necessary for further research and development. In this type of situation, manufacturing and other processes necessary to the development of a final product,

TABLE 6 In vivo tests for tissue compatibility

Sensitization
Irritation
Intracutaneous reactivity
Systemic toxicity (acute toxicity)
Subchronic toxicity (subacute toxicity)
Genotoxicity
Implantation
Hemocompatibility
Chronic toxicity
Carcinogenicity
Reproductive and developmental toxicity
Biodegradation
Immune response

i.e. medical device, have not been carried out. However, the in vivo assessment of tissue compatibility at this early stage of development can be used to evaluate the general tissue responses of the biomaterial, as well as provide additional information relating to the proposed design criteria in the production of a medical device. While it is generally recommended that the identification and quantification of extractable chemical entities of a medical device should precede biological evaluation, it is quite common to carry out preliminary in vivo assessments to determine if there may be unknown chemical entities that produce adverse biological reactions. Utilized in this fashion, early in vivo assessment of the tissue compatibility of a biomaterial may provide insight into its biocompatibility and may permit further development of this material into a medical device. Obviously, adverse reactions observed at this stage of development require further efforts to improve the biocompatibility of the biomaterial and to identify the agents responsible for the adverse reactions. As the in vivo assessment of tissue compatibility of a biomaterial or medical device is focused on the end-use application, it must be appreciated that a biomaterial considered compatible for one application may not be compatible for another.

The second perspective regarding the in vivo assessment of tissue compatibility of medical devices focuses on the biocompatibility of the final product, that is, the medical device and its component materials in the condition in which it is implanted. Although medical devices in their final form and condition are commonly implanted in carefully selected animal models to determine function as well as biocompatibility, it may be not appropriate to carry out all of the recommended tests necessary for regulatory approval on the final device. In these situations, some tests may be carried out on biomaterial components of devices that have been prepared under manufacturing and sterilization conditions and other processes utilized in the final product development.

Specific Biological Properties Assessed by In Vivo Tests

In this section, brief perspectives on the general types of in vivo tests are presented. Details regarding these tests are found in the references. The selection of tests for in vivo biocompatibility assessment is based on the characteristics and end-use application of the device or biomaterial under consideration.

SENSITIZATION, IRRITATION, AND INTRACUTANEOUS (INTRADERMAL) REACTIVITY Exposure to or contact with even minute amounts of potential leachable agents in medical devices or biomaterials can result in allergic or sensitization reactions. Sensitization tests that estimate the potential for contact sensitization of medical devices, materials and/or their extracts are usually carried out in guinea pigs, and should reflect the intended route (skin, eye, mucosa) and nature, degree, frequency duration, and conditions of exposure of the biomaterial in its intended clinical use Emphasis is placed on utilizing extracts of the biomaterials to determine the irritant effects of potential leachables. Intracutaneous (intradermal) reactivity tests determine the localized reaction of tissue to extracts of medical devices, biomaterials or prostheses in the final product form. Irritation and intracutaneous tests may be applicable where determination of irritation by dermal or mucosal irritation tests are not appropriate, for example, albino rabbits are most commonly used.

Because these tests focus on determining the biological response of leachable agents that may be present in biomaterials, their extracts in various solvents are utilized to prepare the injection solutions. Critical to the conduct of these tests is the preparation of the test material and/or extract solution and the choice of solvents, which must have physiological relevance.

SYSTEMIC TOXICITY (ACUTE TOXICITY) AND SUBACUTE AND SUBCHRONIC TOXICITY
Systemic toxicity tests estimate the potential harmful effects of either single or multiple exposures, during a period of less than 24 h, to medical devices, biomaterials and/or their extracts. These tests evaluate the systemic toxicity potential of medical devices, which release constituents into the body. These tests also include pyrogenicity testing.

In these tests, the form and area of the material, the thickness, and the surface area to extraction vehicle volume are critical considerations in the testing protocol. Appropriate extraction vehicles, i.e. solvents, should be chosen to yield a maximum extraction of leachable materials to conduct the testing. Mice, rats, or rabbits are the usual animals of choice for these tests and, depending on the intended application of the biomaterial, oral, dermal, inhalation, intravenous, intraperitoneal, or subcutaneous application of the test substance may be used. Acute toxicity is considered to be the adverse effect, which occurs after administration of a single dose or multiple doses of a test sample given within 24 h. Subacute toxicity (repeat dose toxicity) focuses on adverse effects occurring after administration of a single dose or multiple doses of a test sample per day given during a period of from 14 to 28 days. Subchronic toxicity is considered to be the adverse effects occurring after administration of a single dose or multiple doses of a test sample per day given during a part of the life span, usually 90 days but not exceeding 10% of the life span of the animal.

Pyrogenicity (fever-producing) tests are also included in the systemic toxicity category to detect material-mediated pyrogenic reactions of extracts of medical devices or materials. It is noteworthy that no single test can differentiate pyrogenic reactions that are material-mediated from those due to endotoxin contamination.

GENOTOXICITY In vivo genotoxicity tests are carried out if indicated by the chemistry and/or composition of the biomaterial (see Table 4) or if in vitro test results indicate potential genotoxicity. Initially, at least three in vitro assays should be used, and two of these assays should utilize mammalian cells. The initial in vitro assays should cover the three levels of genotoxic effects: DNA effects, gene mutations, and chromosomal aberrations. In vivo genotoxicity tests include the micronucleus test, the in vivo mammalian bone marrow cytogenetic tests, chromosomal analysis, the rodent dominant lethal tests, the mammalian germ cell cytogenetic assay, the mouse spot test, and the mouse heritable translocation assay. Not all of the in vivo genotoxicity tests need be performed, and the most common test is the rodent micronucleus test. Genotoxicity tests are performed with appropriate extracts or dissolved materials using media as suggested by the known composition of the biomaterial.

IMPLANTATION Implantation tests assess the local pathological effects on living tissue of a sample of a material or final product that is surgically implanted or placed into an implant site or tissue appropriate to the intended application of the device. Evaluation of the local pathological effects is carried out at both the gross level and the microscopic level. Histological (microscopic) evaluation is utilized to characterize various biological response parameters. For short-term implantation evaluation out to 12 weeks, mice, rats, guinea pigs, or rabbits are the usual animals utilized in these studies. For longer-term testing in subcutaneous tissue, muscle or bone, animals such as rats, guinea pigs, rabbits, dogs, sheep, goats, pigs and other animals with relatively long-life expectancy are suitable. If a medical device is to be evaluated, larger species may be utilized. For example, substitute heart valves are usually tested in sheep, whereas calves are usually the animal of choice for ventricular assist devices and total artificial hearts.

HEMOCOMPATIBILITY Hemocompatibility tests evaluate effects on blood and/or blood components by blood-contacting medical devices or materials. In vivo hemocompatibility tests are usually designed to simulate the geometry, contact conditions, and flow dynamics of the device or material in its clinical application. From the ISO standards perspective, five test categories are indicated for hemocompatibility evaluation: thrombosis, coagulation, platelets, hematology, and immunology (complement and leukocytes).

Two levels of evaluation are indicated: Level 1 (required) and Level 2 (optional). Regardless of blood contact duration or time, hemocompatibility testing is indicated for external communicating devices:blood path, indirect; external communicating devices, circulating blood; and blood-contacting implant devices.

Several issues are important in the selection of tests for hemocompatibility of medical devices or biomaterials. While in vivo testing in animals may be convenient, species differences in blood reactivity must be considered, and these differences may limit the predictability of any given test in the human clinical situation. Although blood values and reactivity between humans and non-human primates are similar, European community law prohibits the use of non-human primates for blood compatibility and medical device testing. Hemocompatibility evaluation in animals is complicated by the lack of appropriate and adequate test materials; for example, appropriate antibodies for immunoassays. Use of human blood in hemocompatibility evaluation implies in vitro testing, which usually requires the use of anticoagulants, which are not usually present with the device in the clinical situation, except for perhaps the earliest implantation period. Although species differences may complicate hemocompatibility evaluation, the utilization of animals in short- and long-term testing is considered to be appropriate for evaluating thrombosis and tissue interaction.

CHRONIC TOXICITY Chronic toxicity tests determine the effects of either single or multiple exposures to medical devices, materials, and/or their extracts during a period of at least 10% of the life span of the test animal, e.g. over 90 days in rats. Chronic toxicity tests may be considered an extension of subchronic (subacute) toxicity testing, and both may be evaluated in an appropriate experimental protocol or study.

CARCINOGENICITY Carcinogenicity tests determine the tumorigenic potential of medical devices, materials, and/or their extracts from either single or multiple exposures or contacts over a period of the major portion of the life span of the test animal. Carcinogenicity tests should be conducted only if data from other sources suggest a tendency for tumor induction. In addition, both carcinogenicity (tumorigenicity) and chronic toxicity may be studied in a single experimental study. With biomaterials, carcinogenicity studies focus on the potential for solid-state carcinogenicity, i.e. the Oppenheimer effect. In carcinogenicity testing, controls of a comparable form and shape should be included; polyethylene implants are a commonly used control material. The use of appropriate controls is imperative since animals may spontaneously develop tumors, and statistical comparison between the test biomaterial/device and the controls is necessary.

REPRODUCTIVE AND DEVELOPMENTAL TOXICITY These tests evaluate the potential effects of medical devices, materials, and/or their extracts on reproductive function, embryonic development (teratogenicity), and prenatal and early postnatal development. The application site of the device must be considered, and tests and/or bioassays should only be conducted when the device has potential impact on the reproductive potential of the subject.

BIODEGRADATION Biodegradation tests determine the effects of a biodegradable material and its biodegradation products on the tissue response. They focus on the amount of degradation during a given period of time (the kinetics of biodegradation), the nature of the degradation products, the origin of the degradation products (e.g. impurities, additives, corrosion products, bulk polymer, etc), and the qualitative and quantitative assessment of degradation products and leachable agents in adjacent tissues and in distant organs. The biodegradation of biomaterials may occur through a wide variety of mechanisms that, in part, are biomaterial dependent, and all pertinent mechanisms related to the device and the end-use application of the device must be considered. Test materials comparable to degradation products may be prepared and studied to determine the anticipated biological response of these products in long-term implants. An example of this approach is the study of metallic and polymeric wear particles that may be present with long-term orthopedic joint prostheses.

IMMUNE RESPONSES Immune response evaluation is not a component of the standards currently available for in vivo tissue compatibility assessment. However,

TABLE 7 Potential immunological effects and responses

Effects
Hypersensitivity
Type I-anaphylactic
Type II-cytotoxic
Type III-immune complex
Type IV-cell-mediated (delayed)
Chronic inflammation
Immunosuppression
Immunostimulation
Autoimmunity
Responses
Histopathological changes
Humoral responses
Host resistance
Clinical symptoms
Cellular responses
T cells
Natural killer cells
Macrophages
Granulocytes

ASTM, ISO, and the FDA currently have working groups developing guidance documents for immune response evaluation where pertinent. An example of the need for immune response evaluation is with modified natural tissue implants such as collagen, which has been utilized in a number of different types of implants. The Center for Devices and Radiological Health of the FDA has released a draft immunotoxicity testing guidance document whose purpose is to provide a systematic approach for evaluating potential adverse immunological effects of medical devices and constituent materials (73). Immunotoxicity is any adverse effect on the function or structure of the immune system or other systems as a result of an immune system dysfunction. Adverse or immunotoxic effects occur when humoral or cellular immunity needed by the host to defend itself against infections or neoplastic disease (immunosuppression) or unnecessary tissue damage (chronic inflammation, hypersensitivity, or autoimmunity) is compromised. Potential immunological effects and responses that may be associated with one or more of these effects are presented in Table 7.

Selection of Animal Models for In Vivo Tests

Animal models are used to predict the clinical behavior, safety, and biocompatibility of medical devices in humans (Table 8). The selection of animal models

TABLE 8 Animal models for the in vivo assessment of medical devices

Device classification	**Animal**
Cardiovascular	
Heart valves	Sheep
Vascular grafts	Dog, pig
Stents	Pig, dog
Ventricular assist devices	Calf
Artificial hearts	Calf
Ex-vivo shunts	Baboon, dog
Orthopedic/bone	
Bone regeneration/substitutes	Rabbit, dog, pig, mouse, rat
Total joints—hips, knees	Dog, goat, non-human primate
Vertebral implants	Sheep, goat, baboon
Craniofacial implants	Rabbit, pig, dog, non-human primate
Cartilage	Rabbit, dog
Tendon and ligament substitutes	Dog, sheep
Neurological	
Peripheral nerve regeneration	Rat, cat, non-human primate
Electrical stimulation	Rat, cat, non-human primate
Ophthalmological	
Contact lens	Rabbit
Intraocular lens	Rabbit, monkey

for the in vivo assessment of tissue compatibility must consider the advantages and disadvantages of the animal model for human clinical application. Below, several examples demonstrate the advantages and disadvantages of animal models in predicting clinical behavior in humans.

As described above, sheep are commonly used for the evaluation of heart valves. This is based on size considerations and also the propensity for calves to calcify tissue components of bioprosthetic heart valves. The choice of this animal model for bioprosthetic heart valve evaluation is made on the basis of accelerated calcification in rapidly growing animals, which has its clinical correlation in young and adolescent humans.

The in vivo assessment of tissue responses to vascular graft materials is an example in which animal models present a false picture of what generally occurs in humans. Virtually all animal models, including non-human primates, heal rapidly and completely with an endothelial blood-contacting surface. Humans, on the other hand, do not show extensive endothelialization of vascular graft materials, and the resultant pseudo-intima from the healing response in humans is potentially thrombogenetic. Consequently, despite favorable results in animals, small-diameter vascular grafts (less than 4 mm in internal diameter) yield early thrombosis in humans, the major mechanism of failure, which is secondary to the lack of endothelialization in the luminal surface healing response.

The use of appropriate animal models is an important consideration in the safety evaluation of medical devices that may contain potential immunoreactive materials. The in vivo evaluation of recombinant human growth hormone in poly(lactic-co-glycolic acid)(PLGA) microspheres demonstrates the appropriate use of various animal models to evaluate biological responses and the potential for immunotoxicity. Utilizing biodegradable PLGA microspheres containing recombinant human growth hormone (rhGH), Cleland et al used Rhesus monkeys, transgenic mice expressing hGH, and normal control (Balb/C) mice in their in vivo evaluation studies (77). Rhesus monkeys were utilized for serum assays in the pharmacokinetic studies of rhGH release as well as tissue responses to the injected microcapsule formulation. Placebo injection sites were also utilized, and a comparison of the injection sites from rhGH PLGA microspheres and placebo PLGA microspheres demonstrated a normal inflammatory and wound healing response with a normal focal foreign body reaction. To further examine the tissue response, transgenic mice were utilized to assess the immunogenicity of the rhGH PLGA formulation. Transgenic mice expressing a heterologous protein have been previously used for assessing the immunogenicity of structural mutant proteins. With the transgenic animals, no detectable antibody response to rhGH was found. In contrast, the Balb/C control mice had a rapid onset of high titer antibody response to the rhGH PLGA formulation. This study points out the appropriate utilization of animal models not only to evaluate biological responses but also to evaluate one type of immunotoxicity (immunogenicity).

Future Perspectives on In Vivo Medical Device Testing

As presented above, the in vivo assessment of tissue compatibility of biomaterials and medical devices is dependent on the end-use application of the device under consideration. In this sense, the development and utilization of new biomaterials and medical devices will dictate the development of new test protocols and procedures for evaluating them. Furthermore, it must be understood that the in vivo assessment of tissue compatibility of biomaterials and medical devices is open-ended and new end-use applications will require new tests.

Over the past half-century, medical devices and biomaterials have generally been passive in their tissue interactions. That is, a mechanistic approach to biomaterials/tissue interactions has rarely been used in the development of biomaterials or medical devices. Heparinized biomaterials are an exception to this statement, but considering the five subcategories of hemocompatibility, these approaches have minimal impact on the development of blood-compatible materials.

In the past decade, increased emphasis has been placed on tissue engineering in the development of biomaterials and medical devices for potential clinical application. Rather than a passive approach to tissue interactions, tissue-engineered devices have focused on an active approach in which biological or tissue components, i.e. growth factors, cytokines, drugs, enzymes, proteins, extracellular matrix components, and cells that may or may not be genetically modified, are used in combinations with synthetic, i.e. passive, materials to produce devices that control or modulate a desired tissue response. Obviously, in vivo assessment of the targeted biological response of a tissue-engineered device will play a significant role in the research and development of that device as well as in its safety assessment. It is clear that scientists working on the development of tissue-engineered devices will contribute significantly to the development of in vivo tests for biocompatibility assessment as these tests will also be utilized to study the targeted biological responses in the research phase of the device development.

Regarding tissue-engineered devices, it must be appreciated that biological components may induce varied effects upon tissue in the in vivo setting. Simply put, cell types in the implant site may react differently to the presence of an extrinsic growth factor. Autocrine, paracrine, and endocrine signaling may be different between the same cell types and different cell types in the implant site. Signal transduction systems may be variable depending on the different cells that are present within the implant site. The presence of a growth factor may result in markedly different cell proliferation, differentiation, protein synthesis, attachment, migration, shape change, etc, which would be cell type-dependent. Thus different cell type-dependent responses in an implant site, reacting to the presence of a single exogenous growth factor, may result in inappropriate, inadequate, or adverse tissue responses. These perspectives must be integrated into the planned program for in vivo assessment of tissue compatibility of tissue-engineered devices. Finally, a major challenge to the in vivo assessment of tissue compatibility of tissue-engineered devices is the use of animal tissue components in the early phase of device development, whereas the ultimate goal is the utilization of human tissue components in the final device for end-use application. Novel and innovative approaches in the in vivo tissue compatibility of tissue-engineered devices must be developed to address these significant issues.

LITERATURE CITED

1. Anderson JM. 1993. *Cardiovasc. Pathol.* 2:33S–41
2. Anderson JM. 1988. *ASAIO Trans.* 11:101–7
3. Cotran RZ, Kumar V, Robbins SL, eds. 1999. *Pathologic Basis of Disease*, pp. 50–112. Philadelphia: Saunders. 6th. ed.
4. Gallin JI, Synderman R, eds. 1999. *Inflammation: Basic Principles and Clinical Correlates*. New York: Raven. 2nd. ed.
5. Weissman G, Smolen JE, Korchak HM. 1980. *N. Engl. J. Med.* 303:27–34
6. Salthouse TN. 1976. *J. Biomed. Mater. Res.* 10:197–229
7. Clark RA, Lanigan JM, DellePelle P, Manseau E, Dvorak HF, Colvin RB. 1982. *J. Invest. Dermatol.* 79:264
8. Tang L, Eaton JW. 1993. *J. Exp. Med.* 178:2147–56

9. Tang L. 1998. *J. Biomat. Sci. Polym. Ed.* 9:1257–66
10. Riches DWF. 1989. In *The Molecular and Cellular Biology of Wound Repair*, ed. RAF Clark, PM Henson PM, p.213. New York: Plenum
11. Wahl SM, Wong H, McCartney FN. 1989. *J. Cell Biochem.* 40:193
12. Dvorak HF, Harvey VS, Estrella P, Brown LF, McDonagh J, Dvorak AM. 1987. *Lab. Invest.* 57:673
13. Broadley KN, Aquino AM, Woodward SC, Buckley-Sturrock A, Sato Y, et al. 1989. *Lab. Invest.* 61:571
14. Sporn MB, Roberts AB. 1988. *Nature* 332:217
15. Muller G, Behrens J, Nussbaumer U, Böhlen P, Birchmeier W. 1987. *Proc. Natl. Acad. Sci. USA* 84:5600
16. Madri JA, Pratt BM, Tucker AM. 1988. *J. Cell Biol.* 106:1375
17. Wahl SM, Hunt DA, Wakefield LM, Roberts AB, Sporn MB. 1987. *Proc. Natl. Acad. Sci. USA* 84:5788
18. Ignotz R, Endo T, Massagué J. 1987. *J. Biol. Chem.* 262:6443
19. Spector M, Cease C, Tong-Li X. 1989. *Crit. Rev. Biocompatibility* 5:269–95
20. Wokalek H, Ruh H. 1991. *J. Biomater. Appl.* 5:337–62
21. Marchant R, Hiltner A, Hamlin C, Rabinovitch A, Slobodkin R, Anderson JM. 1983. *J. Biomed. Mater. Res.* 17:301–25
22. Spilizewski KL, Marchant RE, Hamlin CR. 1985. *J. Contr. Rel.* 2:197–203
23. Ganz T. 1988. *Ann. Intern. Med.* 109:127–42
24. Henson PM, Johnston RB Jr. 1987. *J. Clin. Invest.* 79:669–74
25. Malech HL, Gallin JI. 1987. *N. Engl. J. Med.* 317:687–94
26. Jutila MA. 1992. *APMIS* 100:191–201
27. Pober JS, Cotran RS. 1990. *Transplantation* 50:537–544
28. Cotran RS, Pober JS. 1990. *J. Am. Soc. Nephrol.* 1:225–35
29. Hynes RO. 1992. *Cell* 69:11–25
30. Butcher EC. 1991. *Cell* 67:1033–36
31. Henson PM. 1971. *J. Immunol.* 107:1547–57
32. Henson PM. 1980. *Am. J. Pathol.* 101:494–511
33. Weiss SJ. 1989. *N. Engl. J. Med.* 320:365–76
34. Paty PB, Graeff RW, Mathes SJ, Hunt TK. 1990. *Arch. Surg.* 125:65–69
35. Johnston RB Jr. 1988. *N. Engl. J. Med.* 318:747–52
36. Williams GT, Williams WJ. 1983. *J. Clin. Pathol.* 36:723–33
37. Wahl SM, Wong H, McCartney-Francis N. 1989. *J. Cell. Biochem.* 40:193–99
38. Sporn MB, Roberts AB, eds. 1990. *Peptide Growth Factors and Their Receptors I.* New York: Springer-Verlag
39. Fong Y, Moldawer LL, Shires GT, Lowry SF. 1990. *Surg. Gynecol. Obstet.* 170:363–78
40. Kovacs EJ. 1991. *Immunol. Today* 12:17–23
41. Golden MA, Au YPT, Kirkman TR. 1991. *J. Clin. Invest.* 87:406–14
42. Mustoe TA, Pierce GF, Thomason A, Gramats P, Sporn MB, Deuel TF. 1987. *Science* 237:1333–36
43. Maciag T. 1990. In *Important Advances in Oncology*, ed. VT DeVita, S Hellman, S Rosenberg, p. 85. Philadelphia: Lippincott
44. Thompson JA, Anderson KD, DiPetro JM, Zweibel JA, Zmaetta M, et al. 1988. *Science* 241:1349–52
45. Ziats NP, Miller KM, Anderson JM. 1988. *Biomaterials* 9:5–13
46. Rae T. 1986. *Crit. Rev. Biocompatibility* 2:97–126
47. Greisler H. 1988. *Trans. Am. Soc. Artif. Intern. Organs* 34:1051–57
48. Chambers TJ, Spector WG. 1982. *Immunobiology* 161:283–89
49. McNally AK, Anderson JM. 1995. *Am. J. Pathol.* 147:1487–99
50. Kao WJ, McNally AK, Hiltner A, Anderson JM. 1995. *J. Biomed. Mater. Res.* 29:1267–76

51. Möst J, Spötl L, Mayr G, Gasser A, Sarti A, Dierich MP. 1997. *Blood* 89(2):662–71
52. DeFife KM, McNally AK, Colton E, Anderson JM. 1997. *J. Immunol.* 158:319–28
53. McNally AK, DeFife KM, Anderson JM. 1996. *Am. J. Pathol.* 149:975–85
54. Jenney CR, DeFife KM, Colton E, Anderson JM. 1998. *J. Biomed. Mater. Res.* 41:171–84
55. DeFife KM, Shive MS, Hagen KM, Clapper DL, Anderson JM. 1999. *J. Biomed. Mater. Res.* 44:298–307
56. Jenney CR, Anderson JM. 1998. *J. Biomed. Mater. Res.* 44:206–16
57. DeFife KM, Colton E, Nakayama Y, Matsuda T, Anderson JM. 1999. *J. Biomed. Mater. Res.* 45:148–54
58. Jenney CR, Anderson JM. 1999. *J. Biomed. Mater. Res.* 46:11–21
59. DeFife KM, Jenney CR, Colton E, Anderson JM. 1999. *J. Histochem. Cytochem.* 47:65–74
60. DeFife KM, Jenney CR, Colton E, Anderson JM. 1999. *FASEB J.* 13:823–32
61. Honma T, Hamasaki T. 1996. *Virchows Arch.* 428:165–76
62. Roodman GD. 1996. *Endocr. Rev.* 17:308–32
63. Greenfield EM, Bi Y, Miyauchi A. 1999. *Life Sci.* 65:1087–102
64. Merkel KD, Erdmann JM, McHugh KP, Abu-Amer Y, Ross FP, Teitelbaum SL. 1999. *Am. J. Pathol.* 154:203–10
65. Jimi E, Nakamura I, Duong LT, Ikebe T, Takahashi N, et al. 1999. *Exp. Cell Res.* 247:84–93
66. Takahashi N, Udagawa N, Suda T. 1999. *Biochem. Biophys. Res. Commun.* 256:449–55
67. Burgess TL, Qian Y, Kaufman S, Ring BD, Van G, et al. 1999. *J. Cell. Biol.* 145:527–38
68. Lum L, Wong BR, Josien R, Becherer JD, Erdjument-Bromage H, et al. 1999. *J. Biol. Chem.* 75:783–90
69. Clark RAF, Henson PM, eds. 1988. *The Molecular and Cellular Biology of Wound Repair*. New York: Plenum Publishing
70. Hunt TK, Heppenstall RB, Pines E, Rovee D, eds. 1984. *Soft and Hard Tissue Repair*. New York: Praeger Scientific
71. AAMI Standards and Recommended Practices. 1997. Vol. 4, Biological Evaluation of Medical Devices, Association for the Advancement of Medical Instrumentation, 1997. 1998. Vol. 4S, Supplement
72. Chapekar MS. 1996. *J. Biomed. Mater. Res. Appl. Biomat.* 33:199–203
73. Langone JJ. 1998. Immunotoxicity Testing Guidance, Draft Document, Office of Science and Technology, Center for Devices and Radiological Health, Food and Drug Administration
74. FDA Blue Book Memor. 1995. G95-1: FDA-modified version of ISO 10,993-1, Biological evaluation of medical devices - Part 1: Evaluation and testing
75. ISO (Int. Stand. Organ.) 1993. Geneva, Switz. 10,993, Biological evaluation of medical devices. ISO 10,993-1. Evaluation and testing; ISO 10,993-2. Animal welfare requirements; ISO 10,993-3. Tests for genotoxicity and reproductive toxicity; ISO 10,993-4. Selection of tests for interactions with blood; ISO 10,993-5. Tests for cytotoxicity: In vitro methods; ISO 10,993-6. Tests for local effects after implantation; ISO 10,993-7. Ethylene oxide sterilization residuals; ISO 10,993-9. Framework for the identification and quantification of potential degradation products; ISO 10,993-10. Tests for irritation and sensitization; ISO 10,993-11. Tests for systemic toxicity; ISO 10,993-12. Sample preparation and reference materials; ISO 10,993-13. Identification and quantification of degradation products from polymers; ISO 10,993-14. Identification and quantification of degradation products from ceramics; ISO 10,993-15. Identification and quantification of degradation products from metals and alloys; ISO 10,993-16. Toxicokinetic study design for degradation products and leachables

76. ASTM (Am. Soc. Test. Mater., Annu. Book ASTM Stand.) 1999. ASTM F-619–97 Practice for extraction of medical plastics; ASTM F-720-96 Practice for testing guinea pigs for contact allergans: Guinea pig maximization test; ASTM F-748-95 Practice for selecting generic biological test methods for materials and devices; ASTM F-749-98 Practice for evaluating material extracts by intracutaneous injection in the rabbit; ASTM F-981-93 Practice for assessment of compatibility of biomaterials (nonporous) for surgical implants with respect to effect of materials on muscle and bone; ASTM F-1439-96 Guide for the performance of lifetime bioassay for the tumorigenic potential of implant materials; ASTM F-763-93 Practice for short-term screening of implant materials
77. Cleland JL, Duenas E, Daugherty A, Marian M, Yang J, et al. 1997. *J. Control. Release* 49:193–205

Annu. Rev. Mater. Res. 2001. 31:111–37

THIN FILM SYNTHESIS BY ENERGETIC CONDENSATION[1]

Othon R Monteiro
Lawrence Berkeley National Laboratory, University of California, Berkeley, California 94720; e-mail: ormonteiro@lbl.gov

Key Words physical vapor deposition, ion beam-assisted deposition, energetic condensation, diamond-like carbon films

■ **Abstract** The use of energetic particles (ions and atoms) has become increasingly important in physical vapor deposition techniques. These deposition processes can be divided in two main classes: ion beam-assisted deposition and energetic condensation (or deposition). This review focuses on the latter, i.e. processes in which the actual depositing species have energies that far exceed ordinary thermal energies, namely energies greater than 20 eV. The phenomenology of the effect of these high-energy particles on the growth of thin films is first broadly presented, and then specific examples of film deposition are given. The examples drawn here are of films that have been prepared by metal plasma immersion implantation and deposition. The observed microstructures and functional properties of these films are discussed in terms of processing conditions.

INTRODUCTION

The microstructure and properties of films strongly depend on the deposition process and process conditions used (1–5). In physical vapor deposition (PVD) the major process control parameters are substrate temperature, background gas pressure, and energy, flux, and angle of incidence of the depositing particles (atoms and/or ions). Ordinary microstructures of coatings produced by PVD are usually described by maps, as a function of deposition process parameters (also referred to as zone models) (2, 6, 7). Movchan & Demchishin (6) proposed a zone model to describe the microstructure of thick evaporated coatings, according to which three different types of morphologies can result, depending on the substrate temperature relative to the melting temperature of the film material (T/T_m). At the lowest temperatures, the film assumes a columnar structure with a high density of voids in the grain boundaries (zone 1, $T/T_m < 0.3$). At higher temperatures, surface diffusion becomes relevant, resulting in densification of grain boundaries (zone 2, $0.3 < T/T_m < 0.5$). When bulk diffusion becomes important, the film consists of equiaxial grains (zone 3, $T/T_m > 0.5$). Thornton (2) extended this zone model to films produced by magnetron sputtering, and included Ar pressure as another parameter. In addition to the three zones proposed by Movchan & Demchishin (1, 6), Thornton identified a transition zone (T) between zones 1 and 2 consisting of a dense array of poorly defined fibrous grains.

The use of energetic ions or neutral species for the deposition of thin films has become increasingly important in recent years (8, 9). A different zone model was conceived to describe microstructures of films produced by processes that include energetic particles (7, 10, 11). The inclusion of ion energy in this new map led to an increase of the stability domain of the zone T with increasing particle energy (7) (Figure 1). Those zone models were developed for polycrystalline films. However, it is well known that large doses of high-energy ions lead to amorphization. Energetic ions also have a significant effect on the microstructure and properties of amorphous films. For instance, density, intrinsic stresses, and chemical state have all been correlated to the incident energy flux.

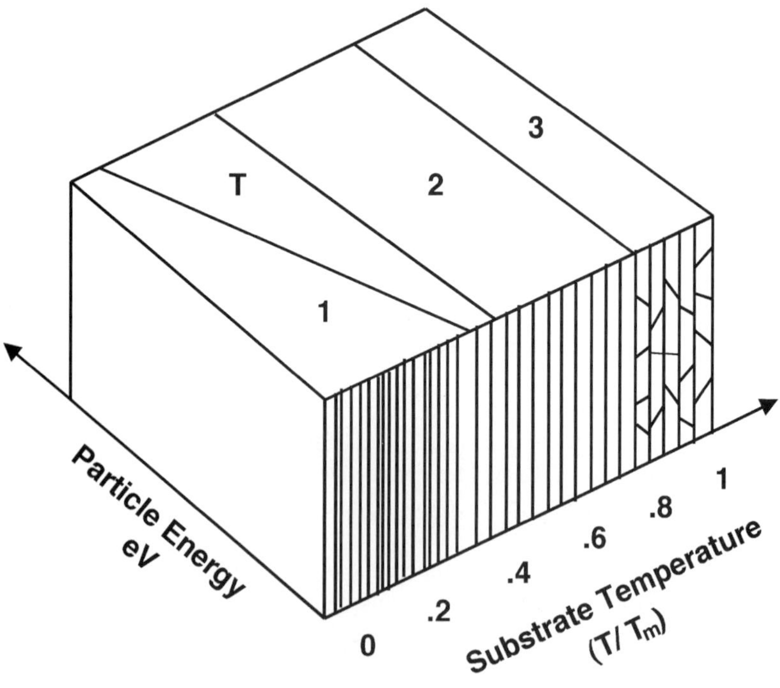

Figure 1 Microstructure zone model for ion beam assisted deposition [after Messier et al (7)].

This review presents and discusses several aspects of the synthesis of films when the depositing species have energy in the range of approximately 20 eV to a few keV. In this energy range, incident ions and atoms can dislodge adsorbed species and penetrate the substrate surface, which may lead to film growth beneath the surface (12–14) in a process that has become known as sub-plantation. Subplantation has been used to describe formation of several materials, in particular non-hydrogenated diamond-like carbon films (DLC). I first focus on processes and films in which the energetic species are constituents of the films. Such processes are generically referred to as energetic deposition and do not include conventional ion beam-assisted deposition (IBAD). Typically, in IBAD, an ion source is responsible for supplying the inert gas ion whose main purpose is to transfer momentum to the growing film. The actual film material originates from a different source. In particular, the films described here are produced by metal plasma immersion ion implantation and deposition (MePIIID) (15, 16). Although I focus on one particular deposition process, the phenomenology discussed here can be extended to other processes that make use of hyperthermal precursors such as pulsed laser deposition (PLD).

The remainder of the review consists of a qualitative discussion of the impact of ion bombardment and quenching on the several steps of the deposition process; a brief introduction of the deposition technique used to prepare the films discussed in subsequent sections; examples of films prepared using this technique with a discussion of the role of energetic ions in each case; and a summary of the beneficial role of energetic ions on thin film deposition.

ROLE OF ENERGETIC DEPOSITING SPECIES

On Adsorption and Adhesion

For a background pressure greater than 10^{-8} torr, the substrate surface in any PVD process consists of adsorbed species that saturate the surface bonds and hinder direct bonding between the deposited film and the substrate (9). A common procedure to remove impurities from the surface, sputter cleaning, when carried out immediately before starting the deposition, is not always effective because the time interval between surface cleaning and starting the deposition is usually long enough to guarantee complete monolayer coverage of the substrate. Moreover, sputter cleaning can cause surface amorphization for some substrates.

This situation may be improved when the deposition is carried out at elevated temperatures, which promotes desorption of impurities, or if the depositing species carry enough energy to break through this adsorbed layer. High-energy species can simply dislodge adsorbed atoms, or they can be incorporated below the substrate surface. Ion energies greater than 25 eV are typically sufficient to promote desorption of physisorbed impurities, as well as some sub-surface implantation. Because adhesion between coating and substrate is greatly affected by the presence of impurities at the interface, the use of energetic ions can greatly enhance adhesion.

During energetic deposition, adhesion can be further improved by using high-energy ions during the early stages of deposition, usually at higher energies than those used during the later stages. At these energies, the incident ions penetrate the depositing film into the substrate. This ion implantation produces dense trails of ionization and direct collisional displacement of the substrate atoms. The random disruption of electronic bonding and short-range atomic displacements produced at the interface can lead to the formation of fresh bonds across the interface. This process is known as ion stitching (17, 18).

An example of an ion-stitched interface between a diamond-like carbon (DLC) film and a silicon substrate is shown in Figure 2*a*. The DLC was deposited from a highly ionized carbon plasma produced by a filtered cathodic arc source. During the early stages of the deposition, a pulsed bias voltage of −2 kV with a 25% duty cycle was applied to the substrate. Duty cycle here is the fraction of the total deposition time that the high voltage is on. The atomic concentration across this interface was calculated using a computer code known as T-DYN (19, 20), and the results are shown in Figure 2*b*. This code is based on the well-known TRIM code (transport and range of ions in matter), which is a Monte-Carlo simulation of ion-solid interactions but corrects for the dynamic change of the solid during the implantation/deposition.

The high resolution image in Figure 2*a* indicates that the observed interface roughness, as measured by the width of the transition region between the crystalline Si and the amorphous C, is much thinner that the projected range for C and

(a)

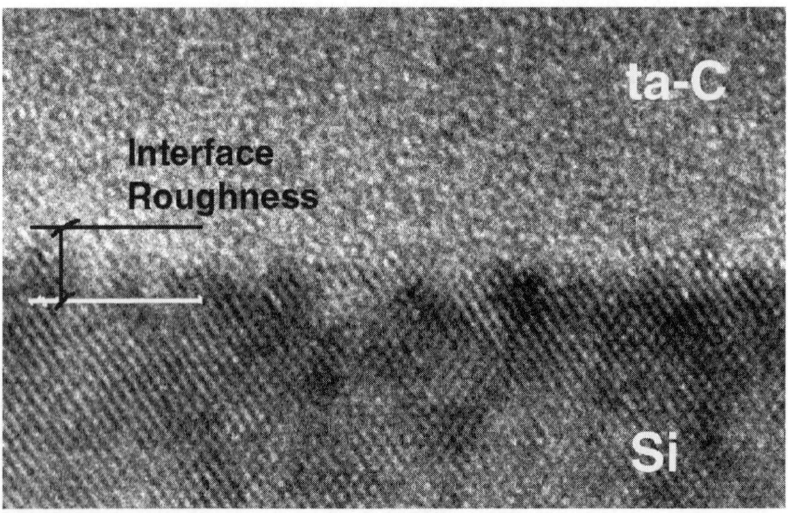

Figure 2 (*a*) Cross-section high-resolution transmission electron image of the interface between Si and DLC prepared with 2 keV C^+ ions and duty cycle of 25%. (*b*) Monte-Carlo simulation of Si and C profiles after deposition of DLC with 2 keV C^+ ions and duty cycle of 25%.

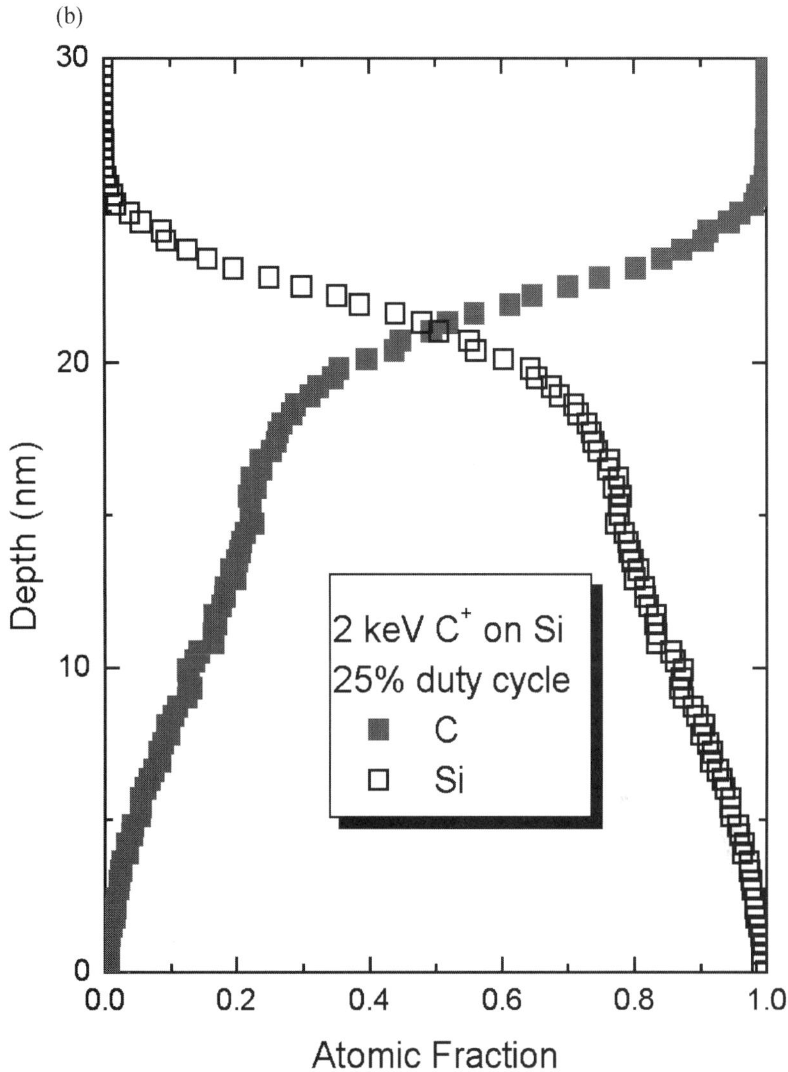

Figure 2 (*Continued*)

extends over only 4 or 5 (001) lattice fringes (i.e. about 2 nm). However, damage areas are observed well into the Si and appear as regions where the fringe contrast is disturbed.

On Surface Mobility

Another important parameter that defines the microstructure resulting from PVD processes is adatom surface mobility. Surface mobility is a thermally activated process. There is a direct correlation between surface mobility and the energy of the impinging ion for two fundamental reasons: the momentum transfer onto the lattice and the formation of surface defects. It is not easy to establish a quantitative description of the effect of depositing energetic particles on adatom mobility or even on sub-surface mobility. Seitz & Koehler (21) have estimated the maximum radial distance r' that an atom can migrate on a surface as a temperature spike is imposed on a surface as being given by

$$r' = \frac{4}{(324\pi)^{1/6}} \left(\frac{E}{Q}\right)^{1/3} r_s, \tag{1.}$$

where E is the energy transferred from the incident atom to the film, Q is the activation energy for adatom diffusion, and r_s is the atomic radius, which can be approximately given by $\alpha k T_m$, where α is a proportionality constant approximately 5, k is the Boltzmann constant, and T is the melting temperature of the material being deposited. The power dependence of r' on E suggests that even for depositing species with energies of hundreds of eV (which is the range of energies that qualify for energetic deposition), the mobility radius should not exceed 10 to 15 times r_s.

On Texturing

Another very important film characteristic that is strongly affected by ion energy is the film texture (8, 22–24). During deposition, texturing (preferential orientation of grains in one or several crystallographic directions) may be thermodynamically or kinetically controlled. Thermodynamically controlled texturing will take place when there is surface energy anisotropy and there is enough adatom mobility so that equilibrium can be attained. Based on the above discussion, ion energy does not have great impact on surface mobility, and therefore thermodynamic control requires elevated deposition temperatures. Kinetically controlled texturing will occur if, for instance, the deposition rate on certain facets is higher than on others owing to difference in sticking coefficients. In ion-assisted deposition or in energetic deposition there is still another cause for preferential growth directions: differences in sputtering yields among the various crystallographic directions. For instance, crystallographic directions that allow channeling will have intrinsically lower sputtering yields, and therefore will be favored in such deposition processes. Another cause for texturing is the presence of stresses during deposition (25–27) and, as discussed below, ion bombardment plays an important role in stress evolution in thin films.

On Stress

Coatings prepared by processes that make use of thermalized particles (e.g. evaporation) are generally under tensile stresses (28). On the other hand, when energetic particles are used, the resulting film is often under compressive stresses. Sputter-deposited films, which make use of atoms with low energy (in general

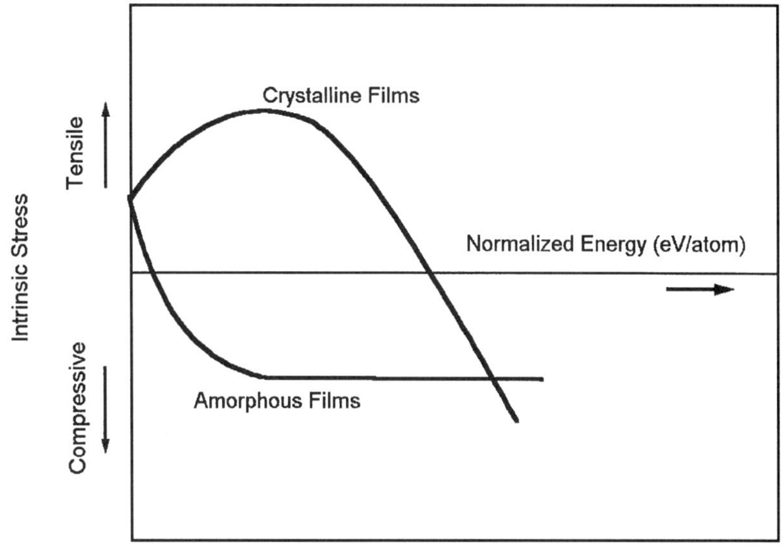

Figure 3 Correlation between intrinsic stress and incident ion energy for a metallic film produced by ion beam-assisted deposition.

2–5 eV), can have either compressive or tensile stresses depending on the background gas pressure. Such films are not discussed here.

In IBAD or in energetic deposition processes, compressive stresses are frequently observed. Typical dependence of intrinsic stresses in polycrystalline and amorphous films are shown in Figure 3. At very low energies, the tensile stresses in polycrystalline films result from coalescence of the film through regions with density less than ideal, such as grain boundaries (29). Increasing energy increases the packing at grain boundaries, eliminating the source of tensile stress. Density fluctuations in amorphous films, e.g. from self shadowing, may account for tensile stresses in amorphous films produced at very low energies. The cross-over point (zero-stress) for amorphous films occurs at small normalized energies, namely a few electron-volts (30).

In energetic deposition, the generation of compressive stress is attributed to the excessive concentration of interstitials, which are produced by direct or recoil implantation, or to ion penning, i.e. lattice distortion resulting from energetic particles striking the film (31, 32). At very low ion energies, interstitial formation can occur without the simultaneous formation of vacancies. As the ion energy increases, Frenkel pairs (vacancy-interstitial) become the predominant defects. As long as the net concentration of interstitials is greater than that of vacancies, the intrinsic stresses will be compressive in nature. [For the sake of clarification, interstitial sites in an amorphous film are those with smaller mean site volume than the sites in an amorphous film having no voids (9).] When the ion bombardment is carried out with ions of non-reactive gases, some incorporation of gas may take place. There is a greater tendency for those ions to be lost to the environment, and therefore the level of intrinsic stresses in such films is expected to be lower (less compressive) than the level of stresses encountered in films produced by energetic condensation.

On Substrate Temperature Evolution

The energy flux onto the substrate during energetic deposition processes can be quite high, and it is not unreasonable to expect that there will be temperature fluctuations at the substrate. Temperature here refers to a macroscopic temperature at the surface of the substrate and not to individual thermal spikes common to ion implantation. Even in actively cooled substrates, the temperature increase may be high enough to cause modification of the films.

The surface temperature of an energetic deposition process consisting of alternating periods of high and low ion energy deposition was estimated by carrying out an energy balance on the substrate, schematically shown in Figure 4. Heat loss through the sides of the substrate is neglected because the substrate is considered thin, as is the case in Si wafers. The heat input is determined by the ion current density i_{ion} and the energy of the depositing ions E_{ion}, which depends on the bias voltage, duty cycle, and ion charge.

The substrate temperature will satisfy

$$\rho c_p \frac{\partial T}{\partial t} = \nabla \cdot k \nabla T, \qquad 2.$$

where ρ is the density, c_p is the specific heat, T is the temperature, k is the thermal conductivity of the substrate (assumed here to be the same as that of the film for

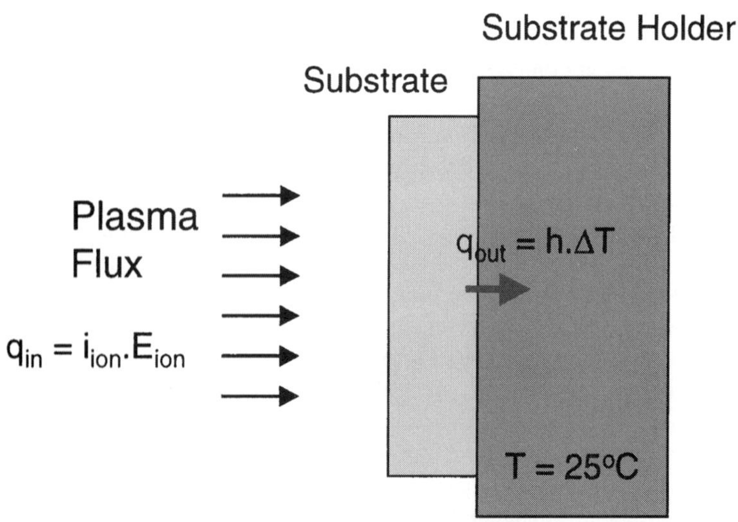

Figure 4 Schematic representation of the thermal balance during energetic deposition.

the sake of simplicity) and t is time. The boundary condition at the backside of the substrate is

$$-k\frac{\partial T}{\partial t} = h_c A(T_b - T_{holder}) \quad 3.$$

where T_b is the temperature at the backside of the wafer, and T_{holder} is the temperature of the holder surface in contact with the substrate, which is assumed to remain constant because the substrate holder usually consists of a metal block directly cooled with water. h_c is the heat transfer coefficient between two solid surfaces in direct contact.

The boundary condition at the film surface is

$$-k\frac{dT}{dx} = q_{plasma}, \quad 4.$$

where q_{plasma} is periodic because of the pulse-biased substrate: q_{on} during the time that the bias voltage is applied to the substrate (t_{on}) and q_{off} during the time that the voltage is not applied (t_{off}). For the sake of simplicity, the temperature gradient in the substrate is assumed negligible. This is a good assumption for thin substrates with high-thermal conductivity, as is the case with thin Si wafers. Under these conditions there is in fact an analytical solution for the temperature at the end of an arc pulse n_p:

$$T_{n_p} = T_{holder} + \left[\frac{q_{on}}{h_c}(1 - e^{-\alpha t_{on}})e^{-\alpha t_{off}} + \frac{q_{off}}{h_c}(1 - e^{-\alpha t_{off}})\right] \times \left[\frac{1 - e^{-n_p\alpha(t_{on}+t_{off})}}{1 - e^{-\alpha(t_{on}+t_{off})}}\right] \times \left[\frac{1 - e^{-m\alpha\left[t_{off}^{arc}+n_p(t_{on}+t_{off})\right]}}{1 - e^{-\alpha\left[t_{off}^{arc}+n_p(t_{on}+t_{off})\right]}}\right] \quad 5.$$

where n_p is the number of pulses, $\alpha = h_c A/\rho c_p V$, A is the surface area, and V the substrate volume.

The treatment of the heat load owing to ion bombardment as a surface heat flux is justified (33) by noticing that the implantation depth for the ordinary energy used in MePIIID (less than 5 keV) is significantly less than a significant thermal depth l, i.e. the length over which there is a significant temperature change in time t, where l is defined by

$$l = \sqrt{\frac{k}{\rho c_p}t}. \quad 6.$$

DEPOSITION PROCESS

The films described below were prepared by MePIIID. In this process, a plasma stream is produced by a cathodic arc plasma source (34) and directed toward the substrate via a quarter torus macroparticle filter. Cathodic arc sources are known for producing large quantities of macroparticles–molten droplets of cathode material–that need to be eliminated in order to obtain high-quality films. Macroparticles that are co-deposited on the film result in phase heterogeneity, and negatively affect the film properties. Compound films (consisting of more than one species) were prepared either by using multiple plasma sources (35, 36) or by carrying out the deposition in the presence of a reactive gas (15, 36).

The substrate is repetitively pulse-biased to a negative voltage. During the periods when the bias voltage is off, plasma ions stream toward the substrate with relatively low energies (up to about 200 eV, depending on the species), where they can condense or be implanted at shallow depths. The native energy of this streaming plasma is a function of the ion species (15, 37) and the gas pressure in the chamber (38, 39). During the periods when the bias voltage is on, a high-voltage sheath rapidly forms at the substrate-plasma boundary, and plasma ions are accelerated through the sheath and into the substrate. Therefore, during film growth there are alternating periods of high-energy ions and low-energy ions or atoms.

A schematic diagram of the apparatus used to generate the plasma beam is shown in Figure 5 with only one plasma source (35). The plasma source was a vacuum arc mini-gun (34), consisting of cathode, anode, and focusing coil. Each mini-gun has a 90° filter to separate the plasma stream from macroparticles and neutrals. Filtering is particularly important during the formation of compound films in order to assure complete reaction between the constituents. This source typically operates with arc current between 150 and 250 A, pulse duration up to 10 ms, and repetition rate of 2 Hz.

The microstructure and composition of the films was characterized by a variety of tools. Rutherford backscattering spectrometry using 1.8 MeV He^+ ions was the primary tool for compositional analysis. High-resolution transmission electron microscopy was carried out on a Topcon 002B TEM to characterize the films and interfaces microstructure. Scanning electron microscopy on a JEOL SEM Model 6400 was used to study the film morphology. Internal stresses in the films were calculated from the change in curvature radius of the Si substrate, measured with a Tencor FLX-2320 stress measurement instrument. Si wafers with 3-inch diameter were used for those measurements. Film hardness was determined by a nanoindentation technique using a Hysitron Picoindenter with loads in the range of 0–250 μN. Surface roughness of the films was measured on (100)Si substrates with a Park T5 SPM in tapping mode.

A combined energy parameter was defined by Musil et al (40) as the average energy carried by arriving ions per condensing particle, which in mathematical terms is expressed by

$$E_P = \frac{E_{on} * J_{on} + E_{off} * J_{off}}{J_{on} + J_{off}}$$

$$= E_{off} * \frac{J_{off}}{J_{off} + J_{on}} + [E_{off} + \langle q \rangle * V_{bias}] * \frac{J_{on}}{J_{on} + J_{off}}, \qquad 7.$$

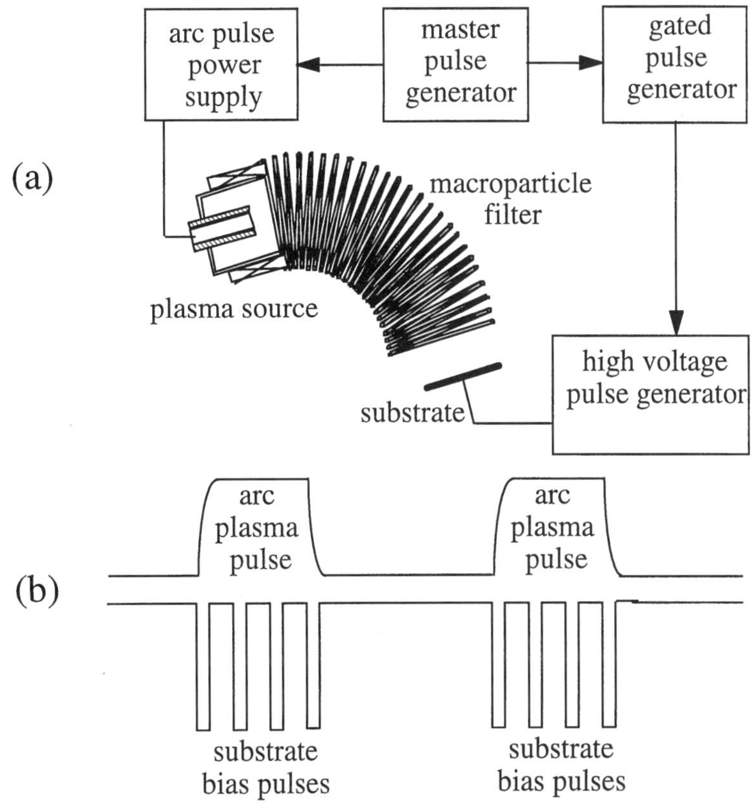

Figure 5 Schematic representation of metal plasma ion immersion implantation and deposition process.

where E_{on} and E_{off} are the energies of the depositing ions during the on time and off time of the pulse bias, respectively; J_{on} and J_{off} are the total ion fluxes arriving at the substrate during the on time and off time of the pulse bias, respectively. V_{bias} is the applied bias voltage, and $\langle q \rangle$ is the mean charge state. By assuming that the integrated fluxes are proportional to the duration of the pulses, this expression can be approximated by

$$E_P = E_{off} + [\langle q \rangle * V_{bias}] * \frac{t_{on}}{t_{on} + t_{off}}. \tag{8}$$

The use of a combined energy parameter has been useful in some events to predict the film microstructure (40). However, it does not include the total rate of deposition and therefore is not appropriate to fully describe changes in microstructures in thin films that may result from energy flux. This is particularly important in the case of DLC prepared by MePIIID, where the rate of arrival of incident ions has a significant impact in the final properties of the film because of localized thermal effects (see below).

FILMS

Non-Hydrogenated Amorphous Carbon Films

Non-hydrogenated amorphous carbon films with high content of sp^3-hybridization (DLC) have been synthesized by a variety of methods including PLD (41–43), filtered cathodic arc (FCA) (16, 44–47), and mass-selected ion beam (MSIB) (48, 49). A common feature in these processes is that the film is produced from incident C^{n+} ions with energies above 20 eV, a threshold for the C^+ ions to penetrate the surface of the growing film (50, 51). When the deposition parameters are optimized, such processes can produce films with sp^3 content as high as 85%, resulting in amorphous carbon films with properties approaching that of crystalline diamond (52, 53). As the energy flux into the growing film increases by increasing the ion energy or the ion flux, conversion of sp^3 into sp^2-bonded carbon occurs through a localized increase in temperature or interaction between cascades.

There is a direct correlation between the density, hardness, level of intrinsic stresses, and the sp^3 content in DLC films (46, 51, 53, 54). The as-deposited DLC films with highest sp^3 content have very high levels of intrinsic stresses, above 10 GPa, which limits the deposition to thickness between 100 and 200 nm before delamination happens (54–56). In order to produce thicker DLC films, Anders and coworkers used an initial step (16) that promotes intermixing between the film and the substrate (see section on adsorption and adhesion). Another approach is the use of a reactive intermediate layer. Nevertheless, both approaches have failed to produce films thicker than 500 nm because of the very high compressive stresses.

Other alternatives suggested to produce thicker films, including annealing of DLC films (57–60), formation of multilayers (54, 61, 62), or incorporation of other elements (35, 63), have focused on relaxing the intrinsic stress. These procedures are discussed below.

Deposition of DLC provides a good example of the difference between energetic deposition or condensation and IBAD-like process. Despite several attempts to produce DLC films using IBAD techniques (64–67), properties such as hardness (max around 20 GPa), intrinsic stresses (max around 2 GPa), and density are significantly lower than those in DLC films with high sp^3 content, which are prepared with energetic C^{n+} ions.

In principle, one could expect that the two deposition processes would be equivalent and therefore produce similar films. In the energetic deposition, the energetic ion arrives and penetrates the surface of the substrate transferring its energy into the lattice. In the IBAD process, energetic ions bombard and penetrate the surface transferring momentum to the atoms on the surface which, by recoil implantation, are also transported into the lattice. In the latter case, the incident ion does not participate in the growth. Whether one accepts the McKenzie model (45) that sp^3 bond stabilization results from high intrinsic stresses or the Lifshitz (48) -Robertson (52) model that such stabilization results from high film density, the phenomenological distinction between those two deposition processes should explain the difference between the products. The most striking difference between the two processes discussed above is the presence of inert gas atoms in the IBAD process. Loss of those atoms to the environment may result in reduction of density and intrinsic compressive stresses, thus reducing the driving force for the formation of sp^3 bonds.

DLC films prepared by FCA and ion beam tend to be very smooth (68) because of the absence of macroparticles, enhanced surface mobility, and self-sputtering. On the other hand, films prepared by PLD tend to be slightly rougher, which may in part be attributed to the presence of particulates. An AFM image of a 50-nm-thick DLC film deposited on (100)Si substrate resulted to a R_{rms} of 0.1 nm, which is comparable to the bare Si surface. A cross-section TEM image of such a film is shown in Figure 2. The image contrast of the amorphous film indicates that no turbostratic structure is present in the DLC film, typical of energetically deposited DLC films.

The film properties are strongly dependent on ion energy and ion flux. A recent compilation of the effect of ion energy on the properties of film prepared with FCA has been published (69), and from those data, I have plotted the film density versus the ion energy in Figure 6. The intrinsic stress in DLC film follows a curve with a

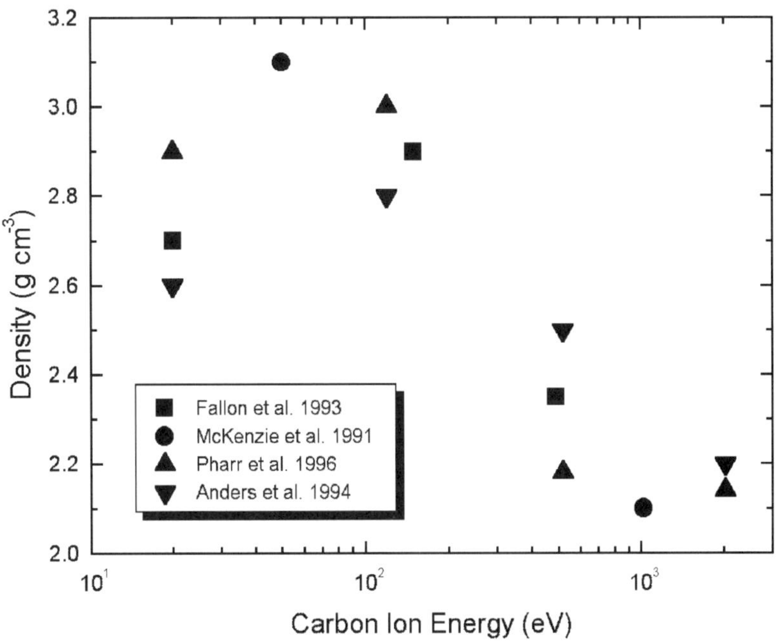

Figure 6 Density of DLC films prepared by filtered cathodic vacuum arc as a function of the C^+ ion energy.

shape similar to that for density, namely, the stress increases up to about 10 GPa by increasing the applied bias voltage; after about −100 V, it begins to decrease. The decrease in stress with increasing voltage beyond 100 V (for FCA) is related to increasing film local (thermal spikes) and global temperatures owing to the energy flux onto the substrate. The latter phenomenon explains the difference between the observed dependence of stress in DLC films on the ion energy and the generic behavior shown in Figure 3.

Early in the development of DLC, Cuomo et al (70) demonstrated the important effect of substrate temperature on the film properties by using substrates with remarkably different thermal conductivities and by carrying out the deposition at ambient and liquid nitrogen temperatures. For the pulsed cathodic arc deposition system, the effect of the arc pulse duration and duty cycle on the film temperature was estimated for the deposition conditions used at Lawrence Berkeley National Laboratory and correlated to the film properties. Equation 5 was used to calculate the temporal evolution of the temperature for different conditions, and the results are plotted in Figure 7. A major uncertainty in this model is the value of the heat transfer coefficient, h_c, between two solid surfaces, but for low substrate temperatures, a sensitivity analysis demonstrated that its impact is not strong. According to the curves in Figure 7, little change is expected in film properties for low applied bias voltages or thin films, but for bias voltages of −2000 V, some change owing to temperature should be expected. The hardness of the DLC films prepared at these same conditions is plotted in Figure 8. The hardness of samples prepared using the

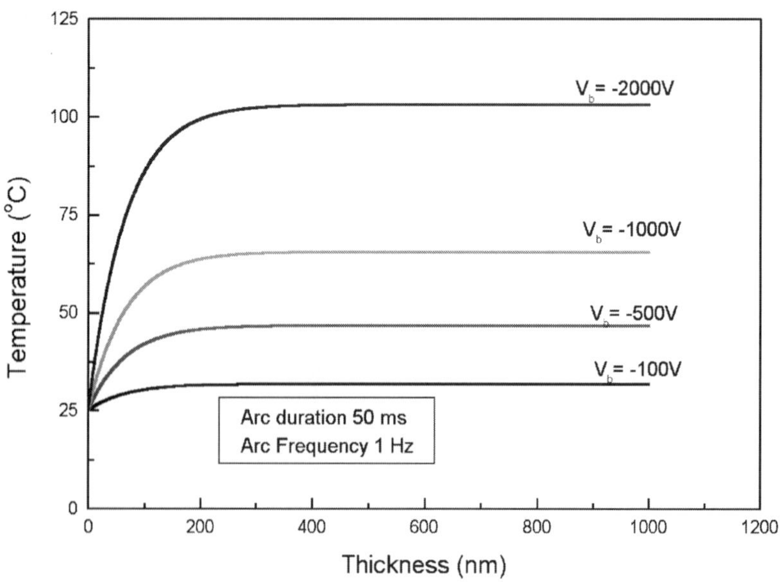

Figure 7 Calculated temperature evolution of a thin Si substrate during deposition of DLC film.

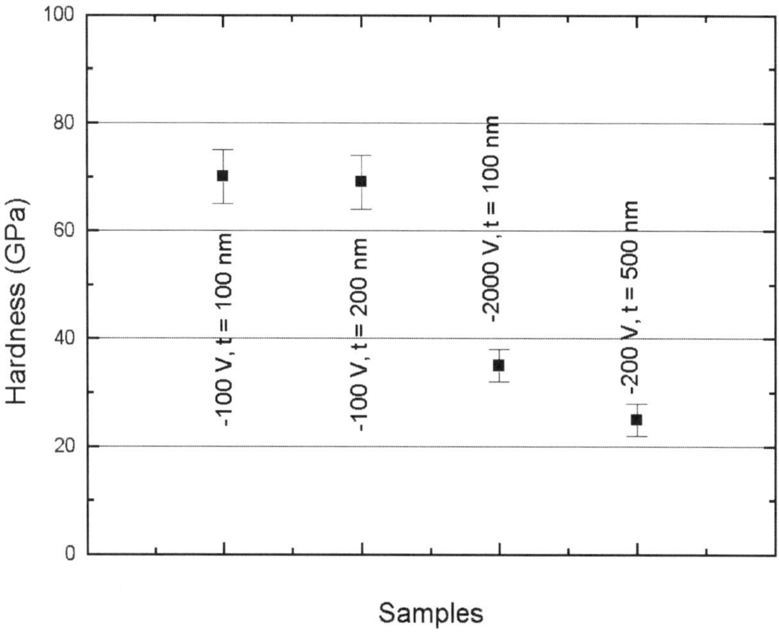

Figure 8 Hardness of DLC films for different total energy deposited onto the films.

low bias voltage was not dependent on the total thickness of the film and therefore on the total deposition time. On the other hand, for the samples prepared at high bias voltage, there was a noticeable decrease in hardness of the thicker films. This dependence fits well with the thermal history of the samples, plotted in Figure 7, which confirms the observations that substrate temperature about 150°C is enough to promote significant reduction of the sp^3 content (51).

Recently it was speculated that surface charge accumulation may lead to a thickness-dependent structure in amorphous carbon films (71). This argument cannot be used here because a decrease in the electric field as a consequence of increasing film thickness would in fact lead to harder DLC instead of softer, and therefore would be the opposite of what is experimentally observed (Figure 8).

Intrinsic stresses of the DLC films can be significantly relaxed by thermally annealing the films without a significant loss of their diamond-like properties (57, 59, 60, 72–74). This provides an important path for the production of thick DLC films (58), which is critical for extending the range of applications of this material.

Metal-Containing DLC and Metal Carbides

Energetic condensation has also been very useful in producing metal-containing DLC films in a very large composition range (35, 62, 75–78), namely from a few ppm to stoichiometric metal carbides. The original motivation for adding refractory metals to DLC films was threefold: to increase the chemical and thermal stability of the films, to modify surface properties of the films, and to decrease internal stresses. Although DLC films can be annealed at temperatures up to 700° in high vacuum (57, 59) without noticeable change in properties, thermal stability in air or in other oxidizing environments is not as good.

FCA was used to prepare silicon-doped DLC films (DLC:Si), using the dual-source approach described above: one Si plasma source and one C plasma source. Films with Si content from 1 to 7 at.% are fully amorphous and show no evidence of segregation at the surface, at the film-substrate interface, or in domains within the film. These films were tested for wear at very high rotation speeds (~35000 rpm) in air and in absence of lubrication. The DLC:Si with 7 at.% silicon showed a remarkable improvement in performance (here measured as the time until failure) at 300°C, when compared with the undoped DLC film or uncoated parts. The exact reason for the enhancement of lifetime is not fully understood. Si incorporation in hydrogenated a-C films increases the sp^3 content (79), but the same reasoning cannot be extrapolated to undoped non-hydrogenated films because the sp^3 content in the latter is already high. However, it is well known that SiC is much more resistant to oxidation at elevated temperatures than DLC. At high temperatures, SiC forms a SiO_2 layer that precludes further diffusion of oxygen into the SiC. A similar mechanism could prevail here because even at such low Si contents, a SiO_2 layer will prevent further oxidation of DLC:Si. If the Si content increases beyond 20 at.%, amorphous silicon carbide, which is softer than DLC:Si, is produced, and the performance suffers (79).

Carbides of transition metals are of great interest in tribological applications because of their hardness and wear resistance (80). Among the carbides, titanium carbide (TiC) is of particular importance. TiC combines the advantages of a high melting point (3140°C) and hardness with those of a relative lightness (density 4.93×10^3 kg m^{-3}). It also exhibits high resistance to both corrosion and high-temperature oxidation and has a relatively low-friction coefficient.

TiC has a NaCl face-centered cubic structure (space group Fm3m) with a lattice parameter of 0.43274 nm (81). This structure can be visualized with the Ti atoms occupying a fcc lattice and C atoms occupying the octahedral interstices. The TiC phase extends through a relatively large solubility range at room temperature, from 40 to 50 C at.%, i.e. a significant amount of C vacancies can be present without disrupting the stability of the compound.

Figure 9 shows a bright-field transmission electron image of the as-deposited DLC:Ti with a low Ti content. The microstructure of DLC:Ti (or Ti_xC, with < 0 x < 1) consists of a mixture of an amorphous carbon with a TiC crystalline phase. The TiC is predominantly nanocrystalline, and the grain size increases with increasing Ti content. The solubility range in TiC is quite large, and therefore the excess C in Ti_xC is either accommodated as interstices in the B1 structure or leads to the formation of amorphous DLC (77). A high-resolution image of a nanocrystalline Ti_xC film with x ~ 1 is shown in Figure 10.

Transmission electron diffraction patterns of TiC films prepared with two Ti concentrations are shown in Figure 11. The films with the highest content of Ti

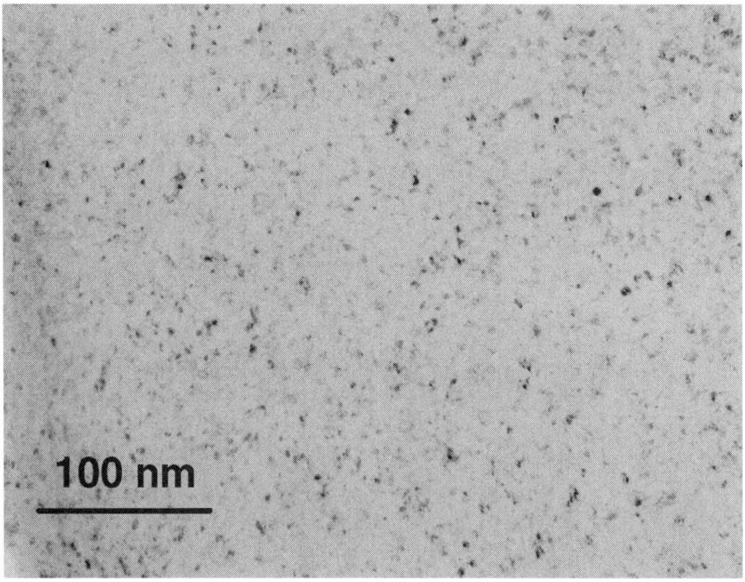

Figure 9 Bright-field image of DLC:Ti.

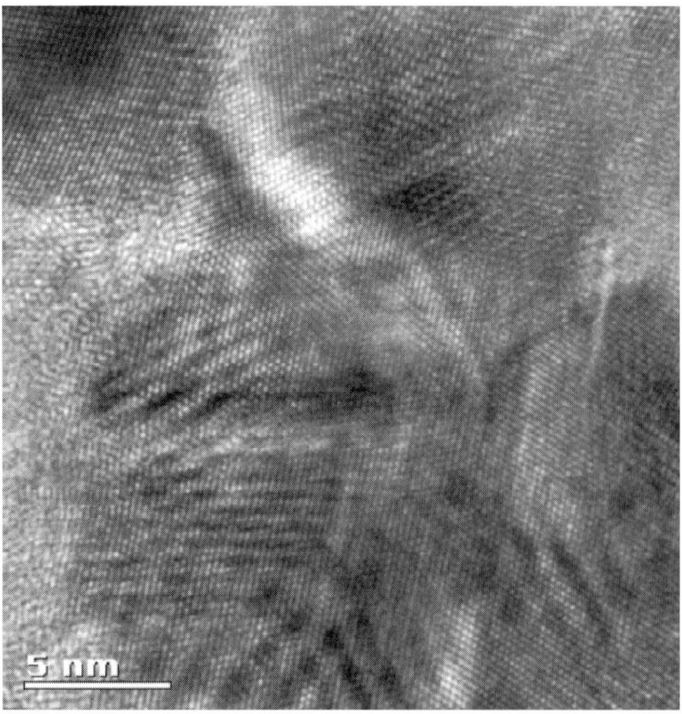

Figure 10 High resolution TEM of TiC.

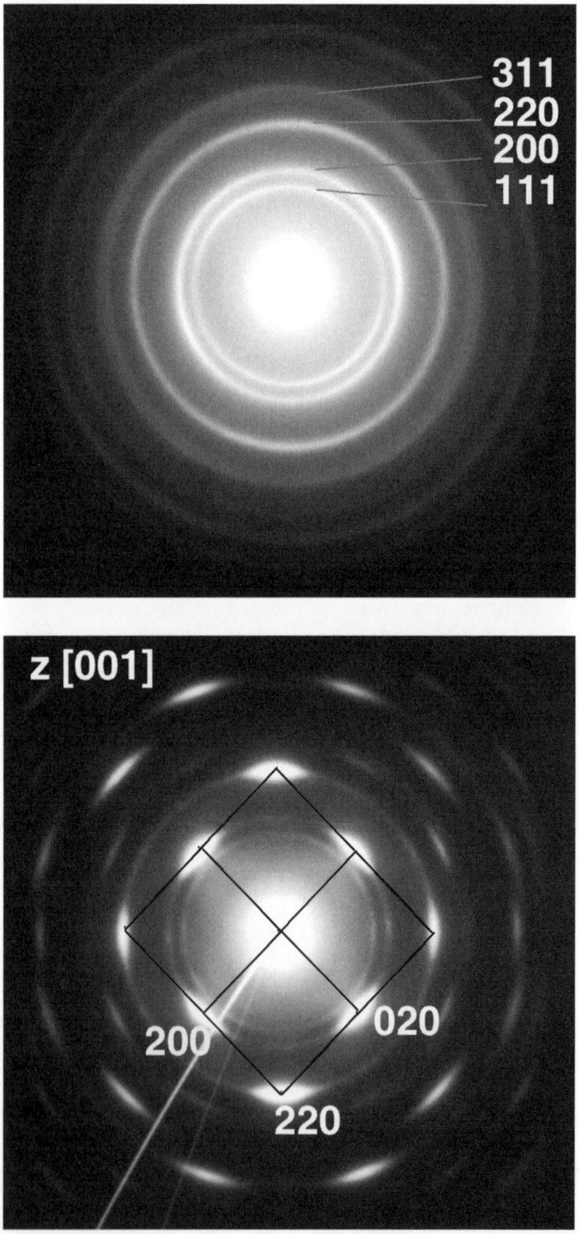

Figure 11 Electron diffraction pattern of titanium carbon films prepared with ion current ratios C:Ti: (*a*) 25:3 and (*b*) 5:3.

were strongly textured with the film normal around a ⟨002⟩ zone axis. As the Ti content decreases, the films become less textured, and the electron diffraction pattern becomes more ring-like, consistent with the decrease in grain size and loss of preferential orientation. Sputter-deposited TiC films have shown strong texturing along ⟨111⟩, instead of ⟨002⟩ observed here (82). This difference may be attributed to differences in the ion energies, because in sputter deposition, the depositing particles have very low energies.

Loss of film texture for increasing carbon content can result by either of two mechanisms. The first is that the formation of an amorphous carbon network (DLC) disperses the titanium carbide grains, favoring constant re-nucleation of carbide particles as the film thickness increases; the second is related to the energy transferred to the growing film by the ions. Vacuum arc plasmas are known to generate multiple ionized species, with an average ion charge state that depends on the cathode elements. The mean ion charge states for titanium and carbon are 2.1 and 1.0, respectively. Therefore, for a given bias voltage, the total deposited energy increases with increasing the Ti content in the plasma stream, which results in greater mobility of the surface atoms when the ion flux is Ti-rich. Also, because of the higher energies, greater sputtering yields of the non-channeling directions are expected with increasing Ti content in the plasma stream. Texturing of ion-assisted e-beam evaporated TiC was recently reported (83). The results indicated that for low temperature a ⟨111⟩ texture prevails, and at elavated temperature the ⟨002⟩ texture prevails.

Another refractory metal carbide of interest is WC (tungsten-carbide) because of its mechanical and tribological properties at ambient and elevated temperatures. Energetic deposition is a successful tool to produce films of WC with a variety of compositions (78, 84, 85), spanning from 1 at.% to 50 at.%.

In a DLC:W film, a C atom may be bonded to another C atom or a W atom. For the same overall W content, the amount of carbidic bonds, i.e. W–C bonds, increases with increasing the incident ion energy (85). This behavior was demonstrated by comparing the intensity of the carbidic peak in X-ray photoelectron spectra of films prepared with voltages from -100 to -2000 V. Carbide formation is favored by high temperature, and here it may result from either an increase in the cascade volume, or an increase in the macroscopic temperature (as described above). Contrary to the observations for TiC, the structure of the DLC:W films with W content below 20 at.% was amorphous. Films with higher W content were primarily nanocrystalline without any specific texture. The WC phase was dispersed in the amorphous carbon phase. Electron diffraction patterns for DLC:W with 20 at.% W and another with 10 at.% are shown in Figure 12 (85). For high W contents, the crystalline phase was β-WC, contrary to that previously reported (85).

Hardness and the intrinsic stresses of DLC:W are plotted as a function of ion energy (bias voltage) and W content in Figures 13 and 14, respectively (85). Small additions of W lead to significant decreases in the intrinsic stresses and a simultaneous increase in hardness for all but the hardest of the DLC films. The

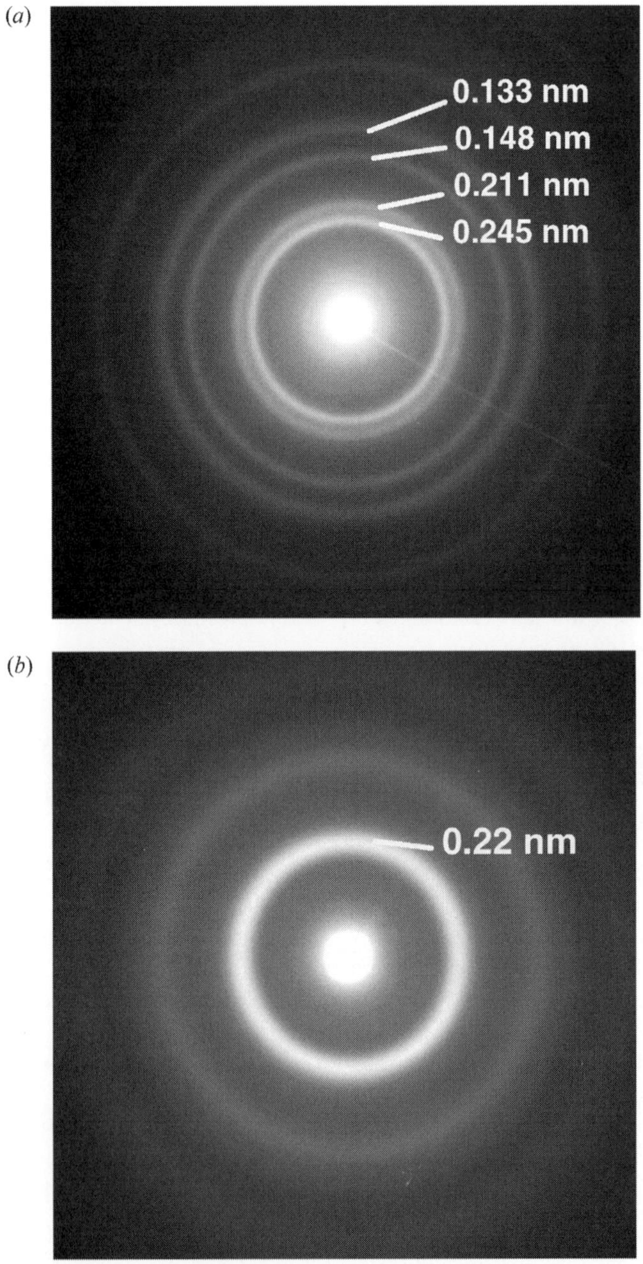

Figure 12 Electron diffraction pattern of tungsten carbon films with 20% W (*a*) and 7% W (*b*).

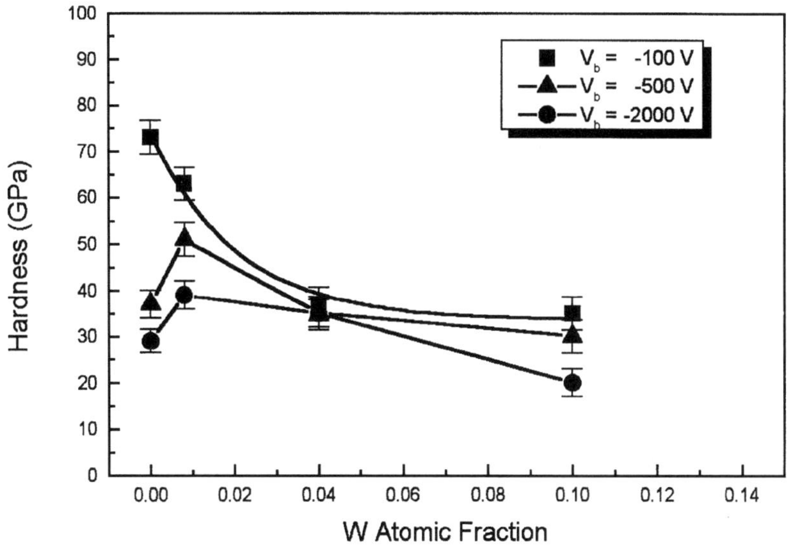

Figure 13 Hardness of DLC:W films prepared by filtered cathodic vacuum arc as a function of W content different bias voltages.

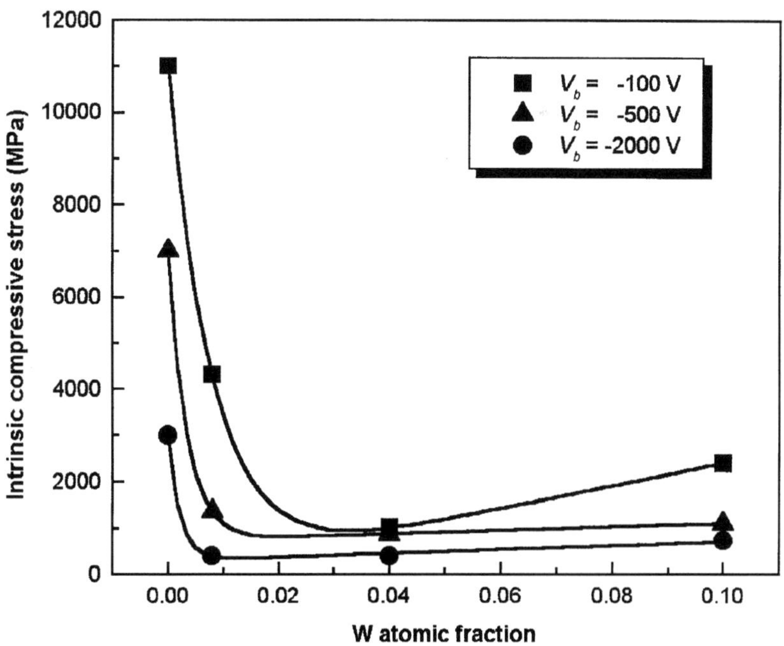

Figure 14 Intrinsic stress of DLC:W films prepared by filtered cathodic vacuum arc as a function of W content different bias voltages.

decrease in intrinsic stress with increasing the W content can be explained on a purely energetic basis. The mean charge state of W is 3.1, opposed to 1 for C, and this results in a net energy flux increase.

The dependence of hardness on the W content is complex. Hardness of undoped non-hydrogenated DLC films results from a balance between the formation of sp^3 hybridization owing to local increases in mass density of the film and lattice relaxation owing to thermal spikes (52, 86). The maximum hardness of these films was found to occur at a bias voltage of around −100 V, which coincides with the maximum sp^3 content and corresponds to C^+ ion energy around 100 eV (46, 52, 53, 86). When a bias voltage of −100 V is applied to a substrate exposed to a C^+ and W^{n+} plasma stream, the average ion energy is approximately 120 and 410 eV, respectively, because of their charge state distribution and native energies. Therefore, addition of W to the plasma stream and to the film results in two competing effects: (*a*) The larger size of W atoms when compared with C atoms leads to greater increases in local density and in intrinsic stress and (*b*) the presence of multiple charged ions in the plasma stream leads to greater energy transfer to film, which favors softening. The final effect on the film hardness is a compromise between these two effects.

Multilayers

Oftentimes a monolithic film will not have a combination of properties that makes it acceptable for a particular application. In these cases, a viable alternative is the use of multilayers, which may result in a combination of properties that best suits the application (87). A typical example is DLC, which has very high intrinsic stresses. In order to build thick DLC films, Anders et al (88) suggested a multilayer of alternating soft and hard phases of amorphous carbon, prepared by filtered cathodic arc, with bias voltages of −2000 and −100 V, respectively. The reduction in intrinsic stress obtained was greater than would be expected by the rule of mixtures, and the final film was still extremely hard.

Energetic deposition is highly suitable for the production of multilayer structures. It can be used at room temperature, which obviates differences in thermal expansion coefficient between layers. It also allows maximizing interlayer adhesion by ion mixing, which creates a graded interface between adjacent layers. The only limitation to the latter is that the minimum thickness of each individual layer must be greater than twice the projected range of the ions during the high-voltage cycle of the process to guarantee differentiation between the layers.

TiC/DLC multilayers prepared by FCA were developed and tested as protective coatings for wear applications (62). A cross-section TEM image of such a multilayer is shown in Figure 15. In addition, the elemental composition map obtained by electron energy filtered imaging is shown in Figure 16, where the concentration of the element indicated is proportional to the brightness in the image. Wear tests performed on such layers showed a remarkable difference depending on whether the first layer on the work-piece was TiC or DLC (89). The one with TiC at the

Figure 15 Cross-section transmission electron micrograph in bright field of a DLC-TiC multilayer.

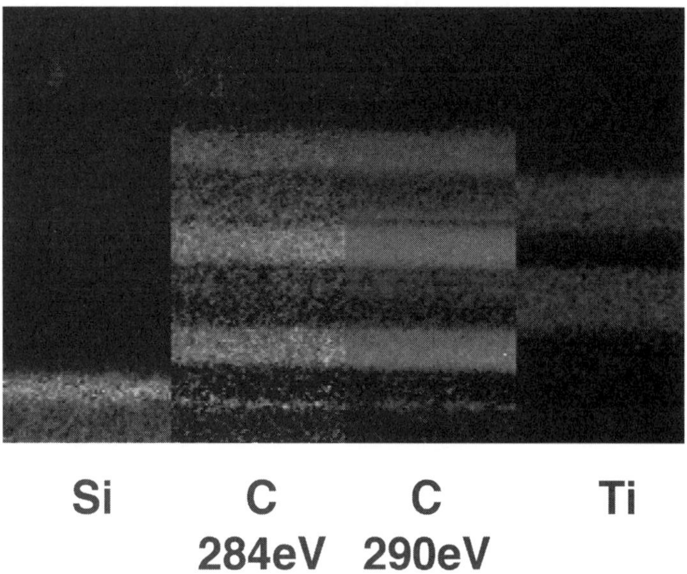

Figure 16 Composition map of DLC:TiC multilayer.

interface failed massively with formation of cracks and debris. The DLC-TiC-DLC showed no sign of degradation beyond 50,000 revolutions.

SUMMARY

The effect of using energetic particles (ions and atoms) during thin film deposition is remarkable. This warrants the use of ion energy and flux as important parameters to tailor the microstructure and properties of thin films. Several materials such as non-hydrogenated amorphous carbon with a high content of sp^3 bonding can only be produced using energetic condensation processes. In addition energetic deposition can lead to the improvement of properties of films that could not be otherwise produced by other techniques.

In particular, the field of superhard materials has greatly benefited from energetic deposition processes. For example, one of the important advantages of energetic condensation is that the substrate temperature can be maintained at a very low level without a negative effect on adhesion. This is particularly important for deposition on substrates such as polymeric materials that are temperature sensitive.

The combined effect of process parameters such as substrate temperature and ion energy is not fully understood, and it is expected that this effect will be material dependent. Therefore, although some broad considerations can be made, there is still much to be investigated for the specific materials systems of interest. MePIIID appears to be a valuable technique for those who are willing to undertake this challenge because it is in fact a truly energetic deposition process with independent control of ion energy and fluxes, as opposed to conventional IBAD.

ACKNOWLEDGMENTS

I thank all my coworkers for the last several years. I am particularly thankful to the Plasma Applications Group staff at LBNL for all the support and fruitful discussions. This work was supported in part by the U.S. Department of Energy under contract number DE-AC03-76SF00098.

Visit the Annual Reviews home page at www.AnnualReviews.org

LITERATURE CITED

1. Movchan BA, Demchishin AV. 1069. *Fiz. Metall. Metall.* 28:653
2. Thornton JA. 1974. *J. Vac. Sci. Technol.* 11: 666
3. Bunshah RF. 1974. *J. Vac. Sci. Technol.* 11: 633
4. Dirks AG, Leamy HJ. 1977. *Thin Solid Films* 47:219
5. Leamy HJ, Dirks AG. 1978. *J. Appl. Phys.* 49:3430
6. Movchan BA, Demchishin AV. 1969. *Phys. Met. Metallogr.* 28:83

7. Messier R, Giri AP, Roy RA. 1984. *J. Vac. Sci. Technol. A* 2:500
8. Cuomo JJ, Rossnagel SM, Kaufman HR. 1989. *Handbook of Ion Beam Processing Technology*. Westwood, CA: Noyes
9. Machlin ES. 1995. *Materials Science in Microelectronics: The Relationship Between Thin Film Processing and Structure.* Croton-on-Hudson: Giro
10. Olbrich W, Kampschulte G. 1993. *Surf. Coat. Technol.* 61:262
11. Kelly PJ, Arnell RD. 1998. *J. Vac. Sci. Technol. A* 16:2858
12. Boyd KJ, Marton D, Rabalais JW, Uhlmann S, Frauenheim T. 1998. *J. Vac. Sci. Technol. A* 16:463
13. Lifshitz Y, Kasi SR, Rabalais JW. 1989. *Phys. Rev. Lett.* 62:1290
14. Lifshitz Y, Kasi SR, Rabalais JW, Eckstein W. 1990. *Phys. Rev. B* 41:10468
15. Brown IG. 1998. *Annu. Rev. Mater. Sci.* 28:243
16. Anders A, Anders S, Brown IG, Dickinson MR, MacGill RA. 1994. *J. Vac. Sci. Technol. B* 12:815
17. Baglin JEE. 1991. Interface design for thin film adhesion. In *Fundamentals of Adhesion*, ed. L-H Lee, pp. 363–82. New York: Academic
18. Baglin JEE. 1994. *IBM J. Res. Dev.* 38:413
19. Biersack JP, Haggmark LG. 1980. *Nucl. Instr. Methods* 174:257
20. Biersack JP. 1987. *Nucl. Instr. Methods* B19/20:32
21. Seitz F, Koehler JS. 1956. Displacement of atoms under irradiation. In *Solid State Physics*, ed. F Seitz, D Turnbull, 2:307. New York: Academic
22. Dobrev D. 1982. *Thin Solid Films* 92:41
23. Ensinger W, Kiuchi M, Satou M. 1995. *J. Appl. Phys.* 77:6630
24. Ensinger W. 1995. *Nucl. Instrum. Methods Phys. B* 106:142
25. McKenzie DR, Marks NA, Guan P, Pailthorpe BA, McFall WD, Yin Y. 1996. In *Surface Science. Principles and Current Applications*, ed. RJ MacDonald, EC Taglauer, KR Wandelt. Berlin: Springer-Verlag. 250 pp.
26. McKenzie DR, Bilek MMM. 1998. *J. Vac. Sci. Technol. A* 16:2733
27. McKenzie DR, Bilek MMM. 1999. *J. Appl. Phys.* 86:230
28. Doerner MF, Nix WD. 1988. *CRC Crit. Rev. Solid State Mater. Sci.* 14:225
29. Doljack FA, Hoffman RW. 1972. *The origins of stress in thin nickel films*. Presented at Int. Conf. Thin Films, Venice, Italy
30. Was GS, et al. 1995. *Residual stress control by ion beam assisted deposition.* Presented at Symp. Ion-Solid Interactions for Materials Modification and Processing. Boston, MA
31. Windischmann H. 1987. *J. Appl. Phys.* 62:1800
32. Windischmann H. 1992. *Crit. Rev. Solid State Mater. Sci.* 17:547
33. Blanchard JP. 1993. *J. Vac. Sci. Technol. B* 12:910
34. MacGill RA, Dickinson MR, Anders A, Monteiro OR, Brown IG. 1998. *Rev. Sci. Instrum.* 69:801
35. Monteiro OR, Wang Z, Hou PY, Brown IG. 1997. *Nucl. Instrum. Meth. B* 127/128: 821
36. Monteiro OR, Zhi W, Brown IG. 1997. *J. Mater. Res.* 12:2401
37. Boxman RL, Goldsmith S. 1989. *IEEE Trans. Plasma Sci.* 17:705
38. Zhitomirsky VN, Kinrot U, Alterkop B. Boman RL, Goldsmith S. 1996. *Influence of gas pressure on the ion current and its distribution in a filtered vacuum arc deposition system.* Presented at 23rd Int. Conf. on Metallurgical Coatings and Thin Films, San Diego, CA
39. Bilek MMM, Martin PJ, McKenzie DR. 1998. *J. Appl. Phys.* 83:2965
40. Musil J, Kadlec S, Valvoda V, Kuzel R, Cerny R. 1990. *Surf. Coat. Technol.* 43–44:259
41. Pappas DL, Saenger KL, Cuomo JJ, Dreyfus RW. 1992. *J. Appl. Phys.* 72:3966

42. Pappas DL, Saenger KL, Bruley J, Krakow W, Cuomo JJ, et al. 1992. *J. Appl. Phys.* 71:5675
43. Voevodin AA, Walck SD, Solomon JS, John PJ, Ingram DC, et al. 1996. *J. Vac. Sci. Technol. A* 14:1927
44. McKenzie DR, Muller DA, Kravtchinskaia E, Segal D, Cockayne DJH, et al. 1991. *Thin Solid Films* 206:198
45. McKenzie DR, Muller D, Pailthorpe BA. 1991. *Phys. Rev. Lett.* 67:773
46. Fallon PJ, Veerasamy VS, Davis CA, Robertson J, Amaratunga GAJ, et al. 1993. *Phys. Rev. B* 48:4777
47. Falabella S, Boercker DB, Sanders DM. 1993. *Fabrication of amorphous diamond films*. Presented at 20th Int. Conf. Metallurgical Coatings and Thin Films, San Diego, CA
48. Lifshitz Y, Lempert GD, Rotter S, Avigal I, Uzan-Saguy C, et al. 1994. *Diamond Rel. Mater.* 3:542
49. Lifshitz Y, Lempert GD, Rotter S, Avigal I, Uzan-Saguy C, et al. 1995. *Diamond Rel. Mater.* 4:318
50. Uhlmann S, Frauenheim T, Lifshitz Y. 1998. *Phys. Rev. Lett.* 81:641
51. Lifshitz Y. 1999. *Diamond Rel. Mater.* 8: 1659
52. Robertson J. 1993. *Diamond Rel. Mater.* 2:984
53. Pharr GM, Callahan DL, McAdam SD, Tsui TY, Anders S, et al. 1996. *Appl. Phys. Lett.* 68:779
54. Anders S, Anders A, Ager JW, Wang Z, Pharr GM, et al. 1995. *Mechanical properties of amorphous hard carbon films prepared by cathodic arc deposition*. Presented at Mechanical Behavior of Diamond and Other Forms of Carbon Symp. San Francisco, CA
55. McKenzie DR, Muller D, Pailthorpe BA, Wang Z, Kravtchinskaia E, et al. 1991. *Diamond Rel. Mater.* 1:51
56. Peng XL, Clyne TW. 1998. *Diamond Rel. Mater.* 7:944
57. Friedmann TA, McCarty KF, Barbour JC, Siegal MP, Dibble DC. 1996. *Appl. Phys. Lett.* 68:1643
58. Friedmann TA, Sullivan JP, Knapp JA, Tallant Dr. Follstaedt DM, et al. 1997. *Appl. Phys. Lett.* 71:3820
59. Anders S, Diaz J, Ager JW, Lo RY, Bogy DB. 1997. *Appl. Phys. Lett.* 71:3367
60. Monteiro OR, Ager JW, Lee DH, Lo Ry, Walter KC, Nastasi M. 2000. *J. Appl. Phys.* 88:2395
61. Ager JW, Anders S, Brown IG, Nastasi M, Walter KC. 1997. *Surface Coat. Technol.* 91:91
62. Delplancke-Ogletree MP, Monteiro OR, Brown IG. 1996. *Preparation of TiC and TiC/DLC multilayers by metal plasma immersion ion implantation and deposition: relationship between composition, microstructure and wear properties*. Presented at Materials Modification and Synthesis by Ion Beam Processing Symp. Boston, MA
63. Dimigen H, Klages CP. 1991. *Surface Coat. Technol.* 543
64. Ullmann J. 1996. *Low energy ion assisted carbon film growth: methods and mechanisms*. Presented at Tenth Int. Conf. on Ion Beam Modification of Materials, Albuquerque, NM
65. Ullmann J, Heger P, Pinkert K, Baba K. 1995. *Ions as a useful tool for carbon film deposition and modification*. Presented at Ninth Int. Conf. Ion Beam Modification of Materials, Canberra, ACT, Australia
66. Ullmann J, Baba K, Martin H, Wolf GK. 1994. *Effect of deposition conditions on the growth and behaviour of thin carbon films prepared by ion-assisted evaporation*. Presented at Fourth Int. Conf. on Plasma Surface Engineering, Garmisch-Partenkirchen, Germany
67. Jun Q, Jianbin L, Shizhu W, Jing W, Wenzhi L. 2000. *Surface Coat. Technol.* 128–129:324
68. Salvadori MC, Galvao RM, Monteiro OR, Brown IG. 1998. *Thin Solid Films* 325:19

69. Sridharan K, Anders S, Nastasi M, Walter KC, Anders A, et al. 2000. In *Handbook of Plasma Immersion Ion Implantation and Deposition*, ed. A Anders, p. 553. New York: Wiley & Sons
70. Cuomo JJ, Pappas DL, Bruley J, Doyle JP, Saenger KL. 1991. *J. Appl. Phys.* 70:1706
71. Lee DH, Misra A, Waiter KC, Nastasi M. 1999. *Phys. Rev. B* 59:12283
72. McKenzie DR, Yin Y, Marks NA, Davis CA, Kravtchinskaia, et al. 1993. *J. Non-Cryst. Solids* 164–166:1101
73. Anders S, Ager JW, Pharr GM, Tsui TY, Brown IG. 1997. *Thin Solid Films* 308–309:186
74. Kalish R, Lifshitz Y, Nugent K, Prawer S. 1999. *Appl. Phys. Lett.* 74:2936
75. Voevodin AA, Capano MA, Safriet AJ, Donley MS, Zabinski JS. 1996. *Appl. Phys. Lett.* 69:188
76. Voevodin AA, Prasad SV, Zabinski JS. 1997. *J. Appl. Phys.* 82:855
77. Delplancke-Ogletree MP, Monteiro OR. 1997. *J. Vac. Sci. Technol. A* 15:1943
78. Monteiro OR, Delplancke-Ogletree MP, Lo RY, Winand R, Brown I. 1997. *Surf. Coat. Technol.* 94/95:220
79. Zhao JF, Lemoine P, Liu ZH, Quinn JP, McLaughlin J. 2000. *J. Phys. Condens. Matt.* 12:9201
80. Pierson HO. 1996. *Handbook of Refractory Carbides and Nitrides: Properties, Characteristics and Applications*. Westwood, CA: Noyes
81. Barret C, Massalski TB. 1980. *Structure of Metal.* New York: Pergamon
82. Delplancke MP, Vassileris V, Winand R. 1994. *Structure and composition of hydrogenated Ti/sub x/C/sub y/ thin films prepared by reactive sputtering*. Presented at 41st Natl. Symp. Am. Vac. Soc. Denver, CO
83. Wolfe DE, Singh J. 2000. *Surface Coat. Technol.* 124:142
84. Voevodin AA, O'Neill JP, Prasad SV, Zabinski JS. 1999. *J. Vac. Sci. Technol. A* 17:986
85. Monteiro OR, Delplancke-Ogletree MP, Brown IGG. 1999. *Thin Solid Films* 342:100
86. Davis CA. 1993. *Thin Solid Films* 226:30
87. Voevodin AA, Erokhin AL, Lyubimov VV. 1994. *Phys. Status Solidi A* 145:565
88. Anders S, Callahan DL, Pharr GM, Tsui TY, Singh Bhatia C. 1997. *Multilayers of amorphous carbon prepared by cathodic arc deposition*. Presented at 24th Int. Conf. Metallurgical Coatings and Thin Films, San Diego, CA
89. Delplancke-Ogletree MP, Monteiro OR. 1998. *Wear behavior of diamond-like carbon/metal carbide multilayers*. Presented at 25th Int. Conf. Metallurgical Coatings and Thin Films, San Diego, CA

Annu. Rev. Mater. Res. 2001. 31:139–69

PHOTOREFRACTIVE LIQUID CRYSTALS

Gary P Wiederrecht
Chemistry Division, Argonne National Laboratory, Argonne, Illinois, 60439-4831; e-mail: wiederrecht@anl.gov

Key Words liquid crystal composites, photo-induced charge separation, gratings, holography, optical materials

■ **Abstract** A review of the field of photorefractive liquid crystals is presented. The first reports of photorefractive liquid crystals occurred in 1994, and the performance of these materials has dramatically improved since that time. Liquid crystalline materials have proven to be highly versatile, showing photorefractive character under a wide range of conditions. For example, new composites based on high-molar-mass liquid crystals are now capable of forming volume (Bragg) gratings with high photorefractive gain coefficients of >600 cm^{-1}. Formation times for photorefractive Bragg gratings of 15 ms with applied fields of only 0.1 V/μm have been reported. Low-molar-mass liquid crystals continue to be developed and show their largest photorefractive character in the thin (Raman-Nath) grating regime. Composites of nonmesogenic polymers and liquid crystals are also discussed. The experiments and theoretical work that have been used to characterize these materials are reviewed.

INTRODUCTION

Photorefractivity was inauspiciously discovered in 1966 as an optical damage mechanism in electro-optic crystals that ultimately destroyed the spatial integrity of laser beams as they passed through the crystals (1). The researchers deduced that at high optical intensities, two-photon absorption of impurities within the crystals produced excited states that shed their energy in the form of charge separation. The more mobile charge, in this case electron holes, migrated out of the illuminated region owing to the photovoltaic effect and ultimately were trapped in the dark regions of the crystal. The inhomogeneity in the charge distribution produced an internal electric field or space-charge field that altered the index of refraction through the electro-optic effect and destroyed the optical quality of the laser beam.

The humble beginnings of photorefractivity as a damage mechanism in electro-optic crystals was soon recognized by scientists as a potentially useful tool for image processing and storage (2, 3). Potential applications included image correlators, novelty filters, and optical limiting. With increasingly demanding photosensitivity and temporal response requirements, the 1990s saw an explosion in

0084-6600/01/0801-0139$14.00

new and easily processed photorefractive organic materials that included polymers (4–17), organic crystals (18), glasses (19–21), and liquid crystals (22–34). The optimization of photorefractive materials is no small feat because the initial charge separation efficiency, charge migration, charge trapping, and electro-optic responses all must be simultaneously optimized. In most cases, optimizing one part of the required photo-induced response forces sacrifices in another. On top of these technological considerations, cost and processability must also be taken into account.

Most applications and quantitative analyses of photorefractive materials occur through photorefractive holography. When two coherent laser beams are overlapped in a sample, a sinusoidal interference pattern of light and dark regions is produced, as illustrated in Figure 1. The distance between the peaks of the illuminated regions is called the fringe spacing (Λ). Charge separation occurs in the illuminated regions, followed by preferential charge migration of the more mobile species into the darker regions of the interference pattern. This produces a sinusoidal space-charge field that modulates the refractive index. Thus a refractive index modulation (grating) results and forms a hologram that can have read-write-erase capability.

Importantly, only those photons that lie within the coherence length of the laser contribute to the interference pattern. This provides the basis for image sensing through highly scattering media, as scattered photons with longer path lengths pass undetected through the material. Furthermore, as shown in Figure 1, the photorefractive grating is typically phase shifted by $\pi/2$ (relative to the interference pattern) (3). This phase shift (or nonlocality of the index modulation relative to the optical interference pattern) is an additional feature of photorefractive holograms that results in asymmetric diffraction, whereby one of the writing beams diffracts light into the other writing beam. The result is noise-free amplification of one of the writing laser beams that potentially carries a weak optical signal. This is referred to as asymmetric two-beam coupling, the magnitude of which is called photorefractive gain. In order to ascertain the photorefractive figure of merit for a particular material, the gain is usually reported as a photorefractive gain coefficient (Γ) in units of inverse length.

In the pursuit of improved photorefractive materials, seminal research by a group at IBM led by Moerner discovered the photorefractive effect in polymers in 1990 (7, 35). Photorefractive polymers are generally composite materials having additives that induce the photorefractive effect (4, 5, 7, 35, 36). They initially consisted of photoconductive polymers doped with a high concentration of a nonlinear optical (NLO) chromophore and a small amount of an optical sensitizer. The sensitizer could be an electron acceptor that formed a charge transfer complex with the polymer so that mobile charges within the composite could be photo-generated. A notable exception to this strategy incorporated all of the required species for photorefractivity onto a polymer backbone (8). Later it was noted that large increases in the photorefractive sensitivity could be achieved by adding a plasticizer for lowering the glass transition temperature (37). This permitted reorientation

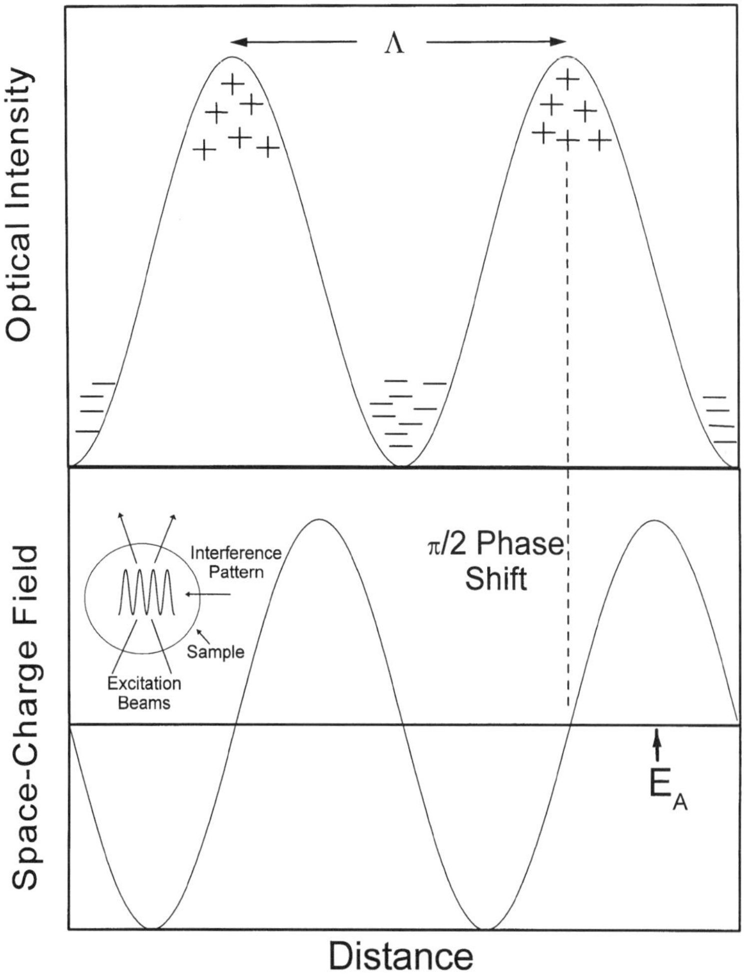

Figure 1 The phase relationship between the optical interference pattern and the space-charge field is illustrated. For liquid crystals, this example illustrates mobile anions migrating into the nulls of the interference pattern. The application of an applied electric field E_A is usually required to observe a phase-shifted photorefractive grating.

of the NLO chromophores in a viscous polymer, thus producing an additional birefringent contribution to the electro-optic effect. Interestingly, the orientational enhancement effect in many polymeric systems was shown to be responsible for the majority of the photorefractive gain (6).

Given the strength of the orientational enhancement effect in polymers, the potential for liquid crystals becomes clear. Photorefractive liquid crystals were first reported in 1994 (22, 23). In many respects, nematic liquid crystals are ideal for observing the orientational photorefractive effect because they are specifically designed to have a strong orientational response, and no nonlinear optical dopant is necessary because the liquid itself is the birefringent component. Thus essentially 100% of the medium contributes to the birefringence rather than the percentage that is a nonlinear optical dopant. Furthermore, the molecules reorient in response to a space-charge field that is orders of magnitude less than that of polymers. The result is that an applied field of only ~0.1 V/μm is required to produce a large index of refraction modulations in liquid crystals, whereas ~50 V/μm is typical for polymers. Also, similar to the photorefractive figures of merit for polymers, those for liquid crystals rapidly improved in just a few years. These improvements resulted from new liquid crystal mixtures and a better understanding of the photo-induced charge transfer processes in liquid crystals.

The widely varying applications of photorefractive media require different response characteristics, defying a "one size fits all" approach to the development (or comparison) of these materials. For example, high-density data storage materials may require very long-charge trapping times at the expense of a large electro-optic response (38, 39), whereas image correlators may require minimal storage times and the largest electro-optic response possible (40–43). Currently, liquid crystalline materials are considered a leading candidate for those applications that do not require long storage times. This review describes the history and present state of photorefractive liquid crystal research.

LOW-MOLAR-MASS NEMATIC LIQUID CRYSTALS

Initial Studies

The first observations and theoretical groundwork of orientational photorefractivity in a nematic liquid crystal were reported in 1994 (22, 23, 44). The low-molar-mass liquid crystalline material 4′-pentyl-4-biphenylcarbonitrile (5CB), doped with small amounts of a sensitizing laser dye rhodamine 6G (R6G) was used. The general experimental geometry that is required to create a photorefractive grating in liquid crystals is illustrated in Figure 2. Two unfocused, coherent laser beams are crossed in the sample. The liquid crystal composite is sandwiched between two indium tin oxide (ITO)-coated glass slides that are treated with a surfactant to induce the liquid crystal director to align perpendicular to the face of the glass slides, i.e. homeotropic alignment (45). The cell thickness is determined by a spacer that is typically 10 to 100 μm thick. A small electric field E_A on the order of ~0.1 V/μm is then applied.

In the initial experiments, self-diffraction of the laser beams was observed resulting from large-fringe spacings and subsequent Raman-Nath (thin) grating behavior. In order to prove that the self-diffraction resulted from the presence of a space-charge field—an essential component of photorefractivity—Rudenko & Sukhov began by assuming a steady-state solution to the current density and

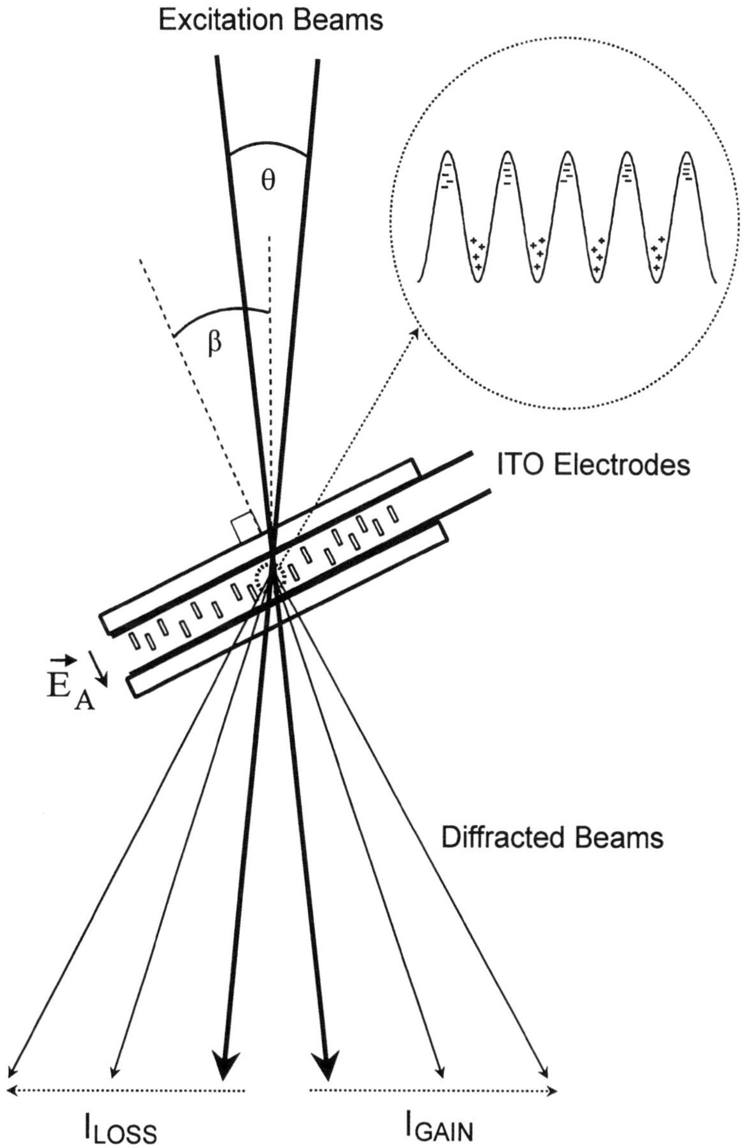

Figure 2 A schematic of the experimental geometry is illustrated. The sample is tilted at an angle $\beta \sim 30^{\circ}$ relative to the bisector of the two beams. This permits charge migration along the grating wavevector, which results in a sinusoidal space-charge field. A phase grating results from the influence of the electric field on the orientational configuration of the birefringent liquid crystal molecules. The beams are polarized in the plane and parallel to the grating wavevector.

showed that the diffusion field in the liquid crystal is given by (23, 46)

$$E_{sc} = \frac{-mk_B Tq}{2e_o} \frac{D^+ - D^-}{D^+ + D^-} \frac{\sigma_{ph}}{\sigma_{ph} + \sigma_d} \sin qx, \tag{1}$$

where the modulation index, m, is given by

$$m = \frac{2(I_1 I_2)^{\frac{1}{2}}}{I_1 + I_2}. \tag{2}$$

Here, σ_{ph} is the photoconductivity, σ_d is the dark conductivity, k_B is the Boltzmann constant, e_o is the charge of a proton, and D^+ and D^- are the diffusion constants for the cations and anions, respectively. Thus the critical factors that determine the magnitude of the space-charge field are the magnitude of the photoconductivity relative to the dark conductivity and the difference in the diffusion coefficients of the cations and anions. These factors allow for one set of charges to preferentially occupy the illuminated regions of the interference pattern and for the opposing charges to occupy the nulls of the interference pattern.

The torque exerted on the liquid crystal molecules has a quadratic dependence on the field. As shown in a simplified illustration of the effect in Figure 3, this means that the presence of a space-charge field owing to the spatial inhomogeneity of cations and anions will produce a grating with twice the wavevector of the optical interference pattern. As a result, no self-diffraction because of a photorefractive mechanism is observed unless a small dc field is applied, which ultimately modulates the index of refraction at the correct wavevector. The need for an applied field was experimentally observed in the initial studies and strongly supports the presence of torque from a space-charge field (22, 23, 44).

From Equation 1, an expression for the diffraction efficiency (η) was derived (23, 46).

$$\eta = \left(\frac{Lmk_b T}{\lambda_{opt} n_e K q e_o} \frac{E_A \Delta\varepsilon_s \Delta\varepsilon_\infty \sin\beta}{1 + \frac{\Delta\varepsilon_S E_A}{2\pi K q^2}} \nu \frac{\sigma_{ph}}{\sigma_{ph} + \sigma_d} \right)^2, \tag{3}$$

where

$$\nu = \frac{D^+ - D^-}{D^+ + D^-} \tag{4}$$

and L is the optical path length, n_e is the index of refraction along the extraordinary axis, K is the single constant approximation of the Frank elastic constant (23, 46), λ_{opt} is the laser wavelength, and $\Delta\varepsilon_s$ and $\Delta\varepsilon_\infty$ are the anisotropies in the static and high-frequency dielectric constants, respectively. The initial studies used Equation 3 to verify that the observed diffraction was from a space-charge field, and similar studies have been used by others to verify that self-diffraction indeed results from a photorefractive mechanism. For example, a linear dependence of optical intensity up to a saturation point was observed, i.e. where the photoconductivity term approaches 1 as $\sigma_{ph} \gg \sigma_d$ (23). A proportionality of $\eta \propto q^{-2}$ and $\eta \propto E_A^2$ was also established in support of Equation 3.

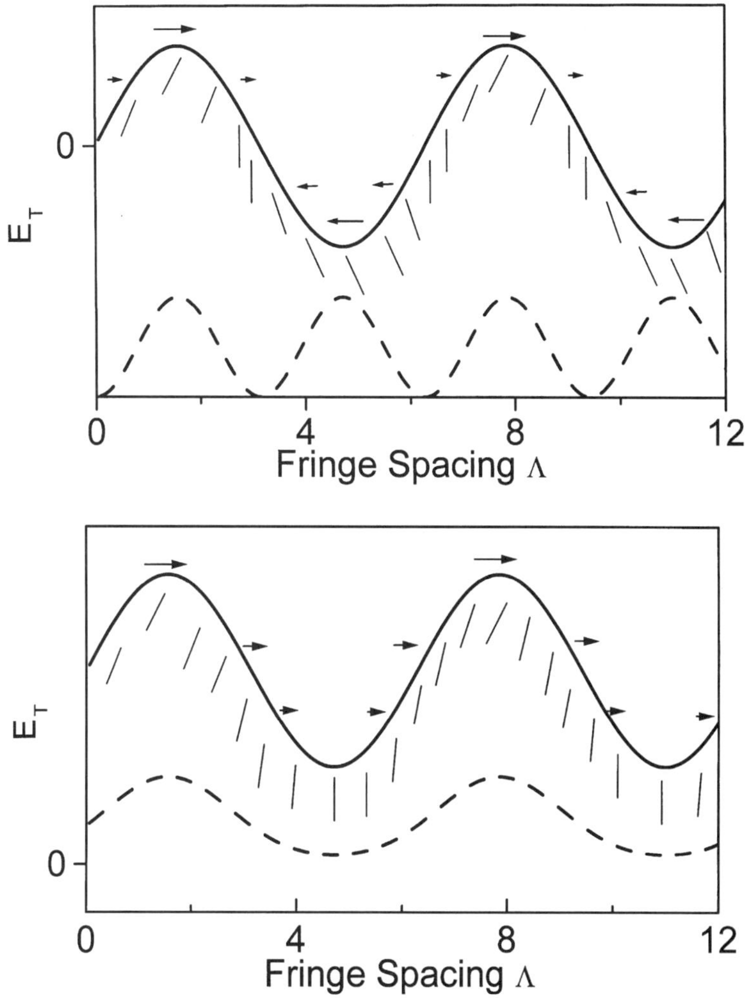

Figure 3 The refractive index change (dashed line) associated with the orientational enhancement effect is illustrated for cases without (*top panel*) and with (*bottom panel*) an applied field. The arrows represent the direction of the total field, E_T, which is the sum of the applied and space-charge fields.

Although these experiments strongly substantiated that diffraction results from a photo-induced space-charge field, further experiments verified that an orientational response was, in fact, the response that modulated the index of refraction. Principle among these was the polarization dependence of the two writing beams on self-diffraction. No self-diffraction in a homeotropically aligned cell was observed unless the beams were *p*-polarized, i.e. parallel to the wavevector of the optical interference pattern. This shows that reorientation was responsible because the director reorientation exposes a portion of the long axis of the liquid crystal molecules to the light wave. If the polarization is perpendicular to the wavevector, reorientation does not lead to interaction changes with the light wave. Furthermore, it was noted that the sample had to be tilted relative to the bisector of the writing beams, a necessary requirement to break the symmetry of the cell and to induce an applied field component along the direction of the wavevector.

Equation 3 has important ramifications beyond allowing the space-charge field mechanism to be verified. Because the only variables that are a function of the dopants are those that make up the diffusion ratio (ν) and conductivity ratio, Rudenko & Sukhov showed that the conductivity ratio saturates at higher light intensities, permitting ν to be determined (23, 44). For example, a value for ν of 0.02 was reported for the experiments of Rudenko & Sukhov (23), which indicates a very small difference in the diffusion coefficients of the positive and negative ions. In these experiments, the charge carriers were not specified. It is believed that R6G functions as the charge generator through a heterolytic cleavage mechanism (44). Such a result implies that a space-charge field nearly two orders of magnitude higher could be formed within the liquid crystal if appropriate cations and anions were utilized. Thus although these initial experiments produced impressive results, the charge generation and transport of the dyes in 5CB were not efficient and much larger photorefractive figures of merit in nematic liquid crystals was possible.

The above studies supply indefatigable proof that dye-doped liquid crystals support a space-charge field and that a director reorientation provides the index of refraction modulation to produce a grating. Khoo et al were the first to report photorefractive effects large enough to observe asymmetric two-beam coupling, reporting a gain coefficient of $\sim$25 cm^{-1}, as given by the equation (37):

$$\Gamma = \frac{1}{L} \cdot \left[\ln \frac{I_{12}}{I_1} - \ln \left(2 - \frac{I_{12}}{I_1} \right) \right], \qquad 5.$$

where I_1 is the intensity of beam 1 in the absence of beam 2, I_{12} is the intensity of beam 1 in the presence of beam 2, and L is the optical path length. The net gain coefficient is cited by subtracting the medium's absorption coefficient (α) and scattering losses.

There is an important caveat to beam-coupling measurements in liquid crystals. Photorefractive gratings in low-molar-mass liquid crystals generally operate in the Raman-Nath (thin) grating regime, producing multiple diffraction orders, as illustrated in Figure 2. This results partially from the larger-fringe spacings in liquid crystals as a result of space-charge field formation through ion diffusion, a process that does not lend itself to small-fringe spacings because of ion recombination issues (23). Also, the collective director orientational response is largest at a high-fringe spacing (47). It is not necessarily true that photorefractive gain increases exponentially with thickness for a thin grating, as is the case with thick gratings. Therefore, there is some controversy regarding the use of gain coefficients in plane gratings, although some groups do report them for plane gratings in other photorefractive media, such as inorganic quantum wells (48, 49). This can be avoided by simply reporting the beam-coupling ratio, I_{12}/I_1 in Equation 5, as a measure of the beam-coupling magnitude.

New Low-Molar-Mass Nematic Liquid Crystals

In 1995, Wiederrecht et al reported a large increase of the orientational photorefractive effect in liquid crystalline materials through two improvements (24). The first was to use a eutectic liquid crystal mixture and the second was to use an organic electron donor and acceptor combination with a well-defined and efficient photo-induced charge transfer mechanism. The eutectic mixture of 35% 4′-(octyloxy)-4-biphenylcarbonitrile (8OCB) and 65% 5CB was utilized. This mixture lowers the liquid crystalline to solid phase transition from 24°C for pure 5CB to 5°C (50) and produces better photorefractive performance because of a greater reorientation angle of the molecules resulting from the lower orientational viscosity relative to 5CB. The birefringence of 8OCB is also slightly higher than that of 5CB (51). Unlike many nonlinear optical chromophores used in photorefractive polymer composites, neither component possesses any visible absorption, reducing unwanted competition with the charge generator for photons.

Perylene was chosen as the dopant that functions as the charge generator for several reasons. First, perylene is a much more ideal solute in the 8OCB/5CB mixture than is R6G, with solubility unlimited for the purpose of these experiments. Second, it possesses a strong visible absorption ($\varepsilon = 40000\ M^{-1} cm^{-1}$ at 442 nm), which has a weak tail that extends farther to the red. At 514 nm, the absorption is 1.5% ($\varepsilon = 610\ M^{-1}\ cm^{-1}$) of the value at 442 nm. Third, perylene is easily oxidized, with an electrochemical oxidation potential of 0.8 V versus saturated calomel electrode (SCE) (52).

The electron acceptor utilized was *N,N′*-dioctyl-1,4:5,8-naphthalenediimide (NI) (53). NI is easily reduced, with an electrochemical reduction potential of −0.5 V versus SCE compared with −1.9 V versus SCE for 5CB. It has no visible absorption in its ground state, but its anion does have a strong absorption ($\varepsilon = 30000\ M^{-1}\ cm^{-1}$) at 480 nm. It also has high solubility in the 8OCB/5CB mixture. The extinction coefficient of the anion at 514 nm is approximately $7000\ M^{-1}\ cm^{-1}$.

The scheme for ion generation with the electron donor and acceptor combination is illustrated in Figure 4. Photoexcitation of the donor molecule (D) is followed by collisional interactions with the neutral electron acceptor (A) molecules. This is followed by charge separation and the generation of an ion pair. The critical final step is for a fraction of the ion pairs to form solvent-separated ions, with rate constant k_{SEP}. Only the solvent-separated ions can diffuse through the sample to create an inhomogeneous spatial charge distribution over macroscopic length scales.

Photocurrent enhancements per unit absorption of more than an order of magnitude were observed relative to R6G. We determined that an excess concentration of NI versus perylene gave the best photoconductivity results, as illustrated in

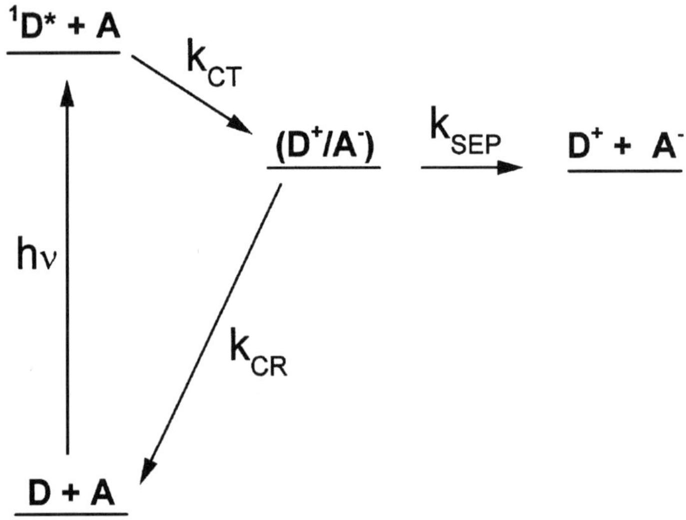

Figure 4 An energy level diagram illustrating the mechanism for production of photoinduced solvent-separated ions in liquid crystals is illustrated, where D is the electron donor and A is the electron acceptor.

Figure 5. The photoconductivity of these composites, normalized for absorption, was found to be 50 times greater than the R6G/5CB mixtures. The beam-coupling experiments were performed with two coherent laser beams from an Ar^+ laser that were crossed in the sample, with a total of 5 mW of *p*-polarized output at 514 nm. The beams were unfocused and had a 1/e diameter of 2.5 mm. The liquid crystal composite was sandwiched between two ITO-coated glass slides that were treated with octadecylsilyl groups to induce the liquid crystal director to align perpendicular to the face of the glass slides, i.e. homeotropic alignment (45). Beam-coupling experiments, shown in Figure 6, for the 37-μm-thick cells, resulted in a beam-coupling ratio of 1.88 for equal intensity beams. Photorefractive rise times as short as 40 ms were observed, although at the expense of photorefractive gain. This was accomplished with very low-light intensities (100 mW/cm^2) and low-applied electric fields (0.04 V/μm) that required only a low-voltage (1.5 V) battery. Additionally, the samples showed no decomposition over more than a one-year period.

Further studies continued to produce a rapid increase in the photorefractive figures of merit of low-molar-mass liquid crystals, especially in regard to the gain coefficient. For example, Khoo et al recently reported studies of 5CB doped with ~0.05% C_{60} by weight (47, 49). The values for the net gain coefficients in the liquid crystals were reported to be nearly 3000 cm^{-1} in the thin grating regime, whereas absorption and scattering losses were only 4 cm^{-1} (49). The huge increase in the gain coefficient was attributed to the improved photoconductivity of C_{60}-doped samples and the thinner cells that permitted a higher E_A. All this was reported using an unfocused 488 nm Ar^+ laser output of less than 5 mW. Applied voltages were 1.5 V. This compares with net gain coefficients for polymers up to ~220 cm^{-1}

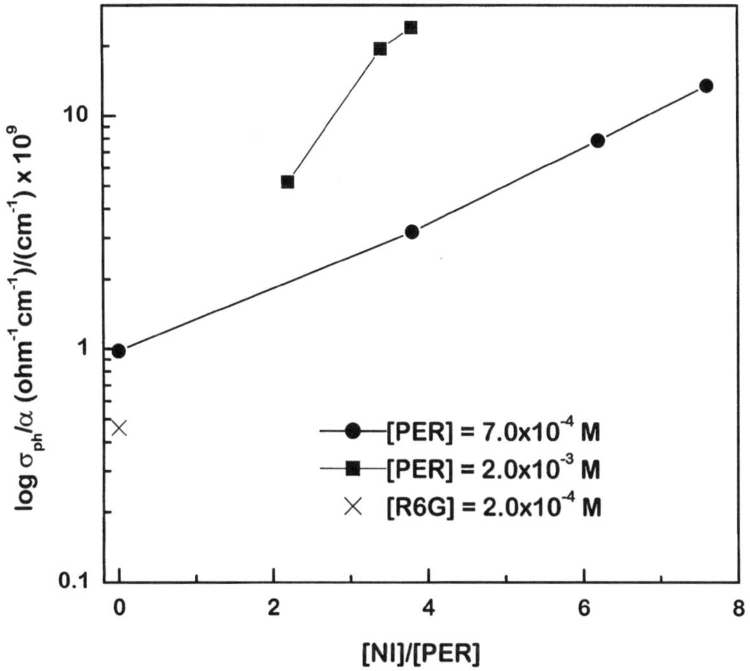

Figure 5 The data illustrate the superior photoconductivity, normalized to an absorption of 1 cm^{-1}, in the perylene and NI-doped liquid crystal versus a R6G-doped sample.

and inorganic ferroelectric materials with gain coefficients of 10–40 cm^{-1} (37). It was noted that some of the drawbacks of nematic liquid crystals were relatively slow response times of >100 ms in the high-gain region and small absolute gains in the amplified beam owing to the thin (10 μm) nature of the film. In other words, a high-gain coefficient does not necessarily imply a high-absolute gain.

The studies of C_{60} in 5CB also revealed one additional important finding—the presence of longer-lived gratings of intermediate and even nearly permanent time scales. As long as the dc voltage was maintained on the sample and the writing beams were turned off, diffraction from a CW He-Ne laser beam occurred that decayed over the course of a few minutes. Normally, gratings in liquid crystals decay on the order of less than 1 s. Thus a contribution to the space-charge field other than ion diffusion (Equation 1) is likely operative. Two possibilities were presented. One is derived from the conductivity anisotropy and is known as the Carr-Helfrich effect (45, 50, 54):

$$E_{SC}^{\Delta\sigma} = -\left(\frac{\Delta\sigma \sin\theta\cos\theta}{\sigma_{\|}\sin^2\theta + \sigma_{\perp}\cos^2\theta}\right)E_A, \qquad 6.$$

where σ is the conductivity, $\Delta\sigma = (\sigma_{\|} - \sigma_{\perp})$, and θ is the director axis reorientation in response to the applied field E_A. Another mechanism to create a

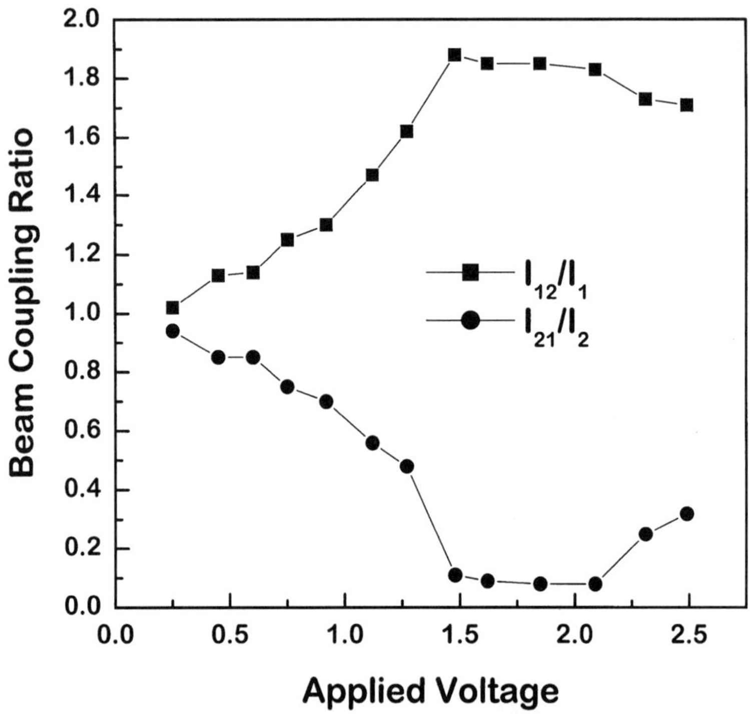

Figure 6 A plot of the beam-coupling ratio versus applied voltage is shown for both the beam that increases in intensity (■) and for the beam that decreases in intensity (●).

space-charge field is due to dielectric anisotropy, given by (45)

$$E_{SC}^{\Delta\varepsilon} = -\left(\frac{\Delta\varepsilon \sin\theta \cos\theta}{\varepsilon_{\|} \sin^2\theta + \varepsilon_{\perp} \cos^2\theta}\right) E_A, \quad 7.$$

where ε is the static dielectric constant and $\Delta\varepsilon = (\varepsilon_{\|} - \varepsilon_{\perp})$. The researchers estimated that these terms were approximately 30% of the total relative to the magnitude of E_{SC} given in Equation 1. These fields are fundamentally different from the photorefractive field described in Equation 1 because they are induced by the applied field and become significant only when the reorientation angle θ is large. Thus they can remain when the writing beams are turned off, unlike the diffusion field in Equation 1.

When the cell was illuminated for longer than 15 min, He-Ne diffraction was observed on a nearly permanent basis when the writing beams were turned off, which is believed to be the result of a surface effect, owing to a irreversible change in the surface director alignment. Surface gratings are of significant interest to the photorefractive liquid crystal community, as the possibility of new mechanisms for inducing space-charge fields in liquid crystals has important ramifications for eventual applications of liquid crystals (32, 55–57). Currently, the resolution of holographic gratings from charge separation effects in neat, low-molar-mass liquid crystals is limited to greater than a few microns. If new mechanisms derived from dielectric anisotropy or the Carr-Helfrich effect can be employed, ultimately these resolutions may be significantly improved. Higher resolution gratings may eventually permit low-molar-mass nematic liquid crystals to operate in the Bragg regime for the first time. Currently, the low resolutions permit only a plane grating response.

As a final discussion on low-molar-mass liquid crystals, recent studies on methyl red doped liquid crystals are particularly intriguing (43, 58, 59). In concentrations of 0.5 to 1%, methyl red–doped liquid crystals show what the authors describe as "extraordinarily large nonlinearity" (58). In fact, for optical intensities as low as 40 μW/cm^2, grating intensities similar to those described in the above research are achieved with two orders of magnitude lower optical intensity, even without the application of an applied field. This provides a refractive index change coefficient, i.e. the index change per unit optical intensity, of ~6 cm^2/W!

For these studies, the authors were careful to rule out other mechanisms through many of the photorefractivity checks described above. It is important to note that while the studies prove that the grating originates from a space-charge field mechanism, the grating seems to form a local, i.e. non-phase-shifted grating. Furthermore, the locality of the grating precludes the observation of asymmetric beam coupling. In this experimental setup, the authors were able to observe the large nonlinearities even without the application of an applied field. Photovoltages up to 12 mV for a 25-μm-thick film were observed that further substantiate the space-charge field mechanism as the origin of the index grating. These cells have subsequently been used to demonstrate a variety of impressive optical signal-processing applications, such as incoherent to coherent image conversion and optical limiting (43).

It should be noted that Raman-Nath gratings are not necessarily a disadvantage for many optical signal-processing applications. In fact, there are significant advantages to thin gratings when multiplexing with the angular or wavelength selectivity of Bragg gratings is not required. Thin gratings do not have angular or wavelength selectivity, and therefore rigorous alignment procedures are avoided. It is clear at this time that large index of refraction changes in low-molar-mass liquid crystals are easily induced under very low–optical powers and applied voltages. For some applications of dynamic holography that are designed to process single images at a time, a Raman-Nath grating may be better suited for the task.

Optimizing Charge Generation in Nematic Liquid Crystals

Charge separation theory can be applied to optimize the magnitude of the photorefractive effect through improved generation of mobile charge. Marcus theory predicts that the rate of charge separation (k_{CS}) will be the greatest when the exothermicity of the reaction is equal to the sum of the reorganization energy of the solvent (λ_o) and the internal vibrational reorganization energy of the ions (λ_i) (60–62). For free energies of charge separation that are more or less than the total reorganization energy, the rates of charge separation will be slower. This can be restated by the following equations (60–62):

$$k = Z \exp(-E_a/kT), \qquad 8.$$

where

$$E_a = \frac{(\Delta G_{CS} + \lambda)^2}{4\lambda} \qquad 9.$$

$$\lambda = \lambda_0 + \lambda_i \qquad 10.$$

$$\lambda_0 = e_o^2\left(\frac{1}{2r_1} + \frac{1}{2r_2} + \frac{1}{r_{12}}\right)\left(\frac{1}{\varepsilon_\infty} - \frac{1}{\varepsilon_s}\right). \qquad 11.$$

Here, Z is the bimolecular collision rate of the uncharged species, ΔG_{CS} is the free energy for charge separation, e_o is the charge of an electron, λ is the total reorganizational energy of the solvent (λ_0) and of the internal intramolecular bonds (λ_i), r_1 and r_2 are the radii of the reactants, and r_{12} is the center to center distance between the two reactants. The free energy for charge separation (ΔG_{CS}) was calculated using the relation

$$\Delta G_{CS} = E_{OX} - E_{RED} - \frac{e_o^2}{\varepsilon_s r_{12}} - E_s, \qquad 12.$$

where E_{OX} is the oxidation potential of the donor, E_{RED} is the reduction potential of the acceptor, E_s is the first excited singlet state of the donor, and ε_s is the static dielectric constant.

As discussed above, the photoconductivity and the difference in diffusion coefficients for the cations and anions (ν) need to be improved in order to increase the magnitude of the space-charge field in liquid crystals. In order to explore these possibilities, we performed studies with four different intermolecular charge transfer dopant combinations (25). The redox potentials and driving forces of the molecules for charge separation and charge return are given in Table 1. The

TABLE 1 The lowest excited singlet state, redox potentials versus the saturated calomel electrode (SCE), and free energies for charge separation and return

Molecule	E_S (eV)	E_{OX} (V)	E_{RED} (V)	ΔG_{CS} (eV)	ΔG_{CR} (eV)
PER, NI	2.8	0.8	−0.5	−1.5	−1.3
PER, PI	2.8	0.8	−0.8	−1.2	−1.6
ANI, NI	2.8	1.0	−0.5	−1.3	−1.5
ANI, PI	2.8	1.0	−0.8	−1.0	−1.8

absorption of the ANI chromophore peaks at 395 nm and extends well into the visible region, permitting the use of the 457-nm line of an Ar^+ laser. The 514-nm line was used for the PER chromophores. The concentration of the donors in all cases was 2.0×10^{-3} M.

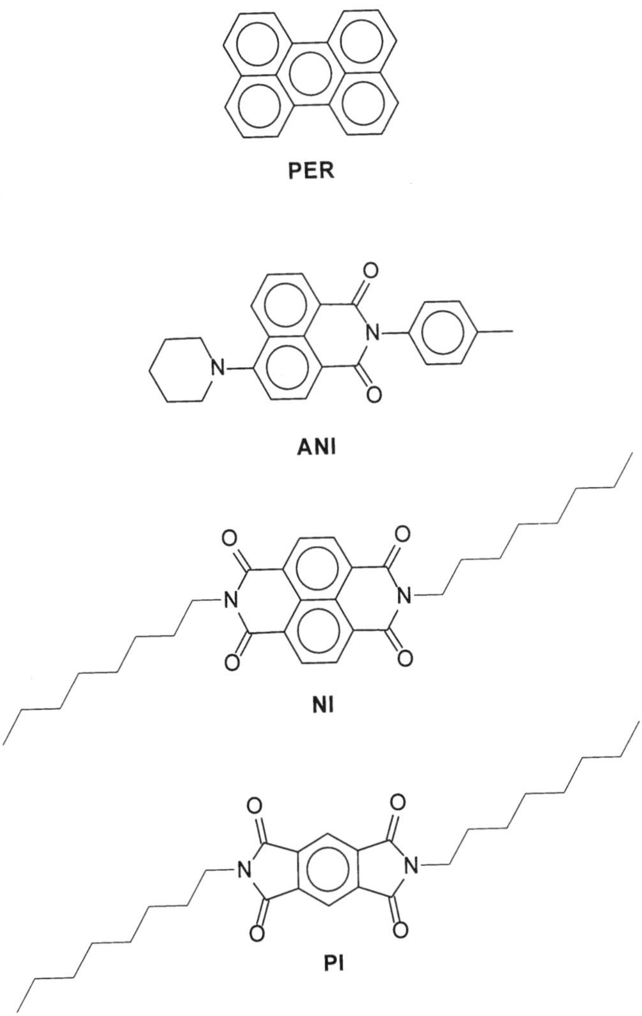

Scheme 1

The results of the conductivity saturation studies that give the value of ν are shown in Figure 7, and the values for saturation diffraction efficiency (η_{SAT}), q, and ν are shown in Table 2. Values for the constants $E_a = 0.04$ V/μm, L $= 37$ μm,

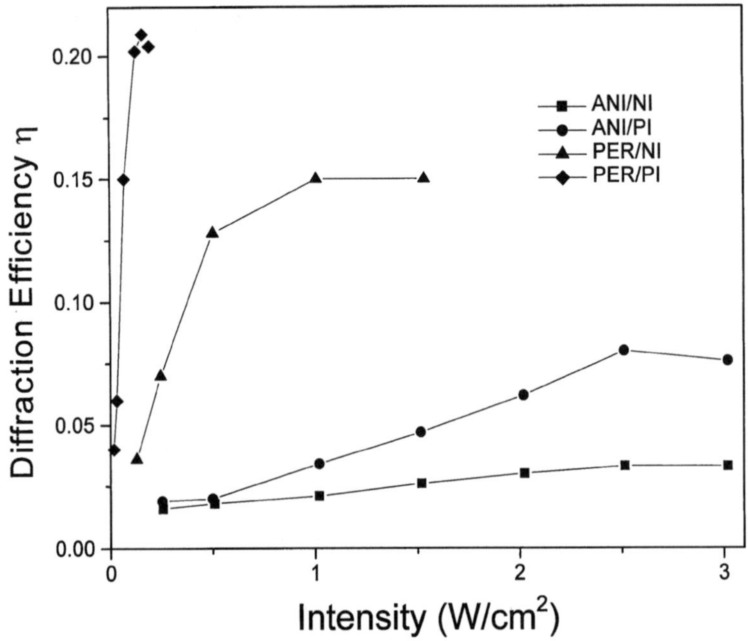

Figure 7 A plot of the diffraction efficiencies versus incident laser power of the composite materials containing ANI/NI, ANI/PI, PER/PI, and PER/NI is illustrated.

$K = 7 \times 10^{-7}$ dyne (45), m $= 1$, ε_∞ of 2.25, ε_s of 15, and 298K for T were used to determine ν from Equation 3. Time of flight measurements were performed to determine the diffusion coefficients for D^- of NI^- and D^+ of ANI^+, which can be utilized in conjunction with Equation 4 to determine the remaining diffusion coefficients.

Clearly, the data indicate that there is a large difference in the values of ν. For the ANI/NI system, the time of flight measurements show that the anion (NI^-) is more mobile than the cation (ANI^+). The results shown in Figure 7 further support

TABLE 2 The saturation diffraction efficiencies (η_{SAT}), wavevectors (q), ν, and diffusion coefficients for each molecular component

Molecule	η_{SAT}	$q \times 10^{-3}$ cm^{-1}	ν	$D^+ \times 10^8$ cm^2/s	$D^- \times 10^8$ cm^2/s
PER, NI	0.15	4.7	0.17	2.5	3.4
PER, PI	0.21	4.7	0.21	2.5	3.8
ANI, NI	0.033	4.2	0.11	2.7	3.4
ANI, PI	0.080	4.2	0.17	2.7	3.8

this conclusion, because the smaller PI has increased mobility that produces a larger value for ν than NI-doped samples. It follows from Figure 7 that PER is less mobile than ANI chromophores because samples doped with PER have superior saturation diffraction efficiencies for a given acceptor. Thus the best sample is the PER/PI combination, which has a value for ν of 0.21, more than 10 times better than the R6G/5CB samples.

The PER/PI sample also saturates at much lower intensities (100 mW/cm^2) than the other composites. This means that the efficiency of mobile charge generation is much greater in this sample, which seems counterintuitive because the more easily reduced NI sample would suggest a better charge separation efficiency. However, as can be seen from Equation 9, the interdependence of ΔG_{CS} and λ in the quadratic exponent indicates that the rate for charge separation is maximized for the PER/PI system in which ΔG_{CS} is -1.2 eV. The larger driving force for the NI/PER composite actually slows the rate of charge separation. Following previous precedent, these values ignore the Coulomb term, which is small in polar environments (62, 63). Although the optimal ΔG_{CS} value at first glance appears rather high, the solvent reorganization energies for solvent separated ion pairs, as opposed to tight ion pairs, are confirmed as higher (62, 63).

In addition to optimizing ΔG_{CS}, the free energy for charge return in the PER/PI system is greater than that for PER/NI. This places ΔG_{CR} farther out in the Marcus inverted regime for the PER/PI system, resulting in slower rates for charge return and therefore increasing the efficiency of mobile charge generation (61, 62, 64). Thus the PER/PI liquid crystal composite is the best of both worlds: It has the largest value for ν, which maximizes E_{SC}, and also has the best efficiency of mobile charge generation within this series of molecules.

HIGH-MOLAR-MASS LIQUID CRYSTALS

Ono et al have recently developed new liquid crystal composites that contain both a low-molar-mass liquid crystal and a liquid crystal polymer, or high-molar-mass liquid crystal (28, 33, 65–71). These composites in many respects have the best photorefractive figures of merit for strictly nematic liquid crystals. Among the standout parameters are net beam-coupling coefficients of greater than 600 cm^{-1} for a high-resolution grating that lies clearly in the Bragg regime. Applied voltages are typically on the order of 1 V/μm, although large gain coefficients are observed in some composites at 0.1 V/μm. Scattering losses are reported to be less than 50 cm^{-1}. All this was accomplished using very low–output powers from a HeNe laser at 632.8 nm of less than 10 mW per beam and a spot size of ~500 μm.

Several different composites have been described that contain C_{60} as a sensitizer in addition to the low- and high-molar-mass liquid crystals, with a typical concentration of 0.02 to 0.05 wt% of C_{60}. Initial studies reported that a low-molar-mass commercial liquid crystal mixture E7 was used along with a side-chain liquid crystalline polymer shown here (28).

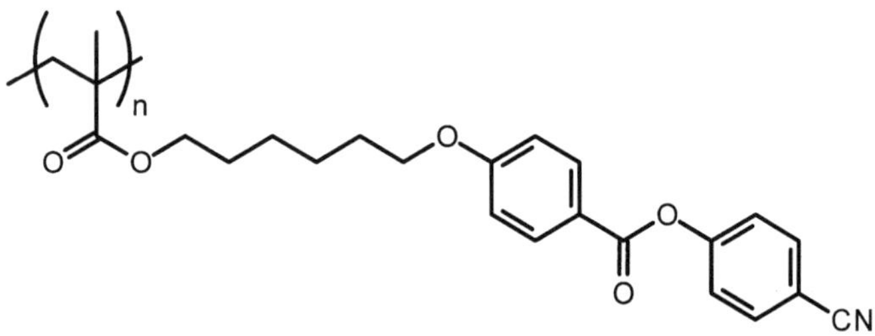

Scheme 2

Depending upon the particular study, the wt% of low- and high-molar-mass liquid crystals varied between 100:0 and 50:50. In all mixtures, a mesophase that did not show phase separation was reported. Exhaustive studies clearly illustrate the tradeoffs that exist as the composition is varied (66). Superior photorefractive gain in the Bragg regime is achieved for higher relative concentrations of the high-molar-mass liquid crystal, i.e. the 50:50 mixtures. This is the ratio in which gain coefficients of over 600 cm^{-1} are reported for a fringe spacing of 2.8 μm, film thickness of 50 μm, and applied fields of 0.7 V/μm. No beam coupling is observed for the 100:0 mixture at this fringe spacing, whereas a gain coefficient on the order of 100 cm^{-1} is observed for the 90:10 mixture. Conversely, faster response times are reported in the Bragg regime for lower concentrations of the high-molar-mass liquid crystal. Response times for the 50:50 mixture are ~3.5 s, whereas the response times for the 90:10 mixture are on the order of 100 ms.

It should also be emphasized that these materials truly operate solidly in the Bragg (thick) grating regime. In order to determine whether the grating is a thin (plane) or volume grating, the following well-known quality parameter (Q) can be used (72)

$$Q = \frac{2\pi L\lambda}{\Lambda^2 n}, \qquad 13.$$

where λ is the wavelength of the light, n is the index of refraction, and L is the thickness of the grating ($L = d/\cos\beta$, where d is the cell thickness). For Q $\ll$ 1, the grating is considered to be a plane grating, and for Q $\gg$ 1, a volume grating is created. Although the literature does not appear to specifically designate an exact value for Q in which the thick-grating regime is reached, the most rigorous treatments designate that Q values of 10 are required to produce a true volume hologram (73). For the parameters described above, the Q values are ~17, and significant beam coupling at much higher Q values has also been reported (65).

More recent studies by Ono et al have explored the effect of molecular weight of the polymer liquid crystal on the temporal and gain coefficients of photorefractive gratings (69). A clear improvement in the response times was found for the lower molecular weight polymers and was attributed to an overall decrease in the viscosity of the composite. Response times as short as 15 ms were reported for composites with gain coefficients of 270 cm^{-1} in the Bragg regime. Furthermore, the required applied voltage was lowered to ~0.1 V/μm. Interestingly, whereas the above results were cited for fringe spacings of 2.5 μm, large beam-coupling coefficients on the order of 100 cm^{-1} were maintained even at $\Lambda = 1.8$ μm in a 50-μm-thick cell.

In addition to studying photorefractivity in new composites, Ono et al has performed measurements that clarify the photorefractive mechanism in liquid crystals, including phase shift, space-charge field, and trap density studies (70). Experiments that further substantiate the phase shift of the index of refraction grating relative to the optical interference pattern are important to further support and understand the photorefractive mechanism in liquid crystals. For example, the magnitude of the phase shift helps to understand the charge transport and trapping characteristics of a composite (74–77). The phase shift studies by Ono et al were performed with a moving grating method, i.e. by translating the sample relative to the optical interference pattern and monitoring the change in the gain coefficient as a function of the translation distance (70):

$$\Gamma(\xi) = \frac{2\pi n}{\lambda \cos\frac{\theta}{2}} \sin(\phi + q\xi), \qquad 14.$$

where ξ is the translation distance, ϕ is the index grating phase shift, and θ is the angle between the writing beams. Such experiments have shown for the low- and high-molar-mass liquid crystal composites described above, that the phase shift is on the order of 70°. By changing the mesogenic polymer to one containing a carbazoyl side group (71), the phase shift drops between 20 to 50°, depending upon the applied voltage. One can determine the trap density limited field (E_q), and ultimately the trap density, by fitting the phase shift to the following expression:

$$\phi = \arctan\frac{E_d^2 + E_d E_q + E_{A||}^2}{E_{A||} E_q}. \qquad 15.$$

Here, $E_{A||}$ is the applied field component along the wavevector direction; E_d is the diffusion field, and can be given by $k_B Tq/e_o$. This is a simplified version of Equation 1 when the photoconductivity is large relative to the dark conductivity, and the difference in diffusion coefficients of the positive and negative charge carriers are also large. The trap density (N_{eff}) can be determined by the relation $E_q = e_o N_{eff}/e_\infty e_{||} q$, where e_∞ is the high-frequency dielectric constant, and $e_{||}$ is the static dielectric constant parallel to the wavevector. The upshot of these calculations is that the trap density limited fields are less than 0.5 V/μm, and N_{eff} is on the order 10^{14} cm^{-3}. These values are considerably smaller than polymeric (nonmesophase) materials, and underscore the fact that the true advantage of liquid crystals is the ease of reorientation to produce very high–gain coefficients.

In other polymer liquid crystal research, Sasaki et al has observed unusual enhancement of photorefractivity in the isotropic phase of polymeric liquid crystals (31). At first glance, the photorefractive effect would be expected to be very weak in the isotropic phase because only an aligned sample with some means of breaking the symmetry of the cell, i.e. oblique incidence of the writing beams, is expected to show self-diffraction from an orientational grating. However, the polymeric liquid crystals reported are capable of withstanding high applied electric fields of 37 V/μm. Thus these voltages are more reminiscent of nonmesophase polymers than those systems described above. They also operate at very small fringe spacings on the order of 1 μm in a 100-μm cell, again comparing well with conventional polymers with strong Bragg character of the grating. The high field enables alignment of the birefringent chromophores in the side chain polymer, even in the isotropic phase. Furthermore, important reductions in scattering losses are reported in the isotropic phase. The side-chain polymeric liquid crystals reported used either 4-nitrobenzylideneaniline or cyanobiphenyl moities as the birefringent chromophore. Electron donor (DEH) and acceptor (TNF) compounds were used to induce charge transport.

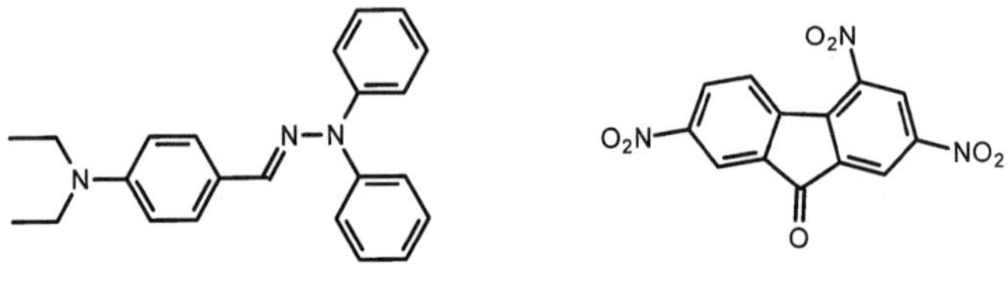

DEH TNF

Scheme 3

Interestingly, the DEH was present in concentrations up to 30% without any phase separation.

The beam coupling and diffraction measurements were performed using the *p*-polarized output of a HeNe laser at 7 mW total power. Unlike low-molar-mass liquid crystals, the laser beams were focused into the sample. Good correlations were performed by varying the length of the side chain to induce greater flexibility of the birefringent chromophore and to study the effect of a nematic phase on photorefractive figures of merit. A clear correlation with longer, birefringent side chains showing a nematic phase also showed higher orientational photorefractivity in the isotropic phase. In other words, shorter side chain polymers that went from a polymer phase directly to an isotropic phase did not show large photorefractivity. For the best composites, gain coefficients of greater than 100 cm^{-1} and diffraction efficiencies of 65% were reported.

LIQUID CRYSTAL/POLYMER COMPOSITES

For many applications, it would be desirable to combine the very large reorientational effects found in low-molar-mass LCs with the longer grating lifetimes and higher resolution of nonmesogenic polymers. Polymers routinely operate with fringe spacings of approximately 1 μm, whereas many of the low-molar-mass liquid crystals with the highest index of refraction changes operate with fringe spacings an order of magnitude larger. Along these lines, researchers have developed composites containing both nonmesogenic polymers that act to alter charge transport and liquid crystals that act as the nonlinear optical chromophore. These materials can be loosely separated into three different classes. The first are those composites that undergo phase separation to produce polymer-dispersed liquid crystals. The second are those composites with a nonmesogenic polymer that do not show phase separation, even with high concentrations of liquid crystals. These systems behave similarly to low-glass transition photorefractive polymers. The third class are those with low concentrations of polymer (i.e. less than 5 wt%) that behave as polymer-stabilized liquid crystals, i.e. much of the low-molar-mass nematic character is preserved in this class of materials. Each is outlined in more detail below.

Polymer-Dispersed Liquid Crystals

Polymer-dispersed liquid crystals (PDLCs) are already of great interest for their electro-optic applications. The dispersions are usually studied with a high concentration of both polymer and liquid crystal molecules that phase separate into droplets. They are opaque owing to the random orientations of the LC directors of the droplets until an electric field is applied. The resulting alignment of the LC directors within each droplet along the electric field produces a transparent material. Such materials are of interest for optical shutters. Orientational photorefractivity was reported in PDLCs by two separate groups at nearly the same time in 1997 (78, 79) and further studies continue to provide material improvements and a better understanding of photorefractivity in PDLCs (80–85).

Golemme et al used PMMA at 45 wt% as the polymer matrix and a eutectic liquid crystal, E49 (from Merck), at 33 wt% (78). The remainder consisted of a hole transport agent, *N*-ethylcarbazole (ECZ), at 21 wt% and TNF at 1 wt%. For an applied field of 22 V/μm, a gain coefficient of 41 cm^{-1} and $\Delta n = 2 \times 10^{-3}$ were reported using a 1 mW *p*-polarized HeNe laser output with a spot size of

250 μm. With the reported fringe spacing at 4.5 μm and film thickness of 53 μm, the grating reasonably fulfills Bragg requirements. However, the high absorption and scattering losses of 735 cm^{-1} precluded net photorefractive gain. In other terms, the authors reported that transmission through the cell with voltage applied was ~20%. Without applied voltage, the sample was opaque because of the disoriented directors of the liquid crystal droplets.

Ono et al also reported the use of PMMA as the polymeric component of PDLCs (79, 86). A slightly different eutectic mixture of liquid crystals, E44 (from Merck) was used along with the C_{60} dopant that functions as a photorefractive sensitizer and charge generator. The PMMA:E44 ratio was 40:60 wt%, and the concentration of C_{60} was 0.05 wt%. A similar Δn ($\sim 3 \times 10^{-3}$), albeit with a higher applied field of 40 V/μm to that of Golemme et al, was reported. However, at fringe spacings of 12 μm and film thicknesses of 25 μm, this composite operated only in the Raman-Nath regime. The nearly identical polymer and liquid crystal mixtures used for the studies of Golemme et al and Ono et al produced completely different photorefractive behavior, highlighting the importance of considering the charge generating and charge transport chromophores in liquid crystal composites.

Shortly after these initial studies, Golemme et al reported dramatically improved PDLC performance (84). One of the principle improvements was the utilization of a different eutectic liquid crystal, TL202, which had both a lower birefringence to reduce scattering losses and higher resistivity to permit higher applied fields. The higher fields improve the alignment (lowering the scattering), charge separation, and quadratic orientational photorefractive effects. Furthermore, a modified electron acceptor, 2,4,7-trinitro-9-fluorenylidene malononitrile (TNFDM) was used that forms a charge transfer complex with ECZ. For samples with the wt% ratio of 40:42:17:1 for TL202:PMMA:ECZ:TNFDM, a beam-coupling coefficient of 136 cm^{-1} was reported with absorption and scattering losses of 99 cm^{-1}. This was reported for a fringe spacing of 3 μm and thickness of 53 μm at an applied field of 22 V/μm. Thus net gain in a Bragg grating was observed for the first time in PDLCs.

An even more impressive aspect of these materials was the observation of total (100%) internal diffraction at low fields for 105-μm-thick samples. The diffraction efficiency of a volume grating varies as the square of $\sin(\pi \Delta nL/\lambda)$; thus it is possible to observe a modulation in the diffraction efficiency at large values of $\pi \Delta nL/\lambda$. In other words, conversely to the initial expectation, larger values of Δn do not necessarily produce larger diffraction efficiencies. This is exactly what was observed in these PDLCs, with total internal diffraction occurring at only 8 V/μm. Similar effects were reported in photorefractive polymers by the same Arizona State group (14).

One reported negative aspect of the PMMA-based PDLCs was the slow speed of the grating formation that was on the order of several minutes (84). This was attributed to the poor charge transport characteristics of PMMA. In response to this, the researchers also tried PVK as a polymer matrix, which is a well-known hole-transporting polymer. Thus ECZ is not required. The electron acceptor (TNF) was used because it is known to form a charge transfer complex with PVK. The wt% ratio of PVK:TL202:TNF was 51:48:1. Although this composite showed diffraction efficiencies of only ~3%, fast grating formation times of 100 ms were observed.

Cipparone et al have published detailed studies that provide new information on the mechanism of orientational photorefractivity in PDLCs; they have also generated new photorefractive PDLC materials (80–83). Detailed optical and scanning electron microscope studies, in addition to two beam-coupling experiments, show that photorefractive gratings form in PDLCs. New materials based on epoxy resins were also reported. Two important observations of these studies were long-lived memory effects and little performance dependence on liquid crystal domain size or shape. A recent report by Golemme et al has also further clarified the orientational response in photorefractive PDLCs (85). This report showed the importance of considering the relative orientation of the total electric field in the PDLC, especially when the threshold for total director reorientation is reached.

Homogeneous Liquid Crystal/Polymer Composites

As described above, the orientational enhancement effect was initially reported in polymers with low glass transition temperatures (4, 37). As a result, new nonlinear optical chromophores continue to be reported for applications in homogeneous photorefractive polymers that have enhanced birefringence and electro-optic properties (13). One possibility is to utilize nematogens as the birefringent donor in photoconductive polymers but in a concentration ratio that maintains a homogeneous (not phase separated) mixture (87, 88). Zhang & Singer reported in 1998 that mixtures of 5CB and PVK could be mixed with up to 65 wt% 5CB without phase separation (88). The 5CB in effect performed a dual role because the high nematogen concentration eliminated the need for a plasticizer to lower the glass transition temperature in addition to functioning as the birefringent chromophore. Diffraction efficiencies of 61% and gain coefficients greater than 100 cm^{-1} were reported for samples with absorption losses of only 18 cm^{-1}. This result was reported using C_{60} (0.2 wt%) and 5CB (40 wt%). The thickness of the cell was 75 μm and the fringe spacing only 3 μm; thus Bragg behavior was observed. Effective trap densities were estimated to be $\sim 3.5 \times 10^{16}$ cm^{-3} or two orders of magnitude higher than for the polymeric liquid crystals described by Ono et al. Here again, photorefractivity is revealed as a game of tradeoffs. Despite the higher space-charge field capabilities of this composite, orientational response is somewhat reduced relative to some other polymer composites because of the relatively high glass transition temperature of 40°C. If one wishes to lower the glass transition for better orientational response by increasing the 5CB concentration, then phase separation is induced.

In other related work, the Moerner group reported the generation of a homogeneous polymer/nematogen composite in studies aimed at establishing the trap density of C_{60}-sensitized polymer composites (87). In these composites, C_{60} acts as the electron acceptor and PVK acts as the hole transporter. The study showed that it is also important to consider the manner in which the nonlinear optical chromophore participates in the charge-trapping process. It was reported that nonlinear optical chromophores with a highest occupied molecular orbital (HOMO) above that of PVK can act as an effective, temporary hole trap that ultimately leads to a higher C_{60} anion concentration (improved space-charge field). To prove this, a homogeneous composite of PVK and 5CB in nearly equal wt% was sensitized with 0.05 wt% C_{60}. 5CB has a HOMO actually below that of PVK, and the results were compared with two other PVK composites that contained differing nonlinear optical chromophores with a HOMO above that of PVK. These two composites contained a plasticizer, whereas the 5CB composite did not. The gain coefficients for the two composites containing nonlinear optical chromophores with HOMOs above that of PVK showed beam-coupling coefficients an order of magnitude above that of the 5CB composite. Thus this work has important ramifications for the development of polymer/nematogen mixtures.

In a different regime of homogeneous polymer/nematogen composites, PPV-doped nematic liquid crystals have been used to induce through bond charge transport in addition to ion diffusion (89). The potential for such systems can be seen by the current density (J) equations (90, 91):

$$J = J^{+} + J^{-}, \tag{16.}$$

$$J^{\pm} = e_o\mu^{\pm}n^{\pm}(x,t)(E_{SC}(x,t) - E_A) \mp e_o D^{\pm}\frac{\partial n^{\pm}[x,t]}{\partial x}, \tag{17.}$$

where $\mu^{\pm}$ is the mobility of the cations and anions, x is the grating wavevector axis, $n^{\pm}$(x, t) is the ion density, E_{SC}(x, t) is the magnitude of the space-charge field, E_A is the magnitude of the applied field, and $D^{\pm}$ is the diffusion constant of the cations and anions, respectively. The first term on the right hand side of Equation 17 describes charge drift, and the second term describes ion diffusion. Previous studies of photorefractivity in liquid crystals indicate that ion diffusion is the charge transport mechanism that creates the space-charge field, a process that limits the efficiency and speed of the effect. The charge drift mechanism has not been a factor because of the short ionic drift length L_E, given by (90)

$$L_E = \frac{\mu^{\pm}\tau^{\pm}V}{d}, \tag{18.}$$

where $\tau^{\pm}$ is the carrier lifetime, V is applied potential, and d is the cell thickness. Typically, $2\pi L_E \ll \Lambda$ in liquid crystals, while the drift mechanism can only contribute to a photorefractive grating if $2\pi L_E \geq \Lambda$ (90).

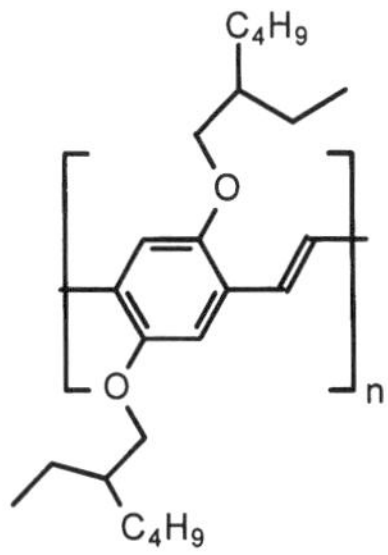

BEH-PPV

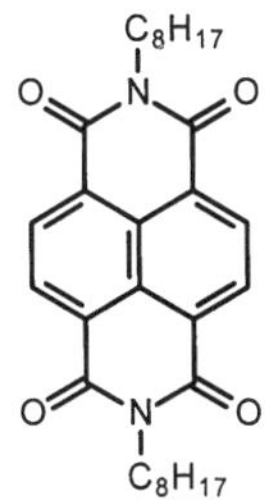

NI

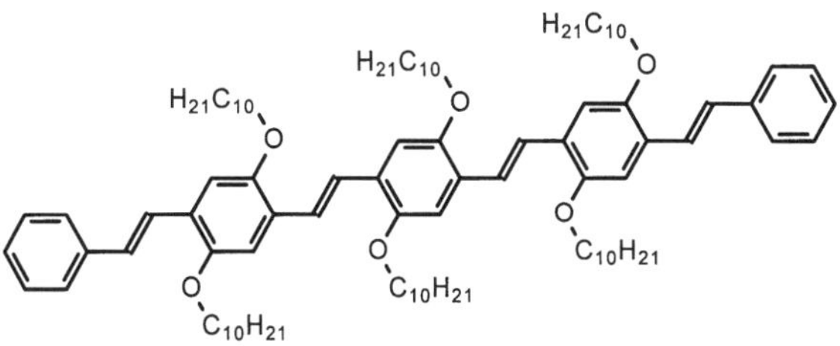

5PV

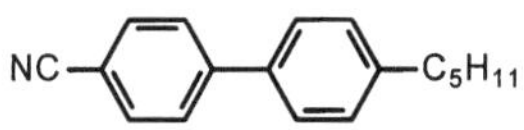

5CB

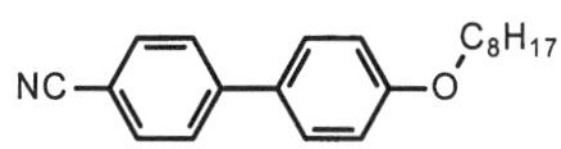

8OCB

Scheme 4

The possibility of inducing drift as a contribution to space-charge field formation was explored by using the conjugated polymer poly(2,5-bis(2′-ethylhexyloxy)-1,4-phenylenevinylene) (BEH-PPV), which shows high through bond electron and hole mobility (92). The 5CB/8OCB eutectic liquid crystal mixture was used, as described above, and doped with 10^{-5} M BEH-PPV (200 kDa by GPC) as the electron donor (93). The molecular weight of the BEH-PPV polymer implies that 500 repeat units of the monomer are present with an extended chain length of 0.35 μm. NI was added as the electron acceptor in a concentration of 8×10^{-3} M (94). Two other liquid crystal composites were also studied as controls. The first control composite contains the 5 unit phenylenevinylene oligomer (5PV) as the electron donor (4×10^{-5} M), along with 8×10^{-3} M NI acceptor, and is designed to avoid the possibility of observing through bond charge transport, i.e. more typical ion diffusion should only be observed in this composite. The second control composite contains 10^{-5} M BEH-PPV polymer with no NI present, designed to show only through bond charge transport, or drift in Equation 17, without ion diffusion.

The results of these studies showed that the direction of asymmetric beam coupling in the BEH-PPV/NI composites inverted at a fringe spacing of 5.5 μm. This suggests that the polarity of the space-charge field inverted at this point owing to a difference in the dominant charge carrier. By comparing the two control composites, neither showed an inversion, but they did show opposite asymmetric beam-coupling directions, which indicates that the dominant charge carrier for the drift contribution to current density were holes, while for diffusion they were anions. Thus for the first time, L_E and $n^{\pm}$ were large enough in the BEH-PPV composites to allow charge drift as a contributor to space-charge field formation in low-molar-mass liquid crystals.

The drift contribution dominates at higher fringe spacings, whereas the diffusion contribution dominates at low fringe spacings because the contribution to the magnitude of the space-charge field derived from ion diffusion is inversely proportional to Λ (23, 54). However, the contribution to the space-charge field magnitude from charge drift in the trap density limited model is independent of Λ (90). The space-charge field due to drift can contribute to the photorefractive grating as long as $2\pi L_E \geq \Lambda$ so that the space-charge field is phase-shifted from the optical interference pattern. Therefore, in the BEH-PPV polymer/liquid crystal composites where L_E is enhanced, the charge drift mechanism dominates space-charge field formation at the larger values of Λ explored in our experiments, but the ion diffusion mechanism dominates at smaller Λ. Furthermore, the fact that the control composite containing only BEH-PPV did not show any beam coupling below 5.5 μm is consistent with the availability of only one charge transport mechanism, charge drift, because the absence of NI^- precludes diffusional space-charge fields at lower values of Λ.

Polymer-Stabilized Liquid Crystals

An entirely different type of phase-separated material results if a small amount (1–2 wt%) of a polymeric monomer is dissolved in an aligned LC and subsequently photo-polymerized with UV radiation in the presence of a photoinitiator. The resulting material is comparable to an anisotropic gel (95–97). Such materials are referred to as polymer-stabilized liquid crystals (PSLCs) and have been discussed most often in reference to the development of shock-resistant LC displays. The birefringence of the nematic LCs is by and large preserved following polymerization because the monomers are rod-like acrylates that align well with the LC director.

PSLCs are a new type of organic photorefractive material (98). These materials are designed to maintain the high birefringence and reorientation characteristics typical of liquid crystals, but permit smaller photorefractive grating fringe spacings by altering the charge transport properties to include trapping effects. The smaller fringe spacing produces a photorefractive grating in liquid crystalline media that operates in the thick grating (Bragg diffraction) regime.

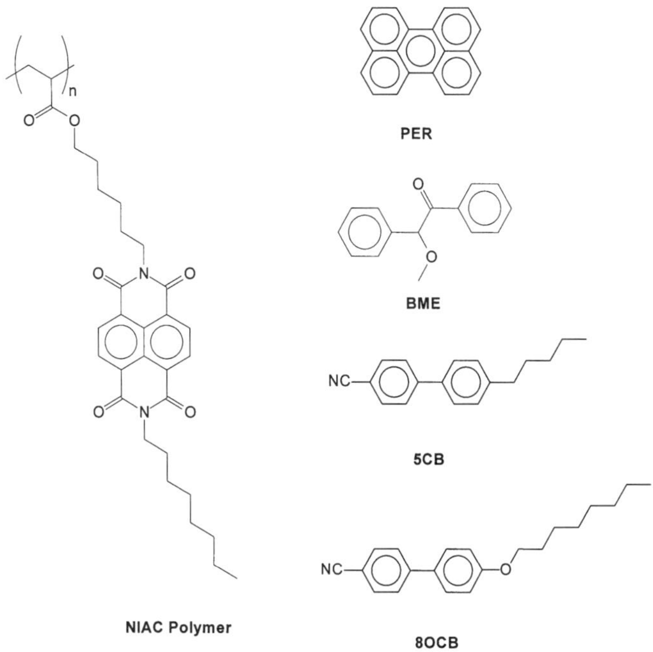

Scheme 5

The nematic LC is the original eutectic mixture of 35% (wt%) 4′-(*n*-octyloxy)-4-cyanobiphenyl (8OCB) and 65% 4′-(*n*-pentyl)-4-cyanobiphenyl (5CB). The sample is doped with perylene and 2% (mol%) of NIAC, which is an acrylate monomer containing the easily reduced 1,4:5,8-naphthalenediimide moiety. Finally, 0.5% (mol%) of benzoin methyl ether (BME) is added to photoinitiate polymerization of the NIAC.

The improved photorefractive grating resolution resulting from polymer stabilization is significant. The nonpolymerized samples do not show any measurable beam coupling for fringe spacings (Λ) below 8 μm, whereas the polymerized samples exhibit beam coupling down to $\Lambda = 2.5$ μm. For our samples, $Q = 10$ is achieved for $\Lambda = 2.6$ μm, when $L = 26$ μm/cos $30^\circ = 30$ μm, and $n = 1.5$. At this fringe spacing, our samples still exhibit a small amount of beam coupling (about $\pm 1\%$), so the PSLCs can be considered to be operative in the thick grating regime. This compares favorably to the best case scenario for low-molar-mass liquid crystals with identical dopants, where beam coupling is observed for fringe spacings no smaller than $\Lambda = 8$ μm, giving $Q = 1$.

CONCLUSIONS

Liquid crystals are proving to be high performance photorefractive materials that are versatile and easily processed. Furthermore, they are outstanding in terms of the very low–applied electric fields and optical fields required to induce large index of refraction changes compared with most other organic photorefractive materials. At present, they are likely to be most useful in dynamic holography applications because of their high nonlinearity and improving response times. Because liquid crystals lack long-lived photorefractive gratings with similarly large index of refraction changes, they are currently not suitable for storage applications. Photorefractive liquid crystals have been proposed for a wide range of applications, including image correlation (pattern recognition) and optical limiting (sensor protection) (43, 45). New composites designed to simultaneously optimize the electro-optic response and space-charge field generation should continue to rapidly improve this young class of optical materials.

ACKNOWLEDGMENT

Support is gratefully acknowledged from the Technology Research Division, Office of Advanced Scientific Computing Research, US Department of Energy, under contract W-31-109-ENG-38.

LITERATURE CITED

1. Ashkin A, Boyd GD, Dziedzic JM, Smith RG, Ballman AA, et al. 1966. *Appl. Phys. Lett.* 9:72
2. Giuliano CR. 1981. *Phys. Today* April:27–35
3. Gunter P, Huignard JP. 1988. *Photorefractive Materials and Their Applications 1: Fundamental Phenomena*. Berlin: Springer-Verlag
4. Moerner WE, Grunnet-Jepsen A, Thompson CL. 1997. *Annu. Rev. Mater. Sci.* 27:585–623
5. Zhang Y, Cui Y, Prasad PN. 1992. *Phys. Rev. B* 46:9900
6. Volodin BL, Kippelen B, Meerholz K, Javidi B, Peyghambarian N. 1996. *Nature* 383:58–60
7. Ducharme S, Scott JC, Twieg RJ, Moerner WE. 1991. *Phys. Rev. Lett.* 66:1846–49
8. Yu LP, Chan WK, Peng A, Gharavi A. 1996. *Acc. Chem. Res.* 29:13–21
9. Bittner R, Daeubler TK, Meerholz K. 1999. *Adv. Mater.* 11:123–27
10. Diaz-Garcia MA, Wright D, Casperson JD, Smith B, Glazer E, et al. 1999. *Chem. Mater.* 11:1784–91
11. Grunnet-Jepsen A, Thompson CL, Twieg RJ, Moerner WE. 1997. *Appl. Phys. Lett.* 70:1515–17
12. Liphardt M, Goonesekera A, Jones BE, Ducharme S, Takacs JM, Zhang L. 1994. *Science* 263:367–69
13. Marder SR, Kippelen B, Jen AK-Y, Peyghambarian N. 1997. *Nature* 388: 845–51

14. Meerholz K, Volodin BL, Sandalph N, Kippelen B, Peyghambarian N. 1994. *Nature* 371:497–500
15. Peng Z, Gharavi AR, Yu LP. 1997. *J. Am. Chem. Soc.* 119:4622–32
16. Wang Q, Wang LM, Yu LP. 1998. *J. Am. Chem. Soc* 120:12860–68
17. Winiarz JG, Zhang L, Lal M, Friend CS, Prasad PN. 1999. *J. Am. Chem. Soc.* 121:5287–95
18. Sutter K, Hulliger J, Schlesser R, Gunter P. 1993. *Opt. Lett.* 18:778–80
19. Lundquist PM, Wortmann R, Geletneky C, Twieg RJ, Jurich M, et al. 1996. *Science* 274:1182–85
20. Hofmann U, Grasruck M, Leopold A, Schreiber A, Schloter S, et al. 2000. *J. Phys. Chem. B* 104:3887–91
21. Zilker SJ, Grasruck M, Wolff J, Schloter S, Leopold A, et al. 1999. *Chem. Phys. Lett.* 306:285–90
22. Khoo IC, Li H, Liang Y. 1994. *Opt. Lett.* 19:1723–25
23. Rudenko EV, Sukhov AV. 1994. *J. Exp. Theor. Phys. Lett.* 59:142–46
24. Wiederrecht GP, Yoon BA, Wasielewski MR. 1995. *Science* 270:1794–97
25. Wiederrecht GP, Yoon BA, Wasielewski MR. 1996. *Adv. Mater.* 8:535–39
26. Wiederrecht GP, Yoon BA, Wasielewski MR. 1997. *Synth. Met.* 84:901–2
27. Wiederrecht GP, Yoon BA, Svec WA, Wasielewski MR. 1997. *J. Am. Chem. Soc.* 119:3358–64
28. Ono H, Saito I, Kawatsuki N. 1998. *Appl. Phys. Lett.* 72:1942–44
29. Miniewicz A, Bartkiewicz S. 1998. *Opt. Commun.* 149:89–95
30. Bartkiewicz S, Miniewicz A, Kajzar F, Zagorska M. 1998. *Appl. Opt.* 37:6871–77
31. Sasaki T, Goto M, Ishikawa Y, Yoshimi T. 1999. *J. Phys. Chem. B* 103:1925–29
32. Tabiryan NV, Umeton C. 1998. *J. Opt. Soc. Am. B* 15:1912–17
33. Ono H, Kawamura T, Frias NM, Kitamura K, Kawatsuki N, Norisada H. 2000. *Adv. Mater.* 12:143–46
34. Miniewicz A, Bartkiewicz S, Kajzar F. 1998. *Nonlinear Opt.* 19:157–75
35. Ducharme S, Scott JC, Twieg RJ, Moerner WE. 1990. Pesented at OSA Annu. Meet. Boston, MA. Postdeadline Pap.
36. Moerner WE, Silence SM. 1994. *Chem. Rev.* 94:127–55
37. Moerner WE, Silence SM, Hache F, Bjorklund GC. 1994. *J. Opt. Soc. Am. B* 11:320–30
38. Psaltis D, Mok F. 1995. *Sci. Am.* November:70
39. Solymar L, Webb DJ, Grunnet-Jepsen A. 1996. *The Physics and Applications of Photorefractive Materials.* Oxford, UK: Clarendon
40. Kippelen B, Marder SR, Hendrickx E, Maldonado JL, Guillemet G, et al. 1998. *Science* 279:54–57
41. Hyde SCW, Barry NP, Jones R, Dainty JC, French PMW, et al. 1995. *Opt. Lett.* 20:1331–33
42. Hyde SCW, Barry NP, Jones R, Dainty JC, French PMW. 1996. *Opt. Commun.* 122:111–16
43. Khoo IC, Wood MV, Shih MY, Chen PH. 1999. *Opt. Exp.* 4:432–42
44. Rudenko EV, Sukhov AV. 1994. *J. Exp. Theor. Phys.* 78:875–82
45. Khoo IC. 1995. *Liquid Crystals: Physical Properties and Nonlinear Optical Phenomena.* New York: Wiley
46. Tabiryan NV, Sukhov AV, Zeldovich BY. 1986. *Mol. Cryst. Liq. Cryst.* 135:1–139
47. Khoo IC. 1995. *Opt. Lett.* 20:2137–39
48. Wang Q, Brubaker RM, Nolte DD, Melloch MR. 1992. *J. Opt. Soc. Am. B* 9:1626–41
49. Khoo IC, Guenther BD, Wood MV, Chen P, Shih M-Y. 1997. *Opt. Lett.* 22:1229–31
50. Collings PJ. 1990. *Liquid Crystals: Nature's Delicate Phase of Matter.* Princeton: Princeton Univ. Press
51. Sen S, Brahma P, Roy SK, Mukherjee DK, Roy SB. 1983. *Mol. Cryst. Liq. Cryst.* 100:327–40

52. Wasielewski MR, Smith RL, Kostka AG. 1980. *J. Am. Chem. Soc.* 102:6923
53. Wiederrecht GP, Svec WA, Niemczyk MP, Wasielewski MR. 1995. *J. Phys. Chem.* 99:8918
54. Khoo IC. 1996. *IEEE J. Quant. Elec.* 32:525–34
55. Tabiryan NV, Umeton C. 1998. *Phys. Rev. E* 58:4619–23
56. Zhang J, Ostroverkhov V, Singer KD, Reshetnyak V, Reznikov Y. 2000. *Opt. Lett.* 25:414–16
57. Zhang G, Montemezzani G, Gunter P. 2000. *J. Appl. Phys.* 88:1709–17
58. Khoo IC, Slussarenko S, Guenther BD, Shih M-Y, Chen P, Wood WV. 1998. *Opt. Lett.* 23:253–55
59. Khoo IC, Shih M-Y, Wood MV, Guenther BD, Chen PH, et al. 1999. *Proc. IEEE* 87:1897–911
60. Wasielewski MR. 1992. *Chem. Rev.* 92:435
61. Marcus RA. 1956. *J. Chem. Phys.* 24:966
62. Gould IR, Ege D, Moser JE, Farid S. 1990. *J. Am. Chem. Soc.* 112:4290–301
63. Gould IR, Noukakis D, Gomez-Jahn L, Young RH, Goodman JL, Farid S. 1993. *Chem. Phys.* 176:439–56
64. Marcus R. 1965. *J. Chem. Phys.* 43:679–701
65. Ono H, Kawatsuki N. 1999. *Opt. Lett.* 24:130–32
66. Ono H, Kawatsuki N. 1999. *J. Appl. Phys.* 85:2482–87
67. Ono H, Kawatsuki N. 1999. *Jpn. J. Appl. Phys.* 38:737–40
68. Ono H, Kawatsuki N. 1999. *Appl. Phys. B* 69:51–53
69. Ono H, Hanazawa A, Kawamura T, Norisada H, Kawatsuki N. 1999. *J. Appl. Phys.* 86:1785–90
70. Ono H, Kawamura T, Frias NM, Kitamura K, Kawatsuki N, Norisada H. 1999. *Appl. Phys. Lett.* 75:3632–34
71. Ono H, Kawamura T, Kawatsuki N, Norisada H, Yamamoto T. 1999. *Jpn. J. Appl. Phys.* 38:L1258–60
72. Eichler HJ, Gunter P, Pohl DW. 1986. *Laser-Induced Dynamic Gratings*. Berlin: Springer-Verlag
73. Ja YH. 1981. *Elect. Lett.* 17:488–89
74. Kondilenko V, Markov V, Odulov S, Soskin M. 1979. *Opt. Acta* 26:239
75. Rupp RA, Maillard A, Walter J. 1989. *Appl. Phys. A* 89:259
76. Sutter K, Gunter P. 1990. *J. Opt. Soc. Am. B* 7:2274
77. Walsh CA, Moerner WE. 1992. *J. Opt. Soc. Am. B* 9:1642
78. Golemme A, Volodin BL, Kippelen B, Peyghambarian N. 1997. *Opt. Lett.* 22:1226–28
79. Ono H, Kawatsuki N. 1997. *Opt. Lett.* 22:1144–46
80. Cipparrone G, Mazzulla A, Simoni F. 1997. *Mol. Cryst. Liq. Cryst* 299:329
81. Cipparrone G, Mazzulla A, Simoni F. 1998. *Opt. Lett.* 23:1505
82. Cipparrone G, Mazzulla A, Nicoletta FP, Lucchetti L, Simoni F. 1998. *Opt. Comm.* 150:297–304
83. Simoni F, Cipparrone G, Mazzulla A, Pagliusi P. 1999. *Chem. Phys.* 245:429–36
84. Golemme A, Kippelen B, Peyghambarian N. 1998. *Appl. Phys. Lett.* 73:2408–10
85. Golemme A, Kippelen B, Peyghambarian N. 2000. *Chem. Phys. Lett.* 319:655–60
86. Ono H, Kawatsuki N. 1997. *Jpn. J. Appl. Phys.* 36:6444–48
87. Grunnet-Jepsen A, Wright D, Smith B, Bratcher MS, DeClue MS, et al. 1998. *Chem. Phys. Lett.* 291:553–61
88. Zhang J, Singer KD. 1998. *Appl. Phys. Lett.* 72:2948–50
89. Wiederrecht GP, Niemczyk MP, Svec WA, Wasielewski MR. 1999. *Chem. Mater.* 11:1409–11
90. Gunter P. 1982. *Phys. Rep.* 93:199–299
91. Maximus B, Ley ED, Meyere AD, Pauwels H. 1991. *Ferroelectrics* 121:103–12
92. Hoofman RJOM, Haas MP, Siebbeles LDA, Warman JM. 1998. *Nature* 392:54–56
93. Wang B, Wasielewski MR. 1997. *J. Am. Chem. Soc.* 119:12–21

94. Greenfield SR, Svec WA, Gosztola D, Wasielewski MR. 1996. *J. Am. Chem. Soc.* 118:6767–77
95. Guymon CA, Hoggan EN, Walba DM, Clark NA, Bowman CN. 1995. *Liq. Cryst.* 19:719–27
96. Hikmet RAM. 1990. *J. Appl. Phys.* 68: 4406–12
97. Rajaram CV, Hudson SD, Chien LC. 1995. *Chem. Mater.* 7:2300–8
98. Wiederrecht GP, Wasielewski MR. 1998. *J. Am. Chem. Soc.* 120:3231–36

Annu. Rev. Mater. Res. 2001. 31:171–81

PHOTOINITIATED POLYMERIZATION OF BIOMATERIALS

John P Fisher[1], David Dean[2], Paul S Engel[3], and Antonios G Mikos[1,4]

[1]*Department of Bioengineering, Rice University, P.O. Box 1892, MS 142, Houston, Texas 77251-1892; e-mail: fisherj@rice.edu*

[2]*Department of Neurological Surgery, Case Western Reserve University, and The Research Institute, University Hospitals of Cleveland, 11100 Euclid Avenue, Cleveland, Ohio 44106-5012; e-mail: dxd35@po.cwru.edu*

[3]*Department of Chemistry, Rice University, P.O. Box 1892, MS 60, Houston, Texas 77251-1892; e-mail: engel@rice.edu*

[4]*Department of Chemical Engineering, Rice University, P.O. Box 1892, MS 362, Houston, Texas 77251-1892; e-mail: mikos@rice.edu*

Key Words photochemistry, photoinitiators, tissue engineered materials

■ **Abstract** Photoinitiated polymerization and polymer crosslinking are viable strategies for biomaterial synthesis because of the mild temperatures and neutral pH environments in which these reactions typically take place. This review summarizes the relevant theories as well as current status of photoinitiated polymerizations in biomaterials. Photoinitiation, photoinitiated polymerization, and photoinitiators are discussed with consideration toward the biological nature of the intended application. Recent investigations into biomaterials, including hydrogels, biodegradable materials, and hard tissue resorbable scaffolds are presented. Lastly, studies of cell interactions with photoinitiated biomaterials are discussed. The work herein illustrates the potential use of photoinitiated polymerization in the development of novel biomaterials for tissue engineering.

PHOTOINITIATION

Photochemistry is concerned with chemical reactions induced by optical radiation (1–3). The radiation is most often ultraviolet (200–400 nm) or visible (400–800 nm) light but is sometimes infrared (800–2500 nm) light. Photochemistry may be used to induce any number of chemical processes, from initiating a chemical reaction to degrading a compound. The focus here is on photoinitiation of polymerization. Two basic steps characterize photoinitiation: the absorption of light to excite a compound and the resulting photochemical reaction of the excited compound (2). These processes are briefly examined, with following sections considering photoinitiated polymerization of biomaterials.

Absorption of a photon of light by any compound causes electronic excitation. The energy causing excitation, E, is described by $E = hc/\lambda$, where h is Planck's constant, c is the speed of light, and λ is the wavelength of the exciting light. Light absorption is described by $a = \varepsilon Cl$, where ε is the molar absorptivity (extinction coefficient), C is the concentration of the species, and l is the light path length. The extinction coefficient, a constant for a compound at a specific wavelength, is an experimental measure of the probability of absorption at that wavelength. The magnitude of the extinction coefficient depends upon the compound's chromophore, the chemical moiety responsible for the absorption of light. Typical chromophores contain unsaturated functional groups such as C=C, C=O, NO_2, or N=N (2, 4). Table 1 lists some chromophores, their wavelength of maximum absorption, and the extinction coefficient at this wavelength (2, 4). These values are qualitative because chromophore absorption is highly dependent upon neighboring substituents. For example, the absorption maximum and extinction coefficient of conjugated dienes are known to be influenced by the number of conjugated double bonds, alkyl substituents, and ring structure (4). Knowledge of the absorption characteristics of the individual species aids in the design and development of pre-fabricated photoinitiated implants, whether of bioinert or tissue-engineered (resorbable) graft materials, by allowing proper choice of initiator concentration and radiation source.

TABLE 1 Local wavelength of maximum absorption and associated extinction coefficient for typical chromophores found in photolabile compounds (2, 4)

Chromophore	λ_{max} (nm)	ε_{max}
C=C	195	10,000
C≡C	195	2000
C=C–C=C	215	20,000
Benzene	185	60,000
	200	8000
	255	200
RRC=O	190	900
	280	15
RCOOH	205	60
N=N	345	10
NO_2	270	18.6
S=O	210	1500

Any wavelength of radiation that falls within the absorption spectrum of a chromophore can effectively excite that chromophore. Excitation is characterized by the promotion of a valence electron from a bonding σ or π orbital or a nonbonding n orbital to an antibonding σ^* or π^* orbital (2, 4). The electronic excitation of photochemical reactions contrasts with thermal reactions, which typically involve translational, rotational, and vibrational excitation (1). Photochemical reactions often localize their excitation on one chromophore and produce (*a*) high levels of electronic excitation, (*b*) alterations in chemical reactivity, and (*c*) relatively low levels of thermal degradation (1). Frequently, the result is to allow photochemical reactions to occur at a mild temperature and neutral pH, often ineffective conditions for thermal systems. These features of photochemistry may provide advantages over thermally cured methods in the production of tissue-engineered scaffolds.

An excited compound (AB^*) can dissociate to fragments (A + B), react with another compound to yield a new compound (ABC), isomerize (BA), luminesce ($AB + h\nu$), ionize ($AB^+ + e^-$), or decay without radiation (AB) (1). All these processes are rapid, often occurring on the nanosecond time scale. Additionally, the primary photochemical step is often a precursor to secondary processes. For most photoinitiated polymerizations, the excited chromophore dissociates into radicals, and these radicals react with monomers to form radical species, which then propagate radical polymerization. A number of different classes of photoinitiatable polymerization reactions as well as photoinitiators are reviewed next.

PHOTOINITIATED POLYMERIZATION

Polymerizations may employ photoinitiation to effectively produce reactive species that initiate polymerization. Specifically, photoinitiated production of a radical is often used to begin radical chain polymerization, whereas photoinitiated production of an ionic monomer may be used in ionic chain polymerizations. In contrast, the photochemically active species can be the monomer itself, leading to polymerization of the monomer alone. Finally, photochemistry can be used in the crosslinking of already formed polymer chains following a radical chain polymerization mechanism. These photoinitiated processes are each briefly examined below.

Photoinitiated radical chain polymerization begins with light irradiation of a properly sensitive compound (2, 5). This activated compound typically fragments into two radicals that attack a monomer, converting it to a radical. The monomer radical attacks a second monomer and this process continues to add monomer units, thereby forming the polymer. The reaction is terminated either by the reaction with another radical (coupling) or by the transfer of a β-hydrogen from one radical to another, producing a saturated and an unsaturated compound (disproportionation). As in thermal polymerizations, the initiator system used in radical photopolymerization may be either a single initiator compound, which dissociates upon light absorption, or an oxidation-reduction reaction that is photoinitiated. The redox strategy within photoinitiated systems involves either (*a*) excitation of the electron donating or accepting species, which then undergoes electron transfer to form a radical pair, or (*b*) excitation of the donor-acceptor pair (often referred to as the charge transfer complex) to also yield a radical ion pair. In either event, redox photopolymerization follows from that radical ion pair, in a solvent-separated form or contact ion form (2).

Photoinitiated ionic chain polymerization is similar to radical chain polymerization but is characterized by the propagation of anionic or cationic species (5). Initiation can be brought about with a single photoinitiator or with a photolabile initiator/co-initiator pair, producing an electron-donating compound (cationic polymerization) or an electron-withdrawing compound (anionic polymerization). Propagation then occurs with the addition of monomers to the initial ion pair. Termination of polymerization can occur by the transfer of the ion pair to a monomer (which terminates the growth of one polymer chain, but starts the growth of another), the transfer of the ion pair to a polymer, combination with a counterion, or by other nonspecific mechanisms. The instability of the ionic species confines ionic chain polymerizations to systems that promote the ion's persistence, such as those run at low temperatures and/or in polar solvents.

Photopolymerization of a pure monomer is restricted to those compounds that are both photolabile and polymerizable. The vinyl monomers styrene, methyl methacrylate, and acrylonitrile meet both criteria, but their polymerization efficiency is low (2). For example, styrene irradiated at 313 nm produces small amounts of polystyrene in addition to larger amounts of styrene oligomers (2). Other compounds require surfaces, photoinitiators, impurities, or oxygen to obtain appreciable rates of monomer photopolymerization (2). Although of interest, photopolymerization of pure monomers severely restricts the choice of components and therefore possible applications.

Photocrosslinking of polymers can be carried out following either radical or ionic polymerization, but this is the reaction of polymers instead of monomers. The product of photocrosslinking is ideally a single, three-dimensional, covalently bonded polymer network; this is in contrast to photopolymerization's production of physical entanglements of many discrete polymer chains. Both photocrosslinking and photopolymerization are valuable, but because the properties of their products are often different, each has specific applications.

PHOTOINITIATORS

A number of issues must be considered when choosing a photoinitiator for the polymerization of a biomaterial. These concerns are not limited to the photoinitiator but extend to interactions between the available light source and other chemical species within the proposed implant production system. Specifically, the peak absorption wavelength and extinction coefficient of the initiator, the reactivity of the initiator toward the other species within the system, and the kinetics of the photoinitiator all determine the applicability of the system. A few classes of photoinitiators are reviewed that have seen use in biomaterials. [A more complete presentation of photoinitiators can be found in Rabek (2) and Fouassier (6).] The final choice of initiator, however, is often based upon similarly reported systems in the literature or prior experience.

Peroxides (ROOR) are effective photoinitiators that absorb in the ultraviolet region from 200 to 300 nm but without a significant absorption peak (2, 6). The absorption of light causes peroxides to degrade by cleavage of the O–O bond, forming alkoxy radicals (RO•) (2). The hydroxyl radicals (HO•) formed from hydrogen peroxide have been used to polymerize methyl methacrylate as well as poly(*N*-isopropylacrylamide) (2, 7). Benzoyl peroxide (C_6H_5–CO–O_2–CO–C_6H_5), a common thermal radical initiator, is also a photoinitiator that has a local wavelength of maximum absorption at 230 and 273 nm. Benzoyl peroxide's benzoyloxy radicals (C_6H_5–CO–O•), and the derived phenyl radicals (C_6H_5•), have been shown to polymerize methyl methacrylate and styrene (2).

Ketones (RCOR) make up an extremely large class of photoinitiators. Some of the most commonly used ketones are related to acetophenone (C_6H_5–CO–CH_3), which has a large absorbance peak at 280 nm and a smaller peak at 330 nm (8–13). These initiators typically form benzoyl radicals (C_6H_5–CO•) after α-cleavage of the C–C bonds; the benzoyl radicals are thought to be primarily responsible for polymerization (6). Benzoin (C_6H_5–CHOH–CO–C_6H_5), a phenylated derivative, has an absorption peak at approximately 315 nm, which often shifts to higher wavelengths with substitutions (2). The frequently used 2,2-dimethoxy-2-phenylacetophenone enhances radical polymerization because of both its additional electron-donating groups and its ability to undergo a secondary light absorption and excitation. The result is a more potent initiator because four radical fragments, instead of the typical two, are produced. Acylphosphineoxides (C_6H_5–CO–PO–$(C_6H_5)_2$) have been reported for dental applications, whereas compounds containing sulfur, such as dibenzoyl disulphides [(C_6H_5)–CO–S_2–CO–(C_6H_5)] or diphenyldithiocarbonate [(C_6H_5)–S–CO–S–(C_6H_5)], have been used in methyl methacrylate photopolymerization (2, 14).

Azo compounds have found use as photoinitiators, as the azo group (−N=N−) absorbs at approximately 350 nm to produce two reactive carbon-centered radicals (2, 15). 2,2′-Azobisisobutyronitrile [AIBN, $(CH_3)_2C(CN)$–N=N–$C(CN)(CH_3)_2$] has been used with significant success (2). However, the low absorptivity and associated low polymerization rate are the chief disadvantages of azo compounds (2).

Finally, a number of longer wavelength systems have been developed. The use of quinones, especially camphorquinone and the dye eosin in the presence of an amine, has been reported with regularity in the biomaterials literature. Both systems utilize blue light (420 to 500 nm) and find significant success in dental applications (16).

PHOTOINITIATED BIOMATERIALS

Many investigations into photoinitiated materials have explored strategies to create a biomaterial that attempts to replace a lost function or to add a new function. Examples of the former include dental and cardiovascular materials; the latter include drug delivery systems and biological adhesives. Photochemistry is an attractive strategy in the production of these materials because it has a controlled reaction initiation and termination, an often short reaction time, and spatial control. Additionally, photochemical reactions are often run at gentle conditions, such as room temperature or under physiologic temperature and pH, allowing both the incorporation of biologically active molecules and in vivo reactions without harm. While photochemistry does offer advantages for biomaterial synthesis, a number of issues must be resolved, such as the incorporation of degradable elements, the identification of water soluble components, the insurance of nontoxic unpolymerized material or initiator residues, and the optimization of cell viability. In this section investigations into photoinitiated biomaterials are reviewed, with an emphasis on the photochemical aspects of each.

Hydrogels

Considerable work has been devoted to the development of photoinitiated hydrogels. The main difficulty lies in the development of water-soluble components that are both functional and photolabile. Three basic strategies have been followed: (*a*) the use of photolabile water-soluble polymers, (*b*) the addition of photolabile groups to hydrophilic polymers, and (*c*) the use of water-soluble photoinitiators with hydrophilic polymers. The first two strategies would employ a photopolymerization of a pure monomer or photocrosslinking of a pure polymer scheme, whereas the latter would typically follow a radical polymerization or crosslinking scheme.

While for a given polymer multiple strategies may produce a hydrogel, few may be viable for the desired application. For example, poly(*N*-isopropylacrylamide) is photocrosslinkable alone at low wavelengths (<290 nm) but forms hydrogels of significantly lower sol fraction when using a hydrogen peroxide/potassium persulfate system and longer UV wavelengths (>290 nm) (7). The crosslinking of the pure polymer is an intriguing approach because it may reduce toxicity by simply reducing the number of components required in the system. However, Kubota & Fukuda (7) show that this often leads to undesirable properties, such as in this case a large sol content. An alternative to an innately photolabile polymer is a hydrophilic polymer modified by the addition of a photolabile group. Poly(ethylene glycol) (PEG) has been modified with photolabile cinnamylidene acetyl groups that, interestingly, are both photolabile and crosslinkable (17, 18). Thus PEG hydrogels can be formed by cinnamylidene acetyl crosslinks. Similarly, chitin has been modified with photolabile azide groups to form a noncytotoxic biological adhesive (15). Upon UV irradiation, the photolabile azide groups release N_2 while becoming highly reactive to nitrogen atoms and thus allowing crosslinking of the chitin. Thus with creative chemical approaches, photoinitiated hydrogels can be formed with considerable success without photoinitiators.

In contrast, hydrogel systems that utilize a discrete photoinitiator possess the advantage of allowing the polymer, or in some cases monomer, to be available for modification. PEG-modified systems crosslinked with a photoinitiator, typically an acetophenone/UV light or eosin Y and triethanolamine/visible blue light system, have been extensively investigated for this purpose (11, 19–25). These systems seek to combine the advantageous properties of a hydrophilic polymer and its resulting hydrogel, a functional group that promotes a desired outcome, and a photoinitiated system. Results have shown that modified PEG can be used as a drug-delivery system (19, 25), lubricant (11), nonadhesive cell surface (20), and an inhibitor to thrombosis (21–24). This strategy is also applicable to other hydrophilic polymers. For example, the hydrophilic monomer 2-hydroxyethyl methacrylate (HEMA) is photopolymerized in the production of hydrogel lenses. It has been shown that poly(HEMA) is effectively produced using the photoinitiator 2,2-dimethoxy-2-phenyl acetophenone (8, 9). Photoinitiated polymerization of HEMA and diethylene glycol dimethacrylate has also been reported in the formation of a hydrogel drug-delivery system (26). Finally, hydrogels have been photoinitiated from modified polymers such as poly(L-lactic acid-*co*-glycolic acid-*co*-L-serine), again using an acetophenone initiator (10). While any of the three strategies outlined above are possible means for the synthesis of photoinitiated biomaterials, the use of photoinitiators probably allows the greatest versatility and therefore widest application.

Biodegradable Materials

Development of biodegradable, biocompatible, photoinitiated compounds is often difficult for it typically requires the incorporation of a biodegradable polymer, such as poly(lactic acid), poly(glycolic acid), or a polyester, and a photoinitiator. Nevertheless, investigators have recently begun development of these materials with success. Copolymers of the degradable polyesters poly(ethylene glycol), poly(propylene glycol), and poly(tetramethylene glycol) and acrylated oligomers of D,L lactic acid were synthesized with the photoinitiator 2,2 dimethoxy-2-phenylacetophenone in *N*-vinylpyrrolidinone to form biodegradable networks (11). Furthermore, Matsuda et al (27) have shown that biodegradable copolymers of ϵ-caprolactone and trimethylene carbonate can be photopolymerized with high-spatial control, allowing stereolithographically prepared surfaces to be synthesized. While a degree of success has been obtained, further work is required to characterize the properties of these materials.

The property many tissue-engineering investigators are often most interested in controlling is the material's degradation rate. The flexibility of photopolymerization offers a variety of schemes to obtain this control. Burkoth et al (12) show that the degradation rate of photoinitiated biodegradable systems (using the acetophenone photoinitiator 2,2-dimethoxy-2-phenylacetophenone) can be tailored by controlling the hydrophobicity of the crosslinked polymer. Similarly, an acetophenone initiator, *N*-vinylpyrrolidinone, and a dimethacrylate monomer can form degradable hydrogels, with degradation controlled by the monomer structure (13). Finally, Gursel et al (28) controlled the degradation rate of azoisobutyronitrile photopolymerized materials by varying the composition of a degrading polymer (polyhydroxybutyrate-co-hydroxyvalerate) and a nondegrading polymer (polyhydroxyethyl methacrylate). Although a variety of biodegradable materials have been successfully synthesized using photoinitiation, considerable room for further developments remains.

Hard Tissue Materials

Since the 1960s, photopolymerized materials have been extensively used in dental applications because they possess advantageous properties such as low shrinkage, low water absorption, and good adhesion (29–31). A typical photoinitiated dental material is comprised of four elements: a monomer, a diluent, a photoinitiator, and an inorganic filler (32). The monomer is often a dimethacrylate, especially 2,2-bis[4-(2′-hydroxy-3′-methacryloyl-oxypropoxy)phenyl]propane (bis-GMA) or 1,6-bis(2′-methacryloyl-oxyethoxycarbonylamino)-2,4,4-trimethylhexane (UDMA) (32). The dimethacrylate monomer bis-GMA has been used clinically, although it is associated with some degradation owing to its hydroxyl groups. Ahn et al examined the effect of substitution of these groups with methacryloyl groups, showing a 90% reduction of water sorption and therefore a likely reduction in degradation (16). The photoinitiator system of choice, camphorquinone (CQ) and an amine such as *N*,*N*,3,5-tetramethyl aniline (TMA), is excited by visible blue light (420–500 nm) (16). The development of orthopedic photoinitiated materials has largely grown from the dental research. A significant deviation, however, is that recent materials proposed for orthopedic applications have been devised so that they will degrade over time, ideally allowing bone to reclaim the defect area. A competing factor is the mechanical strength required by the material in order to withstand the forces carried within load-bearing sites or placed upon protective/supporting bone. Anseth and her associates have developed methacrylated anhydride monomers, based upon sebacic acid, 1,6-bis(p-carboxy phenoxy) hexane, or 1,3-bis(p-carboxy phenoxy) propane, that possess mechanical strength and degrade over periods from weeks to a year (33–37). The methacrylated anhydride monomers may be photopolymerized with a camphorquinone/ethyl-4-*N*,*N*-dimethylaminobenzoate system and visible blue light. Degradation rate can be controlled by alterations in the components and composition that make up the methacrylated anhydride monomer. Additionally, it was shown that hydrolytic degradation occurs from the outer surface toward the interior (surface degradation), in contrast to bulk degradation where there is no spatial dependence owing to the hydrophobicity of the network (37). The development of photopolymerized hard tissue materials may allow for techniques such as in vivo curing and stereolithography printing to better match the patient's imaged anatomy (38–40).

Cell Interactions with Photoinitiated Biomaterials

Recent biomaterial research has shown that photoinitiated materials are suitable platforms for cell growth. As an initial toxicity study, Byrant et al (41) investigated the cytocompatibility of a variety of photoinitiators using a fibroblast cell line. Results showed that a fibroblast cell line is viable with not only direct application of ultraviolet radiation but also in the presence and excitation of a number of photoinitiators, including 1-hydroxycyclohexyl phenyl ketone, 2-methyl-1-[4-(methylthio) phenyl]-2-(4-morpholinyl)-1-propanone, and 2-hydroxy-1-[4-(hydroxyethoxy) phenyl]-2-methyl-1-propanone. In another investigation, chitin modified with both azide and lactose groups was photocrosslinked, through UV excitation of the azide groups and shown to permit cell growth adjacent to the material (15). No cell proliferation, however, was observed on the material itself. Orban et al (42) has studied acrylate functionalized phospholipids on alkylated supports for cardiovascular applications, showing that the visible light initiator system of eosin Y and triethanolamine had little effect upon cell viability.

In addition to cell viability tests, many have investigated the ability of photoinitiated materials to promote or inhibit cell adhesion. Typically, a hydrophilic polymer such as PEG is incorporated in order to inhibit cell adhesion (20). This can be further extended by incorporating a hydrophobic copolymer, such as polystyrene, by photoinitiated copolymerization. The hydrophobic polystyrene allows strong bonding of the copolymer to a poly(ethylene terephthalate) surface, whereas the hydrophilic PEG is highly wettable and cell-adhesion resistant (43). PEG can also be functionalized with bioactive molecules to promote cell adhesion. Hern et al (44) showed that peptide derivitized PEG can be photopolymerized and, depending upon the inclusion of spacer molecules, selective for cell adhesion, as measured by cell spreading. Specifically, PEG without a peptide sequence attached showed little cell spreading, PEG with a peptide attached but without a spacer linkage showed cell spreading nonspecific to the attached peptide sequence, and PEG with a spacer and peptide attached showed cell spreading specific to the peptide sequence. Although the cellular studies discussed to this point have shown that photoinitiated materials can be synthesized to maintain cell viability as well as promote cell adhesion, these studies have not investigated photoinitiated processes in the presence of cells.

Photopolymerized materials for cell encapsulation have been investigated primarily as tissue engineering constructs but have additionally shown the efficacy of photochemistry in biological systems. Islet cell encapsulation for pancreas tissue–engineering and chondrocyte encapsulation for cartilage tissue–engineering have been reported (45–48). Studies with islet cells have shown that photopolymerization of polyethylene glycol diacrylate hydrogels with an argon laser, utilizing eosin Y, triethanolamine, and 1-vinyl-2-pyrrolidinone, is sufficiently gentle to allow approximately 90% of the cells to survive 24 h after the crosslinking reaction (45, 46). Furthermore, it was shown that cell viability could be maintained at this level while increasing the rate of crosslinking that, in turn, improves the efficiency at which the cells are encapsulated. In an extremely effective use of photoinitiation techniques, Elisseeff et al (47, 48) showed that poly(ethylene oxide) hydrogels containing chondrocytes could be formed via transdermal photoinitiation. Ultraviolet light exposure to the surface of the skin polymerized and encapusulated a subdermally injected cell/polymer mixture. Their results show that implants at 2, 4, and 7 weeks exhibited collagen and proteoglycan production, indicating that UV exposure not only permits cell viability but also has no deleterious effect on cell function. These works may guide the future of photoinitiated biomaterials toward more tissue-engineering applications.

SUMMARY

A wide range of biomaterials can be synthesized by photoinitiated polymerization. This review has shown that well-defined hydrogels, degradable materials, and hard tissue materials all have been successfully constructed by using photolabile monomers, photolabile polymers, or photoinitiators. Furthermore, recent studies with tissue-engineering constructs have shown that many of these materials provide suitable substrates for cell encapsulation with maintenance of high levels of cell viability.

ACKNOWLEDGMENTS

This work was partially supported by a National Institutes of Health grant (R01-DE13740) to DD and AGM.

Visit the Annual Reviews home page at www.AnnualReviews.org

LITERATURE CITED

1. Wayne RP. 1988. *Principles and Applications of Photochemistry*. Oxford, UK: Oxford Univ. Press
2. Rabek JF. 1987. *Mechanisms of Photophysical Processes and Photochemical Reactions in Polymers*. New York: Wiley & Sons
3. Calvert JG, Pitts JN. 1966. *Photochemistry*. New York: Wiley & Sons
4. Silverstein RM, Bassler GC, Morrill TC, eds. 1991. *Spectrometric Identification of Organic Compounds*. New York: Wiley & Sons. 5th ed.
5. Odian G. 1991. *Principles of Polymerization*. New York: Wiley & Sons
6. Fouassier JP. 1995. *Photoinitiation, Photopolymerization, and Photocuring*. Cincinnati, OH: Hanser/Gardner
7. Kubota H, Fukuda A. 1997. *J. Appl. Polym. Sci.* 65:1313–18
8. Lai YC, Quinn ET. 1997. In *ACS Symposium Series: Photopolymerization: Fundamentals and Applications*, ed. AB Scranton, CN Bowman, RW Peiffer, 673:35–50. Washington DC: Am. Chem. Soc.
9. Lai YC. 1997. *J. Appl. Polym. Sci.* 66:1475–84
10. John G, Morita M. 1999. *Macromolecules* 32:1853–58
11. Kim BS, Hrkach JS, Langer R. 2000. *Biomaterials* 21:259–65
12. Burkoth AK, Burdick J, Anseth KS. 2000. *J. Biomed. Mater. Res.* 51:352–59
13. Bruining MJ, Edelbroek-Hoogendoorn PS, Blaauwgeers HGT, Mooy CM, Hendriks FH, Koole LH. 1999. *J. Biomed. Mater. Res.* 47:189–97
14. Guggenberger R, Lechner G, Weinmann W. 1997. *Polym. Preprints* 38:105
15. Ono K, Saito Y, Yura H, Ishikawa K, Kurita A, et al. 2000. *J. Biomed. Mater. Res.* 49:289–95
16. Ahn KD, Chung CM, Kim YH. 1999. *J. Appl. Polym. Sci.* 71:2033–37
17. Andreopoulos FM, Deible CR, Stauffer

MT, Weber SG, Wagner WR, et al. 1996. *J. Am. Chem. Soc.* 118:6235–40
18. Andreopoulos FM, Beckman EJ, Russell AJ. 1998. *Biomaterials* 19:1343–52
19. Scott RA, Peppas NA. 1999. *Biomaterials* 20:1371–80
20. Han DK, Hubbell JA. 1997. *Macromolecules* 30:6077–83
21. Hill-West JL, Chowdhury SM, Slepian MJ, Hubbell JA. 1994. *Proc. Natl. Acad. Sci. USA* 91:5967–71
22. Hill-West JL, Dunn RC, Hubbell JA. 1995. *J. Surg. Res.* 59:759–63
23. Hill-West JL, Hubbell JA. 1996. *Proc. Natl. Acad. Sci. USA* 93:13188–93
24. Chowdhury SM, Hubbell JA. 1996. *J. Surg. Res.* 61:58–64
25. An Y, Hubbell JA. 2000. *J. Control. Release* 64:205–15
26. Lu S, Anseth KS. 1999. *J. Control. Release* 57:291–300
27. Matsuda T, Mizutani M, Arnold SC. 2000. *Macromolecules* 33:795–800
28. Gursel I, Balcik C, Arica Y, Akkus O, Akkas N, Hasirci V. 1998. *Biomaterials* 19:1137–43
29. Cook WD. 1992. *Polymer* 33:600–9
30. Venhoven BAM, de Gee AJ, Davidson CL. 1996. *Biomaterials* 17:2313–18
31. Maffezzoli A, Della Pietra A, Rengo S, Nicolais L, Valletta G. 1994. *Biomaterials* 15:1221–28
32. Chung CM, Kim JG, Choi JH. 2000. *J. Appl. Polym. Sci.* 77:1802–8
33. Anseth KS, Shastri VR, Langer R. 1999. *Nat. Biotechnol.* 17:156–59
34. Anseth KS, Svaldi DC, Laurencin CT, Langer R. 1997. In *ACS Symp. Series: Photopolymerization: Fundamentals and Applications*, ed. AB Scranton, CN Bowman, RW Peiffer, 673:189–202. Washington DC: Am. Chem. Soc.
35. Young JS, Gonzales KD, Anseth KS. 2000. *Biomaterials* 21:1181–88
36. Muggli DS, Burkoth AK, Keyser SA, Lee HR, Anseth KS. 1998. *Macromolecules* 31:4120–25
37. Muggli DS, Burkoth AK, Anseth KS. 1999. *J. Biomed. Mater. Res.* 46:271–78
38. D'Urso PS, Earwaker WJ, Barker TM, Redmond MJ, Thompson RG, et al. 2000. *Br. J. Plast. Surg.* 53:200–4
39. Fallahi B, Foroutan M, Motavalli S, Dujovny M, Limaye S. 1999. *Neurol. Res.* 21:281–86
40. Eufinger H, Wehmoller M, Scholz M, Harders A, Machtens E. 1999. *Plast. Reconstr. Surg.* 104:198–203
41. Bryant SJ, Nuttelman CR, Anseth KS. 2000. *J. Biomater. Sci. Polym. Ed.* 11:439–57
42. Orban JM, Faucher KM, Dluhy RA, Chaikof EL. 2000. *Macromolecules* 33:4205–12
43. Nakayama Y, Miyamura M, Hirano Y, Goto K, Matusda T. 1999. *Biomaterials* 20:963–70
44. Hern DL, Hubbell JA. 1998. *J. Biomed. Mater. Res.* 39:266–76
45. Cruise GM, Hegre OD, Scharp DS, Hubbell JA. 1998. *Biotechnol. Bioeng.* 57:655–65
46. Cruise GM, Scharp DS, Hubbell JA. 1998. *Biomaterials* 19:1287–94
47. Elisseeff J, Anseth K, Sims D, McIntosh W, Randolph M, Langer R. 1999. *Proc. Natl. Acad. Sci. USA* 96:3104–7
48. Elisseeff J, McIntosh W, Anseth K, Riley S, Ragan P, Langer R. 2000. *J. Biomed. Mater. Res.* 51:164–71

Annu. Rev. Mater. Res. 2001. 31:183–201

FUNCTIONAL BIOMATERIALS: Design of Novel Biomaterials

SE Sakiyama-Elbert[1] and JA Hubbell[2]

[1]*Department of Biomedical Engineering, Washington University, St. Louis, Missouri 63130-4899; e-mail: sakiyama@biomed.wustl.edu*

[2]*Institute for Biomedical Engineering and Department of Materials, Swiss Federal Institute of Technology and University of Zurich, CH-8044 Zurich, Switzerland; e-mail: hubbell@biomed.mat.ethz.ch*

Key Words drug delivery, cell adhesion, gene delivery

■ **Abstract** The field of biomaterials has recently been focused on the design of intelligent materials. Toward this goal, materials have been developed that can provide specific bioactive signals to control the biological environment around them during the process of materials integration and wound healing. In addition, materials have been developed that can respond to changes in their environment, such as a change in pH or cell-associated enzymatic activity. In designing such novel biomaterials, researchers have sought not merely to create bio-inert materials, but rather materials that can respond to the cellular environment around them to improve device integration and tissue regeneration.

INTRODUCTION

Important advances have recently been made in the development of biomimetic materials that are capable of interacting specifically with their biological environment. This interaction has been accomplished by harnessing the principles involved in biological recognition between protein and proteoglycan receptors on cell surfaces with other biological molecules normally present in the extracellular matrix (ECM), on another cell, or in solution. These biologically active materials can play a pivotal role in tissue engineering by allowing initially cell-free materials to elicit cellular infiltration within an implanted material and to direct the formation of new tissues within the material at the local implant site. In this approach, the implanted materials function both as a mechanical scaffold for tissue regeneration and as a pharmacological agent that alters the course of regeneration via the transmission of biological signals to cells in the surrounding tissue.

For example, biomimetic materials can be used to enhance peripheral nerve regeneration. Synthetic nerve guidance tubes have been used for decades as a mechanical means to reconnect the ends of a severed nerve. However, simply providing a physical connection between the ends of a severed nerve is not sufficient to induce nerve regeneration across a critical size defect. In order for successful nerve regeneration to occur, a tissue-engineered scaffold must be provided that supplies not only the mechanical support for regenerating axons and protection against the ingrowth of scar tissue, but it must also supply direction cues to guide the regenerating nerve back to the appropriate targets of innervation and survival factors to prevent neuronal death during regeneration. Guidance signals in the form of adhesion factors can be incorporated into biomimetic materials to enhance migration on or through the material toward the proper target. In addition, survival and differentiation factors can be covalently or non-covalently incorporated into biomaterials to create delivery systems to modify cellular response during tissue regeneration.

Biological recognition of materials can take several forms. Incorporation of biomimetic adhesion sites, as described for the case of nerve regeneration, can be used to promote cell adhesion and migration on or within bioactive materials. The optimal density and spatial distribution of ligands upon or within materials are important factors to consider when designing biomimetic materials. The selection of which type of cells adhere to a material and their spatial distribution can also be controlled through the selection of the adhesion sites that are incorporated into a bioactive material.

The incorporation and controlled release of growth factors is a second approach for controlling biological recognition within materials. Growth factors are biologically active proteins that act in a highly specific manner and can be used to enhance cell survival, promote cell proliferation, or control cellular phenotype. Delivery systems for growth factors can be incorporated into tissue-engineered scaffolds. The release of growth factors from tissue-engineered scaffolds can occur through a number of mechanisms, including diffusion-based release, degradation of the material, or cell-triggered release. These factors provide a significant degree of control over cells within and near the material by significantly altering the cellular response to the bioactive material during tissue regeneration.

Incorporation of DNA into bioactive materials provides an additional method for controlling cellular response. Plasmid DNA that encodes for polypeptide growth factors can be used to stimulate in situ production of growth factors that promote or enhance tissue regeneration. Bioactive DNA-interacting materials can be used to enhance plasmid DNA uptake and the subsequent protein expression of genes encoded within the plasmid. This approach provides the potential for long-term delivery of bioactive signals from the tissue itself.

An important feature for functional biomaterials is that active material transformation from one state to another state, i.e., from liquid precursor to solid material can occur in the presence of biological systems. Material transformation can be designed to occur in response to an external stimulus, such as light, temperature or chemical composition. Materials capable of such in situ transformation are attractive because they enable new minimally invasive surgical procedures, such as the closure of internal incisions by the injection of a bioactive surgical sealant.

Another class of biomimetic materials that has been developed recently is microfabricated biomaterials. These materials are formed using microfabrication techniques developed for other industries, such as ink jet printing and microchip fabrication. The use of these techniques allows the design and production of materials that have features on the same order of magnitude as a cell diameter.

INCORPORATION OF BIOMIMETIC ADHESION SIGNALS

Biomaterials can be endowed with biological activity through the incorporation of adhesion-promoting oligopeptides. In vivo, cell recognition of traditional biomaterials, such as polytetrafluoroethylene, silicone rubber, or polyethylene, occurs indirectly. Proteins from body fluids adsorb nonspecifically onto the surfaces of these materials, and some of the proteins adsorbed, including fibronectin, vitronectin, and fibrinogen, promote the adhesion of cells via cell-surface adhesion receptors for these proteins. Preventing the nonspecific adhesion of proteins on the material surface and incorporating cell-type-specific adhesion-promoting peptides allows more direct control of cell adhesion on biomaterials (1).

The incorporation of adhesion-promoting peptides into biomaterial surfaces has been studied extensively [for review of adhesion sequences and their incorporation into biomaterials see Yamada (2) and Hubbell (3)]. These peptides are short primary sequences taken from the receptor-binding domains of adhesion proteins such as laminin and fibronectin. The most commonly studied adhesion peptide is the tri-peptide sequence, RGD. Ruoslahti & Pierschbacher first described use of RGD-containing peptides to promote cell attachment (4). Since then RGD has been used extensively to render surfaces and materials biologically active. The advantages of using small peptides rather than whole proteins are that they are relatively inexpensive to make and easy to purify, whereas whole adhesion proteins, such as fibronectin or laminin, are expensive and often not available in a clinically acceptable form.

Functionalization of Biomaterials with Short Peptides Can Incorporate Signals from ECM Adhesion Proteins

In order to use short peptides to render materials bioactive, one must know the amount of peptide required for biological recognition to occur. Massia & Hubbell found that when the tri-peptide RGD was immobilized by its amino terminus via a glycyl spacer, approximately 10^5 copies per cell were required to induce cell adhesion, spreading, focal contact formation, and cytoskeletal organization, whereas many more copies of the complete fibronectin protein were required for focal contact formation. Presumably this was the result of protein unfolding associated with adsorption or absorption on the surface in an orientation such that the receptor-binding domain was not accessible for binding. These results provide a benchmark for the design of biomaterials containing adhesion-promoting peptides and suggest that spacing of about 140 nm between peptides at the minimum peptide density is required for focal contact formation (5).

Other research demonstrated that adhesion-promoting peptides from laminin, such as the penta-peptide YIGSR, could also be used to promote adhesion to biomaterials via the 67-kDa laminin receptor in order to render the materials bioactive. Similar coupling of YIGSR peptides to biomaterial surfaces induced receptor clustering and co-localization with the intracellular cytoskeletal proteins vinculin and α-actinin (6).

Incorporation of cell-type selective adhesion peptides is an important goal in the development of bioactive materials. For example, in the case of vascular graft design, it would be useful to develop materials that support the adhesion and migration of endothelial cells but prevent the adhesion of blood platelets. Already one can target adhesion ligands that interact with cell adhesion receptors expressed selectively on the surface of the target cell population. In the case of a vascular graft the integrin receptor $\alpha_4\beta_1$, which is present on endothelial cells but not on platelets can be targeted. By using the tetra-peptide adhesion ligand from fibronectin, REDV which is specific for this receptor and does not bind to other integrins, one can selectively target the adhesion of endothelial cells, but not platelets, to a biomaterial (7).

Roles of Ligand Density in Cell Migration: Importance in Designing Materials

Several factors play key roles in the biological response to cell adhesion signals displayed on a biomaterial. The presentation of adhesion signals, the average density of the signal, and their spatial distribution all affect cell adhesion and migration. When designing bioactive materials, it is also important to consider that cell migration on or through a material is highly dependent on adhesion ligand concentration and that cell motility speed is often a bimodal function of ligand concentration rather than a linear function. The role of ligand density on cell adhesion and migration has been studied extensively, both experimentally and through mathematical modeling, to determine the optimal parameters for the design of biologically active materials.

Lauffenburger and coworkers have developed critical models for understanding the role of cell adhesion ligands in cell motility. DiMilla et al presented a model to explain the role of cell adhesion receptor-ligand binding and cell mechanical properties in determining the speed of cell migration on a surface (8). This model was later verified experimentally by Palecek et al, who demonstrated that cell migration speed was dependent on three variables: substratum ligand level, integrin expression level, and integrin-ligand binding affinity (9).

Although average ligand density plays an important role in cell adhesion and migration on biomaterials, the spatial distribution of ligand on the surface is also an important consideration in designing biologically active materials. Maheshwar and coworkers studied the role of integrin ligand density and clustering on cell adhesion and motility on thin polyethylene oxide (PEO) hydrogels. It has been shown that the integrin expression level, integrin-ECM binding affinity, and the integrin-ligand density on a surface all affect cell motility on a surface. By tethering the integrin ligand, YGRGD, to surface-immobilized PEO star molecules, both the average surface ligand density and localized clustering of the ligand could be controlled. As the clustering of ligand (controlled through the number of peptides per PEO star) was increased, the average ligand density required to support MR6 (a fibroblast cell line) migration decreased, suggesting that the nanoscale distribution of ligand is important for cell motility (10).

Griffith & Lopina also tested the role of ligand clustering for non-peptide cell adhesion substrates, such as oligosaccharides. Galactose ligands were coupled to linear and star PEO molecules. They observed that the average surface density of ligand and the localized clustering of ligands affected the ability of hepatocytes to spread on surface-bound galactose (11). These results suggest that the role of ligand density in cell motility can be generalized to non-peptide ligands as well. This phenomenon likely depends upon the details of the adhesion receptors and their intracellular mechanics. In the case of the receptor responsible for binding to galactose, the receptor is trimeric. In the case of the receptor responsible for binding to the RGD peptide, the receptor clusters into surface structures that are mechanically integrated via coclustering of intracellular proteins that connect the transmembrane receptors to the cytoskeleton. This phenomenon results in a dramatic enhancement of cell-binding strength (12).

Incorporation of Cell Adhesion Peptides into Three-Dimensional Matrices for Tissue Regeneration

In many physiological situations, a three-dimensional scaffold is required for tissue repair and regeneration. Many researchers have sought to develop biologically active materials for tissue repair through the incorporation of adhesion-promoting peptides into three-dimensional scaffolds. Borkenhagen et al (13) and Bellamkonda et al (14) have developed such materials by coupling short, cell adhesion peptides based on laminin to agarose via a photoactive linker, BP-MAL. This photoimmobilization method permitted patterning of the peptide within hydrogels to provide guidance for regenerating axons back to their target of innervation. Coupling of peptides could also be attained using a carbonyldiimidazole (CDI) linker. The addition of laminin peptides to agarose gels was shown to enhance neurite extension in vitro and nerve regeneration in vivo (13, 14).

Schense & Hubbell developed a novel method for the incorporation of bioactive peptides into three-dimensional tissue-engineered scaffolds. By using a bi-domain peptide containing a substrate for the transglutaminase factor XIIIa in the N-terminal domain, they demonstrated that high levels of bioactive peptides can be incorporated into fibrin matrices (15). The second domain of this bi-domain peptide can serve as a cell adhesion domain by incorporating cell adhesion sequences from fibronectin or laminin. These materials can be used to enhance peripheral nerve regeneration in vivo when placed in nerve guidance tubes (16).

Design of Protein Mimetic Matrices with Cell Adhesion Domains

Another approach to developing bioactive materials for tissue repair is the use of artificial protein polymer as tissue-engineered scaffolds. These protein polymers are expressed in bacteria using recombinant DNA techniques and have the advantage that they can incorporate longer peptide sequences than those synthesized by standard solid-phase peptide synthesis. These proteins can be engineered to contain long repeat domains that mimic the ECM proteins found in nature such as elastin. Cell adhesion–promoting sequences can also be added to these artificial protein polymers to render them biologically active.

Urry and coworkers have developed protein-based polymers that mimic the GVGVP repeats found in elastin. These protein-based polymers were tested in vivo to determine whether they elicit an inflammatory response when implanted subcutaneously. These elastin-like proteins can also be cross-linked though the use of gamma irradiation or, if lysine residues are included in the polymer backbone, lysyl oxidase can be used for the cross-linking of fibers. Such materials may be useful for soft tissue augmentation (17). Through the addition of chemoattractant peptide sequences such as GVGVAP, which attracts monocytes, cellular invasion of the materials made from elastin-like protein polymers can be attained.

Tirrell and coworkers developed protein polymers that consisted of elastin-like repeats and repeating cell adhesion motifs. Building on Urry's work, they incorporated the endothelial cell adhesion site REDV from the fibronectin CS5 domain to promote endothelial cell adhesion to elastin-like artificial proteins. These protein polymers could be useful in the repair and replacement of vascular tissue (18).

BIOMATERIALS FOR DRUG DELIVERY

Polypeptide growth factors are powerful regulators of biological function. They modulate many cellular functions including migration, proliferation, differentiation, and survival. Incorporation of growth factors is an additional approach for rendering biomaterials biologically active. However, the biological activity of growth factors is dependent not only on their presence in solution, but also on their interactions with the ECM. For example, some growth factors are most effective when release occurs over a prolonged period of time, whereas others are more effective when delivered in a bolus (19); some factors are active while tethered to a material whereas others are only active when they have been released from the material and are internalized into a cell. These differences may result from the different pathways that are used to transduce signals from a particular growth factor and how its receptors traffic and are recycled within the cell. All these differences between growth factors must be taken into account when designing a delivery system to be incorporated into a biomaterial, and they often play an important role in selecting what kind of delivery system should be used.

Affinity-Based Delivery from Biomaterials

In nature, the ECM acts as an example of a natural affinity-based drug delivery system by sequestering heparin-binding growth factors via interactions with ECM glycosaminoglycans. The high excess of proteoglycans present in the ECM effectively sequesters the fibroblast growth factors (FGFs) and other heparin-binding growth factors; FGFs that passively dissociate from the matrix are quickly recaptured by unoccupied heparan sulfate sites. FGFs that are in solution must be attached to a heparin-like molecule in order to diffuse through the ECM; otherwise, they are sequestered by the matrix and do not diffuse (20). Thus although the dissociation of FGF from heparin is rapid (k_D equal to approximately 1 min^{-1}), provision of excess immobilized heparin or heparan sulfate can dramatically slow FGF release, both naturally as well as in therapeutic systems. Thus by mimicking nature's approach to drug delivery, affinity-based systems can be designed and incorporated into biomaterials to provide biologically active materials.

USE OF HEPARIN AFFINITY FOR DELIVERY OF HEPARIN-BINDING GROWTH FACTORS By mimicking nature's approach to heparin-binding growth factor delivery, a heparin-based drug delivery system can be developed by incorporating heparin into biomaterials and using the affinity of heparin for heparin-binding growth factors to sequester them within biomaterials. One example of an affinity-based delivery system is the basic fibroblast growth factor (bFGF) delivery system developed by Edelman and coworkers (21). This system consists of heparin-conjugated Sepharose beads that are encapsulated in alginate. Heparin-binding growth factors are immobilized within this system based on electrostatic interactions between basic heparin-binding domains on the growth factors and sulfated groups on heparin. The growth factors are protected from degradation by heparin and are released slowly over time. This delivery system has been tested in a number of animal models and has recently completed Phase I clinical trials successfully (22). Heparin-alginate microcapsules containing bFGF were implanted in ischemic, ungraftable myocardial territories in patients undergoing coronary bypass surgery. After 3 months of implantation, there was significant improvement in the defect size observed by nuclear magnetic resonance imaging in the patients receiving the higher dose of bFGF (100 μg); however, in the placebo-treated group there was a trend toward worsening of the defect size.

In addition to using a microcapsule-based delivery system, functionalizing matrices with covalently immobilized heparin is also a possibility, as demonstrated by Wissink and coworkers. They heparinized collagen matrices by cross-linking collagen with EDC/NHS and then immobilizing heparin with a similar scheme. This heparin could in turn serve to immobilize growth factors such as bFGF within the collagen matrices. The bFGF immobilized in these matrices enhanced endothelial cell proliferation on the gels and reduced the minimum cell-seeding density required for proliferation by four-fold (23, 24).

In a slightly different approach, Sakiyama & Hubbell developed a fibrin-based heparin-containing delivery system for heparin-binding growth factors, where the heparin is not covalently coupled to the matrix, but rather is sequestered via electrostatic interactions. These immobilized heparin sites in the delivery system can sequester heparin-binding growth factors in a manner similar to the way they are sequestered in the ECM. Mathematical modeling was performed to test the feasibility of prolonged release from such delivery systems and to determine the conditions required for slow diffusion-based release of growth factor. These conditions were then tested in a model of in vitro nerve regeneration, and it was found that growth factors delivered from these novel delivery systems were biologically active (25).

In addition to using heparin-based delivery systems to control the release of growth factors, heparin can also be used as a stabilizing agent for some growth factors. Schroeder-Tefft and coworkers used heparin to stabilize transforming growth factor $\beta 2$ (TGF-$\beta 2$) and prevent loss of activity under physiological conditions in vitro and in vivo, when implanted within collagen matrices (26). They demonstrated that the use of heparin to stabilize TGF-$\beta 2$ allows retention of growth factor activity for up to 1 month in vivo, whereas TGF-$\beta 2$ alone loses its activity within 24 h. These results suggest that heparin can be used to stabilize growth factors with low-heparin-binding affinity in vivo and to enhance the delivery of active growth factors from biomaterials.

PROTEIN DELIVERY USING GROWTH FACTOR FUSION PROTEINS WITH EXOGENOUS COLLAGEN-BINDING DOMAINS In addition to a heparin-based delivery system, other types of affinity-based delivery systems have also been studied. For example, instead of functionalizing the base material with a high-affinity binding site for the growth factor of interest, the growth factor itself could be modified to enhance its affinity for a particular biomaterial. Andrades and coworkers used this approach to enhance the ability of collagen matrices to sequester TGF-$\beta 1$. They made recombinant fusion proteins of TGF-$\beta 1$ that contained a collagen-binding domain from von Willebrand factor to allow non-covalent immobilization within collagen-based materials (27). This TGF-$\beta 1$ fusion protein binds collagen with much higher affinity than recombinant TGF-$\beta 1$, which lacks the exogenous collagen-binding domain, and was also shown to promote higher levels of migration, growth, and differentiation of bone marrow mesenchymal cells in collagen gels versus native TGF-β1, bone morphogenetic protein 2 (BMP-2), and bFGF (28).

Enzymatically Regulated Release of Growth Factors

In addition to modifying the growth factor protein with sites for non-covalent immobilization, growth factor proteins could also be modified to contain sites for covalent cross-linking to biomaterials. For example, Sakiyama-Elbert et al have developed modified nerve growth factor (NGF) fusion proteins that contain an exogenous substrate for enzymatic cross-linking via factor XIIIa into fibrin matrices during polymerization (57). These NGF fusion proteins also have a substrate for plasmin cleavage that allows for release of NGF in a relatively native form based on cell-mediated protease activity. They demonstrated the ability of immobilized NGF fusion protein to enhance neurite extension from dorsal root ganglia in vitro, compared with native NGF in the culture medium, and the feasibility of this approach to drug delivery in response to cellular demand.

Another example of enzymatically triggered drug delivery has been developed by Kopecek, who has focused on designing systems that allow drug release in response to enzymes that are present only in specific locations within the cell. For example, he developed delivery systems in which the drugs are covalently linked to the system via an enzymatically degradable linker (29). Short oligopeptide sequences have been identified that are sensitive to lysosomal enzyme cleavage by enzymes such as cathepsin H and cathepsin B. By using these enzymatically degradable linkers, he developed drug delivery systems where release occurs only in the presence of lysosomal enzymes.

Tethering of Growth Factors to Biomaterials

In some cases, growth factors do not need to be released from the material in order to be active. For certain growth factors, simply being presented via a flexible tether molecule allows them to be active while remaining covalently coupled to a material. Kuhl & Griffith-Cima coupled epidermal growth factor (EGF) to aminosilane-modified glass via star PEO molecules. They found that tethered EGF was as effective as soluble EGF in promoting primary rat hepatocyte DNA synthesis and cell rounding responses, whereas physically adsorbed EGF showed no biological activity (30). Tethering growth factors via a spacer arm that provides a significant degree of molecule mobility prevents loss of the protein through diffusion but still presents the growth factor in an active form.

A second example of covalently immobilized biologically active growth factors can be found in the work of Schroeder et al. They coupled TGF-β to injectable fibrillar collagen via a di-functional polyethylene glycol (PEG) linker. The activity of TGF-β2 was retained under physiological conditions in vitro and in vivo for at least 4 weeks when coupled to collagen; however, when the growth factor was not coupled to collagen, activity was lost within 1 week (31). These results demonstrate another technique for rendering materials biologically active.

BIOMATERIALS FOR GENE DELIVERY

The immobilization of DNA is another method for incorporating biological signals into biomimetic materials. Typically, plasmid DNA is presented on or within a biomaterial to enhance the efficiency of its uptake and limit expression of the protein encoded for by the plasmid to the target tissue of interest. DNA incorporation provides an additional degree of control in manipulating the cellular response during tissue repair and regeneration. For example, DNA delivery can be used to express proteins that typically act primarily as an extracellular component in a signaling cascade, such as growth factors; in addition, it can be used for delivering proteins that act intracellularly, such as transcription factors. By incorporating the DNA into the biomaterial, its release from the material and uptake by cells can be regulated through modifications to the material.

Scaffolds for Gene Delivery

DNA incorporated into scaffolds for tissue engineering can be used as a powerful means to alter the cellular wound healing response during tissue regeneration. Several researchers have sought to incorporate DNA into tissue-engineered matrices by simply impregnating the matrix with DNA. For example, Shea and coworkers incorporated plasmid DNA encoding the gene for PDGF into poly (lactide-co-glycolide) (PLGA) matrices. They used the matrices to provide sustained release of the plasmid in subcutaneous implants, which resulted in an increase in granulation tissue thickness and the number of blood vessels, compared with the effect from simple injection of the plasmid alone. Sustained transfection of cells in vivo was measured, using β-galactosidase as a reporter gene, and transfection levels of 40% or higher were observed for 4 weeks (32).

In a related approach, Bonadio and coworkers developed gene-activated matrices to enhance bone regeneration by mixing plasmid DNA and bovine collagen to form collagen matrices. The plasmid used for these studies (pMat-1) encoded the gene for a peptide fragment of human parathyroid hormone. When doses of 40–100 mg of plasmid/matrix were placed in a critical-size bone defect in a dog tibia, bone regeneration of 50–75% was observed, whereas control matrices contained only granulation tissue (33). They demonstrated that delivery of a plasmid encoding the gene for BMP-4 can also enhance bone regeneration in vivo when the gene is expressed by fibroblasts (34).

Materials for Endosomal Escape

DNA delivery is a complex process, requiring not only uptake of the DNA by cells but also often endosomal escape of the DNA following uptake and its subsequent targeting to the nucleus. Only when DNA successfully reaches the nucleus will expression of the encoded protein occur. Presentation of DNA on the surface of a biomaterial often enhances the uptake of DNA by neighboring cells. However, the DNA must then escape the endosome prior to lysosomal fusion in order to prevent degradation within the lysosome. Because endosomal escape is a key and a potentially rate-limiting step in gene expression, many researchers have focused their efforts on developing materials that facilitate endosomal escape.

Pack and coworkers synthesized poly-L-histidine-graft-gluconic acid (pLHis-g-GA) polymers to enhance the endosomal escape of DNA. They chose to graft gluconic acid onto poly-L-histidine (pLHis) in order to solubilize the pLHis at physiological pH. They hypothesized that pLHis polymers could act as proton sponges in the endosome/lysosome and facilitate endosomal escape of the DNA. When pLHis-g-GA was used in a physical mixture with poly-L-lysine (pLLys), transfection was as efficient in the absence of chloroquine as it is with pLLys alone in the presence of chloroquine. Toxicity of pLHis-g-GA is much lower than that of polyethyleneimine (PEI), and relatively little effect on metabolic activity is observed for doses greater than ten times the EC_{50} for PEI (35).

Hoffman and coworkers have developed a series of pH-sensitive polymers for use in intracellular drug delivery and gene therapy, with the purpose of facilitating endosomal escape. This family of polymers, which includes poly(2-ethylacrylic acid) and poly(2-propylacrylic acid), exhibit membrane-disrupting properties when the pH drops below their pK_a. By testing several polymers from this family, Hoffman et al identified polymers that induce red blood cell hemolysis at pH 5–6, similar to what might be found in the endosome or lysosome, but than do not induce hemolysis at pH 7.4. Because endosomal escape is a key step in both intracellular drug delivery and gene therapy, these polymers could provide a critical method to enhance the efficacy of such therapeutics (36–38).

IN SITU TRANSFORMATION OF MATERIALS

Functional biomaterials must also be capable of undergoing an active material transformation from one state to another in the presence of biological systems. This ability to transform from an injectable state to a solid state is highly beneficial for use in minimally invasive surgical procedures. The material transformation can be designed to occur in response to an external stimulus, such as light, temperature, or chemical composition.

One type of material transformation that can be used to modulate the biological activity of materials is the swelling or collapse of materials in response to environmental stimuli. By designing materials that exhibit lower or upper critical solution behavior, gels can be formed at temperatures greater than critical temperature of the material. These types of transitions can be exploited to develop materials that are liquid at room temperature, but upon injection undergo a transition and form nonflowing solids at physiological temperature. In addition, drug-containing biomaterials can be developed that do not release drug until they reach the target tissue for delivery, at which time a shift in the biological environment triggers release.

Artificial-Protein Materials

One approach to developing materials that can undergo transformation while in contact with a biological system is to borrow from nature itself. Many biological proteins assemble in multimeric structures based on a class of protein domains commonly known as coiled-coils. For example, leucine zipper domains, which are found in many proteins, control the assembly of intracellular regulatory proteins such as Jun, a transcription factor that is only active in the dimeric state. Coiled-coil structures, such as the leucine zipper, have been studied extensively. Several researchers have utilized the knowledge derived from these studies to develop materials that assemble based on coiled-coil protein interactions. These materials often undergo a transition in response to either temperature or pH changes.

COILED-COIL SELF-ASSEMBLING MATERIALS Petka and coworkers developed artificial protein polymers that undergo reversible gelation in response to changes in pH or temperature. These proteins contained a central water-soluble polyelectrolyte segment, consisting of the tri-peptide sequence $(PEG)_x$, and terminal leucine zipper domains flank this central domain on either end (39). These terminal leucine zipper domains undergo a transformation near neutral pH that results in the formation of a three-dimensional network owing to coiled-coil aggregation. This transition is reversible, and at higher pHs the coiled-coil aggregates dissociate, resulting in dissolution of the network. These materials may be useful for cell encapsulation or drug delivery applications.

Wang and coworkers developed a stimulus-sensitive hybrid hydrogel that is assembled from water-soluble synthetic polymers and proteins containing coiled-coil protein-folding motifs, which swell or shrink in response to environmental stimuli (40). These hydrogels undergo a reversible, temperature-induced collapse owing to the cooperative conformational transition of the coiled-coil protein domain. The synthetic polymer component is made up of a linear hydrophilic copolymer of *N*-(2-hydroxylpropyl)-methacrylamide (HPMA) and a metal chelating monomer *N*-(*N′*-*N′*-dicarboxymethylaminopropyl) methacrylamide (DAMA) with pendant iminodiacetate (IDA) groups that serve as metal chelating ligands. The IDA groups formed a complex with Ni^{2+}, and these complexes can bind to terminal histidine residues (His tag) of the protein component of the hydrogels. Hydrogels were formed by mixing the chelating copolymers of HPMA charged with Ni^{2+} and the coiled-coil proteins. Homogeneous films of these materials were formed by drying the mixture on Teflon sheets. The dried films were rehydrated and under physiological conditions were stable for at least 24 h, whereas control films lacking polymer, Ni^{2+}, or coiled-coil protein dissolved upon hydration. When heated to 70°C the gels underwent a sudden collapse to 10% of the equilibrium volume at 25°C. Because of the mild conditions under which these hydrogels are formed they may be excellent candidates for the delivery of protein drugs. In addition, the properties of the materials can be varied by tailoring the transition temperature of the coiled-coil domains.

ELASTIN-BASED MATERIALS Another approach to materials design that can be derived from nature is to mimic the structure of naturally occurring proteins to obtain artificial protein polymers with similar mechanical properties. The most commonly mimicked protein is elastin, which has a short-repeating sequence, GVGVP, that can endow artificial proteins with secondary structure and mechanical properties similar to those of elastin.

Urry and coworkers have developed protein-based polymers that mimic the GVGVP repeats found in elastin. These elastin-like proteins can also be cross-linked through the use of gamma irradiation or, if lysine residues are included in the polymer backbone, lysyl oxidase can be used to cross-link the fibers. Such materials may be useful for soft tissue augmentation (17). The incorporation of cell adhesion–promoting peptides to these materials is described above.

Cappello et al developed protein polymers that mimic silk and elastin consisting of GAGAGS and GVGVP repeats, which undergo liquid-to-solid phase transitions in aqueous solution. These protein polymer gels can be used for the delivery of proteins or other high-molecular-weight drugs, without significant reduction of protein activity. The phase transition by which the gels are formed can be controlled by varying the number of silk versus elastin domains, as well as by varying the polymer concentration in solution to alter the gelation temperature and gel time. In addition, this transformation from liquid to solid requires no additional cross-linking agents, which are often toxic and cannot be used for in situ cross-linking (41).

SELF-ASSEMBLING PEPTIDE-BASED SYSTEMS Leon and coworkers have developed materials that form self-assembling matrices based on ionic self-complementary oligopeptides. These oligopeptides self-assemble under physiological conditions to form three-dimensional tissue-engineered matrices capable of supporting cell adhesion. Leon et al analyzed the mechanical properties of matrices formed from the peptide EGFK8, with the sequence KFEFKFEF, and found that the elastic modulus increased with increasing peptide concentration in the matrices (42).

Synthetic Materials

Much research in biomaterials has focused on the design and development of synthetic materials that can undergo transitions in the presence of biological systems. These phase transitions must occur under conditions that do not cause significant damage to the surrounding tissue, but the transitions must also be controllable, for example, so that the precursor material does not become a solid prior to injection. However, because synthetic material design is not constrained by the building blocks supplied by nature, there are more possibilities for material chemistries that will allow activation of material transformation in the presence of biological systems. For example, light can be used in combination with a photo-initiator to trigger material transformation, or material transformation can be triggered by the mixing of two reactive liquids. In addition, transformation stimuli used for artificial protein materials, such as temperature or pH changes, can also be used for synthetic materials.

PHOTOPOLYMERIZED MATERIALS Synthetic materials that can undergo phase transitions in contact with biological systems have a great potential for use in surgery, for example, as adhesives, sealants, and barriers to cell-tissue adhesion. PEG-based materials have been developed that contain a central PEG block with oligo (lactic acid) flanking blocks and terminal acrylates. The PEG block provides water solubility and enhances the biocompatibility of the materials formed from these polymers (43). The oligoesters incorporate degradability into the resulting materials formed, and the terminal acrylates provide functionality for use in polymerization. These polymers can be rapidly transformed into hydrogels by exposure to light in the presence of suitable photo-initiators, such as eosin yellowish. These materials can also be used to block platelet adhesion following balloon angioplasty to prevent restenosis (44) or to prevent scar tissue adhesions formation after gynecological surgery (45).

TEMPERATURE-INDUCED GELATION Temperature-induced material transformation is one of the most widely used methods for the development of biologically active materials. Hoffman and coworkers have extensively studied polymers that exhibit phase transitions in response to stimuli such as a change in pH or temperature. They developed pH-sensitive hydrogels based on the incorporation of poly(acrylic acid) (PAAc) into temperature-sensitive polymers, such as *N*-isopropylacrylamide (NIPAAm). Recently, they focused on the synthesis of graft copolymers, such as NIPAAm-graft-AAc, for rapid thermal responsiveness. These polymers undergo phase separation in response to changes in temperature and pH in a manner similar to individual polymers, NIPAAm (temperature-sensitive) and PAAc (pH-sensitive) (46, 47). Such materials can be used as reversible cell culture substrates upon which cell monolayers or multilayers can be cultured, and then after material transformation, these cell layers can be released in the form of intact sheets for subsequent transplantation (48). These materials can also be used to block a ligand-binding site on proteins in a thermally controlled manner, by utilizing the transformation of an extended chain to a collapsed chain to block binding above a given temperature (49).

Others have also focused on developing biological active materials for drug delivery that undergo temperature-induced phase transitions. Jeong and coworkers synthesized PEO poly(L-lactic acid) (PLLA) block copolymers of AB and ABA architectures. At low concentrations, these polymers form micelles in aqueous solution. At higher concentrations, a sol-gel transition is observed. The sol-gel transition temperature can be manipulated by altering the length of the PLLA block and by varying the concentration of polymer in solution. These polymers can be used to create drug delivery systems that are injected in the sol state (above

37°C) and upon cooling are transformed rapidly to the gel state forming a solid delivery depot. Degradation of the ABA copolymers results in an AB polymer that is in the sol phase at 37°C, thus providing a mechanism for release of drugs from these biodegradable block copolymers (50).

MULTIFUNCTIONAL REACTIVE POLYMERS Another approach to developing materials that can undergo a transformation in the presence of biological molecules is mixing two solutions that when reacted will form elastic gels. This approach can be applied in a manner similar to the clinical use of fibrin glue kits, in which the two species react rapidly upon mixing to form three-dimensional networks in vivo. Harris and coworkers have described a new method for making PEG hydrogels for controlled protein delivery. They use a two-step process, in which an ester-containing, amine-reactive PEG derivative is synthesized and reacted with a branched PEG multiamine to form a hydrogel. Proteins can be incorporated into the gel during polymerization, either by physical entrapment of the protein within the network or by covalently coupling the protein to the hydrogel via a free amine group on the protein. These hydrogels degrade via hydrolysis of the ester linkages to release protein and low-molecular-weight PEG derivatives that can be cleared from the body (51).

HYALURONIC ACID FUNCTIONALIZATION The approach of obtaining sol-gel transitions via the chemical reaction described above can be applied to natural materials, such as hyaluronic acid (HA), as well as purely synthetic materials. For example, Bulpitt & Aeschlimann developed methods for the functionalization of HA in order to make in situ polymerizable hydrogel materials (52). They used carbodiimide-mediated coupling of primary or secondary amines to the carboxyl group of the glucuronic acid moiety using an active ester intermediate. Coupling of primary amines to HA is possible via the formation of a hydrolysis-resistant and non-rearrangable active ester intermediate. Active esters of HA were formed with 1-hydrozylbenzotriazole (HOBt) or *N*-hydroxysulfosucinimide (Sulfo-NHS) using water-soluble carbodiimide EDC for coupling.

These functionalized HA materials can be cross-linked, under physiological conditions, using hetero- or homo-difunctional cross-linkers, to form hydrogels that can be enzymatically degraded by hyaluronidase. These materials have been tested in vivo, and when mixed with the appropriate combination of growth factors, bone or cartilage tissue can be formed within the cross-linked HA hydrogels (52).

ORAL DELIVERY OF PROTEINS VIA pH RESPONSIVE MATERIALS In addition to developing implantable biomimetic materials that incorporate growth factors, developing materials for the oral delivery of proteins such as insulin is another possibility. By taking advantage of the changes in the chemical environment in the gastrointestinal tract, materials can be developed that undergo a transition as they pass from the low pH environment of the stomach to the higher pH environment of the intestine. Brondsted & Kopecek developed a delivery system that incorporates both solvent activation and enzymatic degradation as mechanisms for drug release.

Hydrogels based on *N,N*-dimethylacrlamide, *N*-t-butylacrylamide, and acrylic acid were cross-linked with azoaromatic compounds (53). These materials are protonated at low pH in the stomach, resulting in a low degree of swelling, but when the pH increases farther along the gastrointestinal tract, the polymers are deprotonated and the hydrogels swell. The azo bonds are then vulnerable to cleavage by microbial enzymes that are uniquely present within the colon, allowing colon-specific drug delivery.

Lowman and coworkers have also developed polymers that can be used for oral drug delivery of proteins such as insulin. Protein release was targeted to the intestine using pH-responsive polymers that remain in the collapsed state at low pH, protecting insulin from degradation and slowing release due to the small mesh size of the polymer network. Upon reaching the intestine, an increase in pH triggers dissociation of the complexes found in the collapsed state, and the network pore size rapidly increases, resulting in an increase in the rate of insulin release. The pH-responsive polymers used for these studies were poly (methacrylic-graft-ethylene glycol) p(MAA-g-EG). P(MAA-g-EG) microparticles were formed and the insulin was loaded into the microparticles by equilibrium partitioning in the swollen state. These particles could significantly lower the blood glucose levels of both healthy and diabetic rats for up to 8 h (54).

MICROFABRICATED DEVICES

Another approach for creating biologically active materials is the use of microfabrication techniques to design materials. These materials can have features whose size is the same order of magnitude as a cell diameter, and by using techniques that can be controlled on the micron scale, the cell-sized features created on or within materials can be much more complex in nature because of the enhanced degree of control in the fabrication. For example, microchip devices can be fabricated to contain biologically active features, or features that can become biologically active in response to an external stimulus, such as pulsatile electric fields. Microfabrication techniques can also be applied to develop complex tissue-engineered scaffolds with elaborate features the size of a cell.

Microchip Release Devices

Santini and coworkers have developed a microchip device consisting of a SiO_2 microchip base with multiple reservoirs for drugs. These reservoirs have two openings. A large opening on one side allows filling by microfluidic techniques or inkjet deposition. A smaller opening on the other side of the chip is covered with a gold membrane. This membrane contains trace amounts of chloride ion to allow for gold dissolution when an anode potential is present at the reservoir. These devices could potentially contain hundreds of therapeutic compounds in one device and, through precise activation of release from individual reservoirs, controlled pulsatile release could be attained through the combination of microfabrication and pharmaceutical formulation technology (55).

Three-Dimensional Printing of Scaffolds for Tissue Engineering

Griffith and coworkers have developed a polymer processing technique to allow the three-dimensional fabrication of tissue-engineering scaffolds. They use a solid free-form technique that allows three-dimensional objects to be built as a series of printed layers. Three-dimensional scaffolds were printed from PLA, and the surfaces of these scaffolds were further printed with amphiphilic polymers formed from hydrophilic blocks of PEO and hydrophobic blocks of poly(propylene oxide) (PPO). Those regions that were modified with the PEO-PPO-PEO polymers were rendered non-adhesive to cells, whereas the unmodified PLA surfaces supported nonspecific cell adhesion. The non-adhesive regions could also be rendered selectively adhesive to hepatocytes by covalently coupling a carbohydrate ligand for the hepatocyte asialoglycoprotein receptor to the PEO chain ends. This method may be useful for organizing multiple cell types within an engineering tissue (56).

CONCLUSIONS

The development of modern biomaterials requires that the materials designed are not only bio-inert but also that they interact with and even respond to their biological environment. Bioactive materials must be able to provide biologically active signals that can indicate cell adhesion, migration, proliferation, differentiation, or even apoptosis. Through this type of communication, the materials can share their messages with the surrounding biological environment; for example by providing a morphogenetic signal to promote tissue regeneration. Bioactive materials that are designed to respond to stimuli can also respond to biological activity around them; for example by releasing a drug or degrading in response to a cellular stimulus. The ability of biomaterials to sense biological demand or changes in their environment is critical in the development of "intelligent biomaterials" for the future.

Visit the Annual Reviews home page at www.AnnualReviews.org

LITERATURE CITED

1. Bearinger JP, Castner DG, Healy KE. 1998. *J. Biomater. Sci. Polym. Ed.* 9:629–52
2. Yamada KM. 1991. *J. Biol. Chem.* 266: 12809–12
3. Hubbell JA. 1995. *Biotechnology* 13:565–76
4. Pierschbacher MD, Ruoslahti E. 1984. *Nature* 309:30–33
5. Massia SP, Hubbell JA. 1991. *J. Cell Biol.* 114:1089–100
6. Massia S, Rao S, Hubbell J. 1993. *J. Biol. Chem.* 268:8053–59

7. Hubbell JA, Massia SP, Desai NP, Drumheller PD. 1991. *Biotechnology* 9:568–72
8. DiMilla PA, Barbee K, Lauffenburger DA. 1991. *Biophys. J.* 60:15–37
9. Paleck SP, Loftus JC, Ginsburg MH, Lauffenburger DA, Horwitz AF. 1997. *Nature* 385:537–40
10. Maheshwari G, Brown G, Lauffenburger DA, Wells A, Griffith LG. 2000. *J. Cell Sci.* 113:1677–86
11. Griffith LG, Lopina S. 1998. *Biomaterials* 19:979–86
12. Ward MD, Hammer DA. 1993. *Biophys. J.* 64:936–59
13. Borkenhagen M, Clemence JF, Sigrist H, Aebischer P. 1998. *J. Biomed. Mater. Res.* 40:392–400
14. Bellamkonda R, Ranieri JP, Bouche N, Aebischer P. 1995. *J. Biomed. Mater. Res.* 29:663–71
15. Schense JC, Hubbell JA. 1999. *Bioconjugate Chem.* 10:75–81
16. Schense JC, Bloch J, Aebischer P, Hubbell JA. 2000. *Nat. Biotechnol.* 18:415–19
17. Urry DW, Pattanaik A, Xu J, Woods TC, McPherson DT, Parker TM. 1998. *J. Biomater. Sci. Polym. Ed.* 9:1015–48
18. Pantich A, Tamaoka T, Fournier M, Mason T, Tirrell D. 1999. *Macromolecules* 32:1701–3
19. Dinbergs ID, Brown L, Edelman ER. 1996. *J. Biol. Chem.* 271:29822–19
20. Vlodavsky I, Fuks Z, Ishai-Michaeli R, Bashkin P, Levi E, et al. 1991. *J. Cell. Biochem.* 45:167–76
21. Edelman E, Mathiowitz E, Langer R, Klagsbrun M. 1991. *Biomaterials* 12:612–26
22. Laham RJ, Sellke FW, Edelman ER, Pearlman JD, Ware JA, et al. 1999. *Circulation* 100:1865–71
23. Wissink MJ, Beernink R, Scharenborg NM, Poot AA, Engbers GH, et al. 2000. *J. Control. Release* 67:141–55
24. Wissink MJ, Beernink R, Poot AA, Engbers GH, Beugeling T, et al. 2000. *J. Control. Release* 64:103–14
25. Sakiyama-Elbert SE, Hubbell JA. 2000. *J. Control. Release* 65:389–402
26. Schroeder-Tefft J, Bentz H, Estridge T. 1997. *J. Control. Release* 49:291–98
27. Tuan TL, Cheung DT, Wu LT, Yee A, Gabriel S, et al. 1996. *Connect. Tissue Res.* 34:1–9
28. Andrades JA, Han B, Becerra J, Sorgente N, Hall FL, Nimni ME. 1999. *Exp. Cell Res.* 250:485–98
29. Kopecek J. 1984. *Biomaterials* 5:19–25
30. Kuhl PR, Griffith-Cima LG. 1996. *Nat. Med.* 2:1022–27
31. Schroeder J, Bentz H, Estridge T. 1997. *U.S. Patent No. 5.693,341*
32. Shea LD, Smiley E, Bonadio J, Mooney DJ. 1999. *Nat. Biotechnol.* 17:551–54
33. Bonadio J, Smiley E, Patil P, Goldstein S. 1999. *Nat. Med.* 5:753–59
34. Fang J, Zhu YY, Smiley E, Bonadio J, Rouleau JP, et al. 1996. *Proc. Natl. Acad. Sci. USA* 93:5753–58
35. Pack DW, Putnam D, Langer R. 2000. *Biotechnol. Bioeng.* 67:217–23
36. Lackey CA, Murthy N, Press OW, Tirrell DA, Hoffman AS, Stayton PS. 1999. *Bioconjugate Chem.* 10:401–5
37. Stayton PS, Hoffman AS, Murthy N, Lackey C, Cheung C, et al. 2000. *J. Control. Release* 65:203–20
38. Murthy N, Robichaud JR, Tirrell DA, Stayton PS, Hoffman AS. 1999. *J. Control. Release* 61:137–43
39. Petka WA, Harden JL, McGrath KP, Wirtz D, Tirrell DA. 1998. *Science* 281:389–92
40. Wang C, Stewart RJ, Kopecek J. 1999 *Nature* 397:417–20
41. Cappello J, Crissman JW, Crissman M, Ferrari FA, Textor G, et al. 1998. *J. Control. Release* 53:105–17
42. Leon EJ, Verma N, Zhang S, Lauffenburger DA, Kamm RD. 1998. *J. Biomater. Sci. Polym. Ed.* 9:297–312
43. Sawhney AS, Hubbell JA. 1990. *J. Biomed. Mater. Res.* 24:1397–411
44. West JL, Hubbell JA. 1996. *Proc. Natl. Acad. Sci. USA* 93:13188–93

45. Hill-West JL, Chowdhury SM, Sawhney AS, Pathak CP, Dunn RC, Hubbell JA. 1994. *Obstet. Gynecol.* 83:59–64
46. Chen G, Hoffman AS. 1995. *Nature* 373: 49–52
47. Dong LC, Hoffman AS. 1991. *J. Control. Release* 15:141–52
48. von Recum H, Kikuchi A, Okuhara M, Sakurai Y, Okano T, Kim SW. 1998. *J. Biomater. Sci. Polym. Ed.* 9:1241–53
49. Stayton PS, Shimoboji T, Long C, Chilkoti A, Chen G, et al. 1995. *Nature* 378: 472–74
50. Jeong B, Bae YH, Lee DS, Kim SW. 1997. *Nature* 388:860–62
51. Zhao X, Harris JM. 1998. *J. Pharm. Sci.* 87:1450–58
52. Bulpitt P, Aeschlimann D. 1999. *J. Biomed. Mater. Res.* 47:152–69
53. Brondsted H, Kopecek J. 1992. *Pharm. Res.* 9:1540–45
54. Lowman AM, Morishita M, Kajita M, Nagai T, Peppas NA. 1999. *J. Pharm. Sci.* 88:933–37
55. Santini JT Jr, Cima MJ, Langer R. 1999. *Nature* 397:335–38
56. Park A, Wu B, Griffith LG. 1998. *J. Biomater. Sci. Polym. Ed.* 9:89–110
57. Sakiyama-Elbert SE, Panitch A, Hubbell JA. 2001. *FASEB J.* In press

Annu. Rev. Mater. Res. 2001. 31:203–35

PATTERNED MAGNETIC RECORDING MEDIA

CA Ross
Department of Materials Science and Engineering, Massachusetts Institute of Technology, Cambridge, Massachusetts 02139; e-mail: caross@mit.edu

Key Words patterned media, magnetic recording, single-domain particles, micromagnetics, data storage

■ **Abstract** Continuing increases in the areal density of hard disk drives will be limited by thermal instability of the thin film medium. Patterned media, in which data are stored in an array of single-domain magnetic particles, have been suggested as a means to overcome this limitation and to enable recording densities of up to 150 Gbit cm^{-2} (1 Tbit $inch^{-2}$) to be achieved. However, the implementation of patterned media requires fabrication of sub-50-nm features over large areas and the design of recording systems that differ substantially from those used in conventional hard drives. This review describes patterned media, including the fabrication of arrays of small magnetic particles and their magnetic properties, such as domain structure, reversal mechanisms, thermal stability, and interactions. The practical implementation of patterned media recording schemes is assessed.

INTRODUCTION

The areal density of hard disk drives increased over 100% per annum during the late 1990s, and 8 Gbit cm^{-2} (50 Gb $inch^{-2}$) laboratory demonstrations were reported early in the year 2000 (1, 2). The increases in data density, data rate, and other performance metrics have generally been achieved by scaling to make the read-write head smaller, the medium thinner and higher in coercivity, and the head-medium spacing smaller. The performance of the medium is limited by noise originating from the granular microstructure of the thin film, so there has been a trend to decrease the grain size. However, this trend will eventually be limited by thermal instability (superparamagnetism) of the grains comprising the medium, because if they become too small, thermal energy will be sufficient to allow their magnetization to reverse spontaneously, with consequent loss of the recorded signal. This superparamagnetic limit has been predicted to occur at densities of 6–15 Gbit cm^{-2} (3, 4), depending on the scaling assumptions chosen for extrapolation.

Therefore, considerable effort has been expended in improving media to reduce thermal instability and in designing alternative methods for data storage. Modification of historical scaling trends, for instance reducing the bit aspect ratio (the width of the data track divided by the bit spacing), improving the grain size distribution in the media, and improving signal detection, has allowed densities to exceed early estimates of thermal stability limits (3), but at some point superparamagnetic effects will limit further increases. Patterned media (5–7) (also called discrete media or a quantum disk) have been one of the more widely discussed possibilities for extending data densities. In this review, we first discuss briefly the microstructure of current thin film media and the origin of the superparamagnetic effect. Fabrication methods for patterned media are reviewed, and requirements for patterned media are analyzed, in particular their micromagnetic behavior. Recording devices that incorporate patterned media are assessed.

Thin Film Media

Modern hard disk media incorporate a glass or a NiP-coated aluminum alloy substrate on which a thin film stack is sputtered. The stack consists of one or more underlayers or seed films, a magnetic film, and an overcoat. The magnetic film is a polycrystalline alloy of Co, Cr, and Pt, with additional elements such as B or Ta, and is sputtered at elevated temperatures that promote segregation of non-magnetic elements to the grain boundaries, leading to partial exchange-decoupling of the magnetic grains. Each ~10-nm-diameter grain therefore behaves as a single-domain particle with easy axis parallel to the Co c-axis, usually parallel to the film plane in longitudinal media. During the recording process, small areas of the film (bit cells) are magnetized parallel or antiparallel to the track direction; it is the fringing fields from the magnetization transitions between these areas that are detected by the head during readback. A thin film medium is shown schematically in Figure 1.

In high-density media, each bit cell contains of order 100 grains. The dependence of noise on media parameters has been reviewed recently (1). Transition noise, originating from irregularities or jaggedness in the magnetization transitions, and increased by collective reversal of groups of grains, dominates the overall signal-to-noise ratio (SNR) of the system. Both the SNR and the minimum width of the transition depend on the grain size of the medium. As the down-track linear bit density increases, the grain size must decrease to maintain an acceptable SNR. However, the grains begin to exhibit thermal instability when the ratio of thermal energy kT (k is Boltzmann's constant and T the temperature) to magnetic energy KV (K is the magnetic anisotropy and V the grain volume) exceeds a certain ratio. For isolated grains, stability over a time scale of, for example, 10 years gives a

Figure 1 Schematic of (*a*) conventional thin film medium, consisting of single-domain, exchange-decoupled grains. Bits are represented by the transitions between oppositely magnetized regions. Each bit cell contains tens or hundreds of grains. (*b*) A patterned medium with in-plane magnetization. The single-domain bits are defined lithographically with period p. They can be polycrystalline (indicated by dotted lines) with exchange coupling, or single crystal. (*c*) A patterned medium with perpendicular magnetization. Binary 1 and 0 are shown (from 19) (reprinted with permission, © AIP 1999).

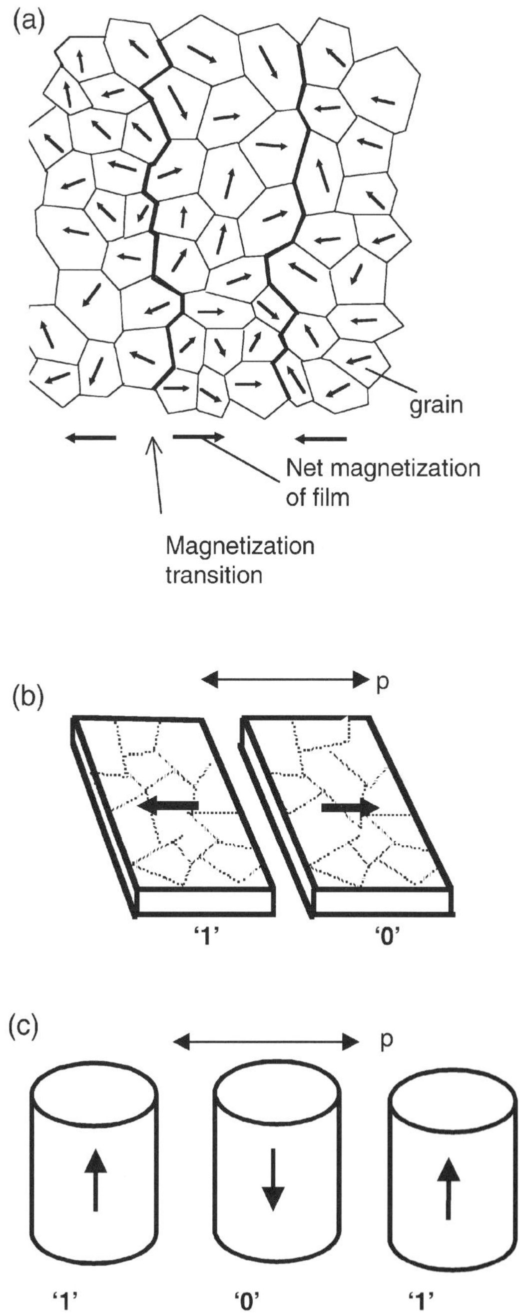
(a)
grain
Net magnetization of film
Magnetization transition
(b)
p
'1'
'0'
(c)
p
'1'
'0'
'1'

stability criterion of KV/kT > 40, but in a hard disk, the presence of demagnetizing fields at the transition lowers the energy for reversal and increases the required stability ratio. Stability ratios of KV/kT > 40–80 have been determined (1, 8–10) depending on the grain size distribution, intergranular coupling, saturation magnetization, and other properties of the medium.

To increase thermal stability, films with higher values of magnetic anisotropy K could be used, but increases in K are limited by the need for the recording head to produce sufficient field to write the medium. The maximum write field is around 400 kA m^{-1}, leading to a minimum grain diameter of approximately 11–12 nm to ensure thermal stability in CoCrPt-based longitudinal media (1, 10). This is not much smaller than the grains used in current media. Improvements in microstructural uniformity, bit aspect ratio, and signal processing will be necessary to increase areal density further, but it is apparent that a significantly different approach will be needed to go well beyond 15 Gb cm^{-2}. Several possibilities exist for achieving ultra-high densities. Perpendicular media are predicted to have higher thermal stability limits (11), perhaps five times greater than longitudinal media (4). One estimate (12) suggests that perpendicular media may be suitable for 150 Gb cm^{-2} recording at 3Gb s^{-1} data rate. Another solution involves the use of high-anisotropy media with thermally assisted writing (13, 14), in which a laser locally heats the magnetic material during writing. For this application, the greater thermal stability of high-anisotropy materials such as ordered (Co, Fe) (Pd, Pt) compounds or rare earth-transition metal compounds (15) could enable an order of magnitude increase in storage density. Patterned media provide a third concept for extending storage densities to very high values without the need for highwrite fields.

Patterned Media

A patterned recording medium, shown schematically in Figure 1*b–c*, consists of a regular array of magnetic elements, each of which has uniaxial magnetic anisotropy. The easy axis can be oriented parallel or perpendicular to the substrate. Each element stores one bit, depending on its magnetization state; for instance magnetization up could represent 1, and down could represent 0. Unlike the thin film medium, the grains within each patterned element are coupled so that the entire element behaves as a single magnetic domain. The medium is incorporated into a system capable of reading and writing data, for example, a spinning-disk system with an ultra-narrow recording head, or a system with an array of scanning probes that address the elements.

For a patterned media system to be viable, it must offer significantly higher data densities than can be achieved in a conventional hard drive. There is confidence that hard drives using longitudinal media will be able to reach 15 Gb cm^{-2} or higher, so patterned media designs must be capable of reaching densities of 30 Gb cm^{-2} and beyond. This implies a periodicity of 50 nm or smaller, for instance 25-nm elements with 25-nm separation. A 50-nm period corresponds to a density of 40 Gparticles cm^{-2}.

The major advantages of such a scheme are first that transition noise is eliminated because the bits are now defined by the physical location of the elements and not by the boundary between two oppositely magnetized regions of a thin film. Second, very high data densities can be obtained because the stability criterion now refers to the volume and anisotropy of the entire magnetic element, not to the individual grains of which it is composed. Elements could therefore be as small as a few nanometers, implying densities exceeding 150 Gbit cm^{-2}. The challenges inherent in patterned media are, however, severe. Fabrication of large-area arrays of elements with dimensions on the sub-50-nm scale requires advanced lithography or accurate self-assembly techniques, and addressing of the array requires the precision application and detection of magnetic fields on a very small scale. In particular, during the writing process the applied field must be synchronized with the physical location of the elements. This differs from recording onto a hard disk, in which the bit positions are defined by the head position at the time when the write field is applied. Most patterned media research has focused on the fabrication and magnetic characterization of media, but read-write mechanisms and system integration are now receiving significant attention.

FABRICATION OF PATTERNED MEDIA

Particle arrays with deep sub-micron feature sizes are too small to fabricate using conventional optical lithography. Instead, a range of nanolithography techniques has been used to make prototype structures, including electron-beam, X-ray, imprint, and interference lithographies, combined with magnetic materials deposition and pattern transfer processes. Self-assembly methods have also been developed in which magnetic particles are synthesized chemically and assembled as a close-packed monolayer on a surface. Since patterned media development is still at an early stage, there is no consensus on the optimum particle size, array geometry, and magnetic properties of the materials, and thus a preferred processing route has not been established. However, it is clear that any commercial patterned medium will need to include an array of magnetic elements with periodicity less than 50 nm, possibly including servo patterns, manufactured over large areas (several square centimeters or more) at low cost.

Lithography

During lithography, the desired pattern is generated by exposure of a resist layer and then transferred into the magnetic material by a sequence of development, etching, and deposition steps. Pattern generation can be accomplished by several methods including electron-beam lithography, X-ray or optical lithography, or imprint lithography. Patterns can then be transferred into magnetic material by additive or subtractive processes. In an additive process, magnetic material is deposited after patterning, for instance by electrodepositing through holes in a lithographically prepared template or by evaporating a film of material over a template followed by liftoff of the template (Figure 2). In a subtractive process,

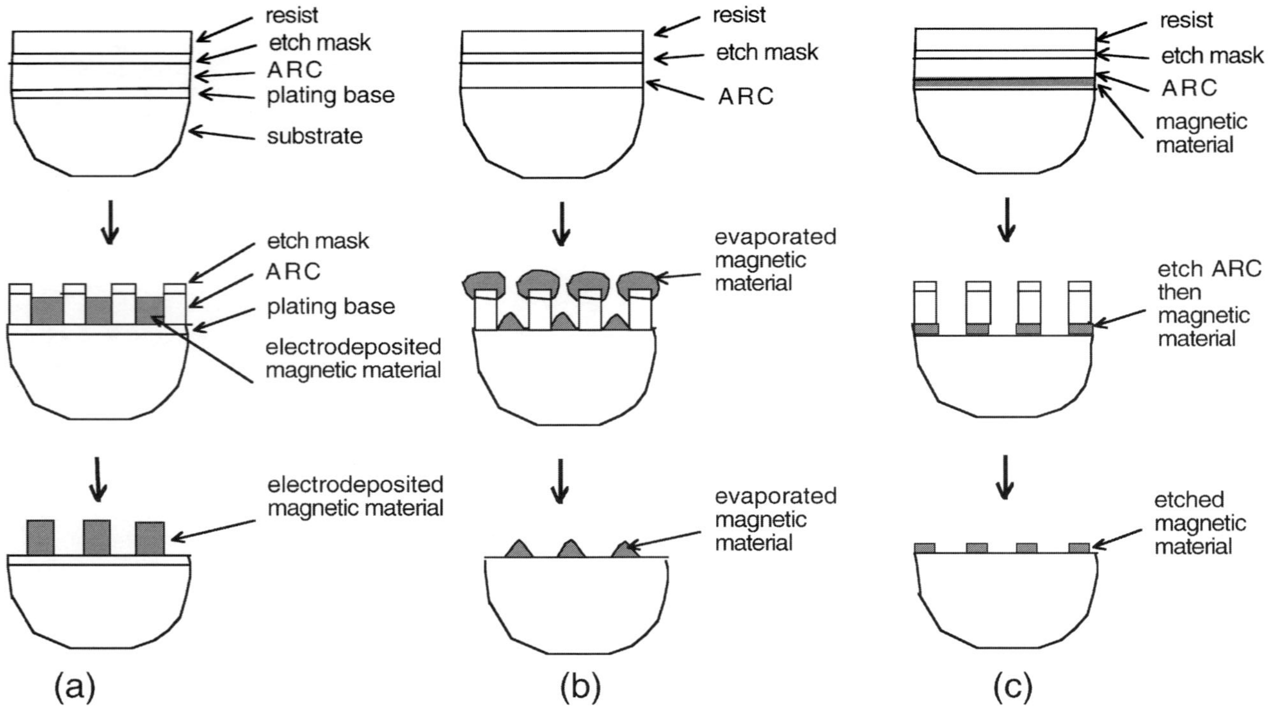

Figure 2 Fabrication steps for making patterned media lithographically. (*a*) A template of holes is prepared, then magnetic material is electrodeposited into the holes. (*b*) A template of holes is prepared, then magnetic material is evaporated on the template. The template is dissolved to leave an array of tapered particles. (*c*) A magnetic material is deposited, then a template of dots is prepared on top of the film to act as an etch mask. The exposed film is then removed by ion milling. These methods are shown using a trilevel resist stack [resist/etch mask/antireflective coating (ARC)] to optimize pattern transfer (from 48) (reprinted with permission, © AIP 1999).

the magnetic material is deposited first, then is etched to form the desired pattern using a lithographically generated mask. Various combinations of lithography and pattern transfer processes have been used to make magnetic arrays. Examples of magnetic arrays are shown in Figure 3.

Electrodeposition, and evaporation followed by liftoff have been the most widely used methods for making magnetic particle arrays. Electrodeposition is well suited to making high aspect ratio particles, with heights limited only by the thickness of the template. Ni, Co alloys such as NiFe, NiCo, and CoPt; multilayers such as Ni/Cu; and granular alloys such as Co-Cu have been electrodeposited (for example, 6, 16–20). Evaporation has been used to produce flat particles or tapered or conical particles from pure metals, from alloys such as NiFe, and multilayer structures (6, 21–25). However, etching of a predeposited magnetic film provides the widest choice of materials and microstructures, including complex alloys, epitaxial or multilayered films. Most magnetic materials are difficult to etch using reactive ion etching, so these materials are usually patterned using ion milling in which the material is selectively masked, then removed by sputtering using high-energy Ar ions (26–32). Ion milling suffers from the drawback that the etching is not highly selective so the mask is also eroded, and sputtered material can redeposit on the substrate thereby leading to undesirable features. The resist pattern can be used as the ion milling mask, but a thin hard mask such as tungsten (30) or aluminum (27) gives better results.

Electron-beam lithography has been the most widely used patterning method for fabricating small structures with dimensions of 15 nm and above. An electron beam is used to write a pattern into a polymethyl methacrylate resist layer, which is subsequently developed to leave a template. Because the writing process is a serial method, it is unsuited for making large area samples but has been used extensively to pattern relatively small (sub-millimeter) areas with fine

→

Figure 3 Examples of regular arrays of magnetic nanoparticles (see also Figure 5). (*a*) Evaporated Ni bar, 15-nm wide, made using electron-beam lithography (6) (reprinted with permission, © IEEE 1997). (*b*) Electrodeposited Ni pillars, 35-nm diameter, 100-nm period, made using electron-beam lithography (6) (reprinted with permission, © IEEE 1997). (*c*) Evaporated Ni-alloy pyramids, 30-nm diameter, 100-nm period, made using achromatic interference lithography (49) (reprinted with permission, © AIP 1999). (*d*) Etched ellipses of a Co/Cu multilayer, 70 × 170 nm, 10-nm thick, made using interference lithography (19) (reprinted with permission, © AIP 1999). (*e*) Topological, and (*f*) corresponding magnetic image of 140 × 250-nm evaporated Co ellipses made using electron-beam lithography (114) (reprinted with permission, © Am. Phys. Soc. 1998). In-plane dipole images (dark and light poles) from the ellipses can clearly be seen in (*f*). (*g*) Arrays of Fe pillars grown by CVD in two separate geometries, and (*h*) a magnetic force micrograph of another array showing that pillars are magnetized as dipoles with vertical moment up or down (dark or light) (58) (reprinted with permission, © AIP 1999). (*i*) and (*j*) FePt particles, 6-nm diameter, self-assembled on a substrate to form regular arrays 4-nm (76) (reprinted with permission, © Am. Assoc. Adv. Sci. 2000).

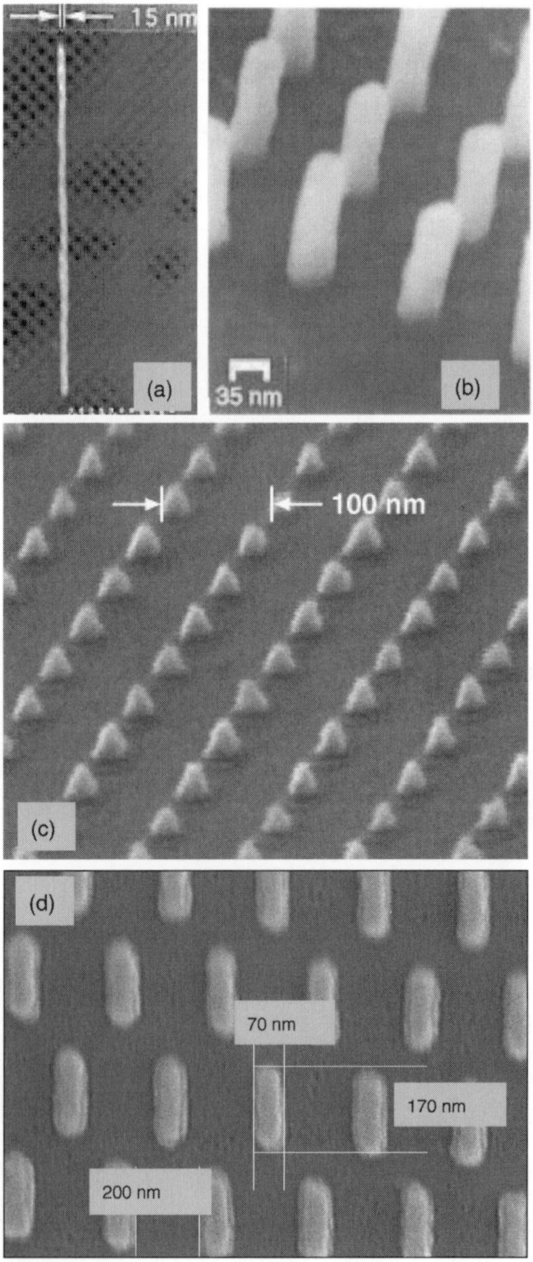
15 nm
(a)
35 nm
(b)
100 nm
(c)
(d)
70 nm
170 nm
200 nm

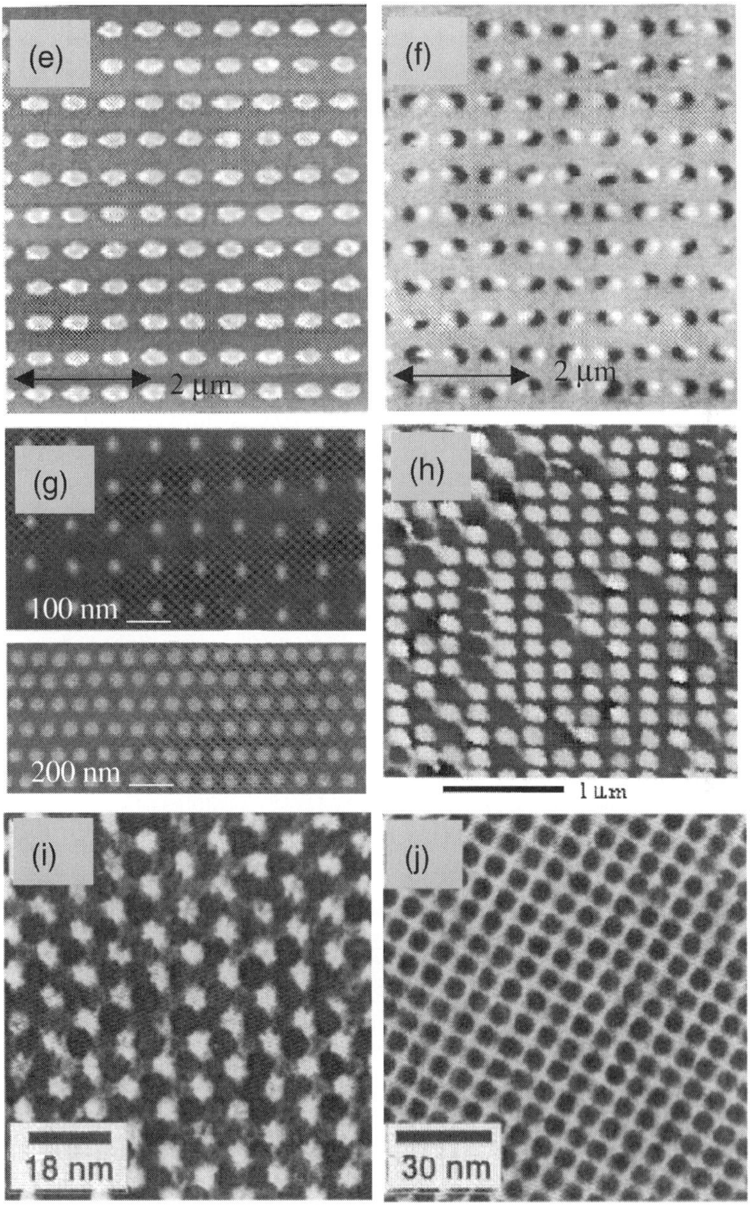

Figure 3 (*Continued*)

features. Rectangles, ellipses, wires, rings, and other shapes of Ni, Co, Fe, permalloy (NiFe), and other alloys have been made by evaporation and liftoff (6, 21, 23, 24, 33–36) including 15-nm-wide Ni wires (6). Electrodeposited columns of 20-nm diameter and wider have been fabricated (5, 37–39). The electrodeposition template can be the electron-beam written resist, or the pattern can be transferred into a more rigid material such as silica (39). Materials made by evaporation or electrodeposition are generally polycrystalline, but single-crystal epitaxial films have also been patterned using electron-beam lithography and an etching process (40, 41). Electron-beam lithography is very convenient for making patterns of arbitrary geometry, but its throughput is limited, and when exposing large areas, stitching errors between the different written fields can distort the long-range coherence of the array. It is excellent for making media prototypes but is unlikely to be suitable for large-scale commercialization unless throughput can be improved.

Three other nanolithography techniques have been developed that can pattern large-area samples. In X-ray lithography, a thin membrane mask is placed in close proximity to a resist-coated substrate and illuminated with soft X-rays of 1–10-nm wavelength (42). Low-intensity laboratory X-ray sources typically require an exposure time of several hours, so in commercial applications, a synchrotron or undulator source would be used to enable high throughput. The mask is typically a silicon nitride membrane patterned with features made from an absorber such as gold or tungsten. Masks can cover areas of several square centimeters and can be used indefinitely, barring mechanical damage. This process has been used to make arrays of thin magnetic dots of diameters 200 nm and greater (27, 43), although feature sizes as small as 20 nm are possible.

In nanoimprint lithography, a rigid mold is used to stamp patterns into a soft polymeric layer, analogous to the stamping of a compact disk. This process has developed rapidly in the last few years (6, 44, 45). Features as small as 10-nm dots on a 40-nm period have been made using evaporation and liftoff, and 70-nm-diameter electrodeposited nickel pillars have been made (44). The mold can be used repeatedly, with its lifetime limited by wear or contamination. In both nanoimprint and X-ray lithography, a master mold or mask must first be patterned, using electron-beam lithography or other methods, but the large number of replications made from each mask or mold would justify this time-consuming step.

Interference lithography, a maskless method, is ideal for patterning large-area periodic structures. A laser beam is split and then recombined on a resist-coated substrate to form a standing-wave pattern of fringes. The period of the pattern is given by $\lambda/(2\sin\phi)$, where ϕ is the half angle of intersection of the two laser beams, and λ is the laser wavelength. After the exposure, the substrate can be rotated in plane through 90° and a second exposure carried out to produce a set of fringes perpendicular to the first. The resulting grid pattern defines an array of holes or dots in the resist that forms the template for deposition or etching of the magnetic material (19). Pattern transfer is greatly improved by the use of a trilayer resist stack consisting of resist, an etch mask, and an antireflective coating layer into which the pattern is transferred to form the template. The minimum period of the pattern is just over $\lambda/2$, typically 180–200 nm, using an Ar-ion laser. Interference lithography has been used to make a variety of dot or ellipse arrays from evaporated Ni, Co (19, 25, 46, 47) or Co/Pt multilayers (22), pillar arrays from electroplated Ni or Co (19), and dots or ellipses etched from single-layer or multilayered sputtered films (19, 31, 48).

Making smaller features requires a shorter wavelength, but the radiation produced from shorter wavelength excimer lasers is insufficiently coherent to produce high-contrast fringes in a conventional interference lithography system. Instead, an achromatic interference lithography system has been developed, in which the radiation is diffracted by a set of three phase gratings. The period of the pattern is now given by half the period of the phase gratings. This system has been used to make 100-nm period patterns, for instance 30-nm evaporated Ni particles or 57–70-nm-diameter electrodeposited Ni pillars on a 100-nm pitch (49). It is expected that this method can be extended to 50-nm period features.

Other lithography processes involving stencils or masks have also been developed. The need for etching can be avoided by depositing magnetic material through a stencil mask into the final pattern (50, 51). This has been used successfully for making 100-nm features, the drawback being that the mask becomes coated with magnetic material that must be removed between uses. Another method involving a stencil mask is to locally modify the properties of a magnetic material by ion irradiation through the mask. For example, the strong perpendicular anisotropy of Co/Pt multilayer films can be destroyed by ion irradiation to give a film with in-plane magnetization (52). Irradiation through a mask with holes, therefore, gives an array of features with in-plane, soft magnetic properties, whereas the remainder of the film retains a perpendicular magnetization. This has been used to make features with 50-nm diameters and greater, with minor changes in surface topography caused by the irradiation (53, 54).

Magnetic particles can also be made by direct writing with a probe tip. For instance, material can be evaporated from a fine tip under pulsed voltage to form sub-100-nm-diameter pillars (55, 56). Iron pillars of heights up to 200 nm and diameters of 10 nm and greater were made by the decomposition of iron pentacarbonyl under a probe tip in a chemical vapor deposition (CVD) process (57, 58). Although these methods are relatively slow, they allow tall structures to be made in precise locations on a substrate.

Templated Growth

Several methods have been developed for making arrays of magnetic particles using self-organized templates. These templates can possess an ordered structure whose period may be smaller than that achievable through lithography. The template is used as a mask for the etching of a magnetic material or as a mold for deposition of magnetic material, using the same techniques as those used in the lithographic processes described above.

One example is the use of anodized alumina films as templates for electrodeposition. When aluminum is anodized, the resulting oxide layer contains a close-packed arrangement of fine pores, with the pore spacing and size controlled by anodization conditions such as voltage, current density, or pH, and a subsequent pore-widening process. The pores can then be back-filled by electrodeposition of cobalt, nickel, or other materials. Pore diameters of 11 nm and periods of 40 nm or more have been reported (17, 59, 60), with good close-packed ordering over areas of up to 100 μm^2. This method is well suited to the fabrication of arrays of long nanowires with lengths of microns or more, although it may be difficult to control the length distribution of short pillars deposited into alumina templates (60).

Block copolymers can also be self-assembled to make structures with various geometries. A block copolymer consists of chains of two chemically distinct, immiscible monomers A and B, joined together to form a polymer A_n-B_m. The polymer phase-separates to form domains of A and B, with a geometry depending on the ratio *n/m*. For instance, if $n \ll m$ the polymer consists of spheres of A in a matrix of B, whereas if $n \sim m$, the polymer forms lamellae. Other structures, such as cylinders or double-gyroid structures, have been described (61). The period of the structure increases with the length of the polymer, i.e. with $n + m$. A nanostructured thin film can be formed by spinning the polymer onto a substrate, then selectively removing one domain to form a template. These polymers have been used as templates for making dense arrays of 20-nm holes or dots in thin films (62, 63), electrodeposition of arrays of nanowires (64), or for patterning of films by etching (65). Magnetic particles can also be selectively grown within one domain of a block copolymer (66) to give a periodic structure.

The drawback of these templating methods is that the structures typically have poor long-range order, even though the short-range order is good. Long-range order could be improved by forming the self-assembled structure on a surface that has larger-scale ordered features, such as variations in surface topography or surface energy, or by growth in a field, in order to align the self-assembled structure over large distances. These processes are still in early stages of exploration.

Self-Assembly of Nanoparticles

Magnetic nanoparticles can be synthesized and assembled on a surface to form a regular array without the need for lithography. This has the potential for making very fine-scale structures over large areas in a rapid process and thus has attracted considerable attention. Structures with period of order 10 nm can be made, a feature size that is beyond that of most lithography techniques. However, it is critical to control the size distribution of the particles to obtain uniform properties and well-ordered self-assembly, and even if particles are monodisperse in size, the tendency of magnetic particles to agglomerate can impede separation and self-assembly. Metal particles also tend to oxidize, which affects their magnetic properties. Oxidation, agglomeration, and self-assembly can be controlled by coating the particles with a passivating layer of precise thickness and chemical functionality to protect the particles and to ensure a uniform separation between the particles once assembled onto a surface. As with templated self-assembled growth methods, these methods suffer from the disadvantage that long-range order may be poor, even though the particles typically form a close-packed array with excellent short-range order. The domain size or correlation length of the structure can be increased by improving the monodispersity of the particles and by careful growth and annealing of the monolayer.

Many methods to grow nanoparticles have been described, including vapor-phase growth by sputtering, laser ablation, arc discharge or evaporation (67–73), and decomposition or reduction of precursors in a solution or a solid matrix (74–77) or in inverse micelles (78, 79). These methods typically produce a distribution of sizes from which highly monodisperse samples can be made by size-selective precipitation. In one case, Co and FePt particles with diameters 3–10 nm, with a standard deviation of 5%, were produced by reduction in solution followed by separation and annealing (75, 76). The FePt initially formed in a disordered face-centered cubic (fcc) structure but changed to a tetragonal structure above 560°C. The Co precipitated as a metastable ε-phase that converted to a hexagonal close-packed structure above 300°C (76). Having made the particles, self-assembly can then be achieved by coating a solution of the particles onto a surface and evaporating the solvent. Some two- or three-dimensional superlattice structures with cubic or hexagonal symmetry have been made and magnetically characterized; an example is shown in Figure 3.

These self-assembly methods are promising for fabricating highly ordered and uniform structures, if the long-range order of the superlattice can be controlled. The magnetic particles themselves have good uniformity, with a microstructure determined by the synthesis and annealing methods. The ultimate method for fabricating a patterned medium with sub-50-nm period may involve self-assembled or templated structures, combined with, perhaps, larger scale lithography to form fiducial marks to align the structure and to provide features such as servo patterns.

MAGNETIC PROPERTIES OF PATTERNED MEDIA

In recent years there has been substantial progress in understanding and modeling the magnetic properties of small particles and particle arrays. However, due to the novelty of patterned media systems, there is so far no agreement on the optimum properties of a practical patterned medium. This can be contrasted with the case of thin film media, whose evolution over the last four decades has been characterized by increasing coercivity, reduction in film thickness and moment, reduced exchange coupling between grains, and increased sensitivity of the read heads. Present-day media have coercivities of around 240 kA m^{-1} and thicknesses of 15 nm to support recording densities of ~2 Gb cm^{-2}.

The magnetic properties of patterned media are driven by different considerations that lead to major differences in microstructure, coercivity, magnetic moment, and anisotropy direction compared with thin film media. The ideal behavior of a patterned medium element is that of a single-domain particle. Each element should have a small number of well-defined remanent states (usually two) to store data. Because the transition width is defined lithographically, high coercivities are unnecessary, and the patterned elements need only to have sufficient coercivity, perhaps below 40 kA m^{-1}, to be stable with respect to demagnetizing fields from their neighbors. It is undesirable for magnetization transitions or domain walls to exist within any element, and if the element is polycrystalline, the grains should be exchange-coupled. In this section, we review the magnetic properties of small particles and discuss design parameters for patterned medium elements.

Size-Dependent Magnetic Behavior

The magnetic microstructure and behavior of a magnetic material are highly size dependent. Exchange coupling aligns the spins parallel to each other in a ferromagnet. The existence of a uniform state of magnetization within the material minimizes the exchange energy. However, the competing tendency to minimize magnetostatic energy favors closed magnetization states in which there is no external magnetic field. This is achieved by the formation of domain walls or other non-uniform magnetization states such as vortices within the material, at the cost of introducing exchange energy. For small particles, the exchange energy term dominates, and the particles have uniform or near-uniform single-domain magnetization states, whereas large particles show multidomain magnetization states at remanence, i.e. in zero applied field (80, 81).

The size at which a particle becomes single domain depends on the balance between exchange coupling and magnetostatic energy and therefore on the geometry, anisotropy, and saturation moment M_s (A m^{-1}) of the particle. The exchange length, which can be defined as $\lambda_{ex} = \surd(2A/\mu_o M_s^2)$ (81), where A is the exchange constant of order 10^{-11} J m^{-1}, and μ_o is the permeability of free space, is a characteristic length scale used to describe magnetization patterns in a magnetic material. This parameter has values of several nanometers for ferromagnetic metals. [Note that in cgs units, λ_{ex} is usually expressed as $\surd(A/M_s^2)$, with A in erg cm^{-1} and M_s in emu cm^{-3}, giving values 2.5 times larger than those calculated using mks units with the definition used here.] Particles with dimensions near λ_{ex} are expected to be uniformly magnetized. As the dimensions increase, the magnetization distribution becomes less uniform, and flower states or magnetization vortices can develop, depending on the shape and anisotropy of the particle (82–86).

Only ellipsoidal bodies (including infinite cylinders, spheres, or plates) can have a truly uniform remanent magnetization state in which the magnetization is parallel at all places within the particle. Analytical methods have been used to calculate the critical size below which the particle is uniformly magnetized (87–89). Non-ellipsoidal or inhomogeneous bodies have non-uniform magnetization patterns at remanence, thus the definition of single-domain is less clear, but for small particles the deviation from uniform magnetization may be minor, and the particles are often described as being uniformly magnetized. Remanent states can be calculated numerically for particles of arbitrary geometry using micromagnetic models (25, 35, 82–86, 90–93). The particle is discretized into small interacting elements that are assumed to behave uniformly. The magnetic moment of each element obeys the Landau-Lifschitz-Gilbert (LLG) equations, in which the moment responds to an effective field derived from exchange coupling from its immediate neighbors, magnetostatic fields from all other elements, magnetocrystalline or other sources of anisotropy, and any applied external field. The equilibrium magnetization structure is then calculated. Figure 4 shows the different remanent states calculated for cubic particles of different diameters, aspect ratios, and anisotropies (84, 85). When the particles are small, a flower state exists that has a high remanence and is stabilized by anisotropy (84–86). As the particle size increases to about $10\lambda_{ex}$, a transition to a vortex state occurs, and multidomain states are predicted at a size of order $50\lambda_{ex}$, depending on the anisotropy. Similar results have been obtained for conical particles (25, 82), and thin particles also show transitions between in-plane flower and vortex states (35).

We have already seen that for a practical patterned media system, particles below 50-nm diameter are necessary to obtain the required high areal data densities. In this size regime, particles made from ferromagnetic metals or oxides are expected to be single-domain, although their non-ellipsoidal shapes and the presence of surface oxides, defects, or other inhomogeneities preclude the existence of a truly uniform remanent magnetization state. A flower state, in which the magnetization is parallel to the axis except near the ends, has high remanence along the axis and would be most suitable as a data storage element, whereas a vortex state has lower remanence as the magnetization twists around the particle axis, leading to a lower readback signal.

Experimentally measured remanent states, deduced from magnetic force microscopy, show a transition from multidomain states to a single-domain state as the particle size decreases, in agreement with the predictions of micromagnetic models. For instance, electrodeposited Ni pillars showed multidomain or vortex states at over 110-nm diameter but had single-domain flower states at 70-nm diameter (45, 94). Evaporated Co rectangles or dots changed from in-plane flower states to vortex or multidomain states with increasing thickness (24, 35), whereas Ni rectangles showed single-domain behavior below 150-nm width (5). The axial remanence of tall, evaporated Ni and Co cones decreased with increasing particle size as a magnetization vortex developed (25).

Contributions to Anisotropy

The magnetic anisotropy, K (J m^{-3}) expresses the tendency of the magnetization to lie along certain directions (easy axes) in a magnetic particle. K represents the energy required to rotate the magnetization perpendicular to the easy axis. In patterned media, a uniaxial anisotropy is preferable, so that at remanence, the

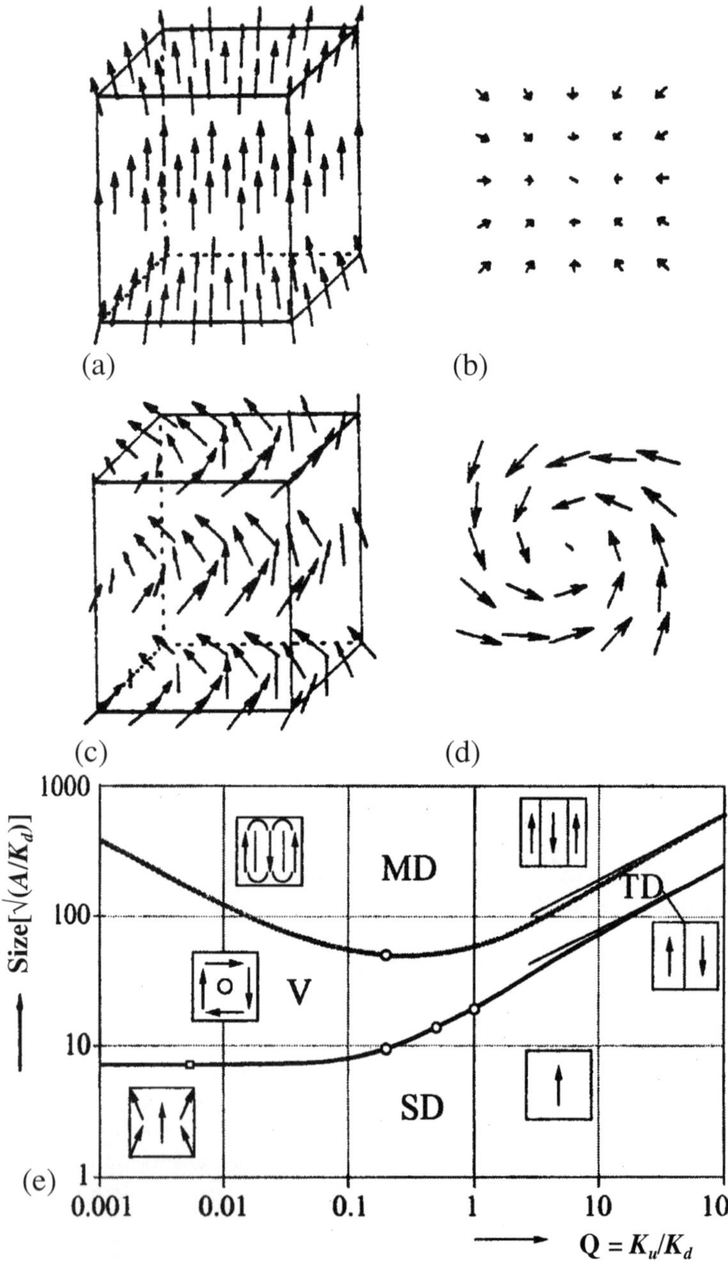

Figure 4 Equilibrium remanent states of a small cubic particles: (*a*) and (*b*) show a flower-state magnetization, (*c*) and (*d*) show a vortex state. In (*b*) and (*d*) the projection of the moments of the bottom plane of the particle is shown (84) (reprinted with permission, © AIP 1988). (*e*) The phase-diagram of remanent states for a cubic particle with uniaxial anisotropy K_u. The anisotropy is expressed as K_u/K_d, where $K_d = \mu_o M_s^2/2$, with M_s the saturation moment SD, single domain; V, vortex; and MD, multidomain states (81) (reprinted with permission, © Springer-Verlag 1998).

magnetization lies in one of two possible directions parallel to the easy axis to store one bit of data. The anisotropy is a key attribute of any particle used in a patterned medium because it influences both the remanent state and the field needed to reverse the magnetization.

There are several possible contributions to K. The magnetocrystalline anisotropy term arises from the preferred magnetization along certain crystallographic axes of the material, for instance [0001] in Co, [100] in Fe, and [111] in Ni at room temperature. Shape anisotropy is a result of the magnetostatic energy associated with a magnetized body, which is smaller if the body is magnetized along its long axis. Other anisotropy terms include magnetoelastic anisotropy, arising from the interaction between strain and magnetization, and interface or surface anisotropies attributed to the asymmetrical environment at interfaces. One important example of interface anisotropy is found in Co/Pt or Co/Pd multilayers, where if the layers are thin, a strong anisotropy exists perpendicular to the layers (54). Magnetoelastic effects are usually neglected, although they may affect the magnetic properties of particles prepared inside a rigid template (20).

Of these terms, the most widely used in controlling the magnetic properties of particles are shape and magnetocrystalline anisotropies. Shape anisotropy provides an easy axis parallel to the long axis of a prolate ellipsoidal particle, such as the oxide particles used in many tapes or floppy disks. (Oblate particles have an easy plane of magnetization perpendicular to the axis, hence such particles do not have two unique magnetization directions as required for data storage.) The uniaxial shape anisotropy $K_s = 0.5\,\mu_o M_s^2(N_x - N_z)$ is expressed in terms of the difference between the demagnetization factors along the axis, N_z, and perpendicular to the axis, N_x. The energy is then given by $K_c + K_s \sin^2\theta$, where K_c is a constant and θ is the angle between the magnetization direction and the axis. Values for N can be determined analytically for ellipsoidal particles (95), and solutions also exist for other geometries such as uniformly magnetized prisms (96–98).

Magnetocrystalline anisotropy is highest for uniaxial materials that include Co; alloys such as the CoCrPt used in hard disks; metallic compounds such as FePt, CoPt, and $SmCo_5$; and some hexagonal magnetic oxides including $BaO.6Fe_2O_3$. In a uniaxial material, the magnetocrystalline anisotropy energy is expressed as $K_o + K_1 \sin^2\theta + K_2 \sin^4\theta$ plus higher order terms. The first term is neglected because it is independent of magnetization direction, and usually $K_1 > K_2$. K_1 varies between 0.2 MJ m^{-3} for CoCrPt alloys, 0.45 MJ m^{-3} for Co, 10 MJ m^{-3} for FePt, and 11–20 MJ m^{-3} for $SmCo_5$ (1, 80).

The net anisotropy of a uniaxial particle is determined by the sum of the anisotropy contributions, taking account of the orientation of the anisotropy axes. For example, consider a small, flat circular particle made from a Co/Pt multilayer film. In this material, the perpendicular anisotropy caused by the layered structure is strong enough to overcome the shape anisotropy of the oblate particle, and such particles are indeed observed to have an out-of-plane uniaxial anisotropy (22). There have been several examples of arrays of particles made from epitaxial Co (41) or Fe (99, 100) thin films, in which the in-plane magnetocrystalline anisotropy is larger than shape anisotropy and dominates the behavior of the particles. In another study (16), the behavior of Co wires was interpreted in terms of the effects of orthogonal shape and magnetocrystalline anisotropies.

Magnetization Reversal

The behavior of uniformly magnetized, non-interacting single-domain particles was identified by Stoner & Wohlfarth in 1948 (101), and this model of coherent or uniform rotation has been widely applied to describe the hysteresis of magnetic particles. The particle is assumed to have uniaxial anisotropy of magnitude K. In zero applied field, the magnetization of the particle lies along the easy axis, but it rotates in response to an applied field. If the field makes an angle of α to the easy axis, and the angle between the particle magnetization and the easy axis is θ, then at equilibrium, $K\sin(2\theta) = \mu_0 M_s H \sin(\alpha - \theta)$, and the magnetization resolved along the field direction is $M_s \cos(\theta - \alpha)$. This model allows the hysteresis loop of the particle to be determined for arbitrary field direction. For fields applied parallel to the easy axis, the particle has a square hysteresis loop with switching field (i.e. the field required to reverse its magnetization) $H_{sw} = 2K/\mu_0 M_s$. For fields perpendicular to the easy axis, the particle shows reversible behavior, and a field of $2K/\mu_0 M_s$ is required to saturate the magnetization parallel to the field. The angular variation of coercivity can therefore be used as an indication of reversal mechanism. The effect of temperature on switching behavior was included by Neel (102) and Brown (103) by considering thermal activation over the energy barrier for reversal.

As the particle size increases, incoherent reversal modes such as curling can occur at lower applied fields than the field require to cause coherent rotation. In incoherent reversal, the magnetization does not remain parallel throughout the reversal process, which leads to an increase in exchange energy but a decrease in the amount of magnetostatic energy. In the case of curling in prolate ellipsoids, the magnetization, which was initially parallel to the long axis of the particle, rotates about an axis perpendicular to the long axis, leading to closed loops of flux perpendicular to the long axis (88, 89, 97). This process occurs more easily than coherent rotation for particles with radii greater than a small multiple of the exchange length, whose value decreases slowly as the particle becomes longer (80, 81).

Numerical micromagnetic methods have also been applied to determine reversal fields for other geometries, such as rectangular prisms or cubes (84, 86, 90, 104), or cylinders with differently shaped ends (105). The presence of surface defects often promotes reversal (106), and in simulations of polycrystalline magnetic elements, the variation in magnetocrystalline anisotropy between grains was found to provide sites at which reversal nucleates (107). In long particles, reversal proceeds by nucleation and propagation of a vortex or domain wall (21, 90, 108, 109).

Experimentally, observations of Stoner-Wohlfarth behavior are rare, either because incoherent reversal occurs more easily (because the particles are relatively large and non-uniformly magnetized, or because defects or surface irregularities exist), or because there are magnetostatic interactions between particles that can either promote or impede reversal. For small particles, thermally induced switching reduces the measured switching field below the coherent rotation value. Particles of dimensions ~100 nm and greater generally show evidence of non-uniform reversal; for example, the energy barrier for reversal or the magnitude or angular dependence of the switching field may be inconsistent with the coherent rotation model (28, 34, 43, 110–115). Switching fields typically increase as the particles become smaller, and their behavior approximates more closely to coherent rotation. For instance, the switching field of 1-μm-long, 35-nm-thick Co bars increased from 47 kA m^{-1} to 240 kA m^{-1} as the width decreased from 300 to 30 nm, whereas for similar Ni samples, the switching field reached 58 kA m^{-1} for 30-nm-wide bars (6).

Data on individual sub-50-nm particles are less abundant. Measurements on an ellipsoidal 25-nm fcc Co crystal made by arc-discharge (72) showed uniform rotation below 6 K, although the behavior diverged from this model if the surfaces were oxidized. The reversal of 10–20-nm-thick barium ferrite crystals has also been measured at low temperatures (116, 117). The Neel-Brown model was followed below 6 K, except for very low temperatures where more complex behavior occurred owing to tunneling or higher-order anisotropy terms. Measurements of 10-nm diameter, 50–250-nm tall iron particles, made using probe-based CVD (57, 58), indicated that curling dominates the reversal at low temperatures, whereas at higher temperatures thermal activation data were inconsistent with a single energy barrier. Ni pyramids of 30 nm showed evidence of non-uniform reversal, attributed to their polycrystalline microstructure (118). Hence, even for small particles, uniform rotation is the exception, and the shape, microstructure, and surface condition of the particle influence the switching field.

The switching speed of particles is of great interest in patterned media applications because it determines how long a field must be applied to ensure that switching occurs. Over long time scales, the magnetization of particles can be calculated from thermal models of reversal over an energy barrier (see below), but on very short, nanosecond time scales, dynamic behavior is determined by magnetization precession and can be modeled using the LLG equations. These equations incorporate a phenomenological damping constant α that controls the evolution of magnetization and therefore the speed of the dynamic magnetization reversal. Values for α are not well known, but the speed of reversal depends critically on the value of α (107) and also on the degree of magnetostatic and exchange coupling between the magnetic particles (119, 120). Calculations of switching speeds for patterned media elements show that tall particles, which switch incoherently, may take 1–2 ns to reverse (121, 122), whereas shorter particles can switch coherently in sub-nanosecond times. If the external field is applied for less than the required switching time, the particle may return to its original magnetization state (6). These considerations will contribute to design of patterned media elements because switching will have to occur within 1 ns to enable GHz writing speeds.

Superparamagnetism

As in the case of conventional media, patterned media are subject to superparamagnetic behavior if the energy barrier for magnetization reversal is small enough. Considering the coherent rotation model for a particle of volume V and net anisotropy

K, the energy barrier for reversal is $E_B = KV$. The rate of switching is given by $1/\tau = f_o \exp(-E_B/kT)$, where f_o is an attempt frequency typically taken as 10^9 s^{-1}. In a laboratory experiment, for the particle to retain its magnetization on average for t = 100 s, $E_B = 25kT$, whereas for data storage over t = 10 years, $E_B = 40$ kT. This allows the minimum KV product to be estimated to ensure thermal stability.

If the particle reverses by an incoherent mechanism, the energy barrier will be lower than that calculated for the uniform rotation model. The energy barrier is often expressed as KV^*, where V^* is an effective switching volume. KV^* can be measured experimentally, for instance, from the dependence of coercivity on the sweep rate of the magnetic field (123) or from the decay of magnetization in an applied reverse field below the coercivity (124, 125). The relationship between V^* and the physical volume of the particles provides some insight into the reversal mechanism.

For interacting particles, the presence of magnetostatic fields (smaller than the switching field) modifies the energy barrier and therefore the thermal stability criterion. For instance, in a transition written on a hard disk, demagnetizing fields can be high [22 kAm^{-1} in one calculation using parameters characteristic of a 1.5 Gb cm^{-2} medium (1)] and lead to a more stringent thermal stability criterion than is derived for isolated particles. In a patterned medium, the proximity between the particles will determine the strength of interactions, but their physical separation will lead to interactions that are smaller than those in thin film media. Patterned media elements made from high anisotropy materials could be as small as a few nanometers without exhibiting thermal instability.

Interactions Between Elements and the Switching Field Distribution

A key issue in patterned media designs is achieving uniform magnetic behavior within the elements of an array. It is commonly observed that magnetic particles in an array exhibit a range of switching fields so that some particles switch at lower applied fields than others (e.g. 24, 32). This spread in switching fields has two origins. There is an intrinsic variability between particles owing to small differences in shape, size, or microstructure, often expressed by a Gaussian distribution of switching fields. There is also an apparent spread in switching fields caused by magnetostatic interactions between particles because interactions contribute to the net field acting on each particle. Intrinsic variability can be minimized by improving the uniformity of the array, which requires good control over lithography or assembly processes and the microstructure of the particles. Magnetostatic interactions can be reduced by lowering the moment of the particles, although this reduces the readback signal they produce, or by increasing the spacing between them, which also reduces the areal density.

To a first approximation, magnetized elements can be treated as dipoles, and nearest neighbor interaction fields (i.e. the magnetic field that an element exerts on its neighbor) can be easily calculated (80). For particles that are closely spaced, the dipole approximation underestimates interaction fields, and the interactions can be calculated by treating the particles as magnetized prisms or cylinders (126, 127). The long-range nature of magnetostatic fields means that contributions from second, third, and further neighbors must also be considered in calculating the total magnetostatic field acting on a particle. As an example, consider a square array of particles with out-of-plane magnetization, in which all the particles are magnetized up. The total magnetic field acting on any particle is equal to $9H_i$, where H_i is the nearest neighbor interaction field. (In the dipole approximation, $H_i = m/4\pi\mu_o r^3$, where m is the moment of each particle and r is the spacing.) The field $9H_i$ acting on each particle is demagnetizing, i.e. it promotes reversal. If the switching field of a particle is given by H_{sw}, the particle will switch when the applied external reverse field reaches $(H_{sw} - 9H_i)$. In contrast, if a particle magnetized up is surrounded by particles magnetized down, an external field of $(H_{sw} + 9H_i)$ will be needed to reverse it.

This suggests that interactions will cause a broadening of the switching field distribution of a perpendicularly magnetized array. Experimentally, this is evident from the shearing of the out-of-plane hysteresis loop. The lowest magnetostatic energy state is a checkerboard pattern, with particles alternately magnetized up and down; this has been observed using magnetic force microscopy in ac-demagnetized arrays of particles with out-of-plane magnetization (49, 94). Other experiments and modeling have demonstrated the effects of interactions; for instance, the switching field of particles is affected by their neighbors (128), and collective switching can lead to steps in the hysteresis loop of a small array of particles (129, 130). It has also been noted that the remanent state of particles can be affected by interactions. For instance, the transition from in-plane to out-of-plane magnetization in cylinders is modified by interactions (130, 131).

Figure 5 shows that a simple model of interacting particles can reproduce hysteresis loops measured for arrays with out-of-plane magnetization (126). In this model, the particles in the array are assigned switching fields from a Gaussian distribution, with mean H_c and standard deviation σ ($\sigma \ll H_c$). The hysteresis loop is calculated iteratively, assuming that each particle switches when the sum of the applied field and the interaction field exceeds its switching field. In the absence of magnetostatic interactions, the collective hysteresis loop is square, except for some rounding of the corners corresponding to the Gaussian distribution in switching fields. However, if interactions are included, then the hysteresis loop becomes sheared, with a slope that is proportional to H_i. An excellent fit to the measured data can be obtained, with one adjustable parameter, σ.

If the maximum interaction field is smaller than the switching field of the particles, then the magnetization state of any particle is stable, irrespective of the magnetization direction of its neighbors. The remanence of the array will ideally be 1, as seen in the sample of Figure 5*a*. However, if interactions are sufficiently large so that $9H_i > H_{sw}$, then the saturated state is unstable at remanence, and particles can spontaneously reverse through interactions from their neighbors. An

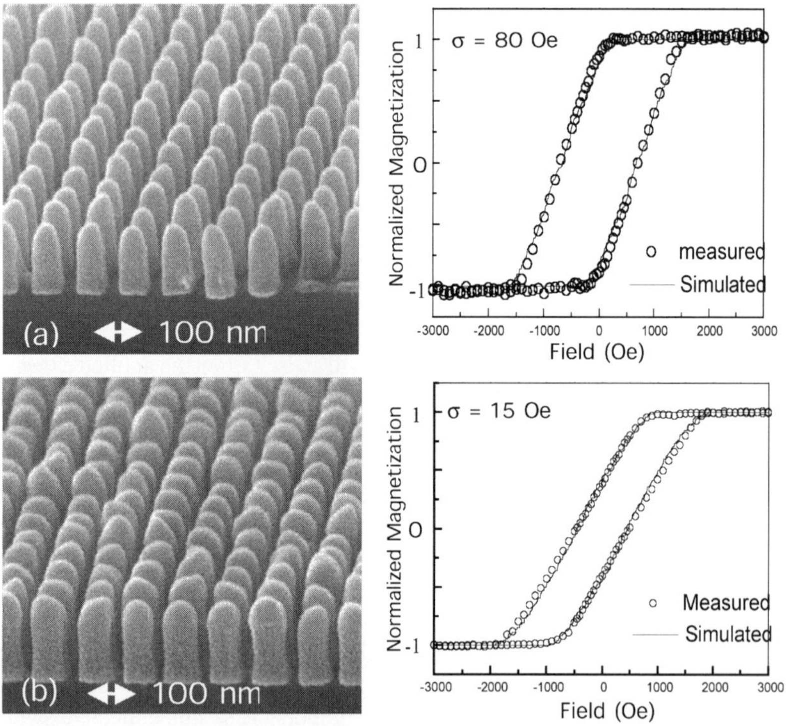

Figure 5 In-plane and out-of-plane hysteresis loops for two samples of 100-nm period electrodeposited Ni pillars. (*a*) 57-nm diameter, 115-nm tall Ni pillars; (*b*) 70-nm diameter, 150-nm tall Ni pillars. The thin line is a fit to the out-of-plane loop from a computational model that treats the pillars as interacting dipoles. The fit has only one adjustable parameter, the standard deviation of the switching fields, σ, which was set to 15 Oe for (*a*) and 80 Oe for (*b*) (1 Oe $=$ 79 A m^{-1}) (126) (reprinted with permission, © IEEE 2000).

array with this characteristic, such as that shown in Figure 5*b*, would be unsuitable for data storage.

In the case of arrays with in-plane magnetization, if all particles are magnetized parallel to each other, then the interaction fields from some neighbors are demagnetizing, while the fields from others stabilize the magnetization state. Unlike the case of perpendicularly magnetized arrays, interactions can actually make the collective hysteresis loops squarer because interactions stabilize rows of particles with aligned magnetization (23, 24). Experiments on magnetized bars show that the switching field decreases as the lateral bar spacing is reduced (5).

The switching field distribution of an array can be characterized by magnetic force microscope (MFM) measurements of the state of the array at various applied fields. By subtracting the magnetostatic interaction field from the applied field, the intrinsic switching field of each particle can be calculated, thus the contributions of intrinsic variability and magnetostatic interactions to switching field distribution can be separated.

Intrinsic variability between particles is generally attributed to differences in microstructure, shape, defects, or surface conditions, such as the presence of an antiferromagnetic surface oxide, which alter the switching field. The variability in magnetic behavior due to polycrystallinity has been calculated for Co (132) and CoCrPt (55) elements, and it was found that random crystal anisotropy could exceed shape anisotropy even for particles containing 50–100 grains. In evaporated polycrystalline conical Ni and Co particles, the intrinsic spread in switching fields for Ni was small, but for Co there was a large spread because the Co grains have a strong uniaxial anisotropy whereas the Ni has a weak cubic anisotropy (25). Another set of experiments on an array of 30-nm diameter evaporated Ni particles (118) showed inhomogeneous reversal with an activation volume similar to the grain size of the Ni. A wide spread in switching fields occurred because the columnar microstructure of the particle dominated the reversal process. In an experiment on single-crystal Ni pillars (133), a standard deviation of $\sigma = 8.3\ \mathrm{kA\ m^{-1}}$ was determined for curling reversal of the particles, owing to fluctuations in particle size and in the direction of the crystallographic axes within the particles. From all this data, it is clear that good control of the particle microstructure will be an essential part of the design of any practical patterned medium.

RECORDING SYSTEMS

The concept of patterning a hard disk originated as early as the 1960s (134) when patterned tracks were proposed as a means to improve noise and tracking. Over 10 years ago, tracks were etched into disks so that the recording properties of narrow tracks could be studied, and the use of patterned features in disks was proposed as a means to detect head misregistration and provide servo information (135, 136). At the same time, the idea of storing data in discrete, nanoscale-patterned elements began to be explored as lithographic patterning techniques improved and as thermal stability effects in conventional media became better understood. These developments have been briefly summarized (6, 7). So far, most research on patterned media systems has focused on the fabrication and properties of the media itself, but there has been some exploration in the last few years of other critical issues in patterned recording systems, such as reading and writing methods, servo systems, synchronization, noise, and signal processing. Patterned recording systems can be broadly divided into two categories: spinning-disk systems, and probe-based systems. Figures 6 and 7 show examples of each category.

Patterned media systems based on spinning disks incorporate a patterned disk that is addressed by an ultra-narrow magnetoresistive (MR) read/inductive write head. The elements can be magnetized longitudinally or vertically. Much of the technology of such systems would be similar to conventional hard drives, with the significant exception that the write signal needs to be synchronized with the

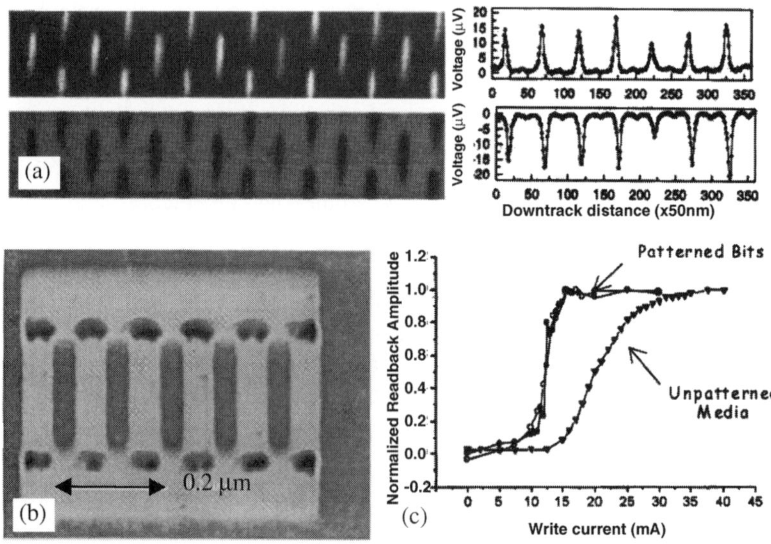

Figure 6 Examples of patterned media written and read using a conventional read-write head. (*a*) Tracks written on an array of nickel pillars with 2-μm spacing using a head dragged over the surface. The dark and light spots indicate pillars with up and down magnetization; the right-hand images show the voltage output from the head (adapted from 141) (*a*) reprinted with permission, © AIP 1999. (*b*) 160 nm × 40 nm bits patterned into a hard disk using a focused ion beam. The dark areas represent material that has been removed. (*c*) Readback voltage versus write current for sample (*b*), showing a sharp response to the write current (139) (reprinted with permission, © IEEE 2000).

physical location of the magnetic elements. Overall, it appears that many of the signal-processing issues will be similar to those encountered in a conventional hard drive system (7). The behavior of these systems, including the origins of noise, have been analyzed (121, 137, 138). Ideally, patterned media systems are free of transition noise, but noise can originate from errors in lithography or from the jaggedness of the edges of the elements. One estimate indicates that a standard deviation of 1 nm or less in pattern placement is necessary in an array with 25-nm pitch to obtain an error rate of 10^{-10} for longitudinal patterned media, and derives an optimum ratio between bit size and array period (137). In another example, both longitudinal and vertical media addressed with single-pole or two-pole heads were analyzed. Error rates of 10^{-9} were calculated for 15 Gbit cm^{-2} longitudinal media with 7% pattern placement errors (138). The write process was also analyzed at 1 GHz; short elements reverse uniformly within 1 ns, so GHz data rates are possible, whereas tall elements reverse more slowly (122, 138). In these recording systems, the pole piece of the write head would be narrow and might act as a single-domain particle capable of writing elements even when no write current is flowing. An anisotropy must be introduced while the write current is switched off

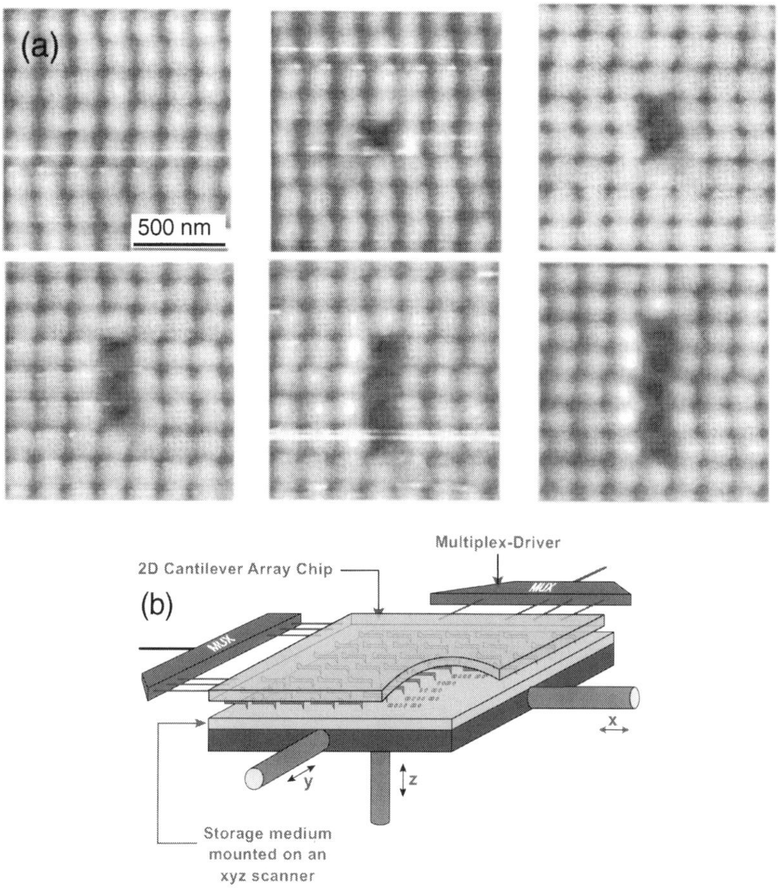

Figure 7 Examples of patterned media written and read using a scanning probe tip. (*a*) A set of 5 bits written sequentially onto a patterned medium using a probe tip, by passing a current that locally heats each dot, allowing it to reverse (29) (reprinted with permission, © AIP 1999). (*b*) The Milipede concept of an array of probe tips addressing a recording medium (155) (reprinted with permission IEEE 1999).

to orient the head magnetization perpendicular to the track to prevent inadvertent writing.

Experimental measurements of patterned hard disks using a spin-stand can be done if the disk surface is smooth enough for the head to fly over it (138, 139). Elements patterned using a focused ion beam (139) show the expected absence of transition noise, and a sharp response of the bit magnetization to the write current. The samples used in these demonstrations are made from conventional media in which the grains are exchange decoupled, but despite this, micromagnetic calculations show a fairly uniform magnetization within each patterned element, and weak demagnetization fields between elements. Increasing the orientation ratio of the media (the ratio of circumferential to radial coercivity) improves the signal by increasing the remanence of the bits. In terms of noise, the media certainly appear to be promising, and the patterned elements provide excellent position error signals, but so far the research has not addressed write synchronization issues or the fabrication of ultra-narrow heads, and tribology issues are unanswered. Synchronization will require much greater control of spindle speed than currently used, or simultaneous readback may be necessary to provide position signals for the write process.

Recording measurements have also been carried out on arrays of magnetic particles such as electrodeposited nickel pillars, by reading and writing using a head that is dragged slowly over the surface. To obtain a flat surface, the pillars are embedded in a rigid substrate such as silica (140) or alumina (39) and smoothed by chemical-mechanical polishing. Individual pillars have been written and read using a scanned MR head (38, 39, 140, 141). It was found that the current through the MR element could cause accidental writing of the pillars (140). Perpendicularly magnetized thin films have also been patterned into 100-nm period single-domain elements, using a focused ion beam, and written using a head dragged over the surface. The results showed that the write signal must be synchronized to the bit positions in order to write the bits cleanly (142).

An alternative to a spinning-disk system is to use a probe-based storage system in which a magnetic probe tip mounted on a cantilever is used to address patterned elements. To write data, a localized magnetic field is applied, for instance, by bringing a magnetic tip in close proximity to a patterned element, possibly with the addition of an external applied field, such that one element is written and not its neighbor. Early demonstrations showed that patterns could be written onto continuous films (143–145). The magnetization state of magnetic bars (146, 147) or ellipses (114) with in-plane moment and pillars with perpendicular moment (19) was reversed using a magnetic tip. A 150-nm period array of perpendicularly magnetized elements was written by passing a current from a MFM tip to heat an element, allowing it to reverse (29). Writing was achieved despite the large spread in switching fields of the elements.

Readback may be accomplished by a method similar to a MFM, in which the magnetization state of the elements is detected from changes in the oscillation of the cantilever that holds the magnetized tip. The tip only needs to detect whether an element has an up or down magnetization state, so the requirements on spatial resolution are relatively modest and could be met by existing MFM technology. In MFM, the detected signal is proportional to the field gradient above the sample. Other detection methods are possible, for instance, by measuring the field directly using a miniature MR element on a probe tip, or by a magneto-optical measurement using an optical probe, but all these systems rely on the mechanical positioning of a probe tip above the surface.

Magnetic systems are just one of several options for ultra-high density probe based storage. Several other types of storage systems have been proposed using atomic force microscopy or tunneling methods, for example, thermomechanical writing (using a heated probe tip to make indentations in a polymer layer), charge injection into a semiconductor layer, oxidation, or ion migration under a localized electric field, and near-field optical storage (148–152). For some of these methods, reading, writing, and erasing have all been demonstrated. In thermomechanical writing, for instance, bit sizes as small as 10 nm have been reported (153). The major drawback of any of these methods is the low data rates achievable from a scanning probe system. Large numbers of tips working in parallel would be necessary to obtain a high data rate. Some of the issues involved in designing systems of multiple probe tips have been investigated. Low-mass cantilevers could reach data rates up to 6 MHz (148, 154), and arrays of up to 1000 tips working in parallel have been developed (151, 155); the combination of these techniques could allow GHz data rates to be achieved in a probe-based system. An integrated array of 32×32 cantilever tips, known as Milipede, has been designed for thermomechanical storage, and is the largest system developed to date (155).

The necessary properties of patterned magnetic media (switching field, anisotropy, magnetic moment, and geometry) have not been well established for either type of recording system. However, some general guidelines can be stated. Maximizing areal density and readback signal favors a densely packed array of high-moment particles with perpendicular magnetization, but this leads to strong interactions between the particles. The array must be sufficiently weakly interacting that interaction fields alone cannot reverse the magnetization of an element. This implies that $9H_i < H_{sw,\,min}$ where $H_{sw,\,min}$ is the smallest switching field of a particle in the array, which can be taken as $H_c - 3\sigma$, where H_c is the average switching field (the coercivity) of the array and σ the standard deviation of switching fields. The writing process imposes a more stringent requirement on particle uniformity. In order to write data, it is necessary for the write element to apply a localized field to one element without writing its neighbor. If the switching fields of neighbors vary by σ, the head field gradient must be well in excess of σ/r to safely write one element but not its neighbor, where r is the array period. From dynamic calculations, switching appears to occur more rapidly for low-aspect ratio particles (122, 136), thus short particles with magnetocrystalline anisotropy may be preferable over tall particles that rely on shape anisotropy.

For readback, the elements need not be uniformly magnetized to saturation, as long as there is sufficient remanence to be detected by the recording system. Readback schemes could detect the magnetization directly (for instance, a near-field magneto-optical probe), the field above the element (as in an MR head), or the field gradient above the element (as in the MFM). Most readback methods described in the literature rely on detection of field or field-gradient, so the fields from neighboring particles will affect the readback and may introduce signal jitter.

Arrays of patterned magnetic elements are also being used in magnetic random access memories (MRAMs) (156). Each data bit is stored in a small element made from a MR spin-valve or spin-tunneling film. Data are written by sending electrical currents through conductor lines to create small magnetic fields near the elements, sufficient to reverse their magnetization, similar to a core memory. Readback is achieved by measuring the resistance of the elements, which depends on their magnetic state. MRAM areal densities will not match those of other patterned media systems because of the need for electrical interconnects, but they have the advantage that there are no moving parts, and they provide a nonvolatile alternative to semiconductor memory. MRAMs have been in development and production for several years. As densities increase, the size of the MR elements has decreased and high-density MRAM chips incorporate sub-micron magnetic elements. The same issues of domain states, switching behavior, inter-element interactions, thermal stability, and the need for high-resolution lithographic patterning are common to both MRAM elements and patterned media based on spinning disks or probe storage.

CONCLUSIONS

We have seen that patterned media present an intriguing possibility for ultra-high density data storage beyond the superparamagnetic limit believed to be approaching for conventional longitudinal media. Whether patterned media are adopted will depend on whether a set of complex fabrication, addressing, and integration problems can be solved economically. To reach small enough feature sizes, unconventional lithography techniques are needed. The ultimate patterned medium may consist of an array of magnetic nanoparticles, self-assembled on a substrate with the aid of lithography to produce an array with long-range order. The particles should have well-defined remanent states, most likely a result of uniaxial anisotropy, so that each one can store a data bit. The geometry of the array, the magnetic moment and the net anisotropy of the particles are chosen to maximize the readback signal and areal density while maintaining sufficiently low interactions between particles such that written data patterns are stable. The key to designing the medium is to minimize intrinsic variability, by control of the particle microstructure (perhaps by using single-crystal or amorphous material, with well controlled anisotropy and surface state), and to minimize magnetostatic interactions by the choice of array geometry and particle moment. One possibility is the use of materials that have strong perpendicular anisotropy, allowing out-of-plane magnetization states to be stable in very low aspect ratio, low moment particles.

The writing and reading of the arrays provide the most difficult challenges. The approach evolved from hard drives is to use a spinning patterned disk, addressed using an ultra-narrow read-write head. This would give high data rates using signal processing schemes derived from current hard drives, but the design of the head still requires considerable development, and write synchronization issues must be solved. A different type of storage system involves the use of parallel probe tips to address a magnetic array. Probe arrays have been fabricated, but the writing process needs to be investigated in much greater detail. In probe-based systems, magnetic storage schemes compete with other methods such as thermomechanical writing and it is not clear which method will ultimately prevail. What is certain is that these systems will be commercialized only if they represent an economical solution to high-density, high data rate storage beyond what is possible with spinning hard disk systems. MRAMs offer another data storage method that is not capable of such high data densities as patterned media but is nonetheless very attractive because it avoids the reliability issues associated with mechanical systems.

The research into lithographically patterned magnetic elements has produced a wealth of information into the magnetization states, switching mechanisms, anisotropy, and interactions of small ferromagnets. This is useful in patterned media designs and also in a range of magnetoelectronic devices, including MRAMs, field sensors, recording heads, and new generations of "spintronic" devices that use magnetic elements to manipulate electron spin. In the future, we expect to see a variety of new hybrid devices emerging that will make use of these nanoscale magnetic elements.

LITERATURE CITED

1. Weller D, Doerner M. 2000. *Annu. Rev. Mater. Sci.* 30:611
2. Proc. Intermag Conf. 2000. *IEEE Trans. Magn.* 36(6): In press
3. Charap SH, Lu L, He Y. 1997. *IEEE Trans. Magn.* 33:978
4. Bertram HN, Williams M. 2001. *IEEE Trans. Magn.* 36:4
5. Chou M, Wei PR, Krauss PB, Fisher J. 1994. *J. Vac. Sci. Technol. B* 12:3695
6. Chou SY. 1997. *Proc. IEEE* 85:652
7. White RL, New RMH, Pease RFW. 1997. *IEEE Trans. Magn.* 33:990
8. Moser A, Weller D. 1999. *IEEE Trans. Magn.* 35:2808
9. Weller D, Moser A. 1999. *IEEE Trans. Magn.* 35:4423
10. Bertram HN, Zhou H, Gustafson R. 1998. *IEEE Trans. Magn.* 34:1845
11. Ouchi K, Honda N. 2000. *IEEE Trans. Magn.* 36:16
12. Wood R. 2000. *IEEE Trans. Magn.* 36:36
13. Katayama H, Sawamura S, Ogimoto Y, Kojima K, Ohta K. 1999. *J. Magn. Soc. Jpn.* 23:233

13a. Katayama H, Hamamoto M, Sato J, Murakami Y, Kojima K. 2000. *IEEE Trans. Magn.* 36:195

14. Ruigrok JJM, Coehoorn R, Cumpson SR, Kesteren HW. 2000. *J. Appl. Phys.* 87:5396
15. Weller D, Moser A, Folks L, Best ME, Lee W, et al. 2000. *IEEE Trans. Magn.* 36:10
16. Ferre R, Ounadjela K, George JM, Pirauz L, Dubois S. 1997. *Phys. Rev. B.* 56:14066
17. Garcia JM, Asenjo A, Velazquez J, Garcia D, Vazquez M, et al. 1999. *J. Appl. Phys.* 85:5480
18. Schwarzacher W, Kasyutich OI, Evans PR, Darbyshire MG, Li G, et al. 1999. *J. Magn. Magn. Mater.* 198:185
19. Ross CA, Smith HI, Savas TA, Schattenburg M, Farhoud M, et al. 1999. *J. Vac. Sci. Technol. B* 17:3168
20. Pignard S, Goglio G, Radulescu A, Piraux L, Dubois S, et al. 2000. *J. Appl. Phys.* 87:824
21. Smyth JF, Schultz S, Fredkin DR, Kern DP, Rishton SA, et al. 1991. *J. Appl. Phys.* 69:5262
22. Thielen M, Kirsch S, Weinforth H, Carl A, Wassermann EF. 1998. *IEEE Trans. Magn.* 34:1009
23. Cowburn RP. 2000. *J. Phys. D Appl. Phys.* 33:R1

24. Evoy S, Carr DW, Sekaric L, Suzuki Y, Parpia JM, Craighead HC. 2000. *J. Appl. Phys.* 87:404
25. Ross CA, Farhoud M, Smith HI, Hwang M, Redjdal M, Humphrey F. 2001. *J. Appl. Phys.* 89:1310
26. New RMH, Pease RFW, White RL. 1994. *J. Vac. Sci. Technol. B* 12:3196
27. Rousseaux F, Decanini D, Carcenac F, Cambril E, Ravet MF, et al. 1995. *J. Vac. Sci. Technol. B* 13:2787
28. Wernsdorfer W, Hasselbach K, Mailly D, Barbara B, Benoit A, et al. 1995. *J. Magn. Magn. Mater.* 145:33
29. Haginoya C, Koike K, Hirayama Y, Yamamoto J, Ishibashi M, et al. 1999. *Appl. Phys. Lett.* 75:3159
30. Walsh M, Hao Y, Ross CA, Smith HI. 2000. *J. Vac. Sci. Technol. B* 18:3539
31. Haast MAM, Schuirhuis JR, Abelmann L, Lodder JC, Popma ThJ. 1998. *IEEE Trans. Magn.* 34:1006
32. Haginoya C, Heike S, Ishibashi M, Nakamura K, Koike K, et al. 1999. *J. Appl. Phys.* 85:8327
33. Smyth JF, Schultz S, Kern D, Schmid H, Yee D. 1988. *J. Appl. Phys.* 63:4237
34. Lederman M, Gibson GA, Schultz S. 1993. *J. Appl. Phys.* 73:6961
35. Cowburn RP, Koltsov DK, Adeye AO, Welland ME, Tricker DM. 1999. *Phys. Rev. Lett.* 85:1042
36. Kirk KJ, Chapman JN, Wilkinson CDW. 1999. *J. Appl. Phys.* 85:5237
37. O'Barr R, Lederman RM, Schultz S, Xu W, Scherer A, Tonicci RJ. 1996. *J. Appl. Phys.* 79:5303
38. O'Barr R, Yamamoto SY, Schultz S, Xu W, Scherer A. 1997. *J. Appl. Phys.* 81:4730
39. Wong J, Scherer A, Todorovic M, Schultz S. 1999. *J. Appl. Phys.* 85:5489
40. New RMH, Pease RFW, White RL, Osgood RM, Babcock K. 1996. *J. Appl. Phys.* 79:5851
41. Kohda T, Otani Y, Novosad V, Fukamichi K, Yuasa S, et al. 1999. *IEEE Trans. Magn.* 35:3472
42. Smith HI. 1995. *J. Vac. Sci. Technol. B* 13:2323
43. Wegrowe JE, Fruchart O, Nozieres J-P, Givord D, Rousseaux F, et al. 1999. *J. Appl. Phys.* 86:1028
44. Wu W, Cui B, Sun X, Zhang W, Zhuang L, et al. 1998. *J. Vac. Sci. Technol. B* 16:3825
45. Cui B, Wu W, Kong L, Sun X, Chou SY. 1999. *J. Appl. Phys.* 85:5534
46. Meier G, Kleiber M, Grundler D, Heitmann D, Wiesendanger R. 1998. *Appl. Phys. Lett.* 72:2168
47. Fernandez A, Gibbons MR, Wall MA, Cerjan CJ. 1998. *J. Magn. Magn. Mater.* 190:71
48. Castano FJ, Hao Y, Haratani S, Ross CA Vogeli B, et al. 2001. *IEEE Trans. Magn* In press
49. Savas TA, Farhoud M, Smith HI, Hwang M, Ross CA. 1999. *J. Appl. Phys.* 85:616
50. Park YD, Caballero JA, Cabbibo A, Childress JR, Hudspeth HD, et al. 1997. *J. Appl Phys.* 81:4717
51. Marty F, Vaterlaus A, Weich V, Stamm C Maier U, Pescia J. 1999. *J. Appl. Phys* 85:6166
52. Chappert C, Bernas H, Ferre J, Kottler V Jamet JP, et al. 1998. *Science* 280:1919
53. Terris BD, Folks L, Weller D, Baglin JEE Kellock AJ, et al. 1999. *Appl. Phys. Lett* 75:403
54. Devolder T, Chappert C, Chen Y, Cambri E, Bernas H, et al. 1999. *Appl. Phys. Lett* 74:3383
55. Hosaka S, Hosoki S, Hasegawa T, Koyanagi H, Shintani T, Miyamoto M. 1995 *J. Vac. Sci. Technol. B* 13:2813
56. Bessho K, Iwasaki Y, Hashimoto S. 1996 *IEEE Trans. Magn.* 32:4443
57. Wirth S, Field M, Awschalom DD, vo Molnar S. 1998. *Phys. Rev. B* 57:R14028
58. Wirth S, von Molnar S, Field M Awschalom DD. 1999. *J. Appl. Phys.* 85 5249
59. Zangari G, Lambeth DN. 1997. *IEE Trans. Magn.* 33:3010
60. Metzger RM, Konovalov VV, Sun M, Xu T

Zangari G, et al. 2000. *IEEE Trans. Magn.* 36:30
61. Bates FS. 1990. *Annu. Rev. Phys. Chem.* 41:525
62. Park M, Harrison C, Chaikin PM, Register RA, Adamson DH. 1997. *Science* 275:1401
63. Lammertink RGH, Hempenius MA, van den Enk JE, Chan VZH, Thomas EL, Vancso GJ. 2000. *Adv. Mater.* 12:98
64. Thurn-Albrecht T, Schotter J, Kastle GA, Emley N, Shibauchi T, et al. 2000. *Science* 290:2126
65. Zhu S, Gambino RJ, Rafailovich MN, Sokolov J, Schwarz SA, Gomez RD. 1997. *IEEE Trans. Magn.* 33:3022
66. Sohn BH, Cohen RE. 1997. *Chem. Mater.* 9:264
67. Chien CL. 1991. *J. Appl. Phys.* 69:5267
68. Hong J, Kay E, Wang SX. 1996. *IEEE Trans. Magn.* 32:4475
69. Christodoulides JA, Shevchenko NB, Hadjipanayis GC, Papaefthymiou V. 1997. *J. Appl. Phys.* 81:5558
70. Li JY, Platt CL, Rudee ML, Berkowitz AE, Cheeks TL. 1996. *J. Appl. Phys.* 79:5072
71. Lee J, Becker MF, Brock JR, Keto JW, Walser RM. 1996. *IEEE Trans. Magn.* 32:4484
72. Wernsdorfer W, Bonet Orozco E, Hasselbach K, Benoit A, Barbara B, et al. 1997. *Phys. Rev. Lett.* 78:1791
73. Gangopadhyay DS, Hadjipanayis GC, Sorensen CM, Klabunde K. 1993. *IEEE Trans. Magn.* 29:2602
74. Leslie-Pelecky D, Zhang XQ, Riecke RD. 1996. *J. Appl. Phys.* 79:5312
75. Sun S, Murray C. 1999. *J. Appl. Phys.* 85:4325
76. Sun S, Murray CB, Weller D, Folks L, Moser A. 2000. *Science* 287:1989
77. Tsang DC, Gafney HD, Sunil D, Rafailovich M, Sokolov J. Gambino RJ. 1996. *J. Appl. Phys.* 79:6025
78. O'Connor CJ, Buisson YSL, Li S, Banerjee S, Premchandram P, et al. 1997. *J. Appl. Phys.* 81:4731
79. Chen JP, Sorensen CM, Klabunde K, Hadjipanayis GC. 1994. *J. Appl. Phys.* 76:6316
80. Cullity BD. 1972. *Introduction to Magnetic Materials*. Reading, MA: Addison-Wesley
81. Hubert A, Schaefer R. 1998. *Magnetic Domains*. Heidelberg: Springer-Verlag
82. Hwang M, Redjdal M, Humphrey FB, Ross CA. 2001. *J. Appl. Phys.* In press
83. Schabes ME. 1991. *J. Magn. Magn. Mater.* 95:249
84. Schabes ME, Bertram HN. 1988. *J. Appl. Phys.* 64:1347
85. Rave W, Fabian K, Hubert A. 1998. *J. Magn. Magn. Mater.* 190:332
86. Newell A, Merrill RT. 1998. *J. Appl. Phys.* 84:4394
87. Brown WR. 1957. *Phys. Rev.* 105:1479
88. Frei EH, Strickman S, Treves D. 1957. *Phys. Rev.* 106:446
89. Aharoni A. 1997. *J. Appl. Phys.* 82:1281
90. Schabes ME, Bertram HN. 1988. *J. Appl. Phys.* 64:5832
91. Fredkin DR, Koehler TR. 1990. *J. Appl. Phys.* 67:5544
92. Fredkin DR, Koehler TR, Smyth JF, Schultz S. 1991. *J. Appl. Phys.* 69:276
93. Zheng Y, Zhu J-G. 1997. *J. Appl. Phys.* 81:5471
94. Shima M, Hwang M, Farhoud M, Savas TA, Tzacyk A, et al. 2001. *Proc. 198th. Electrochem. Soc. Conf.* In press
95. Osborn JA. 1945. *Phys. Rev.* 67:351
96. Joseph RI, Schlomann E. 1965. *J. Appl. Phys.* 36:1579
97. Aharoni A. 1999. *J. Appl. Phys.* 86:1041
98. Aharoni A. 1998. *J. Appl. Phys.* 83:3432
99. Yu J, Rudiger U, Thomas L, Parkin SSP, Kent AD. 1999. *J. Appl. Phys.* 85:5501
100. New RMH, Pease RFW, White RL. 1996. *J. Magn. Magn. Mater.* 155:140
101. Stoner EC, Wohlfarth EP. 1948. *Phil. Trans. R. Soc. London Ser. A* 240:599
102. Neel L. 1949. *Ann. Geophys.* 5:99

103. Brown WF. 1963. *Phys. Rev.* 130:1677
104. Seberino C, Bertram HN. 1997. *IEEE Trans. Magn.* 33:3055
105. Seberino C, Bertram HN. 1999. *J. Appl. Phys.* 85:5543
106. Gadbois J, Zhu JG. 1995. *IEEE Trans. Magn.* 31:3802
107. Schrefl T, Fidler J, Suss D, Scholz W. 2000. *Phys. B* 275:55
108. Richards HL, Kolesik M, Lindgard PA, Rikvold PA, Novotny MA. 1997. *Phys. Rev. B* 55:11521
109. Hinzke D, Nowak U. 1998. *Phys. Rev. B* 58:265
110. Lederman M, Schultz S, Ozaki M. 1994. *Phys. Rev. Lett.* 73:1986
111. Wernsdorfer W, Hasselbach K, Sulpice A, Benoit A, Wegrowe JE, et al. 1996. *Phys. Rev. B* 53:3341
112. Wernsdorfer W, Doudin B, Mailly D, Hasselbach K, Benoit A, et al. 1996. *Phys. Rev. Lett.* 77:1873
113. O'Barr R, Schultz S. 1997. *J. Appl. Phys.* 81:5458
114. Kleiber M, Kummerlen F, Lohndorf M, Wadas A, Weiss D, Wiesendanger R. 1998. *Phys. Rev. B* 58:5563
115. Fruchart O, Nozieres J-P, Wernsdorfer W, Givord D, Rousseaux F, Decanini D. 1999. *Phys. Rev. Lett.* 82:1305
116. Wernsdorfer W, Bonet Orozco E, Hasselbach K, Benoit A, Mailly D, et al. 1997. *Phys. Rev. Lett.* 97:4014
117. Bonet E, Wernsdorfer W, Barbara B, Benoit A, Mailly D, Thiaville A. 1999. *Phys. Rev. Lett.* 99:4188
118. Ross CA, Savas TA, Smith HI, Hwang M, Chantrell R, et al. 2000. *Phys. Rev. B* 62:14252
119. Rizzo ND, Silva TJ, Kos AB. 2000. *IEEE Trans. Magn.* 36:159
120. Stamps RL, Camley RE. 1999. *Phys. Rev. B* 60:12264
121. Hughes GF. 1999. *IEEE Trans. Magn.* 35:2310
122. Boerner ED, Bertram HN, Hughes GF. 1999. *J. Appl. Phys.* 85:5318
123. Sharrock MP. 1994. *J. Appl. Phys.* 76:6413
124. Luo CP, Shan ZS, Sellmyer DJ. 1996. *J. Appl. Phys.* 79:4899
125. Dova P, O'Grady K, Morales MP, Doerner MF. 1997. *J. Appl. Phys.* 81:3949
126. Hwang M, Farhoud M, Hao Y, Walsh M, Savas TA, et al. 2000. *IEEE Trans. Magn.* 36(6):
127. Pardavi-Horvath M. 1996. *IEEE Trans. Magn.* 32:4458
128. Kirk KJ, Chapman JN, Wilkinson CDW. 1999. *J. Appl. Phys.* 85:5237
129. Grundler D, Meier G, Broocks KB, Heyn Ch, Heitmann D. 1999. *J. Appl. Phys.* 85:6175
130. Stamps RL, Camley RE. 1999. *Phys. Rev. B* 60:11694
131. Guslienko KY, Choe SB, Shin SC. 2000. *Appl. Phys. Lett.* 76:3609
132. New RMH, Pease RFW, White RL. 1995. *IEEE Trans. Magn.* 31:3805
133. Abraham MC, Schmidt H, Savas TA, Smith HI, Ross CA, Ram RJ. 2001. *J. Appl. Phys.* Submitted
134. Shew LF. 1963. *IEEE Trans. Broadcast Television Receivers* BTR-9:56
135. Lambert SE, Saunders LL, Patlach LM, Krounbi MT, Hetzler SR. 1991. *J. Appl. Phys.* 69:4724
136. Lambert SE, Saunders LL, Patlach LM, Krounbi MT. 1987. *IEEE Trans. Magn.* 23:3690
137. Nair SK, New RMH. 1998. *IEEE Trans. Magn.* 34:1916
138. Hughes GF. 2000. *IEEE Trans. Magn.* 36:521
139. Zhu JG, Lin X, Guan L, Messner W. 2000. *IEEE Trans. Magn.* 36:23
140. Yamamoto SY, O'Barr R, Schultz S, Scherer A. 1997. *IEEE Trans. Magn.* 33:3016
141. Todorovic M, Schultz S, Wong J, Scherer A. 1999. *Appl. Phys. Lett.* 74:2516
142. Terris B, Lohau J, Moser M, Rettner CT, Best ME, Terris BD. 2001. *IEEE Trans. Magn.* In press

143. Ohkubo T, Kishigami J, Yanagisawa K, Kaneko R. 1991. *IEEE Trans. Magn.* 27:5286
144. Watanuki O, Sonobe Y, Tsuji S, Sai F. 1993. *IEEE Trans. Magn.* 27:5289
145. Manalis S, Babcock K, Massie J, Elings V, Dugas M. 1995. *Appl. Phys. Lett.* 66:2585
146. Kong L, Zhuang L, Chou SY. 1997. *IEEE Trans. Magn.* 33:3019
147. Gibson GA, Schultz S. 1993. *J. Appl. Phys.* 73:4516
148. Mamin HJ, Terris BD, Fan LS, Hoen S, Barrett RC, Rugar D. 1995. *IBM J. Res. Dev.* 39:681
149. Pohl DW. 1995. *IBM J. Res. Dev.* 39:701
150. Sato A, Momose S, Tsukamoto Y. 1995. *J. Vac. Sci. Technol. B* 13:2732
151. Minne SC, Manalis SR, Atalar A, Quate CF. 1996. *J. Vac. Sci. Technol. B* 14: 2458
152. Martin Y, Rishton S, Wickramasinghe HK. 1997. *Appl. Phys. Lett.* 71:1
153. Binnig G, Despont M, Drechsler U, Haberle W, Lutwyche P, et al. 1999. *Appl. Phys. Lett.* 74:1329
154. Ried RP. 1997. *J. Microelectromech. Syst.* 6:294
155. Despont M, Brugger J, Drechsler U, Durig U, Haberle W, et al. 1999. *Presented at 12th IEEE Int. MEMS Conf. Florida.*
156. Tehrani S, Chen E, Durlam M, DeHerrera M, Slaughter JM, et al. 1999. *Appl. Phys.* 85:5822

Annu. Rev. Mater. Res. 2001. 31:237–63

Phospholipid Strategies in Biomineralization and Biomaterials Research

Joel H Collier and Phillip B Messersmith
Northwestern University, Biomedical Engineering Department, Evanston, Illinois 60208-3107; e-mail: j-collier@northwestern.edu; philm@northwestern.edu

Key Words self-assembly, bio-inspired, biomimetics, liposome, vesicle

■ **Abstract** This review summarizes recent advances and future research trends in the field of phospholipid-based biomaterials. Lipids play an important role in biomineralization and countless other biological processes, and they are receiving increasing attention for the synthesis of new biomimetic biomaterials. Several emerging strategies in biomaterials research take advantage of phospholipids to compartmentalize and/or template chemical reactions via self-assembled structures such as liposomes and tubules. Still others exploit the inherent biocompatibility of phospholipids and phospholipid-mimetic materials for use as novel tissue-contacting biomaterials that mimic biological membranes. In the future, phospholipid-based materials may be increasingly utilized as tools for the manipulation of cell and tissue responses to biomaterials, for controlled drug release, for reconstructive surgery, and as tissue-engineered constructs.

INTRODUCTION

Lipids comprise a diverse and essential class of biomolecules. Their involvement in biological processes is ubiquitous and includes the construction of biomembranes, the transduction of cellular signals, their use as a source of energy, and countless other biochemical pathways. One biological process in which lipids play an important role is the synthesis of materials. Through their ability to self-assemble, compartmentalize, and template, lipids enable the formation of a large variety of materials with controlled micro- and nanostructures. Acknowledging this involvement of lipids in natural materials synthesis, a new field is emerging that seeks to use the natural roles of lipids as inspiration for the development of new routes for materials synthesis. This review focuses on the link between how lipid structures are utilized in nature and how these ideas from nature are being exploited to synthesize novel synthetic biomaterials. This review is not intended to be an

0084-6600/01/0801-0237$14.00

exhaustive listing of work being done using lipid-based biomaterials strategies, as several important biomedical applications of lipids, including their use in systemic drug delivery and gene therapy, are not discussed in detail. We begin with a brief overview of lipid structure and self-assembly, along with a discussion of the roles that lipids play in the synthesis of natural biological materials. This is followed by a review of the recent literature on bio-inspired phospholipid-based strategies for developing novel mineralized materials. The latter half of the review focuses on more applied aspects of phospholipid biomaterials, including (*a*) approaches using phospholipid vesicles for the delivery of reactive species, therapeutic agents, or macromolecules, (*b*) tubule, ribbon, and network assemblies of phospholipids for drug delivery and as tissue-engineering scaffolds, and (*c*) membrane-mimetic materials for the enhancement of tissue-material interfaces.

STRUCTURE AND ASSEMBLY OF LIPIDS

The ultimate utility of lipids lies in their ability to self-organize into specific supramolecular aggregates–an attribute conferred by the inherent amphiphilic nature of this diverse class of biomolecules. Because a complete cataloguing of lipid structures is beyond the scope of this review, we instead qualitatively describe the fundamental chemical features of lipids, the major aggregate structures that lipids commonly form, and factors that control phase behavior. Several excellent sources are available for more detailed structural and phase information (1–3).

Chemical Structure

Most lipids are composed of three essential features: a polar head group, one or more hydrophobic tail regions, and a backbone structure that connects the two (Figure 1). These three building blocks are assembled in an extraordinary variety of ways to generate all of the naturally occurring amphiphilic lipids. Tail regions can contain aromatic groups, saturated aliphatic chains, or unsaturated aliphatic chains, and head groups can consist of any number of charged or uncharged polar moieties. Backbone structures occur in similar diversity, from polyol-based structures such as glycerol, to the aromatic polycyclic structures of cholesterol and other isoprenoids. Owing to the large variability in head group and tail chemistries, lipids are often categorized according to backbone structure: isoprenoids, phospholipids, sphingolipids, ceramides, fatty acids, triacylglycerols, eicosanoids, waxes, and glycerophospholipids. Within each of these categories numerous variants exist that are the result of head group or tail modifications. For example, glycerophospholipids, the primary constituents of biomembranes, are composed of two fatty acid chains esterified at C-1 and C-2 of glycerol-3-phosphate (Figure 1). The acyl chains are typically 8 to 24 carbons long, and they can be saturated, unsaturated, or even coupled to form a macrocycle. Typically, an ester linkage couples the glycerophosphate to one of several head grou

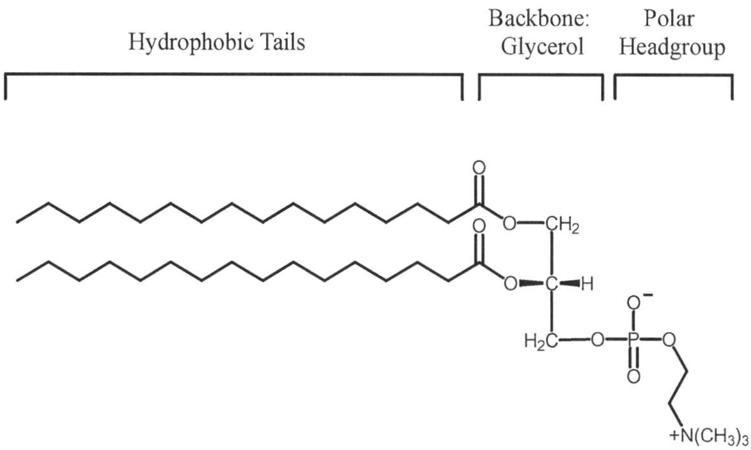

Figure 1 Structural features of lipids, using a glycerophospholipid (phosphatidylcholine) as an example.

compounds such as choline, ethanolamine, or serine to form phosphatidylcholine, phosphatidylethanolamine, or phosphatidylserine, respectively. Such head group modifications may render the lipids either charged, uncharged, or zwitterionic at physiologic pH.

Simple Supramolecular Aggregates of Lipids: Monolayers, Micelles, Bilayers, and Liposomes

When dispersed in a solvent such as water, lipids exhibit a strong tendency to self-assemble into a variety of mesophase structures, including monolayers, micelles, reverse micelles, bilayers, and hexagonal phases (Figure 2). The aqueous phase behavior of lipids is dictated by intrinsic parameters such as lipid polarity, acyl chain length, location and extent of acyl chain unsaturation, acyl chain branching, head group size, head group polarity, and head group charge, as well as extrinsic parameters such as concentration (lyotropism) and temperature (thermotropism) (2–5). For example, increasing acyl chain length results in higher bilayer melting temperatures, whereas the introduction of unsaturated bonds has the opposite effect owing to the disruption of molecular packing and ordering among the fatty chains. Furthermore, in addition to lowering the bilayer melting temperature, the presence of unsaturated bonds can also induce the formation of nonlamellar phases. Head group properties (especially polarity) also impact the phase behavior of lipids by modulating the strength of interaction between head groups and between the head groups and water.

Monolayers are among the simplest structures of lipids. They are nevertheless exceptionally useful as controlled experimental models for the investigation of lipid interactions, lipid surface chemistry, and membrane dynamics. Assembled

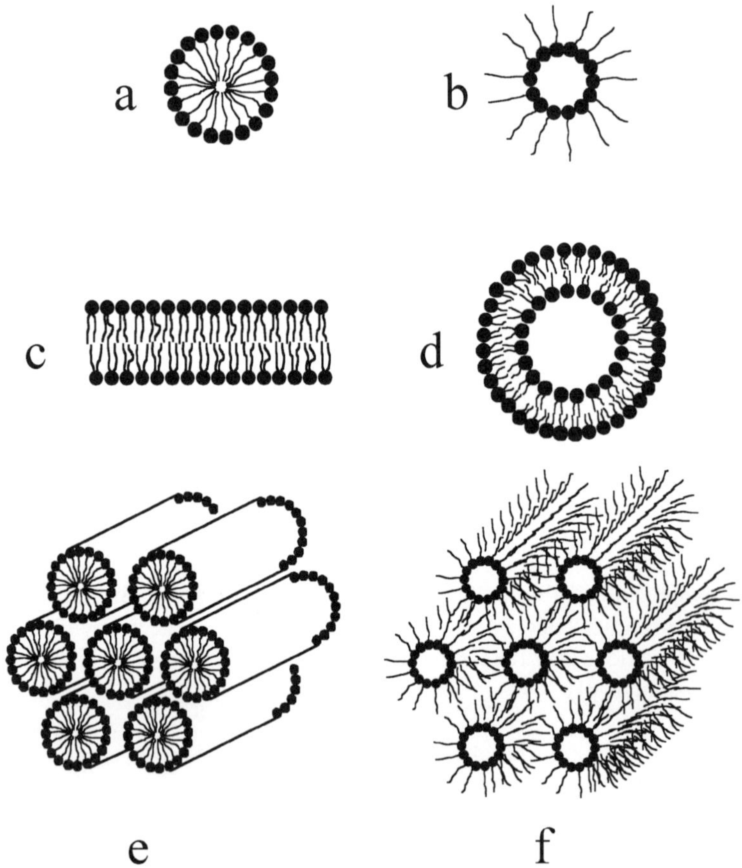

Figure 2 Common lamellar and nonlamellar self-assembled structures of lipids: (*a*) micelle, (*b*) inverse micelle, (*c*) lamellar bilayer, (*d*) bilayer vesicle, (*e*) hexagonal, (*f*) inverse hexagonal.

most commonly at the interface between aqueous media and air, monolayers are formed when an organic lipid solution is spread onto water. Upon evaporation of the organic solvent, the hydrophilic region of the molecule embeds in the aqueous phase to maximize polar interactions, and the hydrocarbon region protrudes into the gaseous phase to maximize hydrophobic interactions. Micelles (Figure 2*a*) are spherical aggregates commonly formed by single-tailed lipids such as bile acids and other surfactants. Formed from 10 to over 100 single molecules in water, micelles minimize their hydrophobic interactions by clustering their hydrocarbon tails in their interiors and exposing their polar head groups to the surrounding aqueous medium. In apolar solvents, inverse micelles are formed, with the hydrophilic head groups clustering in the center (Figure 2*b*). Many lipids, including phospholipids with acyl chains longer than about eight carbons, do not form micelles because their hydrophobic regions are too bulky to fit inside the micellar structure (4). Instead, these lipids form a common and biologically important mesophase, the lamellar bilayer (Figure 2*c*).

Lipid bilayers are the major building blocks of biological membranes that, together with membrane proteins and cholesterol, control cell shape and many cell functions such as the storage of compounds, ion transport, cell fusion, and metabolism. Bilayers exhibit a surprising variety in their phase behavior, forming various shapes and textures such as ripples or saddles (6). Because a planar bilayer sheet dispersed in water would be in the energetically unfavorable position of having its hydrophobic regions exposed to water at its edges, self-organization of phospholipid bilayers usually occurs by the formation of closed vesicles, often referred to as liposomes (Figure 2*d*). Liposomes can be formed in the laboratory using several approaches, perhaps most easily by spontaneous formation when a thin film of lipid is heated above a critical temperature in water. Liposomes range in size from tens of nanometers to tens of microns in diameter and can be made to entrap hydrophilic materials in their inner compartment and/or hydrophobic materials in their membranes. Because of their ability to entrap and compartmentalize, liposomes have received widespread attention as drug-delivery vehicles that can be injected into the blood stream to deliver encapsulated drugs to specific organs, tissues, or cells (7).

MINERALIZATION IN LIPID ASSEMBLIES: A Common Biological Strategy

Biological mineral synthesis, in contrast to conventional mineral-processing techniques, generates materials of very highly controlled size, habit, texture, composition, and structure. Studies of biomineralization processes during the past several decades have yielded extensive information on how nature achieves such exquisite control over mineral formation. These biomineralization processes occur in a diversity of species and tissues, including bone mineral, marine shells, tooth enamel, and marine algae shells. Several excellent reviews of biomineralization have been published (8–11). In many types of biomineralization, macromolecular structures composed of lipids, proteins, and/or polysaccharides are intimately associated with mineral phases and serve vital roles in their crystallization. For example, one ubiquitous biological strategy employed during the formation of mineral phases exploits lipid assemblies to control the location, size, composition, and shape of mineral crystals. Phospholipids, often in the form of bilayer vesicles, are commonly employed in natural biomineralization processes to delineate reaction compartments into which mineral formation takes place (8). Although phospholipid mesophase-mediated mineralization occurs in vastly different species and tissues, the roles that the phospholipid assemblies play in these processes can be generalized as follows:

1. construction of a compartmentalized reaction environment, often through the use of bilayer vesicles;
2. control of the physicochemical conditions inside the reaction environment via transmembrane channels, transporters, or selective ion permeability;
3. control of nucleation kinetics. The organic surface serves as a molecular blueprint for the site-directed formation of the inorganic phase by providing an interface in which the electrostatic, stereochemical, and geometric interactions are conducive to oriented nucleation and growth of the mineral (11);
4. control of the shape and size of the crystal by the reaction compartment. Production of complex crystal shapes is attained by altering the shape of the lipid matrix during crystal growth.

Coccoliths, Matrix Vesicles, and Magnetosomes

To further highlight the ways in which phospholipids play a key role in biomineralization processes, we offer three examples of biominerals that form withi the compartments of lipid assemblies: the coccolith of the marine alga *Emiliania huxleyi*, matrix vesicles found in the mineralizing tissues of vertebrates, an magnetosomes found in magnetotactic bacteria. These examples are from vastl different organisms, yet in each case the formation of the inorganic phases follow several general strategies that can be adapted to the laboratory. The first example is the formation of coccoliths in the alga *Emiliania huxleyi* (12). Coccolith are scales that consist of 30–40 calcite ($CaCO_3$) crystals arranged within an oval shaped structure, and several coccoliths are used by the alga as an exoskeleton t surround and protect itself. The first stage of organized assembly of these structures is the production of a series of phospholipid membrane vesicles that ac as compartments for the subsequent calcite crystallization. Other encapsulatio structures such as proteins or polysaccharide networks are also utilized in biomineralization processes. The second stage of crystallization includes the influx c calcium and carbonate ions. Then, in the third stage, oriented nucleation of calcit crystals occurs within the enclosed vesicles. This is aided by structural, chiral, an chemical complementarity between the organic matrix and the nucleating mineral phase. In the final stage of mineralization, the desired shape of the miner crystal is achieved by control of the vesicle compartment shape via active cellula cytoskeletal elements. Such dynamic control of crystal shape by the phospholipi mesophase leads to the highly elaborate structures found in coccoliths.

Vertebrates also initiate mineral formation within lipid vesicles and lipopeptide aggregates. Noncellular phospholipid vesicles (matrix vesicles) are found i certain actively mineralizing tissues of the human body, including growth pla cartilage and growing dentin, and are the initial sites of calcification in thes tissues (13). Matrix vesicles are unilamellar bilayer vesicles, 100–200 nm i diameter, that are formed by budding from hypertrophic chondrocytes and odontoblasts. The membranes of these vesicles are enriched in acidic (i.e., negatively charged at physiologic pH) phospholipids, ion channels, enzymes, and transport proteins. At the time of release from the cell membrane, matrix vesicles do not contain mineralized crystals. However, as the vesicle becomes embedded within the extracellular matrix secreted by the cell, the interior of the vesicle becomes enriched with Ca^{2+} and PO_4^-, in part by the action of calcium-binding phospholipids and proteins, phospholipase enzymes, and calcium ion channels. When the concentrations of Ca^{2+} and PO_4^- reach supersaturation levels in the vesicle compartment, calcium phosphate mineral then precipitates on the interior surface of the vesicle in the form of amorphous calcium phosphate or poorly crystalline apatite (13). Nucleation at the inner surface of the vesicle membrane appears to be assisted by the chemical and structural complementarity of the inner phospholipid-protein assembly of the matrix vesicle and the mineral phase. Upon further growth, the hydroxyapatite crystals then penetrate the vesicle membrane. Once the membrane has been perforated by the crystal, further mineralization is governed by solution conditions of the extracellular fluid. By this mechanism, matrix vesicles effectively control the initial nucleation kinetics and crystal structure of the biomineral.

Eanes and coworkers utilized lipid vesicles in their work on in vitro models of matrix vesicle mineralization (14–19). In synthetic vesicles of similar lipid composition to matrix vesicles, calcium phosphate crystals nucleated and grew, then ruptured the membranes and seeded further precipitation from the surrounding solution (15). Furthermore, the formation of calcium phosphate could be modulated by altering the permeability of the membrane with cholesterol (17, 18). Much of our current understanding of matrix vesicle mineralization is a result of this work, which employed simulated physiologic conditions to model the sequence of events that occurs during matrix vesicle mineralization.

Another interesting example of phospholipid and organic matrix-mediated mineralization is magnetosomes, which are biologically derived single crystals of magnetite (Fe_3O_4) or greigite (Fe_3S_4) encapsulated in a bilayer vesicle. Magnetosomes are found in certain species of bacteria, and their mechanism of formation shares several common features with matrix vesicle and coccolith formation (20–22). Magnetosomes are highly crystalline, and their sizes and crystal habits are particularly uniform within individual bacterial species. Long chains of interconnected magnetosomes with a common magnetic orientation are often seen in the cytoplasm of magnetotactic bacteria. These organisms are believed to have evolved to utilize these structures, which possess a permanent magnetic dipole, to navigate with respect to the earth's magnetic field. As with coccoliths and matrix vesicles, magnetosomes are formed by the precipitation of mineral within preformed lipid vesicles, which exert a high level of control over crystallization by virtue of the compartmentalization and templating effect of the membrane surrounding the particle. Molecular recognition at the inner vesicle surface is likely to play a role in control of magnetite crystal orientation.

Lipid Strategies in Materials Research

As coccolith, matrix vesicle, and magnetosome mineralization demonstrate, lipid assemblies should not be viewed merely as passive hosts to enclose mineralization reactions, for compartmentalization is only the most basic role that lipids play in the biomineralization process. The formation of a lipid vesicle compartment results in the presentation of an organic matrix to the nucleation environment, the surface of which can strongly influence the growing inorganic phase through the molecular recognition processes of chemical, electrostatic, and chiral complementarity. Lipid monolayers are particularly useful experimental systems for studying these interactions because they are simple and modifiable. Using monolayers, Mann et al and Heywood & Mann have investigated several essential features of the organic-inorganic interface in biomimetic mineralization. For example, they used compressed Langmuir monolayers to control the crystal orientation of nucleating calcium carbonate (23, 24). Initially they used compressed stearic acid monolayers to control the crystallization of $CaCO_3$ from supersaturated solutions, finding that the presence of an organized monolayer gave rise to oriented vaterite phases that formed specifically at the monolayer interface, where the absence of the monolayer resulted in non-uniform rhombohedral calcite crystals that precipitated in multiple locations. Furthermore, the degree of compression of the monolayer dictated the homogeneity of vaterite nucleation (25). In essence, stereochemical and electrostatic matching were the primary contributors to the observed phase changes. To further explore this idea, they then observed crystallization under monolayers of different stereochemistry. Whereas *n*-octadecanoic acid (which forms bidentate motifs) induced the nucleation of the $\{1\bar{1}0\}$ face of calcite, *n*-eicosyl phosphonate and *n*-eicosyl sulfate (which form tridentate motifs) induced the nucleation of the (001) face. Aragonite also showed similar dependence on head group choice. This supports the idea that an interfacial mechanism exists involving stereochemical complementarity between the oxygen atoms of the anionic head group and those of carbonate ions located in lattice positions in the crystal faces of nuclei forming at the organic/inorganic interface. Depending on the stereochemical and electrostatic properties of the phospholipid lattice presented to the nucleating crystal, different faces could be nucleated.

Several laboratories have sought to mimic compartmental biomineralization to synthetically generate inorganic phases with the controlled features of biominerals. In general, these in vitro experiments have attempted to exploit the aqueous inner compartment of liposomes as micro- or nanoscale reaction volumes to form inorganic particles of controlled size and shape. An early approach by Mann and coworkers involved the precipitation of silver oxide inside phosphatidylcholine vesicles (26). In that work, liposomes containing entrapped aqueous silver nitrate were dispersed in a Ag-free alkaline aqueous medium. Under the conditions of the experiment, the Ag^{2+} ions had low permeability across the lipid bilayer membrane and therefore remained entrapped within the vesicles. However, OH^- ions with much higher permeability gradually diffused into the vesicle compartment resulting in an intravesicular pH increase and precipitation of Ag_2O. The crystals that resulted, which formed only inside the liposomes, were essentially of homogeneous size, unlike precipitates formed directly from an analogous lipid-free solution strategy. Similar approaches have been successfully used to generate iron oxide microparticles (27), aluminum oxide particles (29), and nanocomposite particles of aluminum, magnesium, and calcium hydroxides (28, 30). Also, the generation of nanoparticles is of particular interest in semiconductor engineering, and Fendler and coworkers have extensively utilized compartmental synthesis techniques in creating semiconductor particles with unique properties (31–33). The potential applications of such highly controlled inorganic particles include catalysis, recording media, antibody-tagged magnetic particles for molecular biology, and contrast agents for medical imaging.

In addition to vesicles, some bilayer-forming lipids adopt extended nonspherical assemblies such as tubules, filaments, and networks that can be used as templates for materials synthesis. For example, mixtures of galactocerebrosides such as α-hydroxy fatty acid galactocerebroside and non-hydroxy fatty acid galactocerebroside self-assemble into lamellar disks, multilamellar tubules, or fibrous unilamellar tubules in ethylene glycol (34). The inclusion of low levels of anionic, sulfated galactocerebroside into the neutral lipid microstructures gives the microtubules anionic sites for electrostatic binding of cations, and inorganic oxide nucleation can occur on the microtubule surface (35). Microtubules constructed from diacetylene-containing phospholipids have been used as templates to generate solid silica microtubules (36), and nanoparticles of silica (37) and gold (38) have also been selectively arranged on the lipid tubule template to generate patterned three-dimensional objects. Studies have also shown that mineral phases are not the only materials that can be synthesized on these tubule surfaces, as lipid tubules have been utilized for the templating of polymers (39), proteins (40), and metals (41).

PHOSPHOLIPID STRATEGIES IN BIOMATERIALS RESEARCH

We now shift our attention from describing the role of phospholipids in biomineralization to specific applications of phospholipid assemblies as biomaterials. For the purposes of this discussion, we define a biomaterial as any material of natural or synthetic origin that is used for the repair, replacement, reconstruction, or treatment of diseased or traumatized tissue. Although the use of phospholipids as vectors for gene therapy and liposome-based systemic drug-delivery systems certainly falls within this definition, these topics have been reviewed elsewhere and are not the focus of this section. Instead, we focus on phospholipid-based biomaterials strategies that vary widely in approach but can be grouped into three major categories: (*a*) those that utilize lipid vesicles for the delivery of reactive species, therapeutic agents, or macromolecules; (*b*) tubule, ribbon, and network assemblies of phospholipids for drug delivery and as scaffolds for tissue-engineering applications; and (*c*) monolayer, bilayer, or membrane-mimetic materials for the enhancement of tissue-material interfaces. The first two of these categories make use of the unique properties of phospholipids to encapsulate and/or template via self-assembled structures such as liposomes and tubules, whereas the latter takes advantage of the inherent biocompatibility of lipids as surface coatings and membrane-mimetic systems that offer new approaches for engineering novel tissue-contacting biomaterials that mimic biological membranes.

Liposomes and Liposome-Containing Hydrogels

The use of liposomes to encapsulate materials has received widespread attention for the systemic (i.e., intravenous) delivery of drugs and diagnostic reagents (7). Historically, the major barrier to using liposomes for systemic drug delivery has been the rapid clearance of liposomes from the bloodstream by the reticuloendothelial system. Although there is renewed interest in systemic liposome delivery because of recent progress in the use of steric stabilization for increasing the blood circulation lifetimes of liposomes, it is important to note that significant challenges still exist for systemic delivery of liposomes and for the targeting of liposomes to specific tissues or organs. Moreover, some treatments may benefit from a localized approach to sustained delivery of drugs rather than systemic delivery. For example a drug-releasing gel material that could be injected into a solid tumor to deliver chemotherapeutic agents for extended time periods has been widely envisaged.

Several attempts have been made to combine the beneficial properties of liposomes and polymer matrices to develop materials that can sequester and retain liposomes at a discrete tissue site. In most cases, the role of the polymer is to provide a bulk matrix within which the liposomes are entrapped, whereas the function of the liposomes is to provide a reservoir from which drug is released in a controlled manner. Liposome-hydrogel composites have been designed with several applications in mind, including conformal drug-releasing coatings to prevent bacterial adhesions to catheters (42), the prolonged delivery of antigens for enhanced immunization procedures (43, 44), and the delivery of hormones (45). Additional applications of liposome-hydrogel constructs include topical dressings for transcutaneous drug delivery (46) and for the prolonged delivery of substances to cell cultures.

Several different types of matrix materials have been investigated for this purpose, including biopolymeric materials such as collagen (45), chitosan (47), alginate (48–50), alginate-poly(lysine) (43), and gelatin (51). Although sequestration of liposomes in these gels and the evaluation of the liposomes' release profiles is easily achieved, some difficulties have been encountered in the development of these hydrogel-liposome composites. The most significant of these is the tendency of matrix polymers to interact with the liposomal membrane. An example of this is the interaction of alginate with phospholipid bilayers, which results in alteration of the barrier properties of liposomes (52, 53). To prevent such interactions from adversely affecting the release properties of liposomes, more sophisticated strategies involving the introduction of specific functional groups at the liposome surface or within the hydrogel are being investigated. For example, to minimize liposome/matrix polymer interactions, PEGylated lipids have recently been utilized to form a steric barrier around gelatin-embedded liposomes (51). The presence of the PEG chains at the liposome surface prevents membrane disruption by forming a polymeric barrier that sterically inhibits interaction between soluble or matrix polymer and the lipid bilayer.

Triggerable Liposomes and Their Use for In Situ Formation of Biomaterials

In recent years, considerable work has been directed at constructing lipid vesicles that can be induced to release entrapped materials in response to an applied stimulus (54). Typically, the goal has been to trigger site-directed delivery of therapeutic agents. Early work in this area by Yatvin and coworkers utilized temperature-sensitive liposomes in combination with local hyperthermia to deliver a chemotherapeutic agent to a tumor site (55, 56). More recently, photo-, pH-, and other stimulus-responsive systems have been investigated (54, 57–63).

By virtue of their ability to entrap and release reagents under controlled conditions, it is also possible to exploit stimuli-responsive liposomes as a means of controlling chemical reactions that result in the rapid formation of a biomaterial. An ongoing effort in our laboratory is concerned with using stimuli-responsive liposomes for rapid in situ formation of mineral, polymer, and mineral/polymer composite biomaterials, with the goal of developing a versatile approach to inducing rapid in situ formation of biomaterials in response to an applied stimulus (53, 64–67).

Our general strategy (Figure 3) has been to exploit the controllable barrier properties of the bilayer membrane of phospholipid liposomes to encapsulate ions and physically isolate them from reactive species contained within the free (extravesicular) aqueous medium. By using stimuli-responsive lipid vesicles, one is able to create systems in which chemical reactions between the encapsulated species and the extravesicular species are made possible only in response to specific stimuli, such as a change in temperature or pH, or by exposure to light of a specific wavelength. As a result of either a specific physical or chemical (i.e., degradative) mechanism, the applied stimulus induces a change in bilayer barrier properties, resulting in release of the entrapped species. Upon release from the liposome compartment, the entrapped reagent either catalyzes or participates directly in a chemical reaction that results in the formation of a biomaterial.

Much of our work in this area has involved the use of temperature changes to activate liposome-mediated biomaterial formation. This approach exploits the release of entrapped substances from liposomes at temperatures near the lipid chain melting transition (T_m), a temperature at which bilayer permeability is known to

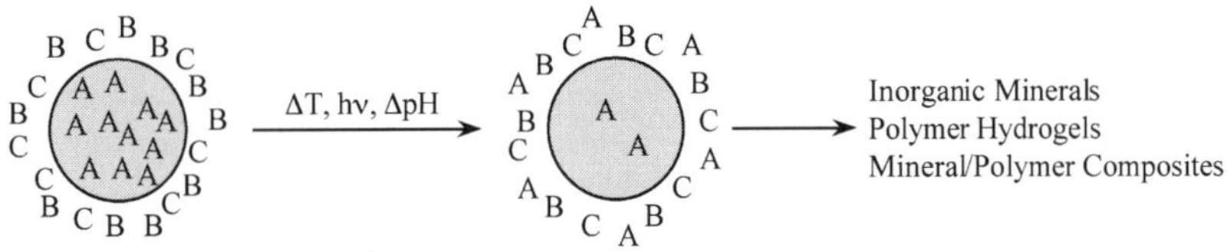

Figure 3 Schematic illustration of general strategy for utilizing stimuli-responsive lipic vesicles for in situ formation of biomaterials. A, B, and C represent reactive solutes, such as organic/inorganic ionic species, low-molecular weight organic molecules, or polymers Exposure of the liposome suspension to an applied stimulus triggers release of entrappe reagent (A), which reacts with components of the free medium (B and/or C) to form solid or semi-solid biomaterial. Alternatively, release of entrapped reagent may be used t catalyze reaction between non-entrapped species.

be significantly enhanced (68). As illustrated in Figure 4, at temperatures belov T_m, phosphatidylcholine bilayers are in the gel phase (L_β), which has low perme ability to encapsulated monovalent and divalent cations. Above T_m, the bilayer are fluid (L_α phase) but still have relatively low permeability. However at T_m, lipi bilayer permeability has been observed to increase by several orders of magnitude a phenomenon attributed to the presence of highly permeable interfacial region between coexisting gel ($<T_m$) and fluid ($>T_m$) bilayer domains (69, 70). Thi fundamental characteristic of phospholipid bilayers permits the design of reagent filled liposomes that are relatively impermeable to encapsulated molecules at on temperature but highly permeable at another temperature. If the initial concentra tion of an ion or low-molecular weight molecule is high inside the liposome bu low outside, heating to T_m will result in diffusion of the entrapped species dow a large concentration gradient, effectively releasing the entrapped species into th extravesicular medium.

Our initial goal was to employ liposomes composed of lipids having T_m betwee ambient and body temperature (37°C), so that the thermal release mechanism de scribed above could be used to trigger biomaterial formation when a fluid precursc

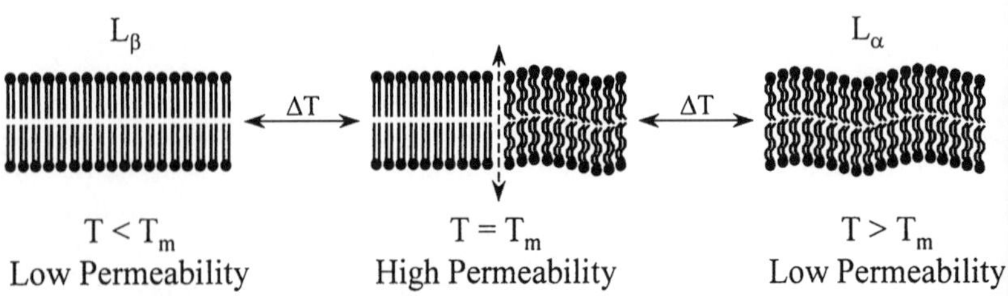

Figure 4 Schematic illustration of the effect of temperature on phospholipid bilayer pe meability. Permeability across bilayers is low in the single-phase regions below and abov the melting temperature (T_m) but is high in the vicinity of T_m owing to the existence defect-rich interfacial regions between coexisting gel and fluid domains.

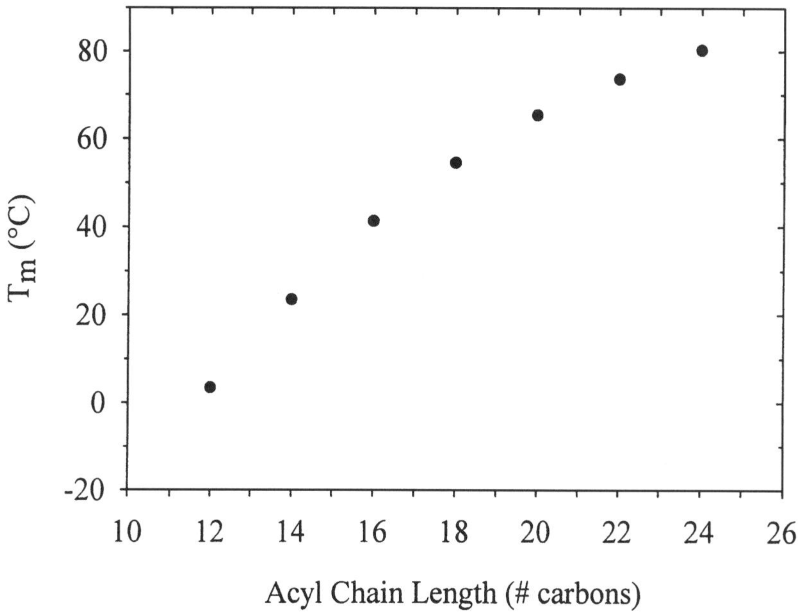

Figure 5 Dependence of bilayer melting temperature (T_m) on fatty acid chain length for saturated diacyl phosphatidylcholines (70a).

suspension was injected into a warm tissue or body cavity. For this purpose we chose saturated phosphatidylcholines, in which T_m is determined primarily by the fatty acid chain length (Figure 5). By selecting phospholipids with appropriate fatty acid chain lengths and ones that exhibit miscibility, such as dimyristoyl phosphatidylcholine (DMPC) and dipalmitoyl phosphatidylcholine (DPPC), we have been able to tailor the T_m of the bilayer in the range of 23 to 41°C. For example, a 9:1 molar ratio of DPPC to DMPC resulted in a T_m that coincides with the body's temperature of 37°C; liposomes of this composition that contain $CaCl_2$ released approximately 90% of entrapped Ca^{2+} between 36 and 38°C (Figure 6). By adjusting the DPPC/DMPC molar ratio, we were capable of tailoring the T_m to less than 37°C (65); such flexibility could be useful in adjusting these systems to lower temperatures, such as might be encountered in a limb or on the external surface of the body.

Initially, we applied this strategy to the formation of calcium phosphate minerals by designing thermally responsive liposomes in which mineral formation was activated when the liposome suspension was heated from ambient to physiologic temperature (65). Separately, calcium- and phosphate-loaded liposomes were prepared and mixed together to form a suspension of equal calcium and inorganic phosphate concentration, and highly supersaturated with respect to calcium phosphate solids such as hydroxyapatite. The suspension could be stored for several

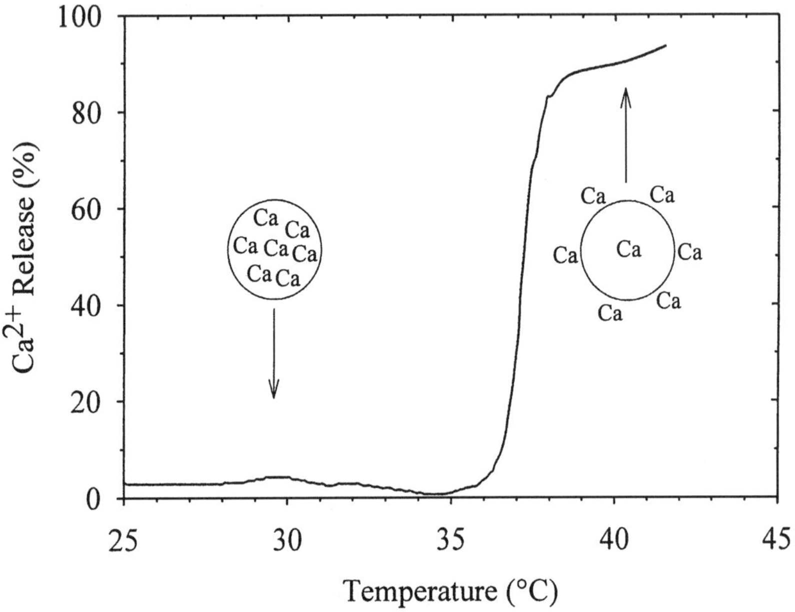

Figure 6 Thermally triggered release of calcium from liposomes at T_m. Liposomes were heated at 1°C/min in the presence of a calcium-sensitive dye as described in text (65).

weeks at room temperature without detectable mineral formation, reflecting the low bilayer permeability of ions in the gel state. However, when the liposome suspension was heated to the T_m of the lipid mixture, calcium and phosphate ions were released into the extravesicular fluid, where, owing to their high concentrations, they rapidly reacted to form calcium phosphate mineral. The mineral phases formed by this strategy included apatite and brushite and could be systematically controlled through manipulation of the pH of the reaction medium (65). In subsequent in vitro experiments we demonstrated how this approach can be used to form calcium phosphate minerals on human enamel and dentin surfaces by exploiting the increased temperature of the oral cavity (Figure 7) (67). The results of these studies suggest that this approach to mineral formation could be used to remineralize tooth surfaces or even possibly for in situ formation of bone repair materials.

We have also used this liposome strategy to elicit rapid in situ formation of polymer hydrogel biomaterials. Having previously developed triggerable Ca^{2+}-loaded vesicles for our mineralization studies, we hypothesized that we could employ the triggered release of calcium to induce hydrogel formation. By utilizing calcium-reactive macromolecules such as alginates and calcium-dependent enzymes that cross-link proteins, we have developed liposomal suspensions that are stable and fluid at room temperature but rapidly form polymer hydrogels at physiologi-

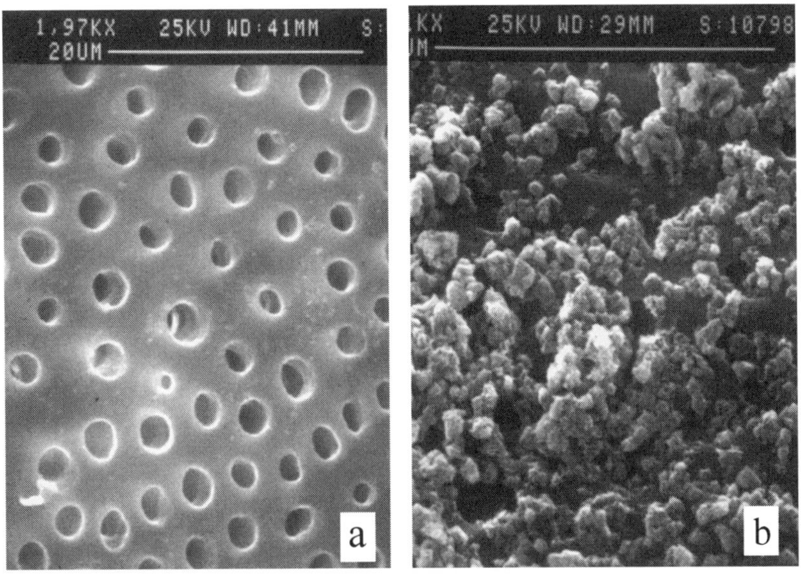

Figure 7 Scanning electron micrographs (SEM) illustrating thermally triggered mineralization of human dentin using calcium- and phosphate-loaded liposomes. (*a*) SEM of untreated human dentin surface showing dentin tubules and smooth intertubular tissue. (*b*) SEM of human dentin treated with mineralizable liposome suspension for 20 min at 37°C. Following treatment the surface was washed to remove excess liposomes as well as unbound mineral, before drying and imaging. Reproduced from (67), with permission from Elsevier Science.

temperature (53, 64, 66). In the alginate system, a precursor fluid consisting of Ca^{2+}-loaded liposomes dispersed in a low viscosity Na-alginate solution remains fluid for several days at ambient temperature. Heating of the precursor fluid to T_m triggers Ca^{2+} release from the liposomes; the calcium in turn cross-links Na-alginate to form a Ca-alginate hydrogel. Characterization of the gelation kinetics reveals that gelation is very rapid in this system (on the order of tens of seconds as shown in Figure 8) when the temperature of the precursor fluid reaches T_m. A consequence of the formation of the Ca-alginate hydrogel is the entrapment of liposomes within the gel matrix; these liposomes can be employed as drug reservoirs for controlled release from in situ formed hydrogels (66).

To create protein-based hydrogels, one can take advantage of biological enzymes to catalyze enzyme-mediated cross-linking of macromolecular substrates. Transglutaminases (TGases) are enzymes that catalyze the formation of cross-links between glutamine- and lysine-containing proteins and have been used to form proteinaceous hydrogels (71, 72). TGase enzymes require Ca^{2+} for activity and therefore can be triggered from an inactive to an active state by release of Ca^{2+} from liposomes. We used thermally triggerable Ca^{2+}-loaded liposomes and TGase

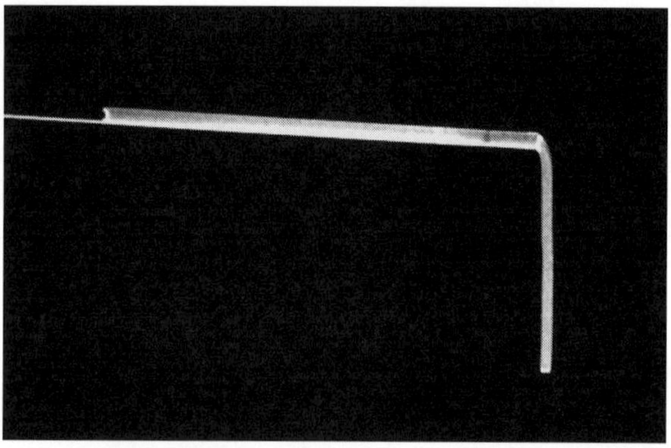

Figure 8 Thermally triggered gelation of alginate using calcium-loaded liposomes. Ca^{2+}-loaded liposomes dispersed in aqueous NaCl were mixed with a Na-alginate solution and loaded into a glass capillary at room temperature. The capillary was then immersed in a 37°C water bath for 20 s, after which the contents of the capillary were partially expelled to reveal the formation of Ca-alginate hydrogel. Reprinted from (53) with permission from Elsevier Science.

in combination with a macromolecular TGase substrate (fibrinogen) to develop in situ gelling proteinaceous hydrogels (53). Figure 9 shows the rheological properties of these alginate and fibrinogen gels as they solidify. Only suspensions that contained Ca^{2+}-loaded liposomes and were heated to 37°C allowed Ca^{2+} release and subsequent gel evolution.

In contrast to the liposome-encapsulated hydrogel systems described above, in which the liposomes act as simple reservoirs for passive release of therapeutic agents, in our approach to rapid in situ formation of biomaterials the liposomes actively participate in the reactions that result in formation of a bulk material. This strategy was designed to take advantage of the clinical versatility of fluid liposome suspensions that are stable during storage at or below room temperature but that rapidly form a hydrogel upon injection into the body. The method of administration (i.e., injection via syringe) of these biomaterials is expected to be much less invasive compared with conventional surgical administration of biomaterials. In the future, this versatile strategy could be utilized for the delivery of biomaterials for tissue repair, reconstructive surgery, and local site-directed drug delivery.

Tubule, Ribbon, and Network Assemblies

In addition to spherical lipid assemblies, lipid tubules have also been explored for encapsulating and releasing bioactive molecules with unique kinetics. Like liposomes, lipid tubules can be loaded with drugs or other bioactive substances for controlled release purposes. In contrast to liposomes, however, drug release from tubules is expected to exhibit zero-order release kinetics under certain conditions. A widely studied tubule-forming lipid is the polymerizable diacetylene-containing

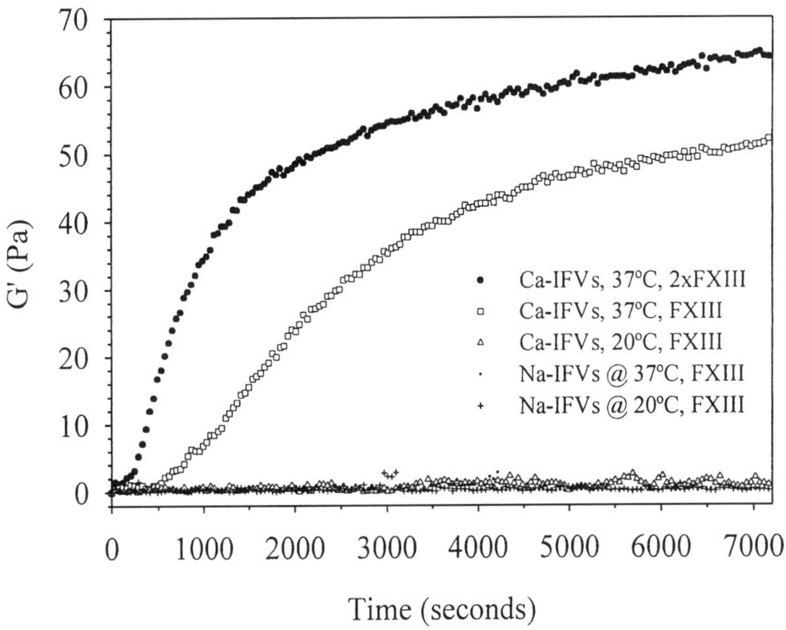

Figure 9 Rheological detection of thermally triggered enzymatic gelation of a fibrinogen gel. Ca^{2+}-loaded liposomes (Ca-IFVs), hrFXIII [either 100 μg (□, °, +, Δ) or 200 μg (•)], and fibrinogen were combined in buffer and monitored by oscillating rheometry. The rheometer was equilibrated at either 20° or 37°C prior to loading the precursor fluid and beginning data collection. Ca^{2+}-free liposomes (Na-IFVs) were used as a control to illustrate the Ca^{2+} dependence of enzyme cross-linking. The rise in storage modulus seen in the samples heated to 37°C is a result of Ca^{2+} release from the liposomes, which activates enzymatic cross-linking of soluble fibrinogen by hrFXIII. Reprinted from (53) with permission from Elsevier Science.

phospholipid 1,2-bis(10,12-tricosadiynoyl)-*sn*-glycero-3-phosphocholine ($DC_{8,9}$ PC) which, under certain conditions, can spontaneously form hollow microtubules (Figure 10) consisting of one or more bilayers (73). Depending on specific processing parameters such as temperature, concentration, and solvent mixture, the dimensions (74–76) and thermodynamic properties (77) of these tubules can be controlled and modified. Zero-order release kinetics were observed over the course of 10 days using $DC_{8,9}PC$ tubules containing radiolabeled growth factors (78). Furthermore, when subjected to hydrolysis by phospholipase enzymes, $DC_{8,9}PC$ tubules degraded in a zero-order fashion as well (79). Zero-order release profiles are highly desired for releasing constant amounts of therapeutic agents for extended periods of time.

$C_{8,9}PC$ tubules, which have been studied extensively by Schnur, Yager, and coworkers (80–82), have been utilized for the templating of a wide variety of

Figure 10 Transmission electron micrograph (TEM) of $DC_{8,9}PC$ microtubules prepared by the method of Yager & Schoen (73). Samples were prepared from aqueous suspension by negative staining with phosphotungstic acid.

materials, and several biological/biomedical applications of controlled release from these tubules have been explored. For example, microtubules have been used as templates for the deposition of metal coatings, resulting in the formation of metal microcylinders (83, 84). The metal microcylinders were then filled with an antibiotic-containing resin and added to marine paint to impart anti-fouling properties to the hulls of ships (85). This composite paint was found to be effective for releasing antibiotics for several months. One medical application of microtubules that has been explored is the use of $DC_{8,9}PC$ tubules for the release of growth factors for nerve regeneration (86). Many existing growth factor encapsulation methods utilize biodegradable polymers, which require the use of biomolecule-damaging organic solvents in their processing. Using phospholipid microcylinders, fragile biomolecules such as peptide growth factors can be encapsulated with mild solvent and temperature conditions. Growth factor-laden $DC_{8,9}PC$ tubules were able to release growth factors in physiologically relevant amounts for at least seven days in vitro (86).

Recently, the minimum diameter of $DC_{8,9}PC$ tubules has been reduced into the sub-100 nm range by the use of a 1:1 mixture of $DC_{8,9}PC$ and 1,2-bis(dinonanoyl)-*sn*-glycero-3-phosphocholine (DNPC) (87). The $DC_{8,9}PC$/DNPC mixture induces formation of lipid nanotubules with diameters in the 50 to 60 nm range

(Figure 11). These tubules are tens of microns in length and are also hollow, with three to four lipid bilayers making up each wall. Unfortunately, it is too early to say with any certainty how, or even if, these lipid nanotubules can be used as controlled release devices. One of the obstacles to their use as biomaterials may be their instability at body temperature. $DC_{8,9}PC$/DNPC nanotubules were found

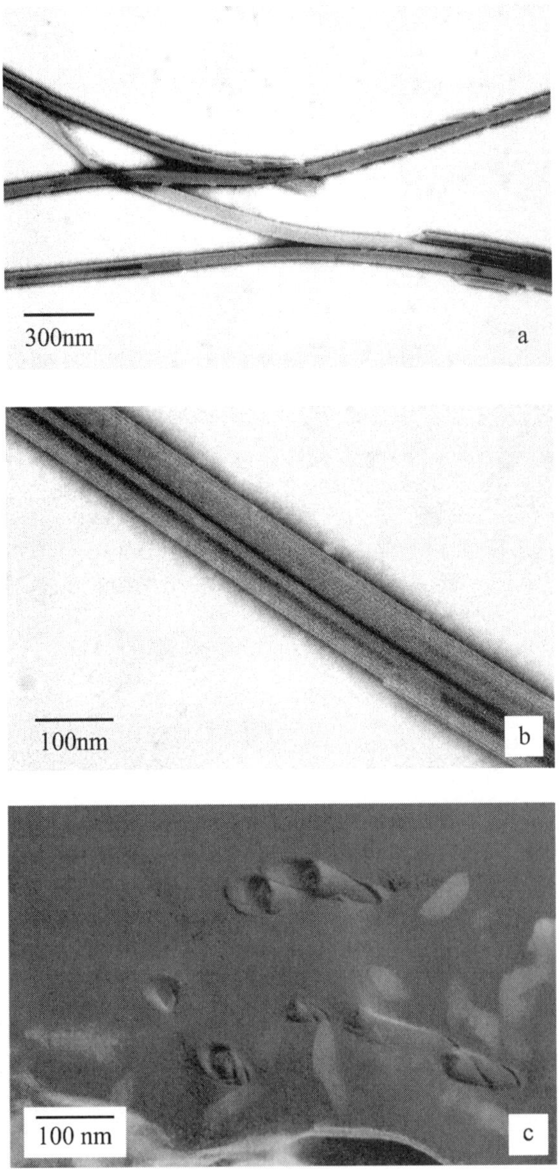

Figure 11 TEM images of $DC_{8,9}PC$/DNPC nanotubules. Samples were prepared from aqueous suspension by negative staining with phosphotungstic acid (*a*,*b*) or by freeze-fracture (*c*).

to undergo morphological changes when stored at or above ambient temperature (87). However, it is interesting to note that this instability may actually give rise to another biomaterial application. Heating of $DC_{8,9}PC$/DNPC nanotubules from sub-ambient to body temperature gives rise to the formation of helical lipid ribbons, which interconnect in three dimensions to form a physical gel (Figure 12). These gels consist of greater than 98% water and could be exploited as biocompatible lipid scaffolds for tissue-engineering and drug-delivery applications. It is also conceivable that the thermally triggered nanotubule-helical ribbon transformation could be exploited to drive a fluid-gel transition of an injectable nanotubule suspension.

Lamellar hydrogels consisting of water-swollen mixtures of saturated phosphatidylcholines and PEGylated lipids (e.g., PEG-DMPE) have been recently developed (88, 89). These gels consist of a network of connected fluid membranes containing phospholipids and small amounts of PEG-lipids that reside in regions of high membrane curvature. Such lipid-based hydrogels are expected to be biocompatible and potentially useful for drug delivery applications and could play host to bioactive membrane-embedded proteins.

Bicontinuous lipid emulsions have been exploited as templates to form three-dimensional inorganic mineral scaffolds whose structure reflects that of the lipid assemblies that were used in their formation (90–93). In these systems, reactions between mineralizing species are confined to the interconnecting water channels of the bicontinuous network, leading to highly complex, interconnected mineral frameworks. Such approaches have significant future potential for constructing mineral, polymer, and composite scaffold structures for biomedical applications. Potential uses of such materials include the design of nanostructured materials for controlled drug delivery, as well as new scaffolds for tissue engineering.

Monolayers, Supported Bilayers, and Membrane-Mimetic Polymers: Biocompatible/Bioactive Surface Coatings

Phospholipids have also been extensively investigated as biomaterials in the form of phospholipid polymers and coatings that take advantage of the hemo-compatible and non-thrombogenetic properties of phospholipids. Several strategies have been developed that utilize either phospholipid monolayers/bilayers (94) or phosphorylcholine groups incorporated into designed polymers (95). These approaches are inspired by the fact that 90% of the phospholipids of the outer membrane of red blood cells possess phosphorylcholine head groups, and it is the phosphorylcholine moiety that is important in promoting thromboresistance (96).

A simple approach for generating such membrane-mimetic surfaces is to create supported lipid bilayers at the surface of bulk materials. These may be constructed in several ways, e.g., the Langmuir-Blodgett technique or fusion of a lipid monolayer onto a self-assembled alkylsilane-treated surface. The more robust systems have a covalent coupling of a phosphorylcholine head group to the substrate surface. Early efforts in this area utilized phospholipids as nonthrombogenic coatings

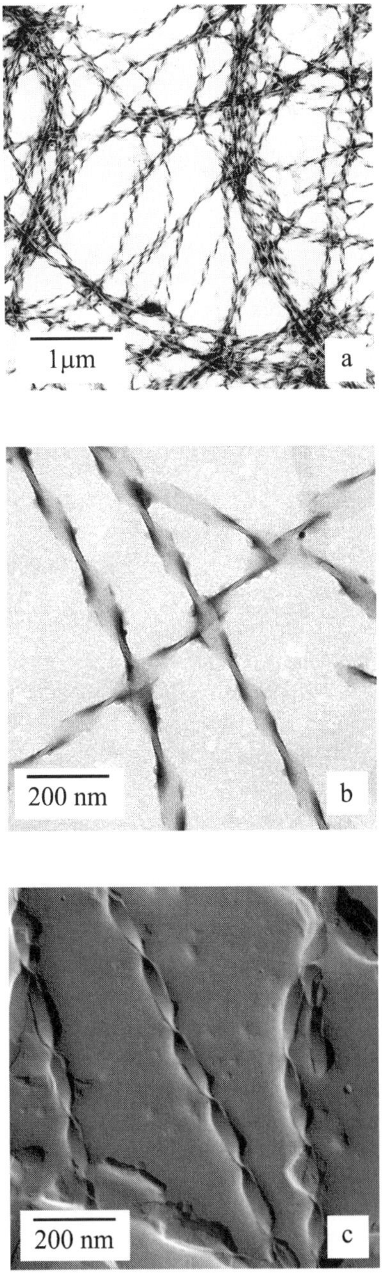

Figure 12 TEM images of $DC_{8,9}PC$/DNPC helical nanoribbons. Samples were prepared from aqueous suspension by negative staining with phosphotungstic acid (*a,b*) or by freeze-fracture (*c*).

on existing implant materials (97–100). Several efforts to coat biomaterials with lipids have used diacetylenic phospholipids in monomeric (97, 101) or polymeric (97, 102) form. In animal studies, these coatings performed as well as surfaces that had been coated with heparin (101), showing the antithrombogenic potential for lipid-coated surfaces. Membrane-mimetic strategies are currently being employed in increasingly complex ways in an effort to more completely imitate the structure of natural biomembranes. The general goal of these studies is to engineer a bioactive surface by incorporating selected proteins/peptides into lipid membrane structures, either by formation of lipid films onto bulk materials or by the chemical conjugation of lipid precursors into a solid matrix. Dori and coworkers recently used the Langmuir-Blodgett technique to form hybrid peptide amphiphile/phospholipid monolayers on hydrophobic supports (103). This approach was designed to yield a biologically active membrane-mimetic surface containing ligands of controlled accessibility to proteins and cells in solution. By controlling accessibility to surface bound ligands, the investigators were able to manipulate the extent of endothelial cell attachment and spreading on hybrid monolayer surfaces. These types of approaches involving functionalized lipid surfaces have significant biomedical potential in the design of medical devices and biosensors with highly controlled surface bioactivity.

Chaikof and coworkers have suggested that supported bilayers could be used to synthesize biomaterials and biosensors with a high degree of similarity to the natural surface of cells and have explored the parameters necessary for controlling the kinetics of multicomponent monolayer formation (104). Furthermore, lipopeptide conjugates (105) are being explored for the modification of such self-assembling supported lipid bilayer surfaces with receptor-activating peptide sequences, and the stability of these supported bilayers has been improved with polymerization of lipid tails after bilayer formation (106). Another possibility that has been explored is patterning individual domains of supported bilayers using microstamping techniques (107). Using such a technique, cells contacting the patterned bilayer could be confined and controlled regio-selectively to study or induce responses that are spatially dependent.

Using the red blood cell membrane as a model, polymers that include phosphorylcholine side-chains have been developed as nonthrombogenic coatings for existing materials. One polymer that has been extensively studied is a copolymer of methacrylate and methacryloyloxyethyl phosphorylcholine (MPC), a monomer used to simulate the polar head groups of phospholipids (108, 109). These copolymers exhibit low protein adsorption characteristics, which are thought to be the basis of their suppression of clot formation. One hypothesis suggested for the low protein adsorption of these materials is that it is a result of the phospholipid-like polymer having a high free-water fraction that disfavors protein conformational change at the surface (110). Several materials have been investigated as substrates for coating with phosphorylcholine-containing polymers, including cellulose dialysis membranes (111), polyurethane (112), polysulfone (113), and polyethylene (114).

In addition to their anti-protein adsorption characteristics, phosphorylcholine-containing polymers can preferentially adsorb natural phospholipids found in the fluid surrounding the implant. Natural synovial fluids contain phospholipids that act as a boundary lubricant, and elastomeric polymers with phosphorylcholine side chains have been investigated as articular surfaces to attract these naturally occurring phospholipids to the wear surface of the implant (115). The authors propose to take advantage of spontaneous lipid adsorption from the synovial fluid to lubricate articular surfaces in total joint replacements.

SUMMARY

In this review we have discussed the structural and self-assembling characteristics of phospholipids, how lipid assemblies are utilized in nature, and current ways in which these biological strategies are being exploited to synthesize novel synthetic biomaterials. Several emerging approaches in biomaterials research take advantage of phospholipids to encapsulate and/or template via self-assembled structures such as liposomes and tubules, whereas others exploit the inherent biocompatibility of lipids for engineering novel tissue-contacting biomaterials that mimic biological membranes.

We have used several examples from the world of biomineralization to illustrate the compartmentalization and templating roles that phospholipids play during formation of biogenic materials with controlled micro- and nanostructures. Phospholipids, often in the form of bilayer vesicles, are commonly employed in biomineralization processes to delineate reaction compartments within which mineral formation takes place. The encapsulation ability of phospholipid vesicles can be exploited for biomaterial applications by entrapment of liposomes within polymer hydrogels for local site-directed delivery of therapeutic agents. A more sophisticated use of phospholipid vesicle compartmentalization involves the use of stimuli-responsive liposomes for rapid injectable in situ-forming biomaterials. In this approach, stimuli-responsive liposomes actively participate in the formation of mineral, polymer, and mineral/polymer composite biomaterials. The use of liposomes in this manner represents a departure from the traditional use of liposomes as delivery vehicles for topical or intravenous administration of entrapped drugs, genetic material, enzymes, and contrast agents. In the future, this versatile strategy could be utilized for minimally invasive administration of biomaterials for tissue repair, reconstructive surgery, and local site-directed drug delivery.

In addition to vesicles, several emerging applications of phospholipid assemblies and membrane-mimetic materials in the form of tubules, monolayers, and lipid gels have been reviewed. Lipid-based tubule structures have potential significance in drug-delivery applications owing to their unique release kinetics and are beginning to be investigated in this capacity. Recent progress on developing lipid-based hydrogels consisting of either fluid polymer-lipid bilayers or interconnected helical lipid ribbons may lead to new applications of lipids as hydrogel matrices for tissue-engineering or drug-delivery applications. Finally, bulk materials coated with lipid monolayers and bilayers offer the promise of biocompatibility, as well as the potential to be optimized for control of surface bioactivity. Many membrane-mimetic strategies employ either modified lipids or lipid-mimetic polymers. As methods for more precise control of surface chemistry are refined and the reproducibility and robustness of these structures improves, they will enable even more sophisticated uses of such structures as biomaterials.

ACKNOWLEDGMENTS

This research was supported in part by National Institutes of Health grants R01 DE13030, R29 DE12599, P60 DE13076, and The Whitaker Foundation. JHC was supported by NRSA grant T32 DE07042.

Visit the Annual Reviews home page at www.AnnualReviews.org

LITERATURE CITED

1. Cevc G. 1993. *Phospholipids Handbook.* New York: Dekker. 988 pp.
2. Jones MN, Chapman D. 1995. *Micelles, Monolayers, and Biomembranes.* New York: Wiley-Liss. 264 pp.
3. Epand RM, ed. 1997. *Lipid Polymorphism and Membrane Properties.* San Diego: Academic. 568 pp.
4. Israelachvili J. 1992. *Intermolecular and Surface Forces.* London: Academic. 450 pp.
5. Seddon JM, Cevc G. 1993. In *Phospholipids Handbook*, ed. G Cevc, pp. 403–54. New York: Dekker
6. MacKintosh FC. 1997. *Curr. Opin. Colloid In.* 2:382–87
7. Gregoriadis G. 1995. *Trends Biotechnol.* 13:527–37
8. Lowenstam HA, Weiner S. 1989. *On Biomineralization.* New York: Oxford Univ. Press. 324 pp.
9. Mann S. 1995. *J. Mater. Chem.* 5:935–46
10. Mann S. 1993. *Nature* 365:499–505
11. Mann S. 1988. *Nature* 332:119–24
12. Young JR, Didymus JM, Brown PR, Prins B, Mann S. 1992. *Nature* 356:516–18
13. Anderson HC. 1995. *Clin. Orthop. Relat. R* 314:266–80
14. Eanes ED, Hailer AW, Costa JL. 1984. *Calcif. Tissue Int.* 36:421–30
15. Eanes ED. 1992. *Bone Miner.* 17:269–72
16. Eanes ED. 1989. *Anat. Rec.* 224:220–25
17. Skrtic D, Eanes ED. 1992. *Calcif. Tissue Int.* 50:253–60
18. Skrtic D, Eanes ED. 1992. *Calcif. Tissue Int.* 50:55–60
19. Eanes ED, Hailer AW. 1985. *Calcif. Tissue Int.* 37:390–94
20. Mann S, Frankel RB, Blakemore RP. 1984. *Nature* 310:405–7
21. Mann S, Sparks NHC, Frankel RB, Bazylinski DA, Jannasch HW. 1990. *Nature* 343:258–61
22. Schuler D, Frankel RB. 1999. *Appl. Microbiol. Biotechnol.* 52:464–73
23. Mann S, Heywood BR, Rajam S, Walker JBA. 1991. *J. Phys. D: Appl. Phys.* 24:154–64
24. Heywood BR, Mann S. 1994. *Chem. Mater.* 6:311–18
25. Mann S, Heywood BR, Rajam S, Birchall JD. 1988. *Nature* 334:692–95
26. Mann S, Williams RJP. 1983. *J. Chem. Soc. Dalton Trans.* 2:311–16
27. Mann S, Hannington JP. 1987. *J. Colloid Interf. Sci.* 122:326–35

28. Bhandarkar S, Bose A. 1989. *J. Colloid Interf. Sci.* 135:531–38
29. Yaacob II, Bhandarkar S, Bose A. 1992. *J. Mater. Res.* 8:573–77
30. Bhandarkar S, Bose A. 1990. *J. Colloid Interf. Sci.* 139:541–50
31. Chang A-C, Pfeiffer WF, Guillaume B, Baral S, Fendler JH. 1990. *J. Phys. Chem.* 94:4284–89
32. Tricot YM, Fendler JH. 1986. *J. Phys. Chem.* 90:3369–74
33. Fendler JH. 1982. *Membrane Mimetic Chemistry*. New York: Wiley
34. Archibald DD, Yager P. 1992. *Biochemistry* 31:9045–55
35. Archibald DD, Mann S. 1993. *Nature* 364:430–33
36. Baral S, Schoen PE. 1993. *Chem. Mater.* 5:145–47
37. Lvov YM, Price RR, Selinger JV, Singh A, Spector MS, Schnur JM. 2000. *Langmuir* 16:5932–35
38. Burkett SL, Mann S. 1996. *Chem. Commun.* 3:321–22
39. Goren M, Qi ZG, Lennox RB. 2000. *Chem. Mater.* 12:1222–28
40. Wilson-Kubalek EM, Brown RE, Celia H, Milligan RA. 1998. *Proc. Natl. Acad. Sci. USA* 95:8040–45
41. Wang GJ, Hollingsworth RI. 1999. *Langmuir* 15:6135–38
42. DiTizio V, Ferguson GW, Mittelman MW, Khoury AE, Bruce AW, DiCosmo F. 1998. *Biomaterials* 19:1877–84
43. Cohen S, Bernstein H, Hewes C, Chow M, Langer R. 1991. *Proc. Natl. Acad. Sci. USA* 88:10440–44
44. Cohen S, Alonso MJ, Langer R. 1994. *Int. J. Technol. Assess. Health Care* 10:121–30
45. Weiner AL, Carpenter-Green SS, Soehngen EC, Lenk RP, Popescu MC. 1985. *J. Pharm. Sci.* 74:922–25
46. Gabrijelcic V, Sentjurc M. 1995. *Int. J. Pharmaceutics* 188:207–12
47. Chandy T, Sharma CP. 1996. *Biomaterials* 17:61–66
48. Monshipouri M, Rudolph AS. 1995. *J. Microencapsul.* 12:117–27
49. Takagi I, Shimizu H, Yotsuyanagi T. 1996. *Chem. Pharm. Bull.* 44:1941–47
50. Takagi I, Nakashima H, Takagi M, Yotsuyanagi T, Ikeda K. 1997. *Chem. Pharm. Bull.* 45:389–93
51. DiTizio V, Karlgard C, Lilge L, Khoury AE, Mittelman MW, DiCosmo F. 2000. *J. Biomed. Mater. Res.* 51:96–106
52. Cohen S, Bano MC, Chow M, Langer R. 1991. *Biochim. Biophys. Acta* 1063:95–102
53. Westhaus E, Messersmith PB. 2001. *Biomaterials* 22:453–62
54. Gerasimov OV, Rui Y, Thompson DH. 1996. In *Vesicles*, ed. M Rosoff, pp. 679–746. New York: Dekker
55. Yatvin MB, Weinstein JN, Dennis WH, Blumenthal R. 1978. *Science* 202:1290–93
56. Weinstein JN, Magin RL, Yatvin MB, Zaharko DS. 1979. *Science* 204:188–91
57. Frankel DA, Lamparski H, Liman U, O'Brien DF. 1989. *J. Am. Chem. Soc.* 111:9262–63
58. Lamparski H, Liman U, Barry JA, Frankel DA, Ramaswami V, et al. 1992. *Biochemistry* 31:685–94
59. You H, Tirrell DA. 1991. *J. Am. Chem. Soc.* 113:4022–23
60. Anderson VC, Thompson DH. 1992. *Biochim. Biophys. Acta* 1109:33–42
61. Thompson DH, Gerasimov OV, Wheeler JJ, Rui Y, Anderson VC. 1996. *Biochim. Biophys. Acta* 1279:25–34
62. Gerasimov OV, Schwan A, Thompson DH. 1997. *Biochim. Biophys. Acta.* 1324:200–14
63. Gerasimov OV, Boomer JA, Qualls MM, Thompson DH. 1999. *Adv. Drug. Deliv. Rev.* 38:317–38
64. Messersmith PB, Han C. 1997. A new approach to in-situ thermal gelation of alginate for controlled release. In *Polym. Prepr. (Am. Chem. Soc., Div. Polym. Chem.)*, pp. 616–17

65. Messersmith PB, Starke S. 1998. *Chem. Mater.* 10:117–24
66. Cui H, Messersmith PB. 1998. Thermally triggered gelation of alginate for controlled release. In *ACS Symp. Ser.*, pp. 203–11
67. Murphy W, Messersmith PB. 2000. *Polyhedron* 19:357–63
68. Papahadjopoulos D, Jacobsen K, Nir S, Isac T. 1973. *Biochim. Biophys. Acta* 311:330–48
69. Marsh D, Watts A, Knowles PF. 1976. *Biochim. Biophys. Acta* 858:161–68
70. Mouritsen OG, Jorgensen K, Honger T. 1995. In *Permeability and Stability of Lipid Bilayers*, ed. EA Disalvo, SA Simon, pp. 137–60. Boca Raton, FL: CRC Press
70a. Koynova R, Caffrey M. 1998. *Biochim. Biophys. Acta* 1376:91–145
71. Sperinde JJ, Griffith LG. 1997. *Macromolecules* 30:5255–64
72. Sperinde JJ, Griffith LG. 2000. *Macromolecules* 33:5476–80
73. Yager P, Schoen PE. 1984. *Mol. Cryst. Liq. Cryst.* 106:371–81
74. Spector MS, Selinger JV, Singh A, Rodriguez JM, Price RR, Schnur JM. 1998. *Langmuir* 14:3493–500
75. Kulkarni VS, Boggs JM, Brown RE. 1999. *Biophys. J.* 77:319–30
76. Thomas BN, Safinya CR, Plano RJ, Clark NA. 1995. *Science* 267:1635–38
77. Spector MS, Selinger JV, Schnur JM. 1997. *J. Am. Chem. Soc.* 119:8533–39
78. Spargo BJ, Stillwell GE, Cliff RO, Monroy RL, Rollwagen F, Rudolph AS. 1992. In *Tissue-Inducing Biomaterials*, ed. LG Cima, ES Ron, pp. 285–92. Pittsburgh: Mater. Res. Soc.
79. Carlson PA, Gelb MH, Yager P. 1997. *Biophys. J.* 73:230–38
80. Schnur JM. 1993. *Science* 262:1669–76
81. Spector MS, Easwaran KR, Jyothi G, Selinger JV, Singh A, Schnur JM. 1996. *Proc. Natl. Acad. Sci. USA* 93:12943–46
82. Spector MS, Price RR, Schnur JM. 1999. *Adv. Mater.* 11:337–40
83. Schnur JM, Price RR, Schoen PE, Yager P, Calvert J, et al. 1987. *Thin Solid Films* 152:181–206
84. Rudolph AS, Calvert JM, Schoen PE, Schnur JM. 1989. In *Biotechnological Applications of Lipid Microstructures, Advances in Experimental Medicine and Biology*, ed. BP Gaber, JM Schnur, D Chapman, pp. 305–20. New York: Plenum
85. Price RR, Patchan M. 1993. *J. Microencapsul.* 10:215–22
86. Yu X, Dillon GP, Bellamkonda RB. 1999. *Tissue Eng.* 5:291–304
87. Svenson S, Messersmith PB. 1999. *Langmuir* 15:4464–71
88. Warriner H, Idziak S, Slack N, Davidson P, Safinya C. 1996. *Science* 271:969–73
89. Warriner HE, Keller SL, Idziak SH, Slack NL, Davidson P, et al. 1998. *Biophys. J.* 75:272–93
90. Walsh D, Hopwood JD, Mann S. 1994. *Science* 264:1576–78
91. Walsh D, Mann S. 1995. *Nature* 377:320–23
92. Mann S, Ozin GA. 1996. *Nature* 382:313–18
93. Eftekharzadeh S, Stupp S. 1997. *Chem. Mater.* 9:2059–65
94. Chapman D. 1993. *Langmuir* 9:39–45
95. Lewis AL. 2000. *Colloids Surf. B Biointerfaces* 18:261–75
96. Chapman D, Charles SA. 1992. *Chem. Br.* 28:253–56
97. Hayward JA, Chapman D. 1984. *Biomaterials* 5:135–42
98. Durrani AA, Hayward JA, Chapman D. 1986. *Biomaterials* 7:121–25
99. Hayward JA, Durrani AA, Shelton CJ, Lee DC, Chapman D. 1986. *Biomaterials* 7:126–31
100. Hayward JA, Durrani AA, Lu Y, Clayton CR, Chapman D. 1986. *Biomaterials* 7:252–58
101. von Segesser LK, Tonz M, Leskosek B,

Turina M. 1994. *Biomaterials* 17:294–300
102. von Segesser LK, Olah A, Leskosek B, Turina M. 1993. *ASAIO J.* 39:43–46
103. Dori Y, Bianco-Peled H, Satija SK, Fields GB, McCarthy JB, Tirrell M. 2000. *J. Biomed. Mater. Res.* 50:75–81
104. Winger TM, Ludovice PJ, Chaikof EL. 1999. *Langmuir* 15:3866–74
105. Winger TM, Ludovice PJ, Chaikof EL. 1996. *Biomaterials* 17:437–41
106. Orban JM, Faucher KM, Dluhy RA, Chaikof EL. 2000. *Macromolecules* 33:4205–12
107. Yang TL, Simanek EE, Cremer P. 2000. *Anal. Chem.* 72:2587–89
108. Ishihara K, Aragaki R, Ueda T, Watenabe A, Nakabayashi N. 1990. *J. Biomed. Mater. Res.* 24:1069–77
109. Yoneyama T, Ito M, Sugihara K, Ishihara K, Nakabayashi N. 2000. *Artif. Organs* 24:23–28
110. Ishihara K, Nomura H, Mihara T, Kurita K, Iwasaki Y, Nakabayashi N. 1998. *J. Biomed. Mater. Res.* 39:323–30
111. Ishihara K, Nakabayashi N. 1995. *Artif. Organs* 19:1215–21
112. Ishihara K, Hanyuda H, Nakabayashi N. 1995. *Biomaterials* 16:873–79
113. Ishihara K, Fukumoto K, Iwasaki Y, Nakabayashi N. 1999. *Biomaterials* 20:1545–51
114. Ishihara K, Iwasaki Y, Ebihara S, Shindo Y, Nakabayashi N. 2000. *Colloids Surf. B Biointerfaces* 18:325–35
115. Williams PF 3rd, Powell GL, Love B, Ishihara K, Nakabayashi N, et al. 1995. *Biomaterials* 16:1169–74

Annu. Rev. Mater. Res. 2001. 31:265–89

EPITAXIAL SPINEL FERRITE THIN FILMS

Yuri Suzuki
Department of Materials Science and Engineering, Cornell University, Ithaca, New York 14853; e-mail: suzuki@ccmr.cornell.edu

Key Words magnetic anisotropy, exchange coupling, cation distribution, ferrite multilayers

■ **Abstract** Recently there have been significant advances in understanding the magnetic properties of epitaxial ferrite films that are not found in bulk ferrites. Much effort has been expended on trying to achieve bulk properties in thin films for a wide range of applications. From a fundamental science perspective, epitaxial thin films and heterostructures have provided model systems in which novel phenomena, such as modified super-exchange interactions, nearly ideal exchange coupling, and perpendicular exchange coupling, have been observed. These magnetic phenomena and other anomalous magnetic properties are interesting in their own right and are highlighted here.

INTRODUCTION

Ferrites constitute a class of materials that has been recognized to have significant potential in applications ranging from millimeter wave integrated circuitry to magnetic recording. In contrast to mature bulk ferrite technology such as transformer cores (1) and microwave magnetic devices (2), these potential applications can only be realized if ferrite films live up to the promise of critical bulk ferrite properties. These properties include strong magnetic coupling, high resistivity, and low loss characteristics at high frequencies. In recognition of their potential, considerable effort has been expended in the synthesis and characterization of ferrite thin film materials optimized to particular applications.

In view of storage applications, some ferrite thin films have been studied as potential perpendicular recording and magneto-optical media material. Barium ferrite and cobalt ferrite films are particularly promising because of their large coercivity, as well as their high corrosion resistance, good mechanical stability, and low noise (3). More recently, high-permeability insulating ferrites have been identified as candidate materials for flux guides and sensors in thin film recording heads (4). While microwave devices using bulk ferrites can be traced back to the 1950s, the development of ferrite thin films will provide several advantages for microwave and millimeter wave devices. For example, the propagation characteristics of magnetostatic wave devices can be significantly improved by films that can provide better uniformity of effective internal field compared with bulk crystals.

Ferrites are magnetic ceramic materials. They are all ferrimagnetic oxides but vary in crystal structure. In the 1950s, ferrites referred to spinel structure crystals. But more recently, ferrite refers to the entire family of iron oxides that includes spinels, garnets, hexaferrites, and orthoferrites. Figure 1 shows the spinel structure in terms of cations in various oxygen coordinations. Spinel structure materials are cubic and have the form AB_2O_4 (A = divalent cations and B = trivalent cations of Fe). The oxygen atoms make up an fcc lattice; an eighth of the tetrahedral (A) sites and half of the octahedral (B) sites are occupied. The exchange interaction between A and B sites is negative and the strongest among the cations so that the net magnetization comes from the difference in magnetic moment between A and B sites. Hexagonal ferrites (e.g. $BaFe_{12}O_{19}$ and Sr $Fe_{12}O_{19}$) have an even more complicated crystal structure. The Fe cations are the sole source of magnetic moment and can be found on any of three crystallographically different sites: tetrahedral, octahedral, and hexahedral sites. The garnets, like the spinels, also have a cubic structure but with trivalent cations (including rare earth and Fe^{3+}) occupying tetrahedral (A), octahedral (B), or dodecahedral (C) sites. The interaction between tetrahedral and octahedral sites is antiparallel, and the resultant magnetic moment is antiparallel to the rare earth ions on the C sites. Rare earth orthoferrites also fall into the family of ferrites, but they are more accurately described as canted antiferromagnets. They have both an orthorhombic crystal structure and the general formula $RFeO_3$, where R is a yttrium or a rare earth.

Given the complexity of the crystal structures of cubic and hexagonal ferrites, it is not surprising that structural disorder, ranging from cation disorder to grain boundaries, has a significant effect on the electronic and magnetic properties. Moreover, these ferrites have open crystal structures so that diffusion of cations and oxygen ions readily occurs. Therefore, in earlier work, microwave ferrite thin film devices did not measure up to their potential (of highly insulating behavior, high permeability, bulk saturation magnetization). By contrast, garnet thin films have been successfully incorporated into magneto-optical applications. This success is based on the single-crystal nature of garnets grown by the liquid

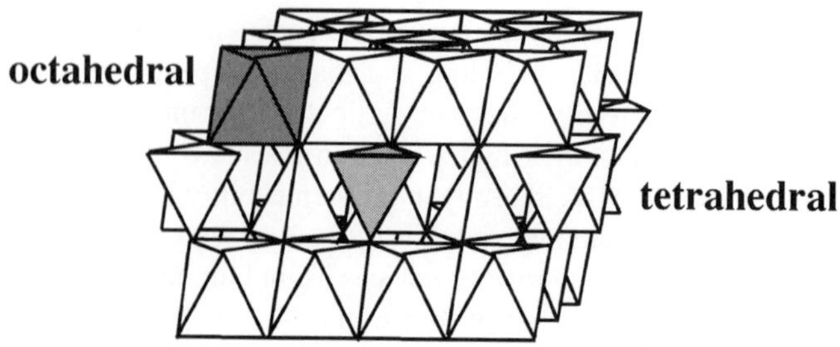

Figure 1 Crystal structure of spinel structure ferrites.

phase epitaxy (LPE) process and the availability of isostructural lattice-matched substrates (5). For magneto-optical applications, in particular optical isolation, micron-thick single-crystal films are necessary to obtain significant Faraday rotation so that LPE is appropriate. Unfortunately, the LPE process does not lend itself easily to the growth of ultrathin films or multilayers.

Recent studies of ferrite thin films have focused on understanding the microstructure and its implications on the electronic and magnetic properties. Studies range from spinel and hexaferrite thin films for perpendicular recording media, integrated inductors, and micromagnetic devices to garnet structure ferrite films for magnet-optical applications. Despite variations in crystalllinity and crystal structure, the fundamental issues of structure-property relationships in ferrite thin films remain the same in these studies. Therefore, in what follows we describe ferrite thin films grown by a variety of methods and with varying crystallinity. Recent progress in the understanding of structure-property relationships in polycrystalline and epitaxial ferrite films is presented. Because many of the issues transcend the different ferrite crystal structure types, we focus, in particular, on epitaxial spinel structure ferrite thin films. Finally, we describe the recent progress in ferrite heterostructures where structural and magnetic ordering across interfaces reveal novel phenomena and provide the foundation for future incorporation of ferrites into technology.

FERRITE THIN FILM GROWTH

Ferrite thin films under study range from nanocrystalline or polycrystalline to epitaxial thin films. Given a particular growth process, it has often been difficult to reproduce results from one laboratory to the next; however, the development and refinement of deposition techniques have resulted in the convergence of results. In polycrystalline and epitaxial ferrite thin films, electroplating, magnetron sputtering, pulsed laser deposition, evaporation, and molecular beam epitaxy have been used extensively.

Of these techniques, electroplating (otherwise known as ferrite plating) has generated much interest because of its ability to generate ferrite crystallites at temperatures below 100°C. Crystallites of varying cation compositions can be deposited from an aqueous solution where Fe ions are oxidized. Due to the low growth temperatures, this technique may allow the fabrication of ferrite film devices on non-heat resistant substrates. Because ferrite plating does not require elevated temperatures or high vacuum, ferrites can be plated on polymer microspheres for toner and carrier particles in xerography and magnetic separation in enzyme immunoassays (6). The use of ultrasound during ferrite plating has been found to improve the coverage of material on the surface. Grain sizes down to 100 nm have been obtained in ultrasound-enhanced ferrite plating (7). Moreover, ferrite-plated films are promising for perpendicular recording media, as well as high performance biasing layers for metallic perpendicular magnetic recording media (8), but also as cladding layers for integrated inductors and micromagnetic devices (9).

Single- and multitarget magnetron sputtering are deposition techniques that can easily be scaled up for large-area production. In the sputtering of oxide thin films, the bombardment of the substrate by negative ions and secondary electrons emitted from the target results in a decrease in the net deposition rate and the preferential sputtering of certain atomic species from the deposited surface (10). Therefore, off-axis sputtering, where the substrate is no longer in the line of sight of the target, negative ions, and secondary electrons, is the geometry of choice for research. Deposition temperatures ranging from 400–600°C and partial pressures of oxygen on the order of 10^{-5} torr or less are typical for spinel structure and hexaferrite thin films. Depending on the choice of substrate and deposition conditions, textured and epitaxial ferrite thin films can be deposited by sputtering.

Pulsed laser deposition (PLD) of oxide thin films has seen remarkable development with the advent of high-temperature superconducting thin films (11, 12). An ultraviolet laser beam is focused on a sintered stoichiometric target of the desired ferrite. The deposition process involves complex laser-solid interactions that depend on the laser wavelength, pulse duration, power density, and the materials specific optical and thermodynamic properties. Unlike in sputtering, the emitted species include neutrals, electronically excited species, ions, electrons, clusters, and macroscopic particles. These species deposit on a substrate facing the target. Deposition temperatures of laser-deposited ferrite films are similar to those of sputtered films, whereas partial pressures of oxygen are typically on the order of 10^{-3} torr or less. Although PLD is not the method of choice for high-volume manufacturing, it is an important research tool because phase space can be investigated quickly. As in sputtering, the ferrite films range from textured to epitaxial depending on composition, substrate temperature, and substrate composition.

Molecular beam epitaxy (MBE) has been employed by some groups to grow oriented ferrite thin films (13–15). Here the precursors decompose into reactive atoms with kinetic energy derived from the high temperatures at the substrate surface. For the deposition of oxide materials in an MBE system with background pressures well below 10^{-9} torr, an oxygen plasma or source is often used to increase the partial pressure of oxygen near the substrate surface. Codeposition of atomic species requires careful monitoring of the evaporation rates of the different sources. In turn, in situ monitoring of the atomic flux onto the substrate surface provides insight into the growth modes and crystallographic order at the surface. This careful control over deposition that MBE affords has resulted in ferrite heterostructure work that is described below. Again, depending on the choice of substrate and deposition conditions, we can obtain textured or epitaxial growth of ferrites using MBE.

In the growth of ferrites, post-annealing treatments must be performed with great care. Divalent Fe ions are easily oxidized to trivalent Fe ions. This oxidation often results in the creation of α- or γ-Fe_2O_3 phases. Although maghemite

γ-Fe_2O_3 is ferrimagnetic with a saturation magnetization slightly less than that of magnetite Fe_3O_4, hematite α-Fe_2O_3 is a canted antiferromagnet and reduces the overall magnetization of the sample.

POLYCRYSTALLINE THIN FILMS

Polycrystalline or textured thin films have been grown by methods ranging from ferrite plating to sputtering. The ability to tune the magnetic properties of these films through the crystallite size and the magnetic coupling among crystallites enables these materials to be incorporated into, for example, perpendicular media applications. Given the high coercivity, large perpendicular magnetic anisotropy, large saturation magnetization, and insulating behavior in some bulk ferrites, ferrite thin films made of $BaFe_{12}O_{19}$, Sr $Fe_{12}O_{19}$, and $(Co_{1-x}Zn_x)Fe_2O_4$ have been the focus of such perpendicular media research (16, 16a–c). Many studies in the literature have been empirical ones of the effects of varying deposition conditions on the magnetic properties. For example, sputtered Co-Zn ferrites that are optimized for high-saturation magnetization exhibit low-perpendicular magnetic anisotropy (17). Studies of ferrite-plated Ni-Zn and Co ferrite film show high-perpendicular coercivities at low-growth temperatures (18, 18a).

In an effort to understand the origin of the deviations from bulk magnetic behavior, techniques such as X-ray (19), neutron diffraction (20), Mossbauer spectroscopy (21, 22), and extended X-ray absorption fine structure (EXAFS) have been employed to probe the local structure. Diffraction techniques are strongly influenced by the element's cross section to X-rays and neutrons, as well as sample mass. Mossbauer spectroscopy is an excellent tool for determining the local environment of Fe ions; however, it is not as successful in elucidating the local structure of dilute or nonmagnetic additives. EXAFS studies have provided both element specificity and local structure sensitivity. As an example, EXAFS studies of ferrite-plated spinel structure Ni-Zn ferrite thin films indicate that while Ni ions are found primarily on the octahedral site, the Zn ions are primarily found on the tetrahedral sites (23). These observations are more or less consistent with the fact that $NiFe_2O_4$ is an inverse spinel (i.e. 8 divalent ions occupy 8 out of 32 octahedral sites, and the 16 trivalent ions randomly occupy the remaining tetrahedral and octahedral sites), and $ZnFe_2O_4$ is a normal spinel (i.e. 8 divalent ions occupy 8 out of 64 tetrahedral sites, and 16 trivalent ions occupy 16 out of 32 octahedral sites). Deviations of cation stoichiometry in ferrite thin films that are grown by a nonequilibrium process are not surprising and may explain some of the observations in, for example, these ferrite-plated Ni-Zn ferrite films.

Details of the microstructure have important implications on the magnetic properties. The crystallite size and the nature of grain boundaries among crystallites are just two of many factors that affect the nature of the magnetic coupling in these polycrystalline films. As is described in the section on epitaxial thin films, small structural deviations can have significant effects on the magnetic properties.

Therefore, determination of local structure and microstructure is key in understanding and improving the quality of ferrite thin films (24, 25).

Most studies of the magnetic properties of ferrite thin films have emphasized the optimization of saturation magnetization, coercivity, or magnetic anisotropy. However, in light of integrated inductor or micromagnetic device applications, the permeability of these ferrite films is also important. Mn-Zn ferrites have been identified as prime candidates given that their bulk permeabilities are as high as 4000. However, the permeability of Mn-Zn ferrite films is very sensitive to the microstructure. In bulk polycrystalline Mn-Zn ferrites, van der Zaag et al found that a nonmagnetic grain boundary model (where a nonmagnetic boundary is thought to modify the effective low-field permeability) is appropriate (26). In Mn-Zn ferrite films, for example, Gillies et al have found permeabilities ranging from 30 to 100, depending on grain size (27). They have performed transmission electron microscopy (TEM) and scanning electron microscopy (SEM) to obtain an accurate measure of grain size and correlated it with low-field permeability, saturation magnetization, and coercive fields. Permeabilities of ferrite-plated Ni-Zn ferrite films are comparable to those of bulk polycrystalline Ni-Zn ferrite with similar grain size (28). Given the high-bulk permeabilities found in Mn-Zn ferrite, we might expect high permeabilities in epitaxial single-crystal ferrite films. The small permeabilities measured in such films are further evidence for the need to understand the materials issues, including epitaxy, microstructure, and strain, that are not present in the bulk, and understand their relation to the electronic and magnetic properties.

EPITAXIAL THIN FILMS

Epitaxial ferrite thin films have provided not only technological promise for the incorporation of magnetically tailored thin film materials in a variety of applications but also an understanding of the structure-property relationships in this family of materials. For example, for short wavelength magnetostatic wave devices and microwave acoustic applications, single-crystal films are necessary to reduce wave attenuation because of scattering at grain boundaries. From a more fundamental perspective, epitaxial ferrite films have provided a model system in which, for example, the effects of modified super-exchange interactions and cation redistribution have been observed. Moreover, epitaxy of the overlying ferrite film on the substrate implies that the film is clamped on one side by the substrate. Therefore, the effects of strain and strain relaxation on the electronic and magnetic properties have to be taken into consideration.

Fe_3O_4 Films

One of the most studied epitaxial ferrite films has been magnetite (Fe_3O_4). It has received renewed interest because of its half metallicity and thus its possible use in the growing field of magnetic multilayer devices. Band structure calculations predict that in Fe_3O_4 the majority spin electrons are semiconducting, with a sizeable energy-gap separating the valence band from empty conduction bands, and that the minority spin electrons are metallic (29). Similar calculations for $NiFe_2O_4$ show a sizeable gap in the majority spins and a smaller one in the minority spin states, thus giving rise to an insulating state, as observed experimentally. Calculations of $CoFe_2O_4$, $MnFe_2O_4$, and $ZnFe_2O_4$ should theoretically give rise to metallicity. Fe_3O_4 is unique among the spinel structure ferrites in that it undergoes what is called a Verwey transition, a transition attributed to a charge ordering of Fe^{2+} and Fe^{3+} ions that occurs below the transition temperature. Below the transition, Fe_3O_4 undergoes a structural transition from cubic to monoclinic (30) and is accompanied by a dramatic increase in resistivity around 120 K (31), with a corresponding decrease in the magnetic moment (32).

In this section, we focus on recent epitaxial work in Fe_3O_4. In the past few years, many groups have fabricated and characterized epitaxial Fe_3O_4 films made by a variety of deposition techniques. Among the recent studies of epitaxial Fe_3O_4 films, the focus has been on their incorporation into device heterostructures where half-metallic Fe_3O_4 provides a reservoir of spin-polarized electrons. Although the Verwey transition is generally accepted to be the result of the ordering of the Fe^{2+} and Fe^{3+} ions, the magnetoresistive and magnetic properties of epitaxial Fe_3O_4 films have only begun to be explained.

Studies of anomalous magnetic properties in Fe_3O_4 films have revealed complicated exchange interactions not found in bulk single crystals. Margulies et al found that their epitaxial Fe_3O_4 thin films did not saturate magnetically even in fields of over 5 Tesla (33, 34). In film studies, the crystalline quality is always called into question. Structural characterization, in the form of X-ray diffraction and high-resolution TEM, showed no evidence for mixed orientations or second phases. From the diffraction intensities, the films had excellent crystallinity. Figure 2 shows one quadrant of a typical magnetization loop with an anomalous approach to saturation. This behavior was observed in films grown by sputtering and evaporation, as well as MBE. Because microstructure varies considerably with deposition technique as well as deposition conditions, the anomalous behavior could not be attributed to grain shape, size, or surface morphology. Furthermore, there was no significant temperature dependence or film thickness dependence on the anomalous magnetic behavior. The lack of temperature dependence suggests that magnetocrystalline anisotropy, which ought to exhibit strong temperature dependence, is not the explanation. In fact, the slight film thickness dependence indicated that the behavior is a volume effect and not a surface effect.

Margulies et al were able to explain this behavior in terms of the exchange interactions across antiphase boundaries found in their Fe_3O_4 films grown on MgO substrates. Although structural characterization showed that the Fe_3O_4 films on MgO were single crystalline, there were antiphase boundaries in the films. The antiphase boundaries are created because the unit cell of MgO ($a = 4.21$ Å) is about half that of Fe_3O_4 ($a = 8.39$ Å). Therefore, when the film nucleates

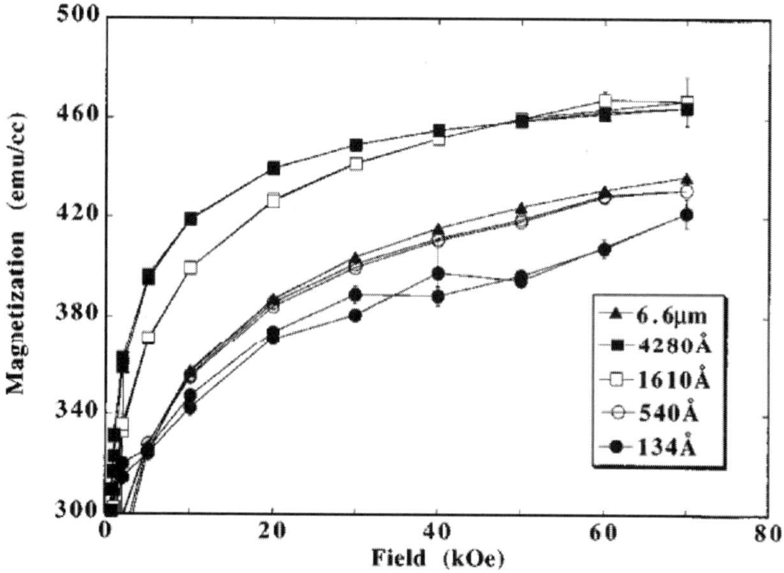

Figure 2 Magnetization loops of Fe_3O_4 thin films versus field show an anomalous approach to saturation. Films are grown on (100) MgO substrates (31).

and grows on the surface of the substrate, neighboring grains may not coalesce with matching cation sublattices. Antiphase boundaries were identified in dark field TEM images (Figure 3*a*,*b*). Varying densities of antiphase boundaries have been found in epitaxial films grown by differing deposition techniques (35).

Only substantial diffusion would allow for improved ordering. Post-annealing treatments in a reducing atmosphere at temperatures up to 600°C were performed with no change in approach to saturation. Annealing at higher temperatures resulted in a reduction of magnetization through the oxidation of Fe_3O_4 into α, γ-Fe_2O_3. At the same time, annealed films on MgO substrates exhibited traces of Mg diffusion into the film (36).

In spinel structure oxides, the superexchange interactions among cations on tetrahedral (A) and octahedral (B) sites are determined by the angle subtended by the cation-anion-cation bond and the cation-anion distances. The more collinear the cation-anion-cation arrangement, the stronger the antiferromagnetic interaction. In Fe_3O_4, the AB antiferromagnetic interaction dominates because the A-O-B angle is closest to 180° and because there are relatively more AB interactions. With an antiphase boundary, (*a*) the A-O-A angle is greatly increased toward 180°, and the number of AA interactions is increased by a factor of two relative to single crystal Fe_3O_4, (*b*) more BB interactions that have B-O-B angles close to 180° are created, and (*c*) the A-O-B angle remains the same but the number of AB interactions decreases. The net result is that across the boundary the usually dominant

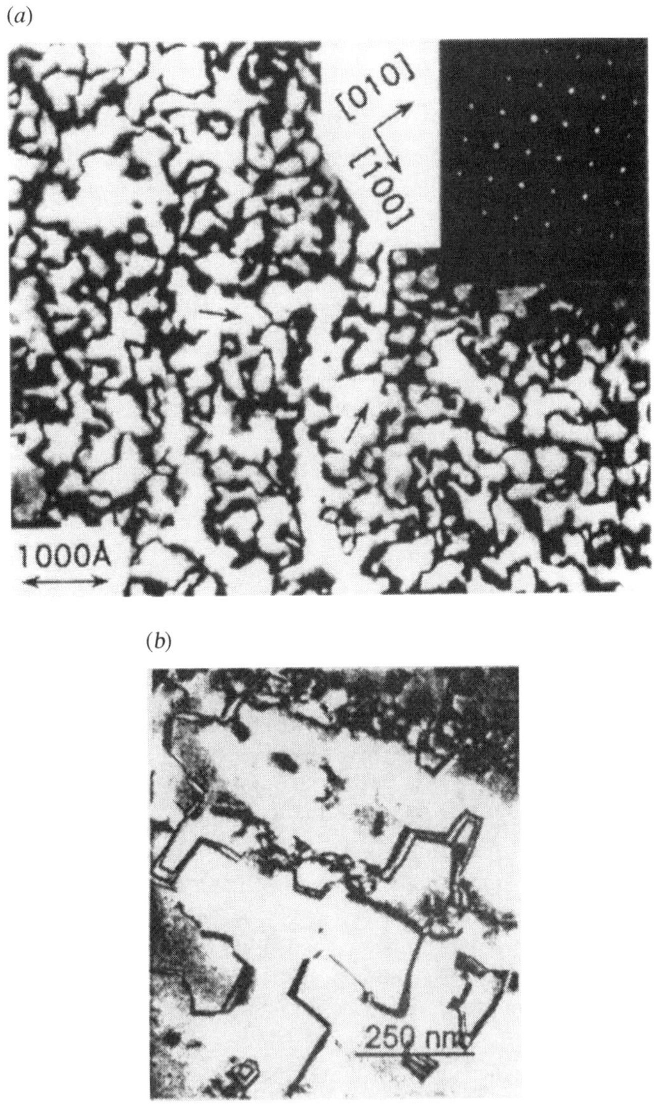

Figure 3 Transmission electron images of antiphase boundaries in (*a*) a 50-nm film grown by sputtering and (*b*) a 200-nm film grown by oxygen-MBE. Neighboring sections are related by shift vectors 1/4a ⟨110⟩ (32, 36).

AB interaction is reduced, whereas the antiferromagnetic interactions between A and B sublattice atoms, respectively, are stronger. Therefore, the intrasublattice superexchange coupling is greatly increased, whereas the intersublattice super-exchange is decreased. In ultrathin films (<5 nm), Voogt et al have found that superparamagnetic behavior stems from the frustration of the interdomain interactions that are dominated by super-exchange interactions across antiphase boundaries (37). More recently, it has been suggested that these same Fe_3O_4 films exhibit room temperature "charge ordering" of the Fe cations on the tetrahedral sites (38). The extra superlattice spots in the high-resolution TEM diffraction images were attributed to a long-range order—a charge ordering. They believe that such charge ordering can only be observed in samples of sufficient crystal quality.

These modified exchange interactions must affect the magnetoresistive properties of epitaxial Fe_3O_4 films. Indeed, magnetotransport studies in epitaxial Fe_3O_4 films have revealed interesting field and temperature dependences and anisotropies (39). As seen in single crystals, the magnetoresistance of the films exhibits a peak around the Verwey transition (40–42). Ogale et al draw similarities to colossal magnetoresistance materials and describe the magnetotransport in Fe_3O_4 films in terms of polaronic states and phonon-magnon effects. Gong et al suggest that the enhanced magnetoresistance they observe is the result of a magnetic field–dependent activation energy for electron hopping transport.

$NiFe_2O_4$ Films

Studies of epitaxial $NiFe_2O_4$ thin films reveal anomalous magnetic behavior similar to that observed in Fe_3O_4 films. However, the origin of the anomalous behavior cannot be entirely explained by modified exchange interactions across antiphase boundaries. Venzke et al investigated the structural and magnetic properties of $NiFe_2O_4$ thin films grown on $SrTiO_3$ (STO) and $Y_{0.15}Zr_{0.85}O_2$ (YSZ) substrates (43). On STO substrates, antiphase boundaries are allowed because the unit cell size is a = 3.905 Å, whereas on (001) YSZ substrates, rotational domains are allowed, given the surface unit cell size of 5.139 Å. Although their as-grown films exhibited good crystallinity, post-annealing at 1000°C in an oxidizing atmosphere improved the crystallinity significantly, as seen in Rutherford backscattering spectroscopy and X-ray diffraction.

Similar to the Fe_3O_4 films, magnetization loops of $NiFe_2O_4$ thin films, grown between 400 and 700°C, exhibited an anomalous approach to saturation. Figure 4 shows typical curves of as-grown $NiFe_2O_4$ thin films on STO. Even at fields as high as 5 Tesla, the magnetization was at two thirds of the bulk saturation magnetization value. The shape of the magnetization loop could not be attributed to a high-anisotropy field from magnetocrystalline or magnetostrictive origin. However, unlike in Fe_3O_4 films, the anomalous approach to saturation decreased with successive annealing steps starting at 600 up to 1300°C. Unlike Fe_3O_4 thin films, $NiFe_2O_4$ contains only Fe^{3+} ion, and Ni^{2+} is not easily oxidized as Fe^{2+}. Therefore, successive high-temperature anneals did not result in second phase formation as in Fe_3O_4 thin films.

A detailed investigation of the microstructure revealed grains on the order of 300–500 Å in diameter. Smooth crystalline areas were surrounded by material that was either amorphous or highly defective. Figure 5 shows a tile-like structure in high-resolution transmission electron micrographs. The crystallites appeared to

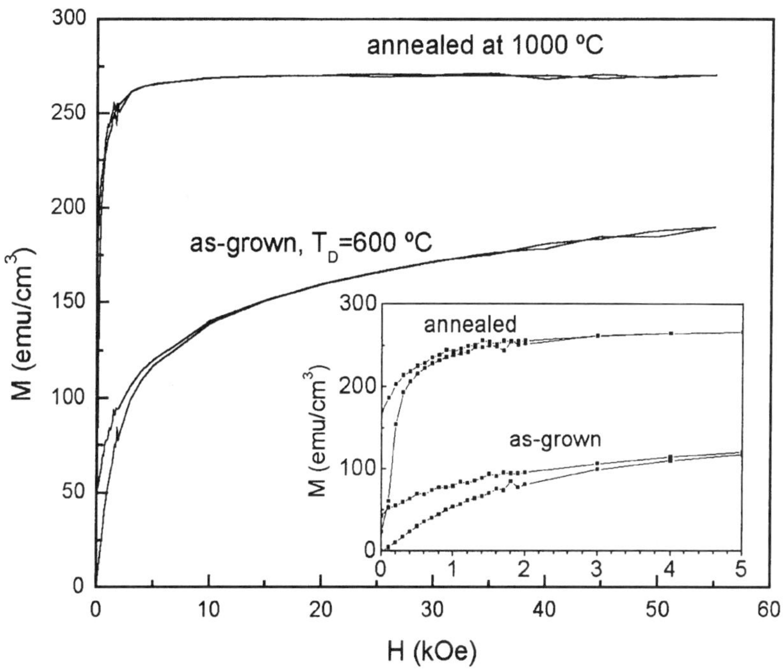

Figure 4 Magnetization loops of $NiFe_2O_4$ films grown on $SrTiO_3$. The bulk saturation magnetization is 270 emu/cm^3. Even in fields of 5 Tesla, the magnetization of the as-grown film is only about two thirds of the bulk value (43).

be well aligned within about a degree due to the influence of the substrate. Upon annealing, the plan view transmission microscopy images showed uniform and smooth micron-size grains with occasional defects threading through the thickness of the film. Therefore, the anomalous approach to saturation in the as-grown films were attributed to magnetic interactions between the crystalline regions and these amorphous or highly defective regions. These interactions were confirmed by the observation of exchange-biased magnetization loops at low temperatures (44).

We can reconcile the results of epitaxial Fe_3O_4 and $NiFe_2O_4$ films. In the case of $NiFe_2O_4$ films, both interactions across antiphase boundaries and among crystalline and highly defective regions may give rise to this anomalous approach to saturation. However, the electron microscope images indicate that the latter is the dominant microstructure. Upon annealing, grain growth occurs by crystallization of the defective regions and eliminates antiphase boundaries at the same time. In the Fe_3O_4 films, the microstructure is dominated by antiphase boundaries and contains minimal defective regions. Annealing of these films in order to eliminate antiphase boundaries requires temperatures that result in the oxidization of the Fe^{2+} ions and the formation of Fe_2O_3.

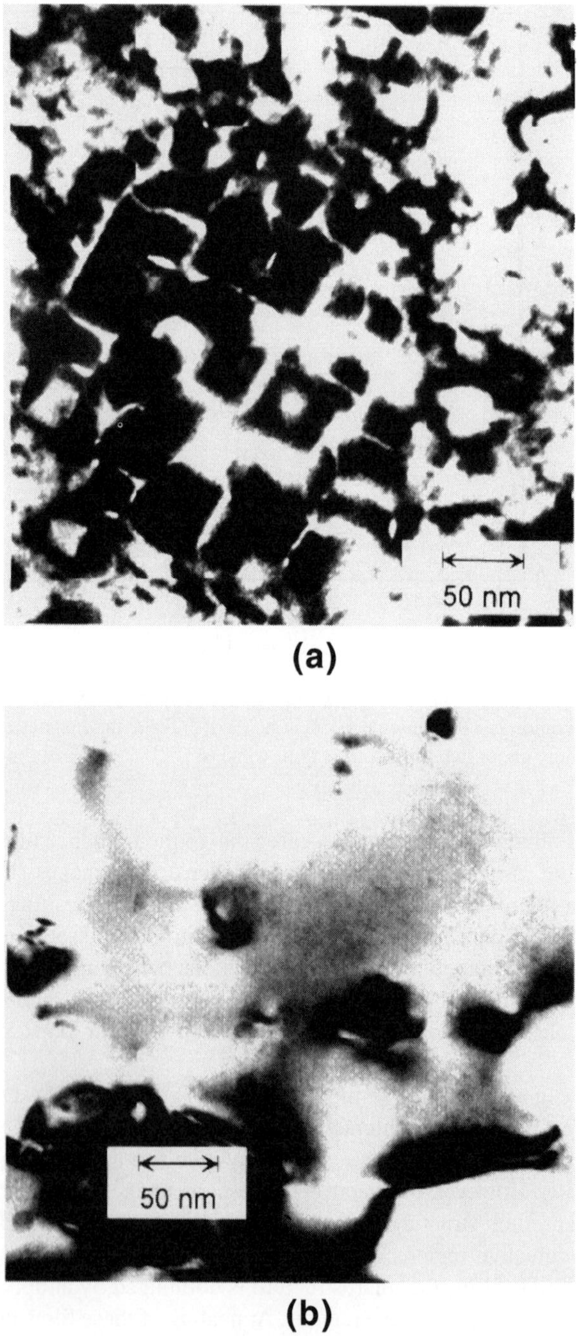

Figure 5 Transmission electron micrographs of typical regions on (*a*) a $NiFe_2O_4$ fil grown on $SrTiO_3$ at 600°C and (*b*) an identical film annealed at 1000°C (43).

(Mn,Zn)Fe_2O_4 Films

Compared with Fe_3O_4 and $NiFe_2O_4$ films, (Mn,Zn)Fe_2O_4 is magnetically softer by two orders of magnitude. As a result, it has been identified as a candidate material for applications where high permeabilities are required. Several groups have studied epitaxial (Mn,Zn)Fe_2O_4 thin films on a variety of substrates.

For studies of (Mn,Zn)Fe_2O_4 thin films on MgO substrates, one would expect anomalous magnetic behavior in the approach to saturation, much as was observed in Fe_3O_4 and $NiFe_2O_4$ films, because antiphase boundaries are inevitable. Williams et al have found that the uniaxial anisotropy, ferromagnetic linewidth, and coercive force are strongly influenced by microstructure, whereas saturation magnetization and magnetocrystalline anisotropy depend greatly on composition and cation distribution (45). When comparing thin films with their bulk counterparts, they have found that saturation magnetization, magnetocrystalline anisotropy, and ferromagnetic resonance linewidth of thin films were comparable to bulk values despite the presence of antiphase boundaries. Dorsey et al have noted that the volatility of Zn made it extremely difficult to substitute Zn^{2+} for Mn^{2+} with accuracy (46). In fact, they find in their studies that on average their film compositions are 42% deficient of Zn, with excess Mn and Fe. The relative ratios of Mn^{2+}: Zn^{2+}: Fe^{2+} determine the saturation magnetization as well as the magnetic softness. In either study, the influence of antiphase boundaries is not apparent in the magnetic data.

In order to eliminate any effect of antiphase boundaries, Suzuki et al have grown (Mn,Zn)Fe_2O_4 on spinel ($MgAl_2O_4$) substrates. Because there is a large lattice mismatch between the spinel structure ferrites and $MgAl_2O_4$, they developed a series of buffer layers that are nonferromagnetic at room temperature (47). Of these choices, $CoCr_2O_4$ and $NiMn_2O_4$ enabled them to grow epitaxial ferrite thin films of high crystalline quality without antiphase boundaries. The idea is to grow the buffered layer thick enough so that misfit dislocations relax strain in the buffer layer. Figure 6 shows a typical example of a (Mn,Zn)Fe_2O_4 film with a square magnetization loop and no trace of an anomalous approach to saturation. Magnetization loops of (100) (Mn,Zn)Fe_2O_4 films measured with fields applied parallel to the film surface and in various directions in the plane of the film revealed no easy axes. Loops of (100) (Mn,Zn)Fe_2O_4 films taken with the field perpendicular to the film showed expected effects due to demagnetization. The measured anisotropy of $H_K \sim 4.5$ kOe agrees with the demagnetizing field of $4\pi M_s \sim 4.5$ kOe (M_s = saturation magnetization) of (Mn,Zn)Fe_2O_4 films. Magnetization loops of (110) (Mn,Zn)Fe_2O_4 films revealed a square loop when the field is applied along an easy axis in the plane.

Films grown directly on $SrTiO_3$ and $MgAl_2O_4$ never reached bulk magnetization values. X-ray fluorescence spectroscopy indicated similar Zn doping values for (Mn,Zn)Fe_2O_4 films grown with and without a buffer layer. Therefore, Zn doping cannot account for the differences in values of buffered and unbuffered (Mn,Zn)Fe_2O_4 films (48). Films grown on $SrTiO_3$ and $MgAl_2O_4$ have lattice mismatches of 8 and 4%, respectively. Therefore, lattice mismatch and thermally

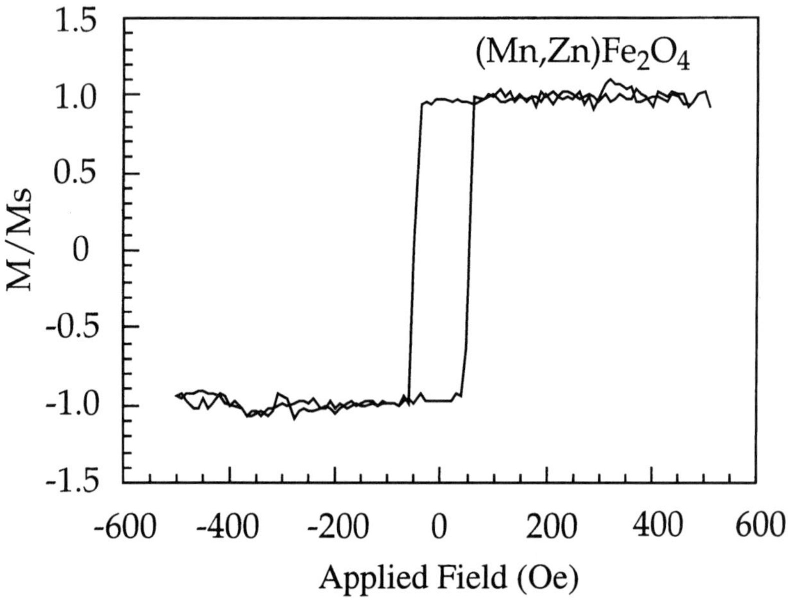

Figure 6 Magnetization loop of (001) $(Mn,Zn)Fe_2O_4$ films versus field shows a square loop with no trace of an anomalous approach to saturation (47).

induced strain, as well as antiphase boundaries, may give rise to disordered regions in $(Mn,Zn)Fe_2O_4$ films during low-temperature growth, thus resulting in a reduced saturation magnetization. When these films were field cooled to low temperatures, magnetization loops showed an offset. Such an offset suggests that there are disordered regions with antiferromagnetic or glassy magnetic behavior and that there is exchange coupling between these disordered regions and crystalline regions of $(Mn,Zn)Fe_2O_4$. For $(Mn,Zn)Fe_2O_4$ films grown on buffered substrates, similar field-cooled magnetization loops showed a much smaller offset, thus indicating a negligible amount of disordered material in the buffered $(Mn,Zn)Fe_2O_4$ films. Therefore, it was concluded that modified exchange interactions are present in the unbuffered $(Mn,Zn)Fe_2O_4$ films but may not be as evident in the magnetization loops as for magnetically harder Fe_3O_4 and $NiFe_2O_4$ films.

$CoFe_2O_4$ Films

$CoFe_2O_4$ is unique among the spinel structure ferrites in that it has a large magnetic anisotropy accompanied by large anisotropic magnetostriction. Therefore, effects of strain and cation arrangement, which may have been masked by other effects such as those of APBs in spinel ferrite thin films grown on MgO substrates, will be enhanced dramatically in the magnetic properties of $CoFe_2O_4$ thin films. Although $CoFe_2O_4$ itself is not technologically promising because of its very short spin lattice relaxation constant, it provides a model system for understanding the magnetic properties of spinel structure ferrite thin films.

Dorsey et al found that increasing deposition temperatures, which in turn affects the microstructure, gives rise to increasing saturation magnetizations and changing sign of the uniaxial magnetic anisotropy (49). Given that their films were grown on MgO substrates, the effect of antiphase boundaries cannot not be ignored. Hu et al have studied $CoFe_2O_4$ films on buffered substrates (see above) in order to determine the effects of strain and cation arrangement that may have been masked by other effects such as antiphase boundaries in spinel ferrite thin films grown on MgO substrates (50). In fact, they observed no antiphase boundaries and low dislocation densities in transmission electron microscopy images. Figure 7 shows a lattice image of a typical (110) oriented $CoFe_2O_4$ film, in cross section, grown under compression. The white line indicates an extra (111) half plane, which forms a Frank dislocation with a Burger's vector of $b = 1/3\langle 111\rangle$. The mismatch strains are relieved by forming these edge dislocations. From images of different regions of the specimen, the dislocation density was estimated to be lower than $1.6 \times 10^{10}/cm^2$.

In Hu et al's study, the $CoFe_2O_4$ films are under compression, as determined from normal and grazing incidence diffraction measurements. The lattice strain

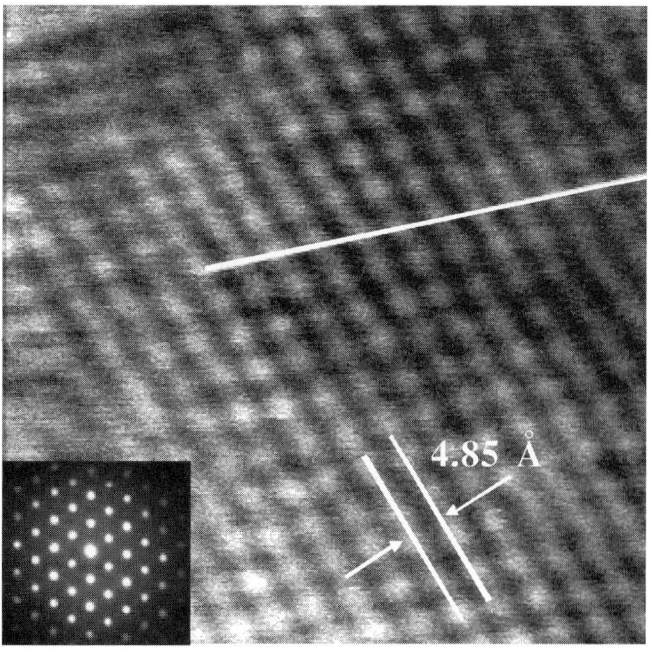

Figure 7 Lattice image of (110) oriented $CoFe_2O_4$ films on $CoCr_2O_4$ buffered $MgAl_2O_4$. The white line indicates an extra half $\langle 111\rangle$ plane (50).

and relaxation effects manifest themselves in the magnetic properties of as-grown and annealed $CoFe_2O_4$ films. The magnetic anisotropy of as-grown compressively strained (110) $CoFe_2O_4$ thin films can be understood self-consistently from magnetoelastic theory. The magnetocrystalline energy terms are dominated by the magnetoelastic energy terms in the expression of the total magnetic anisotropy energy. If, indeed, the biaxial compressive strain from the lattice mismatch between the film and the substrate is predominant, the strain anisotropy should be tuneable via the application of external strain. Magnetization measurements on $CoFe_2O_4$ thin films, while applying external uniaxial tensile strains $[1\bar{1}0]$ along the direction, are shown in Figure 8; the external strains result in shifts of the anisotropy field from 4.3 to 1.4 Tesla, which in turn allows us to estimate the magnetostriction constants $\lambda_{100} + \lambda_{111} = -120 \times 10^{-6}$. These values for magnetostriction confirmed the only measure of magnetostriction obtained in single-crystal $CoFe_2O_4$ (51).

By post-deposition annealing that resulted in strain relaxation and cation redistribution, the magnetoelastic and magnetocrystalline energy contributions to the total energy were varied. The dramatic reversal in the sign of the magnetic anisotropy of the films was explained by the reduction in the magnetoeleastic

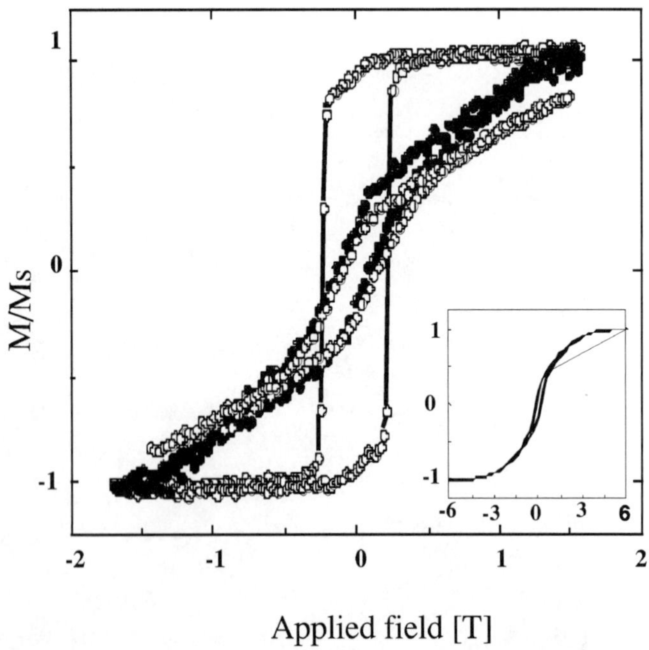

Figure 8 Magnetization loops of $CoFe_2O_4$ grown on $CoCr_2O_4$ buffered $MgAl_2O_4$ without external stress (open circles) along the easy and hard directions and with external tensile stress along the hard direction (solid circles). External stress shifts the anisotropy field from 4.3 to 1.4 Tesla (50).

energy term owing to strain relaxation and the increase in the magnetocrystalline energy term owing to the migration of Co^{2+} ions from octahedral to tetrahedral sites. The migration of the Co^{2+} ions is induced by the preferential migration of Cr^{3+} from the buffer layer into the overlying ferrite thin film and was experimentally verified by electron microprobe, electron energy loss spectroscopy, Rutherford backscattering, and extended X-ray absorption fine structure analyses.

From the structural evidence, Hu et al were able to conclude that (*a*) the large number of tetrahedrally coordinated Co^{2+} ions and (*b*) reduction of the strain are responsible for the evolution of the observed magnetic anisotropy with annealing. By varying cation distribution and intrinsic and external strain, they were able to tune the symmetry and the magnitude of the magnetic anisotropy. This tunability is promising for device applications.

FERRITE HETEROSTRUCTURES

Artificially layering the ferrites with isostructural and non-isostructural material is another route toward tuning the magnetic and electronic properties to incorporate them into planar device and storage applications. On a more fundamental level, modulated structures of ferrite materials have been studied in order to understand the magnetic coupling, ordering, and anisotropy, as well as transport in non-itinerant magnetic systems.

Magnetite Superlattices

Molecular beam epitaxy has been used to create magnetite multilayers with sharp interfaces. Because magnetic coupling between layers is extremely sensitive to the quality of the interface, layer-by-layer growth of complex oxide thin films is indispensable for studying exchange coupling across layers. At the same time, layer-by-layer growth promotes a fully strained state that has significant effects on the magnetic and electronic properties.

Several groups have carried out thorough studies of the growth and characterization of Fe_3O_4/NiO superlattices (52, 53). NiO is an antiferromagnet with a cubic rock salt structure. Single-phase crystal growth in registry with the substrate (rock salt MgO) has been verified with in situ structural monitoring. Chern et al (54) have found that the chemical modulation gives rise to electrical behavior that cannot be simply modeled by a composite model of the materials that comprise the superlattice. The low dimensionality of the modulated structures is believed to give rise to enhanced electrical anisotropy and to suppress the Verwey transition.

Similar structurally perfect magnetite superlattices have been fabricated with CoO that is also an antiferromagnet with a rock salt structure (55, 56). van der Zaag et al studied the exchange anisotropy associated with the Fe_3O_4/CoO superlattices.

In such antiferromagnet/ferromagnet multilayers, when the sample is field cooled through the Neèl temperature of the antiferromagnet, the magnetization loop of the ferromagnet is shifted by the exchange bias field. Similar studies have been performed by Kleint et al on bilayers (57).

There have been many theoretical models proposed to describe the exchange biasing behavior of antiferromagnet/ferromagnet couples, in particular metallic couples such as NiFe/MnFe. However, these metallic systems have a complicated noncollinear spin structure that renders theoretical modeling difficult. Fe_3O_4/CoO superlattices offer a simple collinear spin structure of CoO for exchange biasing. If one assumes that the exchange biasing is caused by a unidirectional anisotropy arising from nearest neighbor exchange coupling across a perfectly flat interface with a magnetically uncompensated antiferromagnetic surface, the calculated exchange bias field is eight times larger than the measured exchange field (58). Although this discrepancy between the model and experiment is certainly smaller than the two orders of magnitude difference in typical as-grown metallic multilayers, more work needs to be done. Alternative models involving planar domain walls (59) or perpendicular domain walls due to interface roughness in the antiferromagnet have been proposed (60). However, Malozemoff's random field model, for example, cannot explain the exchange biasing behavior observed in superlattices with CoO thicknesses up to 300 Å.

More recently, Koon suggested a modified planar domain wall picture where the interfacial exchange coupling is strong with a perpendicular orientation between the magnetization direction of the ferromagnet and antiferromagnet. In this picture, the exchange bias results from the formation of a mainly antiferromagnetic domain wall (parallel to the interface) as a reverse field rotates the ferromagnetic magnetization away from the field-cooled direction (61). Neutron diffraction experiments have demonstrated that in (001) Fe_3O_4/CoO superlattices, with compensated antiferromagnetic surfaces, the CoO antiferromagnetic ordering is long range and propagates coherently through the Fe_3O_4 layers and that these antiferromagnetic CoO spins preferentially align in domains with spins perpendicular to the net Fe_3O_4 moment (62). This was the first direct observation of a perpendicular exchange coupling phenomenon.

In exchange-biased Fe_3O_4/CoO superlattices, finite size scaling of the CoO layers cannot explain the reduction in blocking temperature. This reduction is at odds with the enhancement of the ordering temperature of the CoO layers above bulk values (63). While further studies of the blocking temperature are in order, these results are a good example of how low dimensionality in modulated structures can be used to tune the magnetic properties.

(Mn,Zn)Fe_2O_4/$CoFe_2O_4$ Bilayers

Nearly ideal exchange coupling has been observed in spinel ferrite bilayers (64). This result contrasts with exchange coupling in polycrystalline $Ni_{0.8}Fe_{0.2}$/$Mn_{0.}$

$Fe_{0.5}$ and $Ni_{0.53}Mn_{0.47}$ bilayers, where the exchange interaction constant A_{int} is at best two orders of magnitude smaller than estimates of A_{NiFe}. In ferrite bilayers, the two isostructural ferrimagnetic layers have different magnetic hardness due to the presence of Mn/Zn or Co. Because both layers are spinel structures and the magnetically harder biasing layer is ferrimagnetic, not antiferromagnetic as in the NiFe/-MnFe system, random interface roughness does not affect the strength of the exchange coupling.

Suzuki et al focused on the $(Mn_{0.45}Zn_{0.54})Fe_2O_4/CoFe_2O_4$ bilayer systems to study exchange biasing. Magnetization loops of the bilayers grown showed bulk magnetization values for both layers. Figure 9 is a plot of a major loop of a 1500 Å $(Mn_{0.45}Zn_{0.54})Fe_2O_4$/1600 Å $CoFe_2O_4$ bilayer on $MgAl_2O_4$, when the field is applied in the plane of the bilayer along the easy [001] axis of the $CoFe_2O_4$ layer. The $CoFe_2O_4$ and $(Mn_{0.45}Zn_{0.54})Fe_2O_4$ layers exhibit square hysteretic contributions to the major loop. The schematic diagrams of the soft and hard ferrite layers in Figure 9 indicate the spin directions in four different regions of the loop. In regions (ii) and (iv) of the loop, an increasing fraction of the magnetic moment in $(Mn_{0.45}Zn_{0.54})Fe_2O_4$ reoriented from being parallel to antiparallel to the

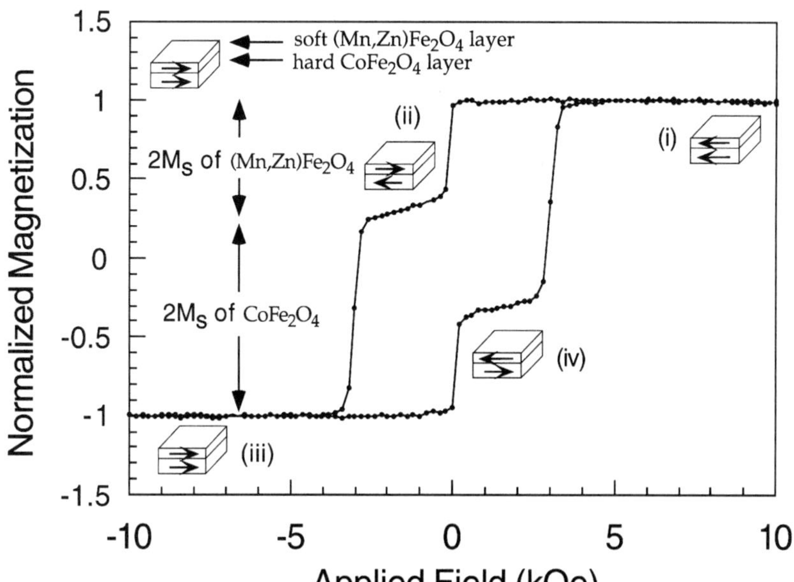

Figure 9 A major loop of a 1500 Å (Mn,Zn)Fe_2O_4/1600 Å $CoFe_2O_4$ bilayer on $CoCr_2O_4$ buffered $MgAl_2O_4$. The spin directions in the hard and soft layers are schematically shown in four different sections of the loop. The magnetizations attributed to the (Mn,Zn)Fe_2O_4 and $CoFe_2O_4$ layers are indicated.

magnetization of the CF layer with increasing field. The spins of $(Mn_{0.45}Zn_{0.54})Fe_2O_4$ farthest away from the $(Mn_{0.45}Zn_{0.54})Fe_2O_4/CoFe_2O_4$ interface reoriented first, while the spins of $(Mn_{0.45}Zn_{0.54})Fe_2O_4$ closest to the interface remained pinned by the adjacent $CoFe_2O_4$ layer until the coercive field of the $CoFe_2O_4$ layer was finally attained. When the field was applied along the hard axis of $CoFe_2O_4$, the major loop revealed a reversible curve with two distinct slopes; the steeper slope near the origin was attributed to the $(Mn_{0.45}Zn_{0.54})Fe_2O_4$; the smaller slope was attributed to the $CoFe_2O_4$.

To isolate exchange-coupled loops of the $(Mn_{0.45}Zn_{0.54})Fe_2O_4$ layer from the $CoFe_2O_4$ contribution, the sample was poled in the plane of the bilayer along the easy [001] axis of the $CoFe_2O_4$ layer to 10 kOe, which is well above its saturation field. Then a minor loop was traced between +0.5 and −1 kOe. In this field range, the magnetization of the $CoFe_2O_4$ was fixed; only the spins in the $(Mn_{0.45}Zn_{0.54})Fe_2O_4$ may reorient. The resulting exchange-biased loop of a 1500 Å $(Mn_{0.45}Zn_{0.54})Fe_2O_4$/1600 Å $CoFe_2O_4$ bilayer is shown in Figure 10. It is

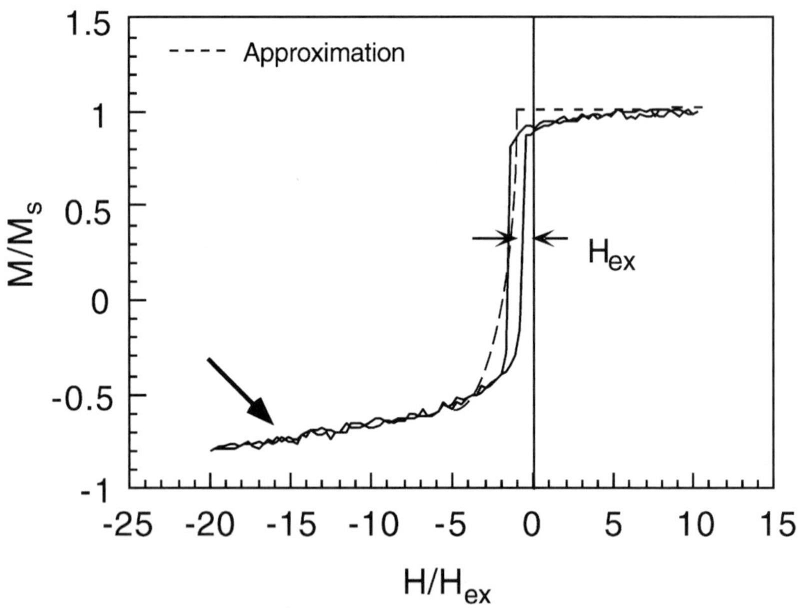

Figure 10 An exchange-biased loop of $(Mn,Zn)Fe_2O_4$ of a 1500 Å $(Mn,Zn)Fe_2O_4$/1600 Å $CoFe_2O_4$ bilayer with an exchange field of 50 Oe. The (fixed) magnetization attributed to the $CoFe_2O_4$ has been subtracted from the data. The asymmetric approach to saturation is characteristic of coherent rotation of the moments in the soft $(Mn,Zn)Fe_2O_4$ layer. The heavy arrow indicates the asymptotic approach to saturation. The dotted line represents a calculation assuming the coherent rotation of spins in the soft layger, as described in the text (64).

important to note that although Figure 10 represents a minor loop for the system overall, it is effectively a major loop for the $(Mn_{0.45}Zn_{0.54})Fe_2O_4$ layer because this layer is being driven from positive to negative saturation and back. We therefore refer to this minor loop as exchange-biased, keeping in mind that in many other systems an offset magnetization loop is not proof of exchange coupling. The exchange coupling of the $(Mn_{0.45}Zn_{0.54})Fe_2O_4$ layer to the adjacent $CoFe_2O_4$ layer is manifest clearly both in the offset, H_{ex}, and in the asymmetry of the loop.

A lower limit for the exchange interaction constant across the bilayer interface was deduced from the exchange-biased loop. This interfacial exchange constant is comparable in magnitude with that of $(Mn_{0.45}Zn_{0.54})Fe_2O_4$ and $CoFe_2O_4$ individually, thus indicating the nearly ideal strength of the exchange coupling across the interface. Indeed, the asymmetry of the exchange-biased loop and the $\sqrt{H}$ form of the approach to saturation are further evidence of strong exchange coupling, according to a simple model involving the coherent rotation of spins in the soft layer (65). In this model, the competition between the magnetostatic energy and the exchange energy of the soft layer gives rise to the rotation of the spins in the soft layer. Because exchange coupling involves the pinning of spins in the soft magnetic layer by the adjacent hard magnetic layer, the effect of exchange coupling must decrease with increasing thickness of the soft layer, everything else being the same. Suzuki et al found that, indeed, the exchange field H_{ex} varied inversely with the $(Mn_{0.45}Zn_{0.54})Fe_2O_4$ thickness (Figure 11). This inverse linear dependence of H_{ex} on $(Mn_{0.45}Zn_{0.54})Fe_2O_4$ thickness is evidence for the interfacial nature of the interaction that gives rise to the exchange-biased loop.

Device Structures

Given the half-metallic nature of Fe_3O_4, spin valves employing Fe_3O_4 as one or both electrodes have been the focus of recent studies. Briefly, a spin valve is a structure composed of two ferromagnetic electrodes separated by a nonmagnetic barrier. Depending on the relative orientation of the magnetic moments in the electrodes, the resistance across this structure is high (magnetic moments in the two electrodes are antiparallel) or low (magnetic moments in the electrodes are parallel). For example, van der Zaag et al and Li et al have fabricated and characterized Fe_3O_4-based all-oxide spin valves (66, 67). Although the observed magnetoresistive effect was on the order of 0.5% at room temperature in their structures, theoretically there is the potential for significantly higher effects given the half metallicity of Fe_3O_4. van der Zaag et al correlated the low magnetoresistive effects with reduced saturation magnetizations of the Fe_3O_4 electrodes. Although the origin of the reduced saturation magnetization is not clear, dead magnetic layers at the interfaces and modified exchange interactions at antiphase boundaries have been cited as possible reasons. Li et al have attributed the reduced magnetoresistance to interface effects (including a disordered spin structure and a reduced FeO layer

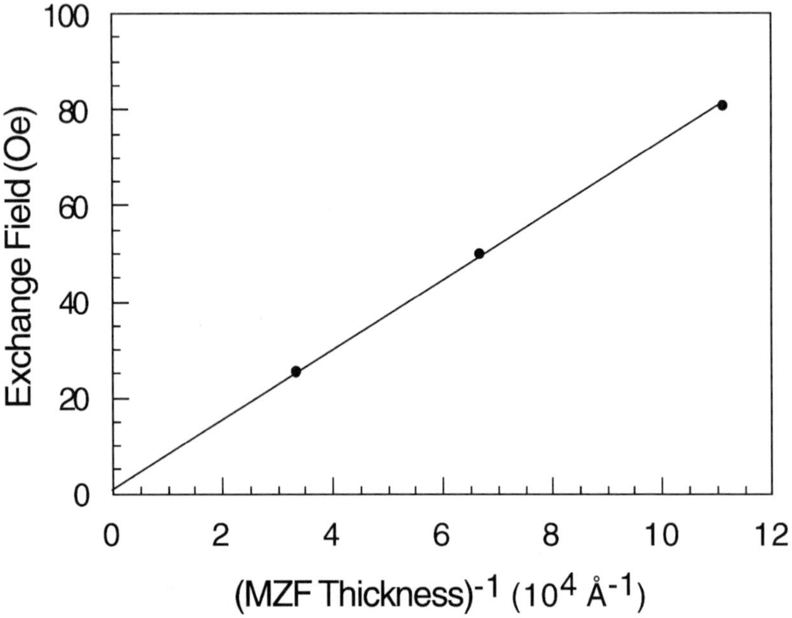

Figure 11 Exchange bias field versus the inverse of the thickness in each of the bilayers. A least-squares fit of the above intercepts the origin, which is the limit that the thickness of the $(Mn,Zn)Fe_2O_4$ layer goes to infinity and the exchange field vanishes (64).

at the interface) and a non-ideal barrier with impurities and defects that can be the source of spin flip scattering. However, the fact that Seneor et al (68) have been able to obtain a 43% magnetoresistance effect in $Co/Al_2O_3/Fe_{3-x}O_4/Al$ structures suggests that the dead interface layer cannot be the primary explanation for the reduced magnetoresistance effect.

There continue to be issues of barrier inhomogeneity, oxygen content, and spin polarization at interfaces that need to be addressed in order to understand the magnetoresistive behavior of ferrite thin film materials in device heterostructures. With continued advances in the understanding of epitaxial ferrite thin films, there will be significant progress in the development of epitaxial ferrite devices.

SUMMARY

Recent work on epitaxial ferrite thin films has revealed magnetic properties that are not found in bulk ferrites. Epitaxial ferrite thin films and heterostructures have provided a model system for studying novel magnetic phenomena such as modified super-exchange interactions, nearly ideal exchange coupling, and perpendicular exchange coupling. With an improved understanding of these film properties, we may now begin to incorporate these materials into applications varying from storage to high-frequency microwave devices. However, outstanding issues in magnetotransport (including the role of interfaces and barrier layers) and magnetism of ferrite thin films remain to be addressed.

ACKNOWLEDGMENTS

The author thanks R Bruce van Dover at Lucent Technologies, Pieter van der Zaag at Philips Research Laboratories, Masanori Abe at Tokyo Institute of Technology, and Guohan Hu at Cornell University. The work was supported by the Office of Naval Research and a David and Lucile Packard Foundation fellowship.

LITERATURE CITED

1. Snoek JL. 1947. *New Developments in Ferromagnetic Materials*. New York: Elsevier
2. Hogan CL. 1952. *Bell Syst. Tech. J.* 31:1
3. Zhang F, Kantake S, Kitamoto Y, Abe M. 1999. *Trans. Mater. Res. Soc. Jpn.* 24:65
4. van der Zaag PJ, Ruigrok JJM, Gillies MF. 1998. *Philips J. Res.* 51:173
5. Winkler G. 1981. *Magnetic Garnets*. Braunschweig, Ger: Friedr. Vieweg & Sohn
6. Abe M. 1999. *Nanocrystalline and Thin Film Magnetic Oxides*, ed. I Nedkov, M Ausloos, p. 27. The Netherlands: Kluwer Acad.
7. Abe M, Kitamoto Y, Matsumoto K, Zhang M, Li P. 1997. *IEEE Trans. Magn.* 33:3649
8. Park JY, Allen MG. 1998. *J. Micromech. Microeng.* 8:307
9. van der Zaag PJ, Lubitz P, Kitamoto Y, Abe M. 1999. *IEEE Trans. Magn.* 35:3436
10. Cuomo JJ, Gambino RJ, Harper JM, Kuptsis JD, Webber JC. 1978. *J. Vac. Sci. Technol.* 15:281
11. Dijkamp D, Venkatesan T, Wu XD, Shaheen SA, Jisrawi N, et al. 1987. *Appl. Phys. Lett.* 51:619
12. Chrisey D, Hubler GK. 1994. *Pulsed Laser Deposition*. New York: Wiley
13. Fujii T, Takano M, Katano R, Bando Y. 1980. *J. Cryst. Growth* 99:606
14. Lind DM, Berry SD, Chern G, Mathias H, Testardi LR. 1992. *Phys. Rev. B* 45:1838
15. Ortiz C, Lim G, Chen MM, Castillo G. 1988. *J. Mater. Res.* 3:344
16. Noma K, Matsushita N, Nakagawa S, Naoe M. 1996. *J. Appl. Phys.* 79:5970

16a. Hiratsuka N, Nozawa M, Kakizaki K. 1998. *J. Magn. Magn. Mater.* 176:31

16b. Chen Y, Kryder MH. 1998. *IEEE Trans. Magn.* 34:729

16c. Zhuang Z, Rao M, White RM, Laughlin DE, Kryder MH. 2000. *J. Appl. Phys.* 87:6370

17. Matshushita N, Noma K, Nakagawa S, Naoe M. 1998. *J. Magn. Soc. Jpn.* 22:17
18. Kitamoto Y, Kantake S, Shirasaki F, Abe M, Naoe M. 1999. *J. Magn. Magn. Mater.* 193:97

18a. Abe M, Tamaura Y. 1984. *J. Appl. Phys.* 55:2614

19. Claasen A. 1925–1926. *Proc. Phys. Soc.* 38:482
20. Shull CG, Wollan EO, Koehler WC. 1951. *Phys. Rev.* 84:912
21. Petitt G, Forester DW. 1971. *Phys. Rev. B* 4:3912
22. Persoons RM, De Grave E, de Bakker PMA, Vandenberghe RE. 1993. *Phys. Rev. B* 47:5894
23. Harris VG, Koon NC, Williams CM, Zhang Q, Abe M. 1996. *J. Appl. Phys.* 79:4561

24. Ortiz C, Lim G, Chen MM, Castillo G. 1988. *J. Mater. Res.* 3:344
25. Gao Y, Kim YJ, Chambers SA. 1998. *J. Mater. Res.* 13:2003
26. van der Zaag PJ, Ruigrok JJM, Noordermeer A, van Delden MHWM, Por PT, et al. 1993. *J. Appl. Phys.* 74:4085
27. Gillies MF, Coehoorn R, van Zon JBA. 1998. *J. Appl. Phys.* 83:6855
28. van der Zaag PJ, Lubitz P, Kitamoto Y, Abe M. 1999. *IEEE Trans. Magn.* 35:3436
29. Penicaud M, Siberchocot B, Sommers CB, Kubler J. 1992. *J. Magn. Magn. Mater.* 103:212
30. Yoshida J, Iida S. 1979. *J. Phys. Soc. Jpn.* 47:1627
31. Shepherd JP, Koenitzer JW, Aragon R, Spalek J, Honig JM. 1991. *Phys. Rev. B* 43:8461
32. Chikazumi S. 1995. *Physics of Magnetism.* New York: IEEE Press
33. Margulies DT, Parker FT, Spada FE, Goldman RS, Li J, et al. 1996. *Phys. Rev. B* 53:9175
34. Margulies DT, Parker FT, Rudee ML, Spada FE, Chapman JN, et al. 1997. *Phys. Rev. Lett.* 79:5162
35. Hibma T, Voogt FC, Niesen L, van der Heijden PA, de Jonge WJM, et al. 1999. *J. Appl. Phys.* 85:5291
36. Shaw KA, Lochner E, Lind DM. 2000. *J. Appl. Phys.* 87:1727
37. Voogt FC, Palstra TTM, Niesen L, Rogojanu OC, James MA, Hibma T. 1998. *Phys. Rev. B* 57:R8107
38. Rudee ML, Smith DJ, Margulies DT. 1999. *Phys. Rev. B* 59:11633
39. Masumoto H, Shirakawa Y. 1941. *Phys. Rev.* 60:835
40. Ogale SB, Ghosh K, Sharma RP, Greene RL, Ramesh R, Venkatesan T. 1998. *Phys. Rev. B* 57:7823
41. Gong GQ, Gupta A, Xiao G, Qian W, Dravid VP. 1997. *Phys. Rev. B* 56:5096
42. Li XW, Gupta A, Xiao G, Gong GQ. 1998. *J. Appl. Phys.* 83:7049
43. Venzke S, van Dover RB, Phillips JM, Gyorgy EM, Siegrist T, et al. 1996. *J. Mater. Res.* 11:1187
44. Korenivski V, van Dover RB, Gyorgy EM, Phillips JM, Felder RJ. 1996. *J. Appl. Phys.* 79:5926
45. Williams CM, Chrisey DB, Lubitz P, Grabowski KS, Cotell CM. 1994. *J. Appl. Phys.* 75:1676
46. Dorsey PC, Lubitz P, Harris VG, Chrisey DB, Horwitz JS. 1995. *IEEE Trans. Magn.* 3455
47. Suzuki Y, van Dover RB, Gyorgy EM, Phillips JM, Korenivski V, et al. 1996. *Appl. Phys. Lett.* 68:714
48. Guillard C, Crevaux H. 1950. *C.R. Acad. Sci.* 230:1458
49. Dorsey PC, Lubitz P, Chrisey DB, Horwitz JS. 1996. *J. Appl. Phys.* 79:6338
50. Hu G, Choi JH, Eom CB, Harris VG, Suzuki Y. 2000. *Phys. Rev. B* 62:R779
51. Sturge MD, Gyorgy EM, LeCraw RC, Remeika JP. 1969. *Phys. Rev.* 180:413
52. Berry SD, Lind DM, Chern G, Mathias H. 1993. *J. Magn. Magn. Mater.* 123:126
53. Terashima T, Bando Y. 1987. *Thin Solid Films* 152:455
54. Chern G, Berry SD, Lind DM, Mathias H, Testardi LR. 1992. *Phys. Rev. B* 45:3644
55. van der Zaag PJ, Ball AR, Feiner LF, Wolf RM, van der Heijden PAA. 1996. *J. Appl. Phys.* 79:5103
56. Ijiri Y, Borchers JA, Erwin RW, Lee S-H, van der Zaag PJ, Wolf RM. 1998. *J. Appl. Phys.* 83:6882
57. Kleint CA, Krause MK, Hohne R, Walter T, Semmelhack HC, et al. 1998. *J. Appl. Phys.* 84:5097
58. Meiklejohn WH, Bean CP. 1957. *Phys. Rev.* 105:904
59. Mauri D, Siegmann HC, Bagus PS, Kay E. 1987. *J. Appl. Phys.* 62:3047
60. Malozemoff A. 1988. *Phys. Rev. B* 37:7673

61. Koon N. 1997. *Phys. Rev. Lett.* 78:4865
62. Ijiri Y, Borchers JA, Erwin RW, Lee SL, van der Zaag PJ, Wolf RM. 1998. *Phys. Rev. Lett.* 80:608
63. van der Zaag PJ, Ijiri Y, Borchers JMA, Feiner LF, Wolf RM, et al. 2000. *Phys. Rev. Lett.* 84:6102
64. Suzuki Y, van Dover RB, Gyorgy EM, Phillips JM, Felder RJ. 1996. *Phys. Rev. B* 53:14106
65. Goto E, Hayashi N, Miyashita T, Nakagawa K. 1965. *J. Appl. Phys.* 36:2951
66. van der Zaag PJ, Bloemen PJH, Gaines JM, Wolf RM, van der Heijden PAA, et al. 2000. *J. Magn. Magn. Mater.* 211:301
67. Li XW, Gupta A, Xiao G, Qian W, Dravid VP. 1998. *Appl. Phys. Lett.* 75:3282
68. Seneor P, Fert A, Maurice JL, Montaigne F, Petroff F, Vaures A. 1999. *Appl. Phys. Lett.* 74:4017

Annu. Rev. Mater. Res. 2001. 31:291–321

DESIGN AND SYNTHESIS OF ENERGETIC MATERIALS[1]

Laurence E Fried, M Riad Manaa, Philip F Pagoria, and Randall L. Simpson
Energetic Materials Center, Chemistry and Materials Science Directorate, Lawrence Livermore National Laboratory, P. O. Box 808, Livermore, California 94551; e-mail: lfried@llnl.gov; manaa1@llnl.gov; pagoria1@llnl.gov; simpson5@llnl.gov

Key Words high explosive, heterocycles, high-energy density material, property prediction

■ **Abstract** Energetic materials are chemical compounds or mixtures that store significant quantities of energy. In this review, we explore recent approaches to property prediction and new material synthesis. We show how the successful design of new energetic materials with tailored properties is becoming a practical reality.

INTRODUCTION

Energetic materials include propellants, explosives, and pyrotechnics. These materials can be thought of as controllable storage systems for chemical energy. Propellants and pyrotechnics release their energy through relatively slow deflagration processes, often taking several seconds to achieve complete combustion. Explosives, on the other hand, release their energy on the microsecond timescale. Despite these differences, explosives, propellants, and pyrotechnics share several chemical similarities. In fact, explosive and propellant mixtures often share the same ingredients, although in different quantities. Energetic materials have numerous military and industrial applications. The vast majority of energetic materials are used by industry.

There is a continuing need for improved energetic materials. In developing new energetic systems, one seeks optimal tradeoffs in energy content, safety, and cost. This is achieved through formulation, that is, combining known chemical compounds, and/or synthesis, producing new compounds. There are significant

[1]The US Government has the right to retain a nonexclusive, royalty-free license in and to any copyright covering this paper.

costs associated with either of these activities. Therefore, it is desirable to try to reduce the amount of trial and error in new material development by predicting energetic material properties.

In this article, we will review current methods to predict energetic material properties. Because this is a very broad area, we focus on the topics of energy content prediction (commonly termed performance) and safety characteristics. An accurate assessment of these material characteristics can be helpful in screening new candidate materials. In the past, these characteristics were predicted with empirical rules and intuition. The development of molecular modeling techniques now makes it possible to predict many, but not all, relevant material properties through much more fundamental means.

To begin we review methods to predict the energy release of an energetic material. This is usually the first step taken in evaluating a target energetic material. In the second section, we examine methods to predict the stability of an energetic material, with particular emphasis on sensitivity to shock or impact. In the final section, we discuss recently synthesized energetic materials, focusing on heterocyclic ring molecules. This class of materials combines favorable material characteristics with interesting synthetic chemistry. A list of material name abbreviations for common energetic materials is given in Table 1. Explicit structures are provided for novel energetic materials.

TABLE 1 Acronyms and abbreviations used for high explosive materials

DATB	1,3-Diamino-2,4,-6-trinitrobenzene
DINGU	Dinitroglycourile
EDNA	Ethylenedinitramine
HMX	Octahydro-1,3,5,7-tetranitro-1,3,5,7-tetraazacine
HNB	Hexanitrobenzene
NG	Nitroglycerine
NTO	3-nitro-1,2,4-triazol-5-one
PETN	Pentaerythritol tetranitrate
RDX	Hexahydro-1,3,5-trinitro-1,3,5-triazine
TATB	1,3,5-triamino-2,4,6-trinitrobenzene
TETRYL	N-methyl-N,2,4,6-tetranitro Benzeneamine
TNA	2,4,6-trinitroaniline
TNAZ	1,3,3-trinitroazetidine
TNB	2,4,6-trinitrobenzene
TNT	2,4,6-trinitrotoluene

PREDICTION OF ENERGY RELEASE

The energy content of an energetic material often determines its practical utility. The useful energy content is determined by the anticipated release mechanism. Since detonation events occur on a microsecond timescale, any chemical reactions slower than this are not relevant when considering a detonation. Another way of looking at energy release mechanisms is through thermodynamic cycles. Detonation can be thought of as a cycle that transforms the un-reacted explosive into stable product molecules at the Chapman-Jouget state (C-J) (1). This is simply described as the slowest steady-shock state that conserves mass, momentum, and energy. Similarly, the deflagration of a propellant converts the un-reacted material into product molecules at constant enthalpy and pressure.

Understanding energy release in terms of thermodynamic cycles ignores the important question of the timescale of reaction. Nevertheless, chemical equilibrium is often nearly achieved when energetic materials react. This is a consequence of the high temperatures produced by such reactions (up to 6000 K). We begin our discussion by examining thermodynamic cycle theory as applied to high explosive detonation. This is a current research topic because high explosives produce detonation products at extreme pressures and temperatures: up to 40 GPa and 6000 K. Relatively little is known about material equations of state under these conditions. Nonetheless, shock experimentation on a wide range of materials has generated sufficient information to allow reasonably reliable thermodynamic modeling to proceed.

One of the attractive features of thermodynamic modeling is that it requires very little information regarding the un-reacted energetic material. The elemental composition, density, and heat of formation of the material are the only information needed. Because elemental composition is known once the material is specified, only density and heat of formation need be predicted.

Material performance depends strongly on density; for example, the detonation pressure is roughly proportional to the square of the density (1). It is very difficult to calculate the density of a molecular crystal (2). Thus the density determination is often the most uncertain step in the prediction of energy release. Density in energetic materials has traditionally been found through group additivity methods (3, 4). The newest methods (5) are of comparable accuracy to more advanced techniques based on intermolecular potentials (6–9). Further improvements in intermolecular potentials may improve the accuracy of atomistic prediction methods (10). The existence of multiple possible polymorphs of varying density but similar energy remains an unresolved difficulty (2).

The heat of formation of each compound in the gas phase can be calculated via quantum chemistry (11). Recent advances in density functional methods make it feasible to determine the heat of formation of large energetic molecules from first principles (12). The calculation of the solid lattice energy, however, is more difficult. Recent techniques relate the surface area of the isolated molecule and electrostatic potential surface to the lattice energy (13).

In spite of these uncertainties, the prediction of densities and heats of formation can yield insight into possible material performance. In Figure 1 we calculate the detonation velocity of several explosives in two different ways, with the Cheetah 1.40 thermochemical code using the BKWC (Backer-Kistiakowski-Wilson/Cheetah) product library (14) at full crystal density. In the first set of calculations, the calculated detonation velocity was based on the experimental heat of formation and density. In the second set, the calculated detonation velocity was based on the predicted density and heat of formation. Densities were predicted with the MOLPAK (molecular packing) program (7). MOLPAK systematically searches through common crystallographic space groups and finds the minimum energy packing of molecules given empirically derived interaction potentials for rigid molecules.

The gas phase heat of formation was calculated using Hartree-Fock theory corrected by group equivalents, a method that is usually accurate to within 4 kcal/mol. The lattice energy was calculated with the MOLPAK program. We see that although differences owing to the use of calculated material properties are apparent, the calculated values are of sufficient accuracy to correctly order the detonation velocity of the various materials. Direct comparison with experimental detonation velocities is not made in the plot because experimental data are available only at densities less than the full crystal.

Chapman-Jouget (C-J) detonation theory (1) implies that the performance of an explosive is determined by thermodynamic states—the C-J state and the connected adiabat. Thermochemical codes use thermodynamics to calculate these states and hence obtain a prediction of explosive performance. The allowed thermodynamic states behind a shock are intersections of the Rayleigh line (expressing conservation of mass and momentum), and the shock Hugoniot (expressing conservation of energy). The C-J theory states that a stable detonation occurs when the Rayleigh line is tangent to the shock Hugoniot.

This point of tangency can be determined, given that the equation of state $P = P(V, E)$ of the products is known. The chemical composition of the products changes with the thermodynamic state, thus thermochemical codes must simultaneously solve for state variables and chemical concentrations. This problem is relatively straightforward because the equation of state of the fluid and solid products is known.

One of the most difficult parts of this problem is accurately describing the equation of state of the fluid components. Despite its simplicity and lack of rigorous derivation, the Becker-Kistiakowski-Wilson (BKW) (15) equation of state (EOS) is used in many practical energetic material applications. There have been a number of different parameter sets proposed for the BKW EOS (16–19). Souers & Kury (20) have critically reviewed these equations of state by comparing their predictions to a database of detonation tests. They concluded that BKW EOS could not adequately model the detonation of a copper-lined cylindrical charge. The BKWC parameter set for the BKW EOS (14) partially overcomes this deficiency through more aggressive parameterization techniques. However, BKWC is not

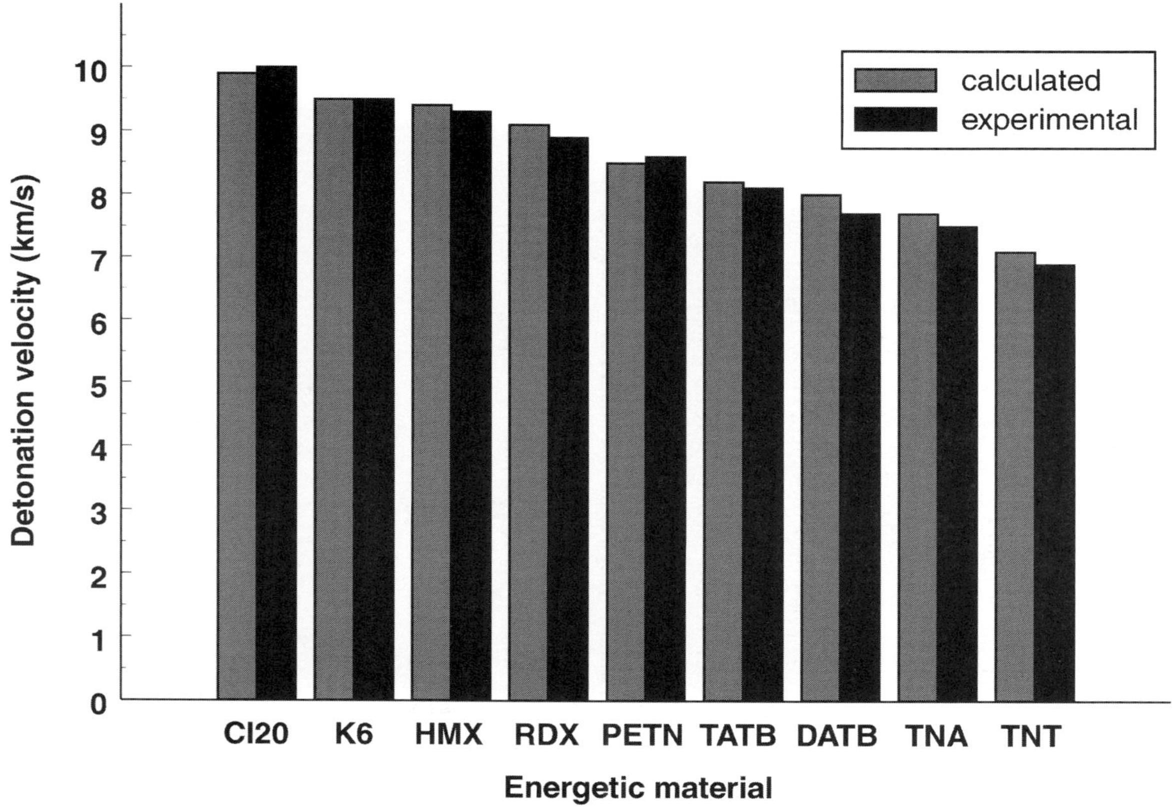

Figure 1 Detonation velocities predicted by Cheetah 1.40, based on calculated densities and heats of formation versus experimental densities and heats of formation.

reliable when applied to explosives very high in hydrogen content. It has long been recognized that the highly empirical nature of the BKW EOS limits the accuracy achievable in detonation calculations. This is particularly important when designing new materials that may have unusual elemental compositions. Efforts to achieve better EOS have largely been based on the concept of model potentials. With model potentials, molecules interact via idealized spherical pair potentials. Statistical mechanics is then employed to calculate the EOS of the interacting mixture of effective spherical particles. Most often, the exponential-6 potential is used for the pair interactions:

$$V(r) = \frac{\varepsilon}{\alpha - 6}[6 \exp(\alpha - \alpha r/r_m) - \alpha(r_m/r)^6]$$

Here, r is the distance between particles, r_{m} is the minimum of the potential well, ε is the well depth, and α is the softness of the potential well.

The JCZ3 (Jacobs-Cowperthwaite-Zwisler/3) EOS was the first successful model based on a pair potential that was applied to detonation (21). This EOS was based on fitting Monte Carlo simulation data to an analytic functional form. Hobbs et al (22) recently reported a JCZ3 parameter set called JCZS (Jacobs-Cowperthwaite-Zwisler/Sandia). JCZS employs some of the parameterization techniques used in the construction of BKWC. It achieves better accuracy for the detonation of common high explosives than BKW EOS. Because JCZ3 is extensively parameterized to detonation, it has difficulty in reproducing reactive shock Hugoniots of hydrocarbons and other liquids (LE Fried, personal communication).

Ross proposed a soft-sphere perturbation theory for the EOS of the exponential-6 fluid (24). Ross, Ree, and others successfully applied this EOS to detonation and shock problems (25–28). Kang et al also derived a fluid perturbation theory designed to work at high density (29). Computational cost is a significant difficulty with an EOS based on fluid perturbation theory. Byers Brown (30) developed an analytic representation of Kang et al's EOS using Chebyshev polynomials. The accuracy of the above EOS has been recently evaluated (31); these authors conclude that Ross's approach is the most reliable. More recently, Fried & Howard have used a combination of integral equation theory and Monte Carlo simulations to generate a highly accurate EOS for the exponential-6 fluid (32).

The exponential-6 model is not well suited to molecules with a large dipole moment. Ree (33) has used a temperature-dependent well depth ε in the exponential-6 potential to model polar fluids and fluid phase separations. Fried & Howard have developed an effective cluster model for the HF (hydrogen fluoride) molecule (34) that is valid to lower temperatures than the variable well-depth model but employs two more adjustable parameters. More progress needs to be made in the treatment of polar detonation product molecules.

Efforts have been made to develop EOS for detonation products based on direct Monte Carlo simulations instead of analytical approaches (35). This approach is promising given recent increases in computational capabilities. One of the greatest advantages of direct simulation is the ability to go beyond van der Waals 1-fluid theory (36). Kerley has advocated the ideal mixing approximation as an alternative to 1-fluid theory (37).

In most cases, interactions between unlike molecules are treated with Lorentz-Berthelot combination rules (38). Non-additive pair interactions have been used for N_2 and O_2 (28). The resulting model for N_2 accurately matches double-shock data, but it is not accurate at lower temperatures and densities (32). A combination of experiments on mixtures and theoretical developments is needed to develop reliable unlike-pair interaction potentials.

Many materials produce large quantities of solid products upon detonation, the most common of which is carbon, although some explosives produce aluminum and aluminum oxide (39). Uncertainties in the EOS and phase diagram of carbon remain a major issue in the thermochemical modeling of detonation. van Thiel & Ree have proposed an accurate Mie-Gruneisen EOS for carbon (40). Fried & Howard (41) have developed a simple modified Murnaghan EOS for carbon that matches recent experimental data on the melting line of graphite. There is considerable uncertainty regarding the melting line of diamond. Fried & Howard argue, based on re-analysis of shock data, that the melting line of diamond should have a greater slope. Shaw & Johnson have derived a model for carbon clustering in detonation (42). Viecelli & Ree have derived a carbon-clustering model for use in hydrodynamic calculations (43, 44).

Thermochemical equilibrium is a first approximation to the chemical kinetic processes controlling detonation. We call an explosive ideal when thermochemical equilibration is nearly achieved in the detonation.We call an explosion non-ideal if the chemical kinetic processes in the detonation wave are poorly approximated by the thermochemical equilibrium, i.e., the chemistry is slower than the hydrodynamics. For most purposes, we define non-ideal explosives as those with a reaction zone of one millimeter or more. These materials have chemical reaction rates that are slow compared with a hydrodynamic time-scale (10^{-6} s) so that the C-J assumption of instantaneous thermodynamic equilibrium breaks down. For example, it is found experimentally that the detonation velocity of non-ideal explosives varies sharply from the C-J value and depends strongly on the cylindrical charge radius.

We are therefore forced to consider the interaction of chemical kinetics with the detonation wave in order to reach an acceptable representation of detonation in non-ideal explosives. Wood & Kirkwood (WK) (45) proposed a two-dimensional steady-state kinetic detonation theory that solves many of the limitations of ZND (Zeldovich-von Neumann-Doring) theory. ZND theory treats a two-dimensional detonation wave. It predicts a high-pressure precursor to the C-J state known as the von Neumann spike (1). WK theory considers a cylindrical charge of infinite length. It solves the hydrodynamic Euler equations in the steady-state limit along the central streamline of the cylinder. Radial expansion is treated as a source term in the one-dimensional flow along the streamline. Bdzil has generalized WK theory to off-axis flow (46). Stewart (47) and co-workers have studied the effect of kinetic rates on the decrease of detonation velocity with decreasing size and on curvature of the detonation wave.

Howard et al (48) have implemented a model of detonation kinetics based on the identification of individual chemical species. The advantage of their treatment is that the same EOS and chemical rate laws can be used on a wide range of explosive mixtures. A mixture of EOS based on thermal, mechanical, and partial chemical equilibrium is used, as implemented in the Cheetah 2.0 thermochemical code (49). Small molecules that are gases at standard conditions are treated with the BKW EOS (14). This framework has been recently extended to the HMSA/MC (hypernetted mean spherical approximation/ Monte Carlo) EOS (32). Solids are treated with a Murnaghan (50) EOS, whereas chemical reaction rates, which depend only on pressure, are employed. These rates represent the consumption of the energetic material by the detonation wave. Fast reaction rates (partial chemical equilibrium) are assumed for species other than the initial material.

Howard, Fried, and Souers solved the WK equations numerically to find the steady-state detonation velocity. The radial expansion is derived from measured radii of curvature for the materials studied. They found good agreement with measured detonation velocities using the same set of EOS and rate laws for each composite. Although the treatment of detonation is by no means exact, the ability to model a wide range of phenomena based on simple EOS and rate laws is encouraging.

PREDICTION OF MATERIAL STABILITY

An energetic material may undergo chemical reaction in response to shock, thermal, or chemical insults. In this review, we concentrate on stability to shock excitation. It is often the case, however, that a material's stability to various forms of loading is highly correlated with one another. Energetic materials exist in a higher energy state than their lowest energy decomposition products. Thus energetic molecules are often termed metastable. Kinetic stability, however, guarantees that decomposition is sufficiently slow at ambient conditions that the species are long lived. Recently, several metastable nitrogen and oxygen compounds have been proposed that contain novel bonding. This has led to recent theoretical studies of hypothetical systems as high-energy density materials (HEDM), such as oxygen ring–strained systems (O_4 and O_8) (51, 52), tetrahedral N_4 (53), and cubic N_8 (54).

From a molecular perspective, the kinetic stability entails two presuppositions. The first is adiabatic stability. Energetic systems must exist at a metastable local minimum on the ground electronic state potential energy surface as a function of nuclear positions. There are significant energy barriers hindering decomposition on this surface. Second, a less commonly realized phenomenon concerns nonadiabatic stability, where processes involving excited electronic states are involved in the decomposition process.

The initial response of energetic molecular systems to shock or impact is a topic of considerable interest and controversy. The goal is to build predictive models

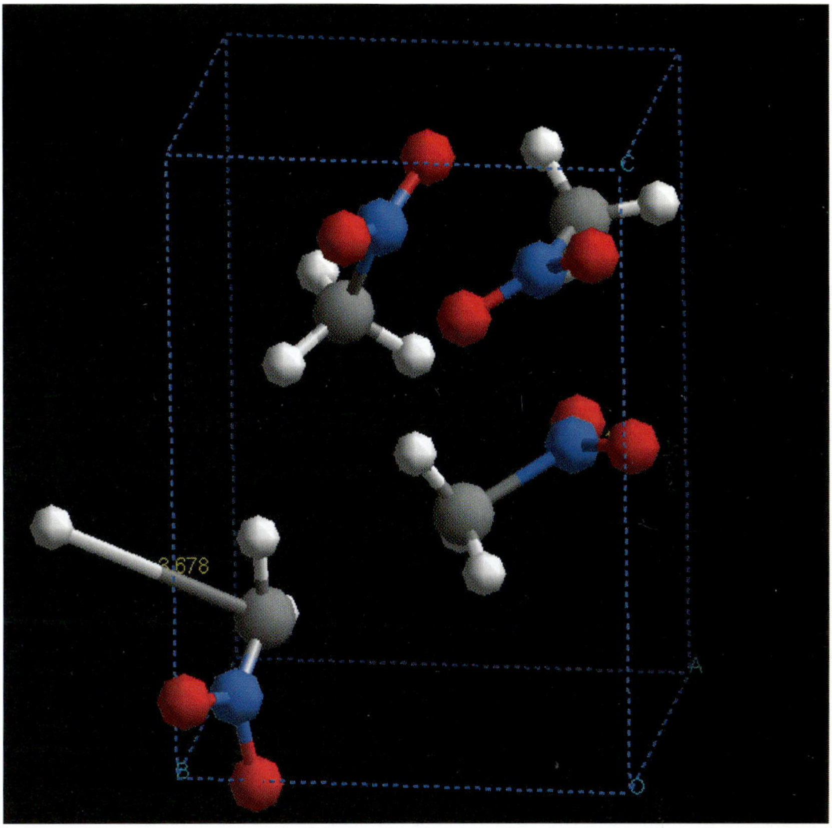

Figure 4 A snapshot of hydrogen abstraction in dissociating nitromethane after 200 fs at a density of 1.5g/cm^3 and temperature of 3000 K. Carbon atoms are gray, nitrogen atoms are blue, oxygen atoms are red, and hydrogen atoms are white.

identifying molecular parameters that can be correlated with sensitivity and performance measures of the energetic molecule. Multiple parameters can be found that correlate with, but do not necessarily cause, energetic material sensitivity (55). Several mechanisms have been proposed that consider either molecular deformations or electronic excitations as the first event under high-pressure conditions.

Microscopic Models of Initiation

Understanding the first step in the response of energetic molecules to a shock is a topic of considerable importance because it ultimately would shed light on the sensitivity properties of energetic materials. It is known that the mechanical energy from compression leads to chemical decomposition and heating. Which bond in the molecule breaks first and what type of chemical reactions (unimolecular versus bimolecular, etc) occur first are still largely unknown, even for the simplest and widely studied systems such as nitromethane. Recent applications of ultrafast spectroscopic methods (56), however, hold great promise for understanding these basic mechanisms at the molecular level through direct spectroscopic measurements.

Several models have been proposed for the initial response of energetic molecular solids to instantaneous high pressure. The vibrational energy up-pumping model (57) suggests that the shock wave produces a bath of excited phonons absorbed by the lowest molecular vibrational modes. Increased phonon absorption and intramolecular vibrational energy redistribution (IVR) lead to excitation of higher-frequency modes. The molecule eventually reaches a transition state and proceeds to chemical reactions. Fried & Ruggiero (58) derived a simple formula for the total energy transfer rate into a given vibron band in terms of the density of vibrational states and the vibron-phonon coupling. The phonon up-conversion rates, based on small anharmonic terms in the Hamiltonian, were shown to correlate with the sensitivity in such explosives as TATB, HMX, and lead styphnate. Coffey (59) postulated that the process of initiation of chemical reactions is the result of tunneling of dislocations in a solid that, at high velocities, could have sufficient energies to directly pump the internal vibrational modes of the explosive molecules.

Other workers considered models associated with the electronic properties. Williams (60) was the first to examine the influence of electronic excited states and electronic transport on the initiation and propagation of detonation waves in solid explosives. He proposed initiation by changes in the Fermi level of solid explosives owing to a double electron and positive hole injection followed by chemical reactions. Dremin et al (61), noting evidence for similarities between shock decomposition intermediates to those of photochemical processes, proposed electronic excitation as the first molecular response in their multiprocess detonation model. Delpuech et al (62) proposed that the minimum polarity affecting an electronically excited state is a unique molecular parameter directly correlated within each family of organic explosives with shock sensitivity. Sharma reported the existence of linear correlation between impact and shock sensitivity with electronic levels, the shake-up promotion energy observed in X-ray photoelectron spectra, of a homologous series of explosive compounds (63, 64). More recently, Gilman (65) postulated that the compression from the wave front causes local metallization, for instance, when bending of covalent bonds decreases the HOMO-LUMO (highest occupied molecular orbital–lowest occupied molecular orbital) gap.

With the aid of ab initio electronic structure calculations on the common explosive, RDX ($C_3H_6N_6O_6$), Kunz and co-workers proposed a mechanism based on electronic excitations induced by an impact wave propagating through the crystal (66). They showed that edge dislocations cause a dramatic reduction of the optical gap owing to the splitting of local electronic states from both the valence and conduction bands. The band gap is further reduced to near zero by the pressure inside the impact wavefront, resulting in the breakage of the N-NO_2 molecular bond.

Conclusive experimental evidence is still lacking in support of any of the models described above. The complexity of the process under shock or impact for solid materials is, at best, a multiscale event in time and space. Intuitively, electronic excitation is plausible, considering the fact that electrons are light, fast, and quantal, thus responding first to exterior perturbations. Following electronic excitation, however, several processes, including dissociation on the excited state surface, radiative processes, and nonradiative energy deactivation, could be at play to induce chemical reactions.

In a recent study, Manaa & Fried (67) have considered the topology of the ground and first excited states of nitromethane, a widely studied prototypical model of energetic materials. The equilibrium geometry of the triplet state of

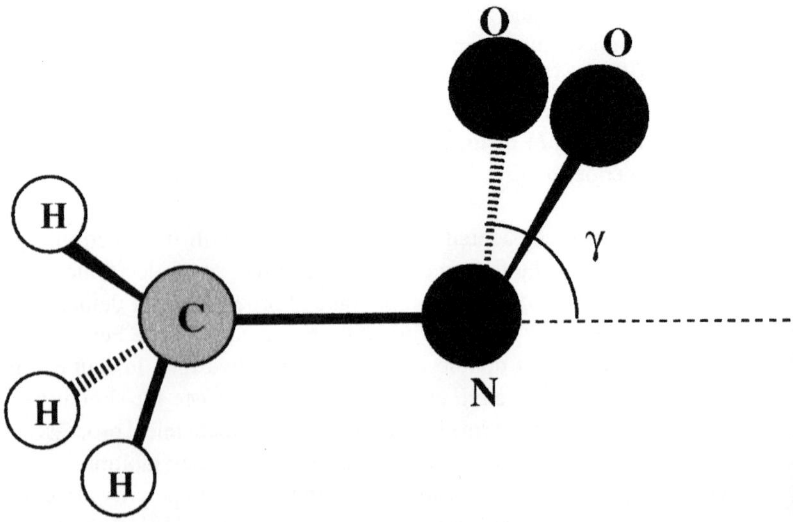

Figure 2 Geometric structure of nitromethane molecule at the equilibrium of the excite[d] triplet state.

nitromethane is such that the ONO bond makes an angle of about 50° relative to the C-N bond (Figure 2), which is markedly different from the equilibrium geometry of the ground state where δ is zero. An adiabatic barrier to dissociation on this first excited state with respect to the C-N bond (see Figure 3) was determined (using highly correlated ab initio multireference configuration interaction methodology) to be about 33 kcal/mol. The existence of this barrier should presumably allow the triplet state to support a few vibrational levels and, owing to a different spin multiplicity that forbids decaying, to be long lived. An intersystem crossing, however, provides a nonradiative deactivation channel. Such a crossing has been located by determining the minimum energy crossing point (MECP) on the singlet-triplet surface of intersection. This point, as indicated in Figure 3, is located 13 kcal/mol above the equilibrium geometry of the triplet state, thus establishing a second barrier for energy redistribution that is more than half of the adiabatic one. It is expected, therefore, that thermal activation along the nitro-bending mode permits efficient energy transfer from the triplet state to higher vibrational levels of the ground surface. There are no experimental measurements on the triplet formation for nitromethane. Nevertheless, a recent study on nitrobenzene (68) showed that the triplet formation is very efficient with a quantum yield of $\geq$80% and that this state has a lifetime of $\leq$500 ps. An earlier study (69) established a short nonradiative lifetime of about 1 ps for the excited singlet state of nitromethane, with efficiency of 76% in energy deactivation. Vibrationally hot ground-state molecules could thus be generated by fast internal conversion (singlet-singlet) and intersystem (singlet-triplet) crossings. The excess nonradiative (kinetic: heat) energy could then be supplied to near shell molecules and speed up the decomposition process.

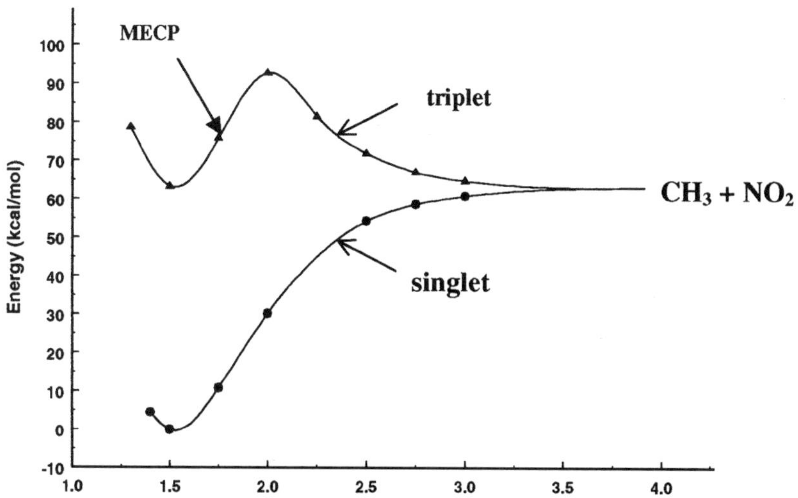

Figure 3 The optimized potential energy curves for the ground singlet and the excited triplet states of nitromethane.

Reed et al (70) have studied electronic excitations in solid crystalline nitromethane under conditions of shock loading and static compression. Bandgap lowering under uniaxial strain owing to molecular defects and vacancies was considered. Ab initio molecular dynamics simulations were done of all possible nearest neighbor collisions at a shock front and of crystal shearing along a sterically hindered slip plane. In all cases, the bandgap lowering was not sufficient to produce a significant population of excited states in the crystal.

The reaction chemistry under detonation conditions for most explosives, including materials as simple as diatomic nitric oxide, NO, is still largely unknown. The most widely studied example is nitromethane, a clear liquid with a density of 1.13 g/cm^3 at 298 K. Static high-pressure experiments (71, 72) show that the time of explosion for deuterated nitromethane is approximately ten times longer than that for protonated materials, suggesting that a proton or hydrogen atom abstraction is involved in the rate-determining step. Isotope-exchange experiments, using diamond cell methods, also give evidence (73) that the aci ion ($H_2CNO_2^-$) concentration increases with increased pressure. Other studies (74, 75) also suggest that reactions occur more rapidly and are pressure enhanced when small amounts of bases are present, giving further support to the aci ion production. Shock wave studies of the reaction chemistry are still inconclusive and at odds with mass spectroscopic studies. Mass spectroscopic studies suggested condensation reactions (76); time-resolved Raman spectroscopy pointed to a biomolecular mechanism (77); UV-visible absorption spectroscopy indicated either no sign of chemical reaction (78) or the production of $H_3CNO_2^-$ intermediate for amine-sensitized nitromethane (79). It was noted, however, that part of the discrepancy might be due to the fact that the ring-up (multiple shock) experiments are mapping lower temperature regimes (~1000 K) than experienced under detonation (76).

Molecular dynamics studies, using semi-empirical or preferably ab initio potentials for the molecular forces play an important role, at least in elucidating the initial steps of reaction mechanism. In the case of nitromethane, a preliminary fully ab initio molecular dynamics simulation was conducted recently (MR Manaa, manuscript in preparation). The calculation employed constant volume and temperature conditions similar to the C-J of fully reacted materials (density of 1.5 g/cm^3 and temperature of 3000 K). The initial step in the chemical reaction appears after 200 fs. As is shown in Figure 4 (see color insert), this first step is clearly identified with the hydrogen abstraction, and thus with the formation of aci ion, in accordance with experimental work noted above. This bond specificity is remarkable because in the gas phase, the C-N bond is the weakest, whereas the C-H bond is the strongest in the nitromethane molecule. In the condensed phase however, vibrational energy is most significant in the C-H mode, in part owing to a caging effect that eventually leads to a proton extraction. Later snapshot of this simulation show that the proton transfers to an oxygen atom of a nearby nitromethane molecule. Similar simulations can be useful for characterizing the first events in the decomposition process, although the molecular size might still hinder such undertaking.

Because of the extended time of the reaction zone of explosive materials (10^{-9}–10^{-6} s), ab initio methods are cumbersome and perhaps unrealistic for use in exploring decomposition kinetics. However, semi-empirical methods, such as tight-binding and bond-order type calculations, now offer a more realistic alternative, as recent simulations of shocked hydrocarbon materials indicate (80, 81). These methods can, in principle, be used for such extended times when implemented on massively parallel computer platforms.

The dissociation energy of the weakest bond of the explosive molecule is thought to play an important role in initiation events. However, previous attempts to correlate bond strengths to impact sensitivity were not successful (82). Politzer & Murray have calculated C-NO_2 and N-NO_2 bond dissociation energies in several small and moderately sized high explosives. They concluded that the correlation between bond strength and impact sensitivity is not general but limited within certain classes of molecules (82).

Given the complexity of detonation chemistry of high explosives, it is not surprising that bond dissociation energy alone is not enough to capture high explosive sensitivity. In order to understand this phenomenon, we herein examine the energy content of high explosives. Wu & Fried (83) have shown that the bond dissociation energy scaled by energy content is a promising indicator for predicting high explosive sensitivity.

We have listed the the calculated (using density functional methods) strengths of the weakest bond (N-NO_2 or C-NO_2) in Table 2. Impact sensitivity is usually characterized through a drop hammer test. In this simple experiment, a weight is dropped on the explosive from varying heights. The height at which violent reaction occurs 50% of the time is referred to as the drop hammer height (H_{50}) The impact sensitivity measurements are taken from the *LLNL Explosives Handbook* (84), the *Navy Explosives Handbook* (85), and Kohler & Meyer (86). Note that these types of measurements are typically rough. Measured impact sensitivities depend strongly on the apparatus used and the experimental protocol.

There appears to be some relation between the bond dissociation energy D_e and explosive sensitivity, but it is clearly not the only controlling factor. For instance, HNB is the most sensitive material in the comparison study, but has a D_e value similar to the much less sensitive DINGU. Nitrobenzene compounds, which are the least sensitive compounds studied here, are found to have the highest D_e values.

It is clear that the energy content of the material must also play a role in determining the impact sensitivity. It is well known that high-energy materials tend to be more sensitive. We have determined the energy content of the explosives studied here with the Cheetah 2.0 thermochemical code (49). The Cheetah code is used to determine E_d, the energy of decomposition, into equilibrium products at standard state, which is a measure of the total energy content of the material. We

TABLE 2 Bond strength (D_e) of the weakest bond, energy content (E_d, kJ/cc) and impact sensitivity H_{50} (cm) of high explosives

Material	Weakest bond	D_e (kJ/mol)	E_d (kj/cc)	H_{50} (cm)
TATB	C-NO_2	323	8.6	>320
DATB	C-NO_2	312	8.6	>320
TNA	C-NO_2	300	8.1	177(N)[a]
NTO	C-NO_2	284	7.7	>280
TNB	C-NO_2	283	8.6	100(N)[a]
TNT	C-NO_2	261	7.7	148
EDNA	N-NO_2	207	9.2	35 (K)[b]
HNB	C-NO_2	183	14.3	8.5
DINGU	N-NO_2	180	8.5	24 (K)[b]
HMX	N-NO_2	179	11.1	32
RDX	N-NO_2	174	10.4	28
TNAZ	C-NO_2	167	11.2	29
PETN	O-NO_2	167	10.5	14
NG	N-NO_2	157	10.0	20
TETRYL	N-NO_2	120	8.8	37

[a](N) indicates values from the *Navy Explosives Handbook* (85).
[b](K) indicates values from Kohler & Meyer (86).
All other data were taken from LLNL measurements.

display the results of these calculations in Table 2. Energy content can be used to explain some of the apparent anomalies in the D_e results. For instance, although HNB has a typical D_e value, it has the greatest energy content.

In Figure 5 we show scatter plots to help visualize the results of Table 2. Note that a logarithmic axis is used to represent the drop hammer values (H_{50}). D_e is seen to be useful in separating the very insensitive materials such as TATB, DATB TNA, and NTO. Within the group of more sensitive materials, however, there seems to be little apparent relation between D_e and H_{50}. The opposite conclusion appears to be true of the energy of decomposition E_d. The relationship between E_d and impact sensitivity appears to be strongest for materials that are very high in energy and becomes more constant at larger H_{50} values.

The ratio between D_e and E_d could be an important quantity in determining the impact sensitivity. The physical motivation for this is that D_e plays the role of an activation barrier. Once a microscopic region ignites in response to mechanical deformation, E_d controls the energy released by the reaction. This in turn will control the local temperature and the likelihood of a propagating chemical reaction.

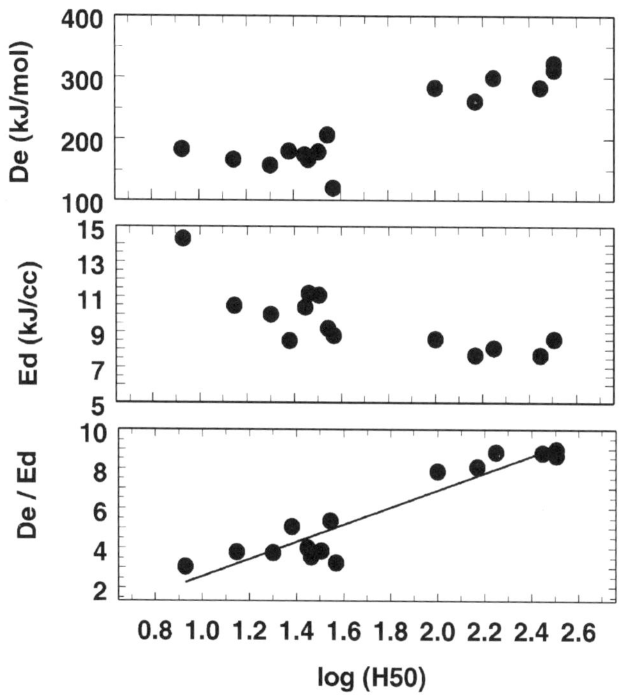

Figure 5 Bond dissociation energies (D_e), Decomposition energies (E_d) and the ratio D_e/E_d are plotted versus the log of the drop hammer height.

Arrhenius kinetics depends on the ratio of activation barrier to temperature, thus motivating the observed correlation between D_e/E_d and H_{50}.

High-Energy Density Materials: Polynitrogen

In contrast to the same group elements (phosphorus and arsenic), nitrogen has thus far resisted transformation from its basic molecular diatomic phase. The possible existence of polymeric forms of nitrogen, in which the strong triple bond of the diatom is replaced by single or single and double bonds, is of a particular interest because of the energetic properties associated with these materials. As potential high-energy density materials from an abundant natural source, they become leading candidates for high-energy density materials. To date, the only known forms of nitrogen are N_2 and N_3^-. Recently, the synthesis of the ion N_5^+ (C_2v symmetry) in a complex with AsF_6 anion (87) is an encouraging step toward the realization of these materials.

The prospect of destabilizing the strong N≡N triple bond under high pressure invited several experimental (88, 89) and theoretical (90–92) attempts to locate and characterize bulk polymeric phases. Hemley et al (89a) have recent papers showing evidence for a polymeric phase. Calculations suggest polymeric regimes to exist below 100 GPa, at 70 and 50 GPa, whereas diamond-anvil-cell experiments failed to record a transition from the diatomic form for pressures of up to 180 GPa. In a non-static, shock-compressed nitrogen experiment (93) an anomaly at pressure above 30 GPa was recorded that was later interpreted as a dissociation to dense atomic phase (94).

Quantum chemical-based computational studies at various levels of theory (53, 54, 95, 96) have shown that N_4 and N_8 are metastable species, having (several) local minima on their respective potential energy surfaces. For tetrahedral N_4 (Figure 6*a*), Lee & Rice determined (53) the energy difference ΔE (N_4– $2N_2$) to be 778 kJ/mol and calculated an energy barrier to decomposition of 255 kJ/mol. Later work examined reaction pathways to other structures (95) and determined the lowest-energy conformation of N_4 to be an open-chain triplet (97) and a dissociation barrier of 264 kJ/mol (98). The results suggest that tetrahedral N_4 has a reasonably long lifetime and can be trapped and studied by matrix isolation techniques.

Computational studies on polyazacubane (99, 100) indicate that the high-energy content of these species, ΔE (N_8–$4N_2$) = 1770 kJ/mol for cubic N_8 (Figure 6*b*), is not the result of strain energy but rather of a weak N-N single bond. Five isomers of N_8 have also been predicted to be metastable (96, 101), using highly correlated methods. In order to guide laboratory synthesis, protonated structures of N_4 and N_8 have been optimized and harmonic vibrational frequencies calculated (102). Recent studies reported ten isomers of N_8 corresponding to analogous CH

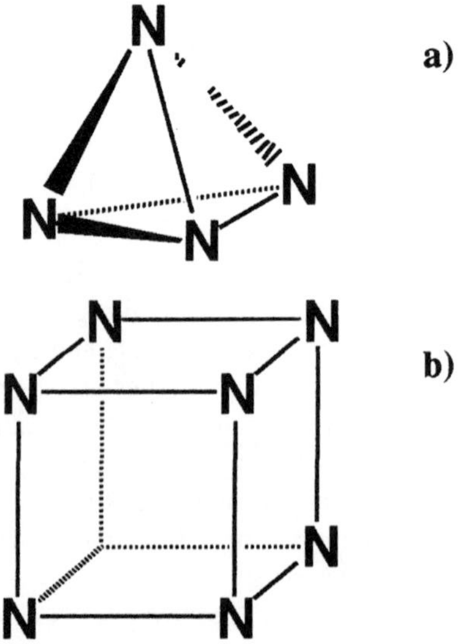

Figure 6 (*a*) Tetrahedral N_4 and (*b*) cubic N_8.

structures (103) and the isomerization reaction of N_8 cubane to planar bicyclic (pentalene-like) structure.

Recently (104), these provocative ideas were suggested: (*a*) Could a nitrogen fullerene analog, N_{60}, be a possible candidate, and (*b*) if so, what is a reasonable building block for such a cluster. In this vein, several computational methods were employed to locate extrema of bicyclic N_{10}. A geometry search located a planar structure (D_2h symmetry) as a saddle point and an orthogonal structure (D_2d symmetry) as a local minimum, with only 13–25 kJ/mol separating the two structures. The equilibrium structure was determined to be remarkably stable with respect to ring-connecting bond homolysis. Furthermore, it was postulated that this molecule could serve as a building block for yet a super-high-energy content N_{60} cluster, which was shown to have a local minimum with purely single-bonded nitrogen atoms. Harmonic vibrational frequencies of N_{10} and infrared active modes of N_{60} were reported to aid in future identification of these species.

For nitrogen, one triple bond is stronger than three single bonds, whereas for carbon the opposite is true. This is perhaps why nitrogen analogs of conjugated carbons are yet to be discovered. Computational tools are currently efficient, however, and should eventually indicate if there is a viable pathway for the experimental characterization and production of some form of polymeric nitrogen.

SYNTHESIS OF NOVEL ENERGETIC MATERIALS

In recent years, the synthesis of energetic heterocyclic compounds has received a great deal of interest. Heterocycles generally have a higher density and oxygen balance, important parameters for increasing the performance of an energetic material, than their carbocyclic analogues. Heterocycles high in nitrogen content also tend to have high heats of reaction. In this review, we discuss recent advances in the synthesis of heterocycles as energetic materials. Other types of novel energetic materials have been reviewed by Agrawal (105).

Pyrazoles

Molecular modeling and thermochemical codes have been used to guide the synthesis of new energetic materials containing the pyrazole ring. This is one of the few examples in which molecules were designed with modern tools, synthesized, and characterized. Caged molecular compounds and fused ring systems typically have higher densities than their single-ring analogues. For this reason, we investigated several fused ring systems computationally. Density predictions were made with the MOLPAK program, whereas predictions of heats of formation were based on Hartree-Fock calculations in combination with empirical corrections calibrated to energetic and non-energetic materials.

It was found that ring systems containing two fused 5-member rings consistently yielded higher densities than two fused 6-member or 4-member rings. The ring

TABLE 3 Predicted densities and heats of formation for target energetic materials

Material	Density[a] (g/cc)	Density[b] (g/cc)	Heat of formation[c]	Heat of formation[d]
DNPP	2.0	1.9	339	287
LLM-119	1.9	2.0	510	584
LLM-121	2.2	2.0		500

[a]MOLPAK 1 computer program used.

[b]Group additivity method of Ammon & Mitchell.

[c]Corrected Hartree-Fock calculations.

[d]DFT calculations (P. Politzer, private communication).

systems chosen for synthesis were predicted to be flat, thus enhancing packing. We also chose heterocyclic ring molecules as a way of improving the energy content of the molecule. Unsaturated bonds in the ring system also improved the calculated energy content and density. Densities and heats of formations were predicted for the target materials by several methods. The results of these calculations are shown in Table 3. Version 1 of the MOLPAK density prediction code was used because potential parameters used in more recent versions were not available for fused heterocyclic ring systems.

In Figure 7, we show pyrazoles that were developed at Lawrence Livermore National Laboratory. Energy release from the C-J state is a more reliable indicator of detonation performance than the detonation velocity. The energy release at a relative volume expansion of 2.2 is a good indicator of the energy usable for rapid metal acceleration. Based on the MOLPAK densities and Hartree-Fock heats of formation shown in Table 3, the metal acceleration energy of three compounds—DNPP, LLM-119, and LLM-121—were predicted to be 0.95, 1.1, and 1.4 times that of HMX, respectively.

Efforts to synthesize these materials began with 3,6-dinitropyrazolo[4,3-c]pyrazole (DNPP). In 1993, Shevelev and co-workers (106) reported its synthesis. Pagoria et al (107) developed an alternative synthesis of DNPP that eliminated the need for an extraction step, thus making the synthesis more amenable to scale-up. DNPP has a measured ΔH_f of +272 kJ/mol, in good agreement with the simple Hartree-Fock calculation in Table 3. The crystal density was determined

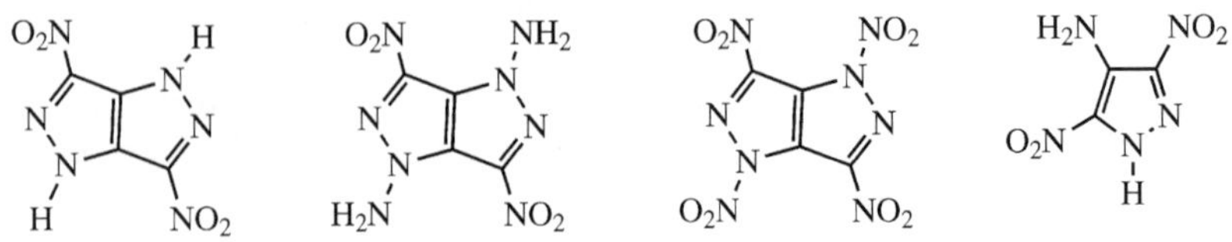

DNPP **LLM-119** **LLM-121** **LLM-116**

Figure 7 The molecules DNPP, LLM-119, LLM-121, and LLM-116.

by X-ray crystallography to be 1.865 g/cc (108). The crystal density is less than the density predicted by MOLPAK but agrees with the group additivity prediction. DNPP is a thermally stable [peak exotherm on differential scanning calorimetry (DSC) at 603 K] energetic material relatively insensitive to shock. The good thermal stability and performance of DNPP make this compound an attractive alternative to RDX, TATB, and TNT. DNPP has two acidic protons with $pK_a = 6$ and 3, respectively, that may be problematic in some applications.

DNPP was also used as a precursor to 1,4-diamino-3,6-dinitropyrazolo [4,3-c]pyrazole (LLM-119). N-amino derivatives of nitrohetrocycles, in comparison to the parent compounds, have an increased predicted performance and possess no acidic protons. LLM-119 was synthesized by amination of DNPP. LLM-119 has a density of 1.845 g/cc as determined by X-ray crystallography (108), which is significantly lower than predicted in Table 3. The low density may be attributed to the fact that the crystal structure shows the amino groups are orthogonal to the plane of the molecule. This is contrary to the predicted ab initio geometry, where the amino groups are in the same plane as the rest of the molecule. The ab initio geometry maximizes the delocalization of their lone pair of electrons with the heterocylic π–electron system and maximizes hydrogen bonding with the nitro groups. With the lower than anticipated density, LLM-119 has a predicted performance 90% that of HMX, based on the experimental density and predicted heat of formation. In this case, the underprediction of density led to a significant decrease in the expected performance of the material. LLM-119 has a peak exotherm at 526 K as determined by DSC; the thermal stability is good for an N-amino compound.

Efforts to synthesize the target materials with performance significantly greater than HMX have not been fully successful. Attempts at further nitration of DNPP to LLM-121 yielded a highly sensitive product. IR spectral analysis and ^{2}H-NMR suggest that LLM-121 is produced. LLM-121 is highly reactive and is itself an effective nitrating agent. LLM-121 may spontaneously revert to DNPP in most common solvents. It appears that very high-energy heterocycles are synthetically achievable but may not be sufficiently stable for routine handling.

Pagoria et al (109) reported the first use of 1,1,1-trimethylhydrazinium iodide (TMHI) as a nucleophilic aminating reagent in the amination of a series of 3-substituted nitrobenzenes. This method uses chemistry coined by Makosza & Winiarski (110) as the "vicarious nucleophilic substitution of hydrogen" in which an amino-group formally replaces a hydrogen atom on an electrophilic aromatic ring. Pagoria used this method to synthesize 4-amino-3,5-dinitropyrazole (LLM-116) (111). LLM-116 has a density of 1.90 g/cc (108), a decomposition point of 451 K, and is relatively insensitive to shock.

Pyrazines and Pyridines

The difficulty of synthesizing nitrated heteroaromatic systems can be attributed to their electron deficiency, making electrophilic aromatic substitution problematic. By the addition of electron-donating substituents to the heteroaromatic ring, nitration may proceed more readily. This is illustrated in the next few examples in which activated pyridine and pyrazine precursors are nitrated to yield the desired dinitro-substituted heterocyles. These examples also illustrate the concept of increasing density and thermal stability by the use of an alternating array of amino- and nitro-groups.

Pagoria et al (107) reported the synthesis of 2,6-diamino-3,5-dinitropyrazine-1-oxide (LLM-105) by the oxidation of 2,6-diamino-3,5-dinitropyrazine (ANPZ) (112). LLM-105 has a density of 1.913 g/cc (108) and a decomposition point of 627 K. This work also illustrates another method to increase density and oxygen balance in heterocyclic systems, i.e., through the conversion of tertiary amines to their corresponding N-oxides. The N-O bond of a tertiary N-oxide is relatively strong, possessing significant double-bond character owing to π-back bonding by the lone pair of electrons of the oxygen (113). The formation of a heterocyclic N-oxide also changes the charge distribution of the heterocyclic ring and leads to, in some cases, an increase in the aromaticity of the heterocyclic ring, thus stabilizing the ring system (113). It should be noted that ANPZ has a crystal density of 1.84 g/cc (108), whereas LLM-105 has a crystal density of 1.913 g/cc. Thus the N-oxide functionality not only increases oxygen balance but also allows better crystal packing. Ritter & Licht (114) reported the synthesis of 2,6-diamino-3,5-dinitropyridine-1-oxide (ANPyO). ANPyO has a density of 1.878 g/cc and a melting point (mp) > 613 K with decomposition (dec). Hollins et al (115, 116) extended this work and synthesized 2,4,6-triamino-3,5-dinitropyridine-1-oxide (compound 1) by the amination of ANPyO. Compound 1 is an insensitive energetic material with a density of 1.876 g/cc and a mp 581 K(dec) (see Figure 8).

Tetrazines

Coburn et al (117) reported the synthesis of 3,6-diamino-1,2,4,5-tetrazine-1,4-dioxide (LAX-112), an example of a cycloaromatic energetic material without a nitro-group as an oxidizing group. LAX-112 has a relatively high detonation velocity and good density. Using a more aggressive final oxidation step, 3-amino-6-nitro-1,2,4,5-tetrazine-2,4-dioxide was made, a sensitive, energetic material that decomposes at 383 K (117) (see Figure 9).

Chavez & Hiskey (118) have continued the research on 1,2,4,5-tetrazine-based explosives and synthesized a number of derivatives that are interesting as propellant

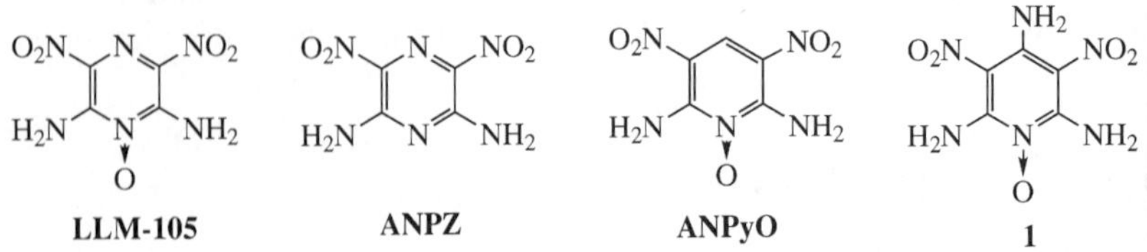

Figure 8 The molecules LLM-105, ANPZ, ANPyO, and compound 1.

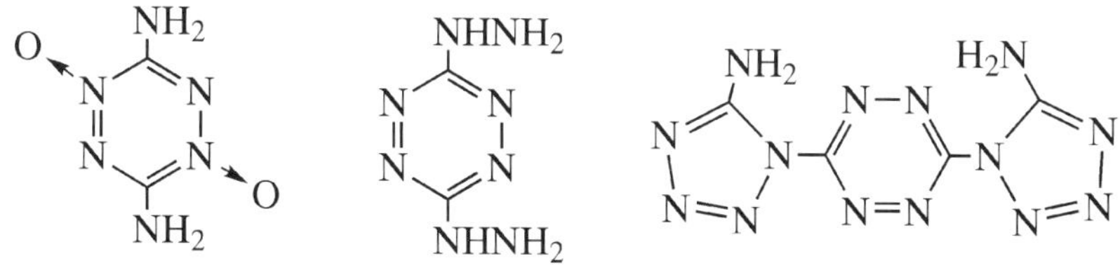

LAX-112 **DHT** **2**

Figure 9 The molecules LAX-112, DHT, and compound 2.

and pyrotechnic ingredients because of their low carbon content. An example is 3,6-dihydrazino-1,2,4,5-tetrazine (DHT) (119), an energetic fuel with a density of 1.61 g/cc and a $\Delta H_f = 536$ kJ/mol. DHT was further converted to 3,6-bis (1H-1,2,3,4-tetrazol-5-ylamino)-1,2,4,5-tetrazine (2), which has a mp $= 537$ K and a measured $\Delta H_f = 883$ kJ/mol. Chavez and co-workers (120) recently reported the synthesis of 3,3′-azobis(6-amino-1,2,4,5-tetrazine (3), a high-nitrogen propellant ingredient with a density of 1.84 g/cc, a mp $= 546$ K, and a $\Delta H_f = +862$ KJ/mol (see Figure 10).

Furazans

3,4-Diaminofurazan (DAF), first synthesized by Coburn (121), has been an important precursor to a series of furazan-based energetic materials that are necessary as propellant ingredients and explosives. Solodyuk et al (122) reported the oxidation of DAF to yield 3-amino-4-nitrofurazan (ANF); 4,4′-diamino-3,3′-azoxyfurazan (DAAF); or 4,4′-diamino-3,3′-azofurazan (DAAzF). Chavez et al (120) scaled-up the synthesis of DAAF and performed measurements of its explosive properties. DAAF has a higher energy content than TATB. DAAF has a crystal density of 1.747 g/cc, a $\Delta H_f = 444$ kJ/mol, and is not sensitive to impact. Sheremetev and co-workers (123) have exploited the high reactivity of the nitro- groups of DNF, DNAF, and DNAzF to nucleophiles in the synthesis of a large number of 3-substituted-4-nitrofurazan derivatives (Figure 11).

Gunasekaran & Boyer (124) synthesized a highly energetic liquid [boiling point (bp) $= 433$–438 K], 5-[4-nitro-(1,2,5)oxadiazolyl]-5*H*-[1,2,3]triazolo[4,5-c][1,2,5]oxadiazole (NOTO) from DAAF.

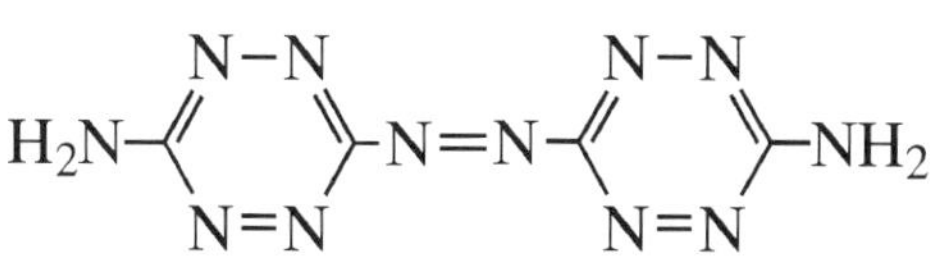

3

Figure 10 Compound 3.

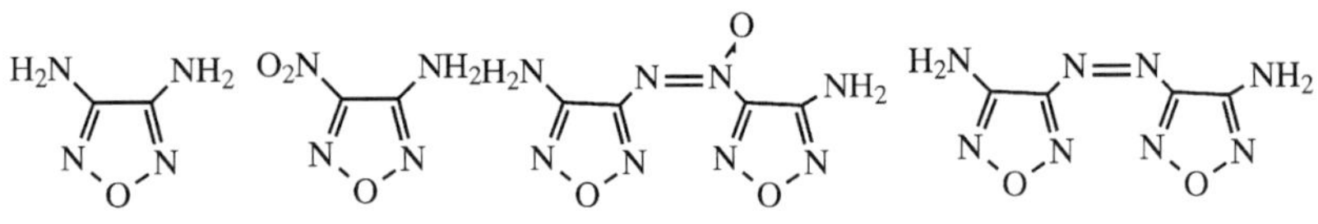

DAF **ANF** **DAAF** **DAAzF**

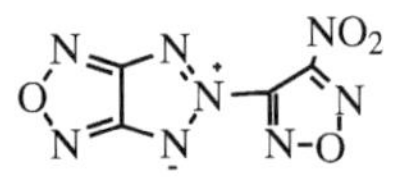

NOTO

Figure 11 DAF, ANF, DAAF, DAAzF, and NOTO.

Trinitroazetidine (TNAZ) and Small-Ring Energetic Materials

Highly nitrated small-ring heterocycles and carbocycles are interesting energetic materials because of the increased performance from the additional energy release upon opening of the strained ring system during decomposition. The most widely studied small-ring energetic material to date is 1,1,3-trinitroazetidine (TNAZ), a melt-castable explosive. TNAZ has a mp of 376–377 K, a crystal density of 1.84 g/cc, and thermal stability >513 K. TNAZ was first synthesized by Archibald and co-workers (125). Coburn and co-workers (126) improved the synthesis making it more amenable to scale-up. Other alternative synthetic pathways to TNAZ have been investigated (127, 128) (Figure 12).

Hiskey et al (129) have synthesized 3,3-dinitroazetidine (DNA), an energetic material with a pK_b of 6.5. They (129) exploited this basicity and synthesized a number of energetic salts. All these salts have low differential thermal analysis (DTA) exotherms (413–433 K) and impact sensitivities ranging from sensitive to moderately sensitive.

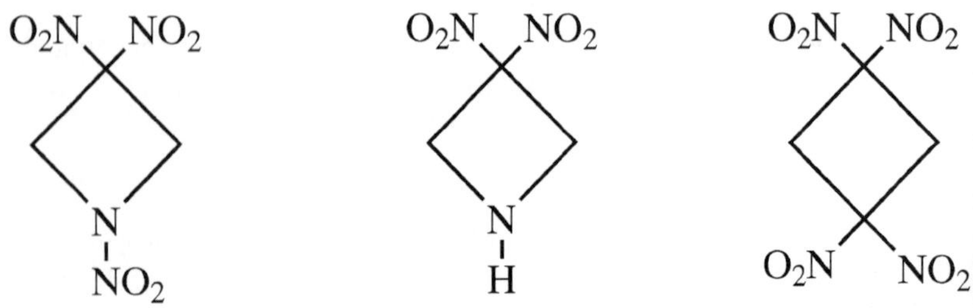

TNAZ **DNA** **TNCB**

Figure 12 TNAZ, DNA, and TNCB.

Earlier, Archibald et al reported the synthesis of 1,1,3,3-tetranitrocyclobutane (TNCB) (130) by the oxidative nitration of 1,3-dinitrocyclobutane with $AgNO_3$ and $NaNO_2$. TNCB has a crystal density = 1.83 g/cc and a mp = 438 K.

Nitrotriazoles

The synthesis of nitrotriazoles as energetic materials and as intermediates to energetic materials has received a great deal of attention in the past ten years (131). The most studied nitrotriazole explosive, 4,6-bis(5-amino-3-nitro-1,2,4-triazolyl)-5-nitropyrimidine (DANTNP), was reported by Wartenberg and co-workers (132). The conception and synthesis of DANTNP is an example of models guiding the synthesis of new energetic materials. Sensitivity prediction was performed by the method of Delpuech et al (62). DANTNP is an insensitive explosive with a crystal density of 1.865 g/cc, $\Delta H_f = 431$ kJ/mol, mp = 603 K, and performance similar to TATB (Figure 13).

3-Nitro-5-amino-1,2,4-triazole (ANTA) was first prepared by Pevzner et al (133). Lee and co-workers (134) subsequently reported an improved synthesis of ANTA. Simpson et al (135) used this method of synthesis, with some small modifications, to scale-up the synthesis of ANTA. ANTA was found to be an insensitive energetic material with a density of 1.819 g/cc, $\Delta H_f = 88$ kJ/mol, a mp = 511 K, and performance similar to TATB.

Ammonium Dinitramide

Bottaro et al (136) reported in 1991 the synthesis of ammonium dinitramide (ADN) $[(NH_4)^+ N(NO_2)_2^-]$. ADN is a very strong oxidizer that may have potential uses in environmentally benign rocket propellant ingredient and as a cationic phase transfer agent. Following this paper, Tartakovsky et al (137–143) published a number of articles on their research on the synthesis and use of dinitramide salts. A significant number of salts of the dinitramide anion have been synthesized, including the alkali salts and the guanidinium; hydroxylammonium; aminoguanidinium; cubane-1,4-diammonium; cubane-1,2,4,7-tetraammonium; biguanidinium; and 1,2-ethylenediammonium salts (136).

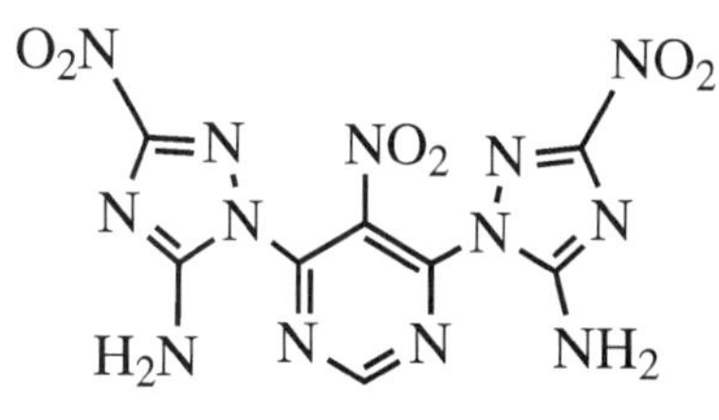

DANTNP

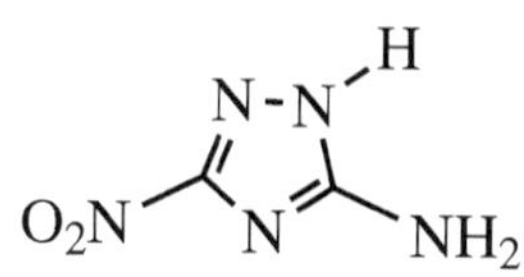

ANTA

Figure 13 DANTNP and ANTA.

ADN is a very strong acid with a $pK_a \sim -5$, is stable between pH 0–15, but slowly decomposes in concentrated acid. It has a mp = 365 K and a DTA exotherm that leaves the baseline at 403 K and peaks at 669 K, and a density of 1.801 g/cc (144).

Highly Nitrated Cage Compounds

Highly nitrated cage compounds hold great promise as energetic materials, based on the premise that the strained rings of cage compounds and the rigid, highly compact cage structure should result in a highly dense, more powerful explosive. A drawback has been the corresponding increase in the difficulty in synthesis of these caged structures. The most studied example of highly nitrated cage compounds is 2,4,6,8,10,12-hexanitro-2,4,6,8,10,12-hexaazatetracyclo[5.5.0.05.9.03,11]dodecane (CL-20), first synthesized by Nielsen and co-workers (145) at the Naval Air Warfare Center-China Lake. CL-20 (see Figure 14) is the most powerful explosive currently being investigated at the pilot plant scale or larger (146). CL-20, in its ε-crystal polymorph, has a density of 2.04 g/cc and a DSC decomposition temperature of 501 K.

Highly nitrated cubanes and azacubanes were predicted to be very dense, highly energetic compounds with explosive performance greater than CL-20 (147–152) (see Figure 14). Recently Zhang and co-workers (153) reported the synthesis of heptanitrocubane (compound 4) and octanitrocubane (compound 5), the culmination of an ongoing project in the synthesis of nitrocubanes as energetic materials. Previously, Eaton et al (154) reported the synthesis of 1,3,5,7-tetranitrocubane (compound 6) from the oxidation of the tetramino derivative with dimethyldioxirane. Tetranitrocubane has a density of 1.814 g/cc and a mp = 475 K. The more highly nitrated species proved to be more difficult to synthesize. The pentanitro- (ρ = 1.959 g/cc) and hexanitrocubanes (155) were synthesized by the treatment of the anion of tetranitrocubane with N_2O_4 at the interface between frozen THF and N_2O_4. Heptanitrocubane was synthesized by the treatment of tetranitrocubane with $NaN(TMS)_2$ and frozen N_2O_4 in THF/isopentane. Octanitrocubane was synthesized by the treatment of heptanitrocubane with $LiN(TMS)_2$ in CH_2Cl_2 at 195 K.

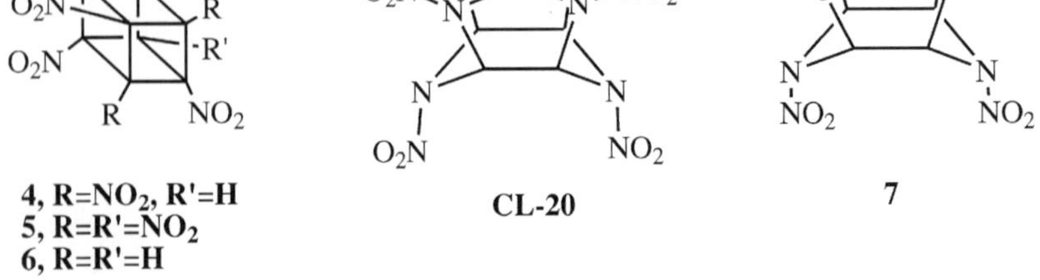

Figure 14 Compounds 4–7 and CL-20.

with NOCl followed by ozonation. Heptanitrocubane has a density of 2.028 g/cc whereas octanitrocubane has a density of 1.979 g/cc (Figure 14).

Vayalakkavoor and co-workers (156) reported the synthesis of 4,10-dinitro-2,6,8,12-teraoxo-4,10-diazatetracyclo[5.5.0.05,9.03,11]dodecane (7) (Figure 14). Compound 7 has the surprisingly high density of 1.99 g/cc, considering it possesses only two nitramine moieties. The high density was attributed to the caged structure. It is thermally stable with an mp > 523 K and has been investigated as an insensitive energetic material.

Polynitrogen Compounds

Polynitrogen compounds are of significant interest as high-energy density materials (see above). Recently, Christe and co-workers (157) synthesized $N_5^+AsF_6^-$, the first example of a new homoleptic polynitrogen ion since the discovery of the azide ion in 1890. $N_5^+AsF_6^-$ was synthesized by the condensation of $N_2F^+AsF_6^-$ with HN_3 at 195 K in anhydrous HF. $N_5^+AsF_6^-$ is a white solid that is marginally stable at room temperature but can be stored for weeks at 195 K.

Nanostructured Energetic Materials

Alternative approaches to monomolecular energetic materials have been formulations of separate fuels and oxidizers. These composites have the potential for significantly greater energy density, but energy release rates have been dramatically slower than monomolecular counterparts owing to mass transport rates. Recent developments have reduced the scale of mixing to the nanometer range. Aumann et al (158) developed a gas condensation method to produce 20–40 nm particles of MoO_3 and passivated Al. When stoichiometrically mixed, an energy density greater than 16 kJ/cm^3 can be attained. Very fast burn rates have been reached, making these nanostructured materials attractive alternatives to monomolecular structures. Other methods to prepare energetic nanostructures include chemical syntheses and epitaxial deposition. One technique uses sol-gel chemistry to produce nanometer scale gels made of either a fuel or oxidizer (159, 160). Both organic and inorganic gels may be made. Figure 15 shows a Fe_2O_3 gel (the oxidizer) with 30-nm aluminum particles (the fuel) entrained. The primary particle cluster sizes are 2 to 5 nm. For all these nanostructured materials there are few property data, and the effect on reactivity of increased surface area, high surface energy, and higher compressive stresses in nanoparticles is largely unknown.

CONCLUSION

The design of energetic materials is currently in a transitional stage. While certain properties can be predicted with confidence, others elude current simulation methods. Thermochemical calculations of energetic material energy delivery are probably the most mature. Detonation properties of mixtures of known compounds

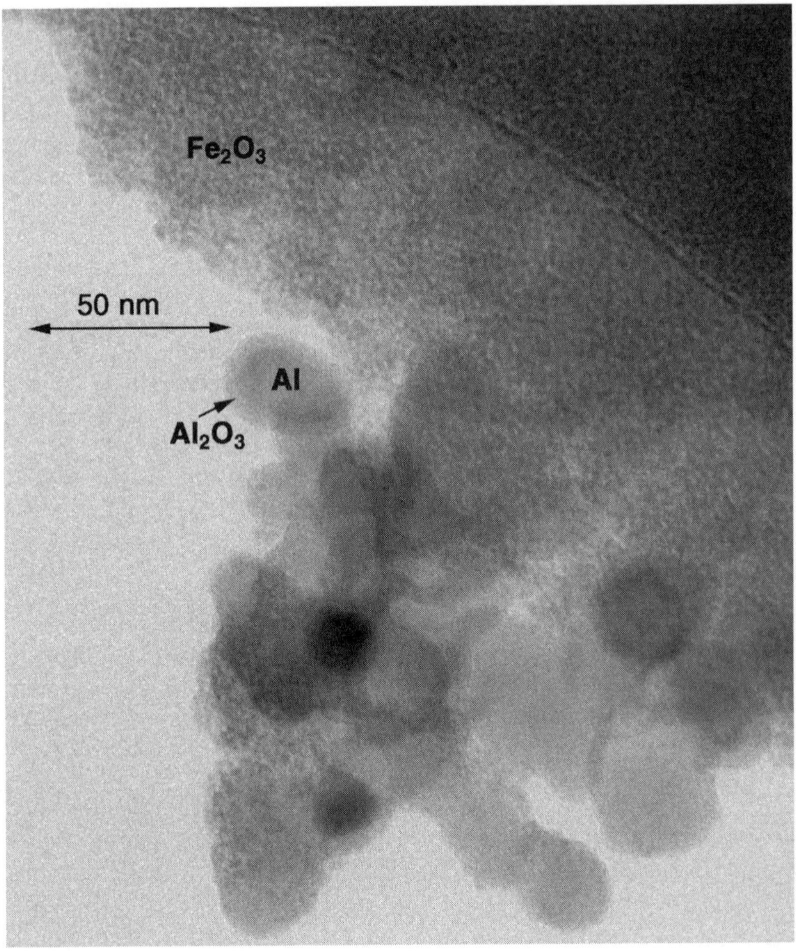

Figure 15 LA Fe_2O_3 gel (the oxidizer) with 30-nm aluminum particles (the fuel) entrained

can be calculated with errors of roughly 2–3%. Higher uncertainties occur, however, with composite materials including high concentrations of metals and other slowly reacting components. In this case, the development of kinetic detonation schemes shows promise for the future.

Significant progress has been made in the last several years in the prediction of the gas phase heat of formation. The calculation of the solid phase heat of formation, however, is less well developed. The lattice energy now represents the greatest uncertainty in estimating the energy content of molecular solids. Density prediction schemes are not yet mature. Although several atomistic methods have been developed, group additivity is also a competitive practical method. The prediction of energetic material stability is hindered by a lack of basic theoretical understanding. Simply calculating the bond dissociation of an isolated molecule may provide some insight, given that the molecular energy content is also taken into account.

The above uncertainties in the design of energetic materials can lead researchers into inadvertently overestimating material stability and performance. The overestimation is produced by simple selection; researchers select from a large set of candidates, i.e., those materials with the most outstanding predicted properties. Sometimes, however, what is selected is merely a molecule poorly treated by the computational method. The development of more robust performance and stability predicting tools will help to alleviate this difficulty.

Synthesis of new energetic materials with novel combinations of properties is continuing to proceed at a rapid pace. Most of these materials, however, live within a rough envelope of tradeoffs between energy and stability. New material classes, such as nano-compositie materials, may offer tradeoffs that are more favorable. More sophisticated synthetic techniques can optimize the tradeoff between energy and stability but usually at an increased material cost. The importance of material cost diminishes, however, as energetic materials find new, technically sophisticated, applications. In the future, we expect a wider range of energetic materials to be tailored to specific applications.

ACKNOWLEDGMENTS

This work was performed under the auspices of the US Department of Energy by the Lawrence Livermore National Laboratory under contract No. W-7405-ENG-48.

LITERATURE CITED

1. Fickett W, Davis WC. 1979. *Detonation*. Berkeley: Univ. Calif. Press
2. Lommerse JPM, Motherwell WDS, Ammon HL, Dunitz JD, Gavezzotti A, et al. 2000. *Acta Crystall. B* 56:697–714
3. Tarver CM. 1979. *J. Chem. Eng. Data* 24:136
4. Stine JR. 1981. *A method for predicting the density of organic molecular crystals. Rep. LA8920*, Los Alamos, NM: Los Alamos Sci. Lab.
5. Ammon HL, Mitchell S. 1998. *Propellants Explosives Pyrotechnics* 23:260–65
6. Karfunkel HR, Gdanitz RJ. 1992. *J. Comput. Chem.* 13:1171–83
7. Holden JR, Du ZY, Ammon HL. 1993. *J. Comput. Chem.* 14:422–37
8. Piacenza G, Legsai G, Blaive B, Gallo R. 1996. *J. Phys. Org. Chem.* 9:427–32
9. Dzyabchenko AV, Pivina TS, Arnautova EA. 1996. *J. Mol. Struct.* 378:67–82
10. Sorescu DC, Rice BM, Thompson DL. 1999. *J. Phys. Chem. A* 103:989–98
11. Rice BM, Pai SV, Hare J. 1999. *Combustion Flame* 118:445–58
12. Habibollahzadeh D, Grice ME, Concha MC, Murray JS, Politzer P. 1995. *J. Comput. Chem.* 16:654–58
13. Politzer P, Murray JS, Grice ME, Desalvo M, Miller E. 1997. *Mol. Phys.* 91:923–28

14. Fried LE, Souers PC. 1996. *Propellants, Explosives, Pyrotechnics* 21:215–23
15. Kistiakowsky GB, Wilson EB. 1941. *Report on the prediction of detonation velocities of solid explosives. Rep. OSRD-69*, Washington, DC: Off. Sci. Res. Dev.
16. Finger M, Lee E, Helm FH, Hayes B, Hornig H, et al. 1976. *The effect of elemental composition on the detonation behavior of explosives*. Sixth Int. Symp. Detonation, Coronado, CA
17. Mader CL. 1979. *Numerical Modeling of Detonations*. Berkeley, CA: Univ. Calif. Press
18. Gubin SA, Odintsov VV, Pepekin VI. 1985. *Sov. J. Chem. Phys.* 3:1152
19. Hobbs ML, Baer MR. 1993. *Calibrating the BKW-EOS with a large product species data based and measured C-J properties*. Tenth Int. Detonation Symp., Boston, MA
20. Souers PC, Kury JW. 1993. *Propellants, Explosives, Pyrotechnics* 18:175
21. Cowperthwaite M, Zwisler WH. 1976. *The JCZ equation of state for detonation products and their incorporation into the TIGER code*. Sixth Int. Symp. Detonation Coronado, CA
22. Hobbs ML, Baer MR, McGee BC. 1999. *Propellants, Explosives, Pyrotechnics* 24:269–79
23. Deleted in proof
24. Ross M. 1979. *J. Chem. Phys.* 71:1567
25. Ross M, Ree FH. 1980. *J. Chem. Phys.* 73:6146–52
26. Ree FH. 1984. *J. Chem. Phys.* 81:1251
27. van Thiel M, Ree FH. 1987. *J. Appl. Phys.* 62:1761–67
28. van Thiel M, Ree FH. 1996. *J. Chem. Phys.* 104:5019–25
29. Kang HS, Lee CS, Ree T, Ree FH. 1985. *J. Chem. Phys.* 82:414
30. Byers Brown W. 1987. *J. Chem. Phys.* 87:566
31. Charlet F, Turkel ML, Danel JF, Kazandjian L. 1998. *J. Appl. Phys.* 84:4227–38
32. Fried LE, Howard WM. 1998. *J. Chem. Phys.* 109:7338–48
33. Ree FH. 1986. *J. Chem. Phys.* 84:5845–56
34. Fried LE, Howard WM. 1999. *J. Chem. Phys.* 110:12023–32
35. Shaw MS. 1991. *J. Chem. Phys.* 94:7550–53
36. Leland TW, Rowlinson JS, Sather GA. 1947. *Trans. Faraday Soc.* 64:1447
37. Kerley GI. 1985. *Theoretical equations of state for the detonation products of explosives*. Eighth Int. Symp. Detonation, Albuquerque, NM
38. Reed TM, Gubbins KE. 1973. *Statistical Mechanics*. New York: McGraw-Hill
39. Jones HD, Zerilli FJ. 1991. *J. Appl. Phys.* 69:3893–900
40. van Thiel M, Ree FH. 1989. *Int. J. Thermophys.* 10:227–36
41. Fried LE, Howard WM. 2000. *Phys. Rev. B* 61:8734–43
42. Shaw MS, Johnson JD. 1987. *J. Appl. Phys.* 62:2080–85
43. Viecelli JA, Ree FH. 1999. *J. Appl. Phys.* 86:237–48
44. Viecelli JA, Ree FH. 2000. *J. Appl. Phys* 88:683–90
45. Wood WW, Kirkwood JG. 1954. *J. Chem Phys.* 22:1920–24
46. Bdzil JB. 1981. *J. Fluid Mech.* 108:195–206
47. Stewart DS, Yao J. 1998. *Combustion Flame* 113:224–35
48. Howard WM, Fried LE, Souers PC. 1998 *Kinetic modeling of non-ideal explosives with Cheetah*. 11th Int. Symp. Detonation Snowmass, CO
49. Fried LE, Howard WM, Souers PC 1998. *Cheetah 2.0 User's Manual. Rep UCRL-MA-117541 Rev. 5*, Livermore, CA Lawrence Livermore Natl. Lab.
50. Murnaghan FD. 1944. *Proc. Natl. Acad Sci. USA* 30:244–47
51. Seidel ET, Schaefer HF. 1988. *J. Chem Phys.* 88:7043–49
52. Kim KS, Jang JH, Kim S, Mhin B, Schaefe HF. 1990. *J. Chem. Phys.* 92:1887–92
53. Lee TJ, Rice JE. 1991. *J. Chem. Phys* 94:1215–21

54. Engelke R, Stine JR. 1990. *J. Phys. Chem.* 94:5689–94
55. Brill TB, James KB. 1993. *J. Phys. Chem.* 97:8752–58
56. Dlott DD. 1999. *Annu. Rev. Phys. Chem.* 50:251–78
57. Dlott DD, Fayer MD. 1990. *J. Chem. Phys.* 92:3798–812
58. Fried LE, Ruggiero AJ. 1994. *J. Phys. Chem.* 98:9786–91
59. Coffey CS. 1993. *Energy localization and the initiation of explosive crystals by shock or impact.* Structure and Properties of Energetic Materials, Boston, MA
60. Williams F. 1971. *Adv. Chem. Phys.* 21:289–302
61. Dremin AN, Klimenko VY, Davidova ON, Zoludeva TA. 1989. *Multiprocess detonation model.* Ninth Int. Symp. Detonation, Portland, OR
62. Delpuech A, Cherville J, Michaud C. 1981. *Molecular electronic structure and initiation of secondary explosives.* Seventh Int. Symp. Detonation, Annapolis, MD
63. Sharma J. 1991. In *Shock Compression of Condensed Matter*, ed. SC Schmidt, RD Dick, JW Forbes, DG Tasker. Williamsburg VA: Elsevier
64. Sharma J, Beard BC, Chaykovsky M. 1991. *J. Phys. Chem.* 95:1209–13
65. Gilman JJ. 1995. *Philos. Mag. B* 71:1057–68
66. Kuklja MM, Stefanovich EV, Kunz AB. 2000. *J. Chem. Phys.* 112:3417–23
67. Manaa MR, Fried LE. 1999. *J. Phys. Chem. A* 103:9349–54
68. Takezaki M, Hirota N, Terazima M. 1997. *J. Phys. Chem. A* 101:3443–48
69. Rajchenbac C, Gediminas J, Rulliere C. 1995. *Chem. Phys. Lett.* 231:467–75
70. Reed EJ, Joannopoulos JD, Fried LE. 2000. *Phys. Rev. B* 62:16500–9
71. Shaw R, Decarli PS, Ross DS, Lee EL, Stromberg HD. 1979. *Combustion Flame* 35:237
72. Shaw R, Decarli PS, Ross DS, Lee EL, Stromberg HD. 1983. *Combustion Flame* 50:123
73. Engelke R, Schiferl D, Storm CB, Earl WL. 1988. *J. Phys. Chem.* 92:6815
74. Brasch JW. 1980. *J. Phys. Chem.* 84:2084
75. Naud DL, Brower KR. 1992. *High-Pressure Res.* 11:65
76. Blais NC, Engelke R, Sheffield SA. 1997. *J. Phys. Chem. A* 101:8285–95
77. Winey JM, Gupta YM. 1997. *J. Phys. Chem. B* 101:10733–43
78. Winey JM, Gupta YM. 1997. *J. Phys. Chem. A* 101:9333–40
79. Gruzdkov YA, Gupta YM. 1998. *J. Phys. Chem. A* 102:2322–31
80. Kress JD, Bickham SR, Collins LA, Holian BL, Goedecker S. 1999. *Phys. Rev. Lett.* 83:3896
81. Bickham SR, Kress JD, Collins LA. 2000. *J. Chem. Phys.* 112:9695
82. Politzer P, Murray JS. 1996. *J. Mol. Struct.* 376:419
83. Wu CJ, Fried LE. 1998. *First principles study of decomposition energetics.* Proceed. 11th Int. Symp. Detonation, Snowmass, CO
84. Dobratz BM, Crawford PC. 1985. *LLNL Explosives Handbook.* Rep. UCRL-52997
85. Hall TN, Holden JR. 1985. *Navy Explosives Handbook: Explosion Effects and Properties Part III. Rep. NSWC MP88–116*, Dahlgren, VA: Naval Surface Warfare Center
86. Kohler J, Meyer R. 1993. *Explosives.* Weinheim, Germany: VCH
87. Christe KO, Wilson WW, Sheehy JA, Boatz JA. 1999. *Ang. Chem. Int. Ed.* 38:2004–9
88. Reichlin R, Schiferl D, Martin S, Vanderborgh C, Mills RL. 1985. *Phys. Rev.Lett.* 55:1464
89. Bell PM, Mao HK, Hemley RJ. 1986. *Physica B* 139:16
89a. Goncharov AF, Gregoryanz E, Mao H, Zhenxian L, Hemley RJ. 2000. *Phys. Rev. Lett.* 85:1262

90. McMahan AK, LeSar R. 1985. *Phys. Rev. Lett.* 54:1929
91. Martin RM, Needs RJ. 1986. *Phys. Rev. B* 34:5082
92. Mailhiot C, Yang LH, McMahan AK. 1992. *Phys. Rev. B* 46:14419
93. Radousky HB, Nellis WJ, Ross M, Hamilton DC, Mitchell AC. 1986. *Phys. Rev. Lett.* 57:2419
94. Ross M. 1987. *J. Chem. Phys.* 86:7110
95. Lauderdale WJ, Stanton JF, Bartlett RJ. 1992. *J. Phys. Chem.* 96:1173
96. Leininger ML, Sherrill CD, Schaefer HF. 1995. *J. Phys. Chem.* 99:2324
97. Glukhovtsev MN, Schleyer PR. 1993. *Int. J. Quantum Chem.* 46:119
98. Dunn KM, Morokuma KJ. 1995. *J. Chem. Phys.* 102:4904
99. Alkorta I, Elguero J, Rozas I, Balaban AT. 1990. *J. Mol. Struct.* 206:67
100. Engelke R. 1993. *J. Am. Chem. Soc.* 115:2961
101. Tian A, Ding F, Zhang L, Xie Y, Schaefer HF. 1997. *J. Phys. Chem. A* 101:1946
102. Leininger ML, Huis TJV, Schaefer HF. 1997. *J. Phys. Chem. A* 101:4460
103. Gagliardi L, Evangelisti S, Roos BO, Widmark P. 1998. *J. Mol. Struct.* 428:1
104. Manaa MR. 2000. *Chem. Phys. Lett.* 331:262–68
105. Agrawal JP. 1998. *Prog. Energy Combustion Sci.* 24:1–30
106. Shevelev SA, Dalinger IL, Shkineva TK, Ugrak BI, Gulevskaya VI, Kanishchev MI. 1993. *Russ. Chem. Bull.* 42:1063–68
107. Pagoria PF, Mitchell AR, Schmidt RD, Simpson RL, Garcia F, et al. 1998. *Synthesis, scale-up and experimental testing of LLM-105 (2,6–diamino-3,5–dinitropyrazine-1–oxide)*. Insensitive Munitions and Energetic Materials Technology Symp., Meet. No. 956, San Diego, CA
108. Gilardi R. 2000. *X-ray crystallographic determination*. Washington, DC: Naval Res. Lab.
109. Pagoria PF, Mitchell AR, Schmidt RD. 1996. *J. Organic Chem.* 61:2934–35
110. Makosza M, Winiarski J. 1987. *Acc. Chem. Res.* 20:282–89
111. Pagoria PF, Mitchell AR, Schmidt RD. 1996. *Vicarious amination of nitroarenes with trimethylhydrazinium iodide*. 211th Am. Chem. Soc. Natl. Meet. New Orleans, LA
112. Donald DS. 1974. *U.S. Patent No. 3,808,209*
113. Albini A, Pietra S. 1991. *Heterocyclic-N-oxides*. Boca Raton, FL: CRC Press
114. Ritter H, Licht HH. 1995. *J. Heterocyclic Chem.* 32:585–90
115. Hollins RA, Merwin LH, Nissan RA, Wilson WS, Gilardi R. 1996. *J. Heterocyclic Chem.* 33:895–904
116. Hollins RA, Merwin LM, Nissan RA, Wilson WW, Gilardi R. 1996. *Aminonitroheterocyclic-N-oxides–a new class of insensitive energetic materials*, pp. 31-36. Pittsburgh, PA: Mater. Res. Soc. Symp.
117. Coburn MD, Hiskey MA, Lee KY, Ott DG, Stinecipher MM. 1993. *J. Heterocyclic Chem.* 30:1593–95
118. Chavez DE, Hiskey MA. 1999. *J. Energetic Mater.* 17:357–77
119. Chavez DE, Hiskey MA. 1998. *J. Heterocyclic Chem.* 35:1329–32
120. Chavez D, Hill L. Hiskey M, Kinkead S. 2000. *J. Energetic Mater.* 18:219–36
121. Coburn MD. 1968. *J. Heterocyclic Chem.* 5:83–87
122. Solodyuk GD, Boldyrev MD, Gidaspov BV, Nikolaev VD. 1981. *J. Organic Chem. USSR (Engl.)* 17:756–59
123. Sheremetev AB, Kharitonova OV, Mantseva EV, Kulagina VO, Shatunova EV et al. 1999. *Z. Organic. Khimii* 35:1555–66
124. Gunasekaran A, Boyer JH. 1993. *Heteroatom Chem.* 4:521–24
125. Archibald TG, Gilardi R, Baum K George C. 1990. *J. Organic Chem.* 55:2920–24

126. Coburn MD, Hiskey MA, Archibald TG. 1997. *Waste Management* 17:143–46
127. Hayashi K, Kumagai T, Nagao Y. 2000. *Heterocycles* 53:447–52
128. Marchand AP, Rajagopal D, Bott SG, Archibald TG. 1995. *J. Organic Chem.* 60:4943–46
129. Hiskey MA, Coburn MD, Mitchell MA, Benicewicz BC. 1992. *J. Heterocyclic Chem.* 29:1855–56
130. Archibald TG, Garver LC, Baum K, Cohen MC. 1989. *J. Organic Chem.* 54:2869–73
131. Yuxiang O, Boren C, Jiarong L, Shuan D, Jianjuan L, Huiping J. 1994. *Heterocycles* 38:1651–64
132. Wartenberg C, Charrue P, Laval F. 1995. *Propellants Explosives Pyrotechnics* 20:23–26
133. Pevzner MS, Kulibabina TN, Povarova NA, Kilina LV. 1979. *Khim. Geterotskil* 8:1132–35
134. Lee KY, Storm CB, Hiskey MA, Coburn MD. 1991. *J. Energetic Mater.* 9:415–28
135. Simpson RL, Pagoria PF, Mitchell AR, Coon CL. 1994. *Propellants Explosives Pyrotechnics* 19:174–79
136. Bottaro JC, Penwell PE, Schmitt RJ. 1997. *J. Am. Chem. Soc.* 119:9405–10
137. Lukyanov OA, Gorelik VP, Tatakovsky VA. 1994. *Russ. Chem. Bull.* 43:89–92
138. Lukyanov OA, Konnova YV, Klimova TA, Tartakovsky VA. 1994. *Russ. Chem. Bull.* 43:1200–2
139. Lukyanov OA, Anikin OV, Gorelik VP, Tartakovsky VA. 1994. *Russ. Chem. Bull.* 43:1457–61
140. Shlyapochnikov VA, Cherskaya NO, Lukyanov OA, Gorelik VP, Tartakovsky VA. 1994. *Russ. Chem. Bull.* 43:1522–25
141. Shlyapochnikov VA, Oleneva GI, Cherskaya NO, Lukyanov OA, Gorelik VP, et al. 1995. *Russ. Chem. Bull.* 44:1449–53
142. Lukyanov OA, Shlykova NI, Tartakovsky VA. 1994. *Russ. Chem. Bull.* 43:1680–83
143. Lukyanov OA, Agevnin AR, Leichenko AA, Seregina NM, Tartakovsky VA. 1995. *Russ. Chem. Bull.* 44:108–12
144. Martin A, Pinkerton AA, Gilardi RD, Bottaro JC. 1997. *Acta Crystall. B* 53:504–12
145. Nielsen AT, Chafin AP, Christian SL, Moore DW, Nadler MP, et al. 1998. *Tetrahedron* 54:11793–812
146. Simpson RL, Urtiew PA, Ornellas DL, Moody GL, Scribner KJ, Hoffman DM. 1997. *Propellants Explosives Pyrotechnics* 22:249–55
147. Reed LH, Jayasuria K, Koovaakat SK, Allen LC. 1991. *J. Phys. Org. Chem.* 4:714–20
148. Chapman DA, Kaufman JJ, Buenker RJ. 1991. *Int. J. Quantum Chem.* 40:389–403
149. Eremenko LT. 1997. *Chem. Phys. Rep.* 16:1675–83
150. Astakhov AM, Stepanov RS, Babushkin AY. 1998. *Combustion, Explosion, Shock Waves* 34:85–87
151. Kortus J, Pederson MR, Richardson SL. 2000. *Chem. Phys. Lett.* 332:224–30
152. Jursic BS. 2000. *J. Mol. Struct. Theochem.* 530:21–30
153. Zhang MX, Eaton PE, Gilardi R. 2000. *Ang. Chemie Int. Ed.* 39:401–4
154. Eaton PE, Xiong YS, Gilardi R. 1993. *J. Am. Chem. Soc.* 115:10195–202
155. Lukin KA, Li JC, Eaton PE, Kanomata N, Hain J, et al. 1997. *J. Am. Chem. Soc.* 119:9591–602
156. Vayalakkavoor RT, Vedachalam M, Boyer JH. 1990. *Heterocycles* 31:479–80
157. Christe KO, Wilson WW, Sheehy JA, Boatz JA. 1999. *Ang. Chemie. Int. Ed.* 38:2004–9
158. Aumann CE, Skofronick GL, Martin JA. 1995. *J. Vac. Sci. Technol. B* 13:1178–83
159. Tillotson TM, Hrubesh LW, Simpson RL, Lee RS, Swansiger RW, Simpson RL. 1998. *J. Non-Cryst. Solids* 225:358–63
160. Tillotson TM, Simpson RL, Hrubesh LW, Gash AE, Thomas IM, Poco JF. 2001. *J. Non-Cryst. Solids.* In press

Annu. Rev. Mater. Res. 2001. 31:323–55

BLOCK COPOLYMER THIN FILMS: Physics and Applications[1]

Michael J Fasolka
Optical Technology Division, National Institute of Standards and Technology, Gaithersburg, Maryland 20899; e-mail: mfasolka@nist.gov

Anne M Mayes
Department of Materials Science and Engineering, Massachusetts Institute of Technology, Cambridge, Massachusetts 02139; e-mail: amayes@mit.edu

Key Words confinement, self-consistent field, wetting, nanoarray, nanopattern, self-assembly

■ **Abstract** A two-part review of research concerning block copolymer thin films is presented. The first section summarizes experimental and theoretical studies of the fundamental physics of these systems, concentrating upon the forces that govern film morphology. The role of film thickness and surface energetics on the morphology of compositionally symmetric, amorphous diblock copolymer films is emphasized, including considerations of boundary condition symmetry, so-called hybrid structures, and surface chemical expression. Discussions of compositionally asymmetric systems and emerging research areas, e.g., liquid-crystalline and A-B-C triblock systems, are also included. In the second section, technological applications of block copolymer films, e.g., as lithographic masks and photonic materials, are considered. Particular attention is paid to means by which microphase domain order and orientation can be controlled, including exploitation of thickness and surface effects, the application of external fields, and the use of patterned substrates.

INTRODUCTION AND SCOPE

Block copolymers (BCs) consist of two or more chemically distinct polymer fragments, or blocks, covalently bonded together to form a larger, more complex macromolecule. If the constituent polymers are immiscible, phase separation is induced on a scale that is directly related to the size of the copolymer chains, which results in morphologies typified by a pattern of chemically distinct domains of periodicity L_0 in the 10–100 nm range (1, 2). The order/disorder transition (ODT)

[1]The US Government has the right to retain a nonexclusive, royalty-free license in and to any copyright covering this paper.

temperature and the specific pattern motifs formed by a given BC are functions of the polymer molecular weight, the segmental interactions, and the volumetric composition (1, 2). The BC composition in particular strongly determines the microphase morphology. For example, diblock copolymers having blocks of comparable volume exhibit a lamellar motif. Increasing the degree of compositional asymmetry leads to the gyroid, cylindrical, and finally, spherical phases (2).

BC microphase separation has been the focus of intense research activity—both fundamental and applied—spanning the last 30 years. These efforts have produced a solid foundation of theoretical and experimental understanding of the rich behavior of these systems and provided insight into how this spontaneous nanoscopic pattern formation might be harnessed for use in a variety of technological schemes. While the majority of this work has concentrated upon bulk systems, more recent efforts have striven to understand BC microphase separation in thin films. This latter research has, in large part, been concerned with the understanding and exploitation of the enhanced role of surface/interfacial energetics, as well as the interplay between the copolymer's characteristic length scale, L_0, and the film thickness, t, in dictating film structure. Indeed, as film thickness decreases, these effects become increasingly consequential.

Our review of the current state of BC thin film research is organized into two sections First, the major ideas that constitute our understanding of BC thin film morphology are outlined. This encompasses a discussion of the experimental and theoretical treatments of the effects of confinement and surface energetics on microphase separation. This is followed by a review of research concentrating on the application of BC thin films to future technologies, with discussion centering largely on the control of thin film morphology.

As with all such reviews on extensively researched subjects, the scope of this article is limited by space. Accordingly, much of the following focuses on the simplest and most studied system, cast films of amorphous diblock copolymers that are compositionally symmetric. The topic of adsorbed block copolymer films—of relevance to colloidal and biomaterials applications—is not discussed here. Finally, the current discussion does not deal with BC thin film kinetics. A brief review of this subject can be found in an article by Matsen (3).

PHYSICS OF BC THIN FILM MORPHOLOGY

Bulk BC morphologies are typified by grains of ordered domains oriented randomly with respect to each other, much like the grain structure of small molecule or atomic crystals (2, 4, 5). In contrast, BC thin films are often characterized by highly oriented domains (6–8). This orientation is a direct result of surface and interfacial energy minimization, as illustrated by a film of a bulk-lamellar diblock with block segments denoted A and B bounded by surfaces 1 and 2 shown in Figure 1*a*,*b*. The overriding trend exhibited by films of thickness $t > L_0$ is full lamellae oriented parallel to the film and substrate surfaces (6) as shown in

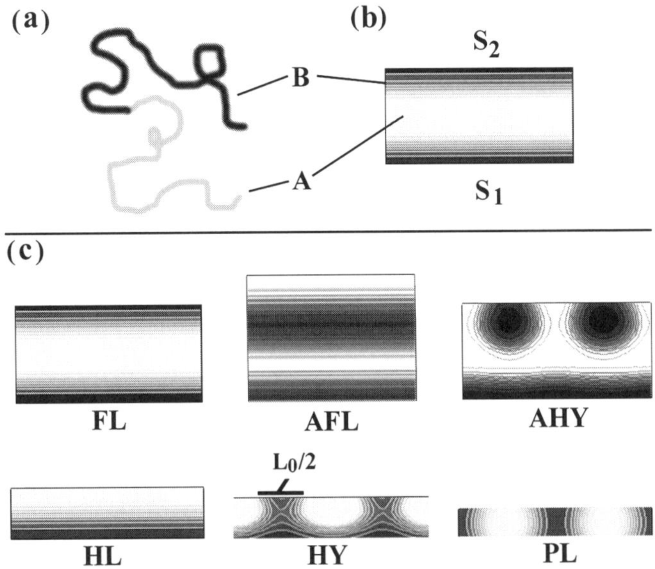

Figure 1 Diblock copolymer thin film morphologies. (*a*) Schematic representation of volume symmetric diblock with A (light) and B (dark) type segments. (*b*) Diagram of BC film system in cross section indicating the bottom (1) and top (2) surfaces, with surface interaction energies S_1 and S_2, respectively. (*c*) Summary of diblock thin film morphologies, generally organized by their appearance as film thickness decreases from L_0. These calculated cross sections indicate the density of B-type segments, i.e., Black = 100% B, white = 100% A. These structures are referred to in the text with the abbreviations included under each diagram. FL: symmetric surface-parallel full lamella; AFL: anti-symmetric surface-parallel lamella; AHY: anti-symmetric hybrid structure; HL: half-lamella; HY: symmetric hybrid structure; PL: surface-perpendicular lamellae.

Figure 1*c* (FL). This morphology develops through the surface and substrate boundary conditions that demand the most energetically compatible block be expressed at each of the surfaces. These surface-parallel lamellae also optimize the interfacial energetics of the system by minimizing the amount of A/B interface while maintaining L_0 periodicity. By convention, if the same block, for example B, is found at each boundary, the copolymer is said to exhibit symmetric wetting. Alternately, BC films that express different blocks at each surface (6, 9, 10) are termed anti-symmetric [see Figure 1*c* (AFL)]. At equilibrium, symmetric film systems exhibit a series of stable films when $t = nL_0$ $(n = 1, 2, 3, 4)$, whereas anti-symmetric films exhibit a similar series of stable films when $t = (n + 1/2)L_0$.

Symmetric Boundary Energetics

The wetting and domain orientation a given BC exhibits depends upon the energetic/chemical nature of the surfaces that bound the film, i.e., the magnitude and type of interaction (attraction/repulsion) these boundaries have with each of the block species. Surface boundary energetics can be classified into two general categories: symmetric boundary conditions, in which the energetics imposed by each surface is identical, and asymmetric boundary conditions consisting of surfaces with different energetic qualities. Research considering symmetric boundary conditions is discussed below.

THEORY Several researchers have used theoretical techniques to examine the prospect of morphological shifts as a function of film thickness for the case of lamellar diblock films bounded by identical or neutral surfaces, i.e., symmetric surface energetics. This problem was first approached by Turner (11), who used a scaling analysis to determine the relative stability of symmetric versus antisymmetric lamellae as a function of film thickness and surface potential. Walton et al (12) extended this effort to first consider the case of lamellae oriented perpendicular to the film surfaces. This lateral arrangement [or perpendicular lamellae (PL) seen in Figure 1*c*] relaxes the entropic penalty imposed on chains when surface-parallel lamellae are constrained to film thicknesses incommensurate with integer multiples of L_0. Comparing this advantage with the surface and interfacial penalties associated with lamellar reorientation, these researchers (12) outlined the following trends. First, the domain orientation was film thickness dependent. In particular, PL gained stability when $t \neq nL_0$, and especially for $t < L_0$. Second, for neutral surface energetics, PL was stable for all film thicknesses. As will be seen, this neutrality condition in surface energetics has technological implications.

Using computational approaches, other researchers have verified and expanded upon the concept of thickness-dependent domain orientation under symmetric boundary conditions (13–17). Pickett & Balazs (13) employed two-dimensional self-consistent mean-field (SCF) calculations to test the neutrality condition by comparing the equilibrium morphologies found for both neutral and attractive boundaries. They found, like Walton et al, that segment-selective surfaces stabilized parallel morphologies and gave thickness-dependent domain orientation, whereas PL was stable for all film thicknesses in the case of neutral walls. Matsen (14) and Tang & Witten (15) also used SCF calculations to examine this system while additionally including the possibility of mixed-orientation or hybrid morphologies that exhibit both surface-parallel and surface-perpendicular components (see, for example, the AHY morphology of Figure 1*c*). While both studies concluded that these mixed forms were metastable only for the case of symmetric boundary conditions and volume symmetric BCs, stable hybrid morphologies were observed in Monte Carlo simulations of similar systems by Kikuchi & Binder (18) (and as discussed below in experiments).

Compositionally asymmetric BCs—those that form spheres, columns, or the gyroid structure in bulk—have garnered less theoretical attention regarding their thin film structure, primarily owing to the extra complexity they pose to analytical treatments and the increased computing power needed to capture the extra dimensionality inherent to these materials. Three studies that consider the cylindrical morphology under symmetric boundary conditions are notable. An analytical treatment of cylinder orientation in thin films, following the formalism of [Turner (11) and Walton et al (12)] was presented by Suh et al (19). This followed an earlier effort by Turner et al (20) who computed the surface affinities required to impose a cylinder to FL transition in thin films using a Ginzburg-Landau formalism. Most recently, using calculations based on so-called dynamic density functional theory, Huinink et al (21) have produced an extensive analysis of the cylinder morphology that outlines both domain orientation transitions and motif shifts as a function of surface energetics and film thickness. In particular, these researchers found that symmetric boundaries with a strong affinity for the cylinder-forming (minority) BC component could induce formation of FL, which seemingly captures the behavior observed in the experiments of Radzilowski et al (22), discussed below.

EXPERIMENTS Symmetric boundary conditions can be realized experimentally in free-standing films and when films are sandwiched between two substrates. Sandwiched systems were first used by Lambooy et al (23) and Koneripalli et al (24) to probe confinement/surface effects on BC film morphology. In these studies, hard boundaries (evaporated SiO_2 in the former case, glassy polymer in the latter) introduced to the system were strongly attractive to one of the copolymer blocks, a condition that greatly stabilizes surface parallel morphologies (12, 25). Accordingly, each team observed contraction and expansion of the lamellar period as the film thickness was ramped through values incommensurate with multiples of L_0, despite the entropic penalties inherent in these structures. While surface-perpendicular domain orientations in diblock films had been previously seen experimentally (26), Kellogg et al were first to induce lamellar reorientation through symmetric boundary conditions (27). Neutral boundary conditions for polystyrene-b-poly(methyl methacrylate) (PS-b-PMMA) were created by sandwiching this diblock between films of an equimolar random copolymer (RC) of styrene and MMA. This technique was later refined by Huang et al (28–30). Using a better estimate of the RC composition required for surface-energy neutrality in the PS/PMMA system as reported by Mansky and coworkers (31, 32), highly oriented PL (28, 29) and surface-perpendicular cylindrical domains (33) were observed. The innovative techniques these researchers used to localize the RC layers at the BC film surfaces are discussed below. A notable, more quantitative discussion concerning the design of neutral surfaces for a given BC system can be found in Reference 30.

Freestanding films offer the other experimental opportunity to observe symmetric boundary conditions in constrained BC systems. Although the inherent fragility of such films makes examination of these specimens formidable, Radzilowski and coworkers (22), through elegant sample preparation and cross-sectional TEM, tracked the thickness-dependent morphology of cylinder-forming PS-b-poly(butadiene) (PS-b-PB). They noted several trends. Thicker films showed surface-parallel layering, with surfaces composed of half-lamellar layers of PB, despite the entropic expense, as a means of producing the lower surface tension (and here, minority) component at the surface. Karim et al observed this near-surface structure previously via reflectometry (34). For the thinnest films ($t < L_0$), the consequence of these surface layers is found in an observed change in motif from surface-parallel PB cylinders to volume-asymmetric FL, i.e., thin PB layers sandwiching a PS core. Given the entropic expense associated with this latter morphology, this transition indicates the dominance of surface energetics on the system. As mentioned above, these observations likely find a theoretical framework through the recent calculations of Huinink and coworkers (21). It would be interesting to see these calculations expanded to include asymmetric energetic boundary conditions, discussed next, and thus used to describe substrate-supported films.

Asymmetric Boundary Conditions—Substrate-Supported Films

Although identical film surface boundaries simplify some analyses, many film systems involve asymmetric boundary conditions, i.e., where the block-segment/surface interactions, S_1 and S_2, are different in strength and/or sign. This is certainly the case for substrate-supported films that form the majority of technologically relevant and experimentally tractable specimens. In supported film systems, the surface energy of a given monomer can differ from its substrate interfacial energy by an order of magnitude (25). Furthermore, as discussed below, the presence of asymmetric boundary conditions can result in specific morphological trends—for example, the formation of hybrid morphologies that combine surface-parallel and surface-perpendicular components—that cannot be accounted for with symmetric surface energetics. Accordingly, the understanding of substrate-supported films requires analysis that explicitly includes boundary condition asymmetry.

THEORY Several published studies have employed boundary condition asymmetry in analytical and computational analyses of the morphology of volume-symmetric diblock copolymers. Walton et al (12) included this case in their analysis of lamellar domain orientation and defined conditions for the stability of the PL morphology analogous to that supplied for symmetric boundary conditions. Matson also considered this case using SCF methods (14). This latter work centered generally on film thicknesses greater than L_0. Mixed-orientation morphologies were included in the analysis, making possible predictions comparable to recent observations of such structures by Huang et al (35) in this thickness regime. However, Matson's conclusion was that these hybrid forms are metastable. In contrast, another recent study by Tang (36) considered the case of L_0-thick films under asymmetric boundary conditions. While that work did not consider thickness effects, it succeeded in demonstrating the stability of a particular hybrid morphology (AHY in Figure 1*c*), which had been observed in the PS/PMMA system via transmission electron microscopy (TEM) by Morkved & Jaeger (37) and inferred from reflectivity measurements by Russell et al (38). Tang's study also predicts a number of new morphologies that have yet to be observed.

We recently completed a comprehensive SCF analysis of the morphological behavior of compositionally symmetric, substrate-supported diblock films in the thickness regime $t \leq L_0$ (25). Predicted morphology trends were compared with concurrent experimental observations on a series of PS-b-poly(*n*-alkyl methacrylate) materials supported by silicon substrates. A discussion of our model and main results are presented next.

Two-dimensional SCF calculations were performed using the diblock thin film model of Pickett & Balazs (13). Compositionally symmetric diblock copolymers were modeled on a lattice as chains N statistical segments in length, each segment occupying one lattice space, with an A/B segmental repulsion quantified by the Flory interaction parameter χ_{AB}. Thin films were modeled by confining chains to a rectangular lattice, t units high (x) and $r > t$ units long (y). The substrate boundary (1) at $x = 0$ was energetically selective of the B segments with an interaction energy of S_1^B, whereas the (impenetrable) free surface boundary (2) at $x = t$ was selective of either the B segments (for most of the calculations performed) or the A segments (representing anti-symmetric boundary conditions), but always with an interaction energy (S_2) of lower magnitude than the substrate interaction, i.e., $|S_1| \geq |S_2|$. The conditions $S_1^A = -S_1^B$ and $S_2^A = -S_2^B$ were further imposed to simplify the calculation. Two-dimensional morphologies (analogous to film cross sections) were calculated from a variety of initial conditions (39), and phase diagrams were compiled from graphical and free energy data as described in Reference 25.

The stable morphologies calculated in our study, as summarized in Figure 1*c*, are presented in the general order (with decreasing film thickness) in which they occur and identified by abbreviations defined below. The images shown are cross-sectional views of the calculated film morphologies. The gradients depict the calculated density of B segments. Black indicates 100% B segments, white 100% A segments. In addition to the FL, AFL (anti-symmetric, surface-parallel lamellae) and PL morphologies discussed above, three other film structures are indicated. The HL, or half lamella, is the first ($n = 0$) stable morphology in the classical anti-symmetric thickness series and consists of a half layer of B segments next to the substrate capped by a half layer of A. The HY or hybrid morphology is a mixed-orientation structure consisting of a surface-parallel wetting layer of B at the substrate with surface-perpendicular protrusions approximately $L_0/2$ in width, extending through a half layer of A to the free surface. The HY structure resembles the mixed-orientation lamellar phases predicted by Kikuchi & Binder (18, 40) and by other later studies (14, 15, 36). In contrast to those studies, however, the HY structure is predicted to be a stable morphology for thicknesses below L_0. The AHY, or anti-symmetric hybrid morphology mentioned above with respect to the work of Tang (36), was also found. The relationship between these hybrid forms, HY and AHY, is discussed below.

Figure 2*b* shows these morphologies in the context of an example phase diagram. Here, the ordinate gives the reduced film thickness t/L_0. The abscissa $R = S_2^B/S_1^B$ gives a measure of the asymmetry of the surface energetics. The sign of R distinguishes symmetric (+) and anti-symmetric (−) wetting conditions. The

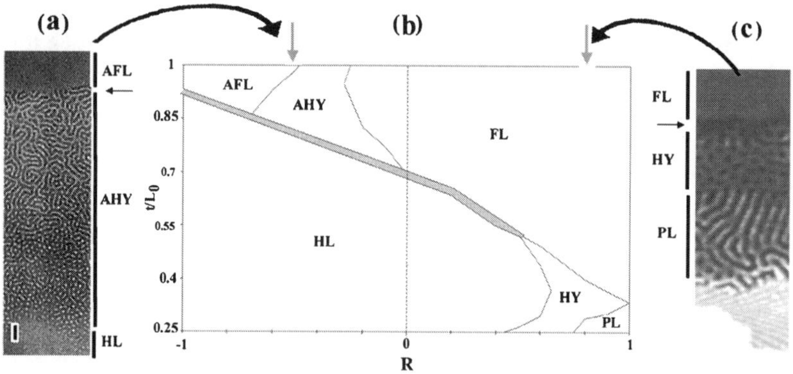

Figure 2 Comparison of predicted morphologies with experimental observations. (*a*) TEM micrograph of anti-symmetric wetting PS-b-PMMA thin film as observed by Morkved et al (reprinted with permission from 37, copyright 1997 EDP Sciences). This film section decreases in thickness from $3/2L_0$ to $1/2L_0$ moving from the top. PS domains appear dark. The morphologies that appear are labeled on the right using abbreviations from Figure 1*c*. The horizontal arrow indicates where a drop in film thickness occurs as described in the text. (*b*) Phase diagram of thin film morphologies calculated with the following parameters: $N = 200$, $S_1^B = -0.3$ kT, $\chi_{AB} = 0.1$. $R = S_2^B/S_1^B$. Phase fields are labeled with abbreviations from Figure 1*c*. The gray area marks where the HY morphology is metastable. Vertical gray arrows (*top*) indicate behavior of PS-b-PMMA (*left*) and PS-b-PLMA (*right*). (*c*) AFM phase micrograph of PS-b-PLMA droplet edge. The film thickness decreases from the top from L_0 to 0. Softer PLMA domains appear dark. Morphologies observed are indicated on the left and the horizontal arrow denotes a sharp drop in film thickness [sections (*b*) and (*c*) are reprinted with permission from 25, copyright 2000 Am. Chem. Soc.].

phase fields are labeled with abbreviations found in Figure 1*c* and designate conditions for which each morphology is predicted to be stable. The hatched area (along the FL/HL border) is a region of HY metastability.

The positive (symmetric wetting) values of R are considered first. For symmetric surface energetics ($R = 1$) a single morphological transition occurs from FL to PL, as other studies have predicted (12, 14, 41). With increasing energetic asymmetry ($0 < R < 1$), the PL phase field splits, giving way first to a region in which the HY structure gains stability, then to another new phase field, the HL morphology. Thus as the surface interactions become more asymmetric, morphologies that lack a horizontal plane of symmetry (i.e., HY and HL) are favored. On the whole, the system behavior can be divided into two regimes based upon the existence of a stable HL phase. For more asymmetric values of R, a stable HL phase intervenes in the transition from FL to the perpendicular structures (HY and PL). As R becomes larger, a direct transition from FL to HY (or PL) occurs.

As seen in Figure 2*b*, the HY phase field is typically situated between the surface parallel morphologies (FL or HL) and the PL phase. With both surface-parallel and surface-perpendicular elements, this structure may be seen as a compromise morphology that serves as an intermediate between fully parallel and fully perpendicular states. Moreover, the metastable existence of the HY morphology between the FL and HL phase fields suggests that such hybrids are possible intermediates between parallel-symmetric and anti-symmetric morphologies as well. This possibility is pursued further in the discussion of experimental observations (below). A more complete discussion of the effects of the molecular (N, χ_{AB}) and surface-energy (S_1^B) parameters on the morphological behavior of this system can be found in Reference 25.

Different behavior is noted for the anti-symmetric ($R < 0$) portion of the phase diagram. Here, no perpendicular phases are found in the sub-$L_0/2$ thickness region. Indeed, thinner films are all predicted to have the HL morphology, consistent with the classical view of anti-symmetric systems. Overall, three regimes of behavior are observed in our calculations. Generally, these trends are reflected in the morphology of the film near L_0, which is based on the degree of surface-energy asymmetry, i.e., the magnitude of R. This is consistent with the work of Tang (36) whose analysis focused on morphological behavior in this thickness regime. For nearly equal A and B affinities for the surface and substrate, respectively ($R \approx -1$), the AFL morphology is stable, as predicted previously (11, 12, 14). When R is negative but small, there exists a region of FL stability for $t > 0.5L_0$. In this case, the entropic drive toward L_0 periodicity overcomes the penalty of localizing B segments at the S_2 surface. For intermediate values of R, a window of stability appears for the AHY structure. Similarities exist between this morphology and the HY structure found for $R > 0$. First, the AHY appears, like the HY, to be dependent upon boundary condition asymmetry. Also, like the HY, the AHY morphology seems to be an energetic compromise but between surface-parallel states. While our calculations do not extend into the $t > L_0$ thickness regime, the analysis of Matsen (14) strongly indicates that the AHY phase field is bounded above by an anti-symmetric surface-parallel lamellar structure (AFL). Furthermore, the position of the AHY phase field between the AFL and FL indicates its status as intermediary between anti-symmetric and symmetric structures as R is changed. Indeed, the mixed composition of AHY at the free surface (2) surface provides a plausible transition between the B-wetting FL and the A-wetting AFL.

To elaborate further, the AHY may be seen as a fusion of two "half-period" films, each responding to local asymmetric surface energetics. The HL morphology, constituting the bottom half of the film, optimizes S_1^B energetics, as the B block wets the substrate. [This condition was implicit in previous calculations by Tang & Witten (15) who "quenched" this section of film into a static HL morphology.] In the top half of the film, the A segments are favored at both the free surface and at the HL substrate but with differing affinities, now determined by the differential surface energies and the attraction of A for itself. The consequence is the formation of the hybrid morphology, where A segments now form the contiguous surface-perpendicular features. Hence, the HL and HY structures combine to form the AHY morphology.

We see this deconstruction of the AHY structure as interesting in ways that transcend the particulars of anti-symmetric wetting systems. For when considered in this light, the behavior of such films is identical to the symmetric wetting systems, only shifted $0.5L_0$ in thickness. The latter shows a hybrid morphology near $0.5L_0$, whereas the former exhibits these structures near L_0. A conjugate trend is already known to govern the sequence of stable surface-parallel states for each case (6, 10, 42), i.e., symmetric systems exhibit a surface-parallel state at L_0, whereas anti-symmetric systems find these at $0.5L$ and $3/2L_0$. Thus $t < L_0$-thick films are unified under the nL_0 and $(n + 1/2)L_0$ paradigm widely used to describe the morphology of thicker films (6, 10, 42). Framing the question in this manner also suggests that the structure of thin lamellar films can be understood through a general half-period principle of morphological development. Indeed, this half-period principle seemingly applies to the structure of thicker lamellar films also and may well be a way of re-examining between period lateral morphologies as well as explaining hybrid morphologies in more complex copolymer systems as they are observed. Some of these possibilities are considered below in the discussion of experimental observations on substrate-supported systems.

EXPERIMENTS Before continuing, an important difference between real supported films and those considered in the majority of model systems discussed above must be distinguished. Theoretical analyses of thin films most often place the specimen between hard, impenetrable boundaries. Real films supported by simple substrates, however, have only one such boundary. The existence of a free surface (indeed, a form of boundary asymmetry in itself) has important consequences on the film morphology. This can be illustrated by considering a substrate-supported lamellar BC film that exhibits symmetric wetting. Recall that at equilibrium, this system exhibits a series of stable films for $t = nL_0$ $(n = 1, 2, 3, 4 \ldots)$. Now, consider the case of an as-cast substrate-supported film of thickness t_i, such that $nL_0 < t_i < (n + 1)L_0$ with $(n \geq 1)$, i.e., the film thickness is not commensurate with the equilibrium period of the copolymer. Upon annealing, the morphology of this system develops such that this thickness mismatch with the copolymer periodicity is resolved. This is accomplished by a bifurcation into film regions having two distinct thicknesses, which represent the closest conditions of stability, i.e., nL_0 and $(n+1)L_0$. [Excellent more quantitative illustrations of this phenomenon can be found in the analyses of Matsen (14) and Huinink et al (21).] Accordingly, the free surface forms plateaus, of $(n + 1)L_0$ height on a foundation of film that is nL_0 thick (9, 10). The film area fractions of these quantized domains mirror the degree of mismatch so that the amount of thicker film is almost exactly equal to $t_i/(\mathrm{n} + 1)L_0$, whereas the thinner film fraction is $1-t_i/(n + 1)L_0$. (9) In common nomenclature, when the area of thicker film is less than one half, the higher regions are isolated from each other and are thus called islands. Likewise, isolated lower regions (when the area of thinner film is less than one half) are called holes. In film systems that exhibit a thickness gradient, e.g., the edges of droplets, this effect manifests itself as a series of step-like L_0-high terraces that accommodate the change in

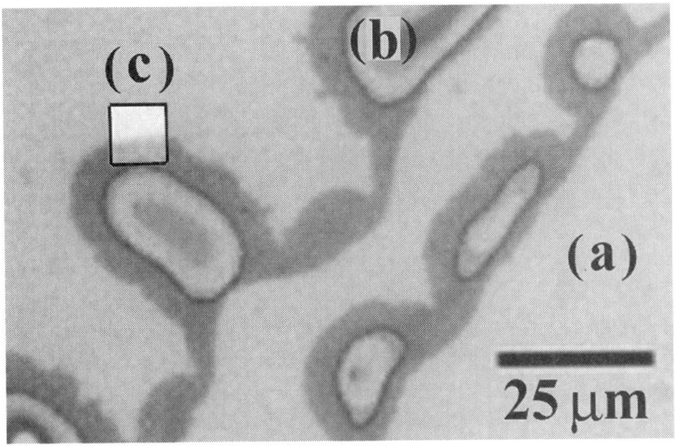

Figure 3 Optical micrograph of lamellar PS-b-PnBMA droplets exhibiting FL morphology and the terracing phenomenon. (*a*) Silicon substrate, (*b*) terraced droplet. Bands of contrast indicate specific optical interference conditions caused by quantized L_0-sized (43 nm) jumps in film thickness. (*c*) Example area (light box) for AFM analysis. An exposed region of bare substrate allows film thickness determination.

thickness. An example of a terraced BC droplet can be seen in Figure 3. In this optical micrograph of a compositionally symmetric PS-b-poly(butyl methacrylate) (PS-b-PBMA) system, the discrete jumps in film thickness are observable because of an optical interference condition with the substrate that selects for the reflection of a specific wavelength of light. Terracing has been most studied for lamellar systems but has been seen for off-symmetric BC compositions as well (22, 43).

Indeed, it was observations of this terracing phenomenon with optical microscopy that led Kämpf and coworkers (7) to first conclude that BC films organize into discrete layers and hypothesize the FL morphology. The verification of this hypothesis, however, came through a series of groundbreaking experiments by Russell and coworkers (6, 38, 42, 44, 45). These researchers probed thin lamellar films with X-ray and neutron reflectivity, demonstrating the ABBA-patterned and L_0-periodic internal structure that characterizes the FL morphology. This same group also established that even when the BC was well above its bulk order/disorder transition temperature, surface effects could induce FL layering (42, 45, 46), as predicted theoretically by Fredrickson (8).

Domain reorientation (formation of PL) owing to thickness constraints were first observed in supported films by Henkee et al (26) through TEM analysis of BC droplet edges. Another study conducted by Carvalho et al (47) showed that terrace edges in ordered BC droplets exhibited surface-perpendicular domains. These edge regions are necessarily incommensurate with L_0, thus demonstrating that intermediate film thicknesses induce lamellar reorientation (12, 14, 25). The authors hypothesized that these perpendicular domains are PL that extend from the film surface to the substrate. However, in light of the theoretical analysis of supported films presented above, an explanation invoking a topmost-layer HY (or AHY) morphology may be more appropriate.

Mixed-orientation morphologies have also been observed in supported films. Fasolka et al (48) observed the HY morphology in symmetric wetting systems. Morkved & Jaeger (37) first noted the AHY structure [inferred by Russell et al (38)] in anti-symmetric BC films. Thicker ($t > L_0$) films supported by an energetically neutral substrate have also exhibited mixed-orientation lamellae as shown by Huang and coworkers (35).

We have recently completed an extensive empirical study of BC film morphology focusing on the $t < L_0$ thickness regime (25). This work employed a series of lamellar PS-b-poly(*n*-alkyl methacrylate) copolymers supported by silicon substrates with native oxide surfaces. For all but PS-b-PMMA, this homologous series exhibits symmetric wetting (of the methacrylate block) but also offers a range of free-surface and substrate interactions. Hence these systems are comparable to our model calculations described above. A useful AFM (atomic force microscopy) technique facilitated the study of film structure at the edges of ordered BC microdroplets. (In Figure 3, the light box shows a typical area of analysis). These droplet edges exhibit a profile of film thickness from 0 (substrate) to L_0. Simultaneous AFM height and phase imaging of these areas allowed for the concurrent measurement of film thickness (height, when an area of bare substrate is included in the image) and local morphology (phase, which is modulated by elasticity differences between the PS and polymethacrylate domains). Through this technique, accurate correlations between domain orientation and film thickness can be made. Figure 2*c* (*right*) shows a typical AFM phase micrograph collected from an annealed droplet of PS-b-poly(lauryl methacrylate) (PS-b-PLMA). In this image, darker areas are the softer PLMA domains; the brighter areas are PS. The lightest region at the bottom of the image is the substrate. The film thickness ranges from L_0 to 0 going from top to bottom, as noted to the right of the image. Immediately apparent is the thickness-dependent morphological behavior. At L_0, the lack of phase contrast denotes the homogeneous surface layer of the FL morphology. This is followed by a sharp drop in film thickness (arrow), constituting a terrace edge analogous to those discussed above, but with a magnitude of approximately $L_0/2$. All the specimens in the series we examined had this feature. Below the terrace edge, a shelf (approximately $L_0/2$ thick) of film exhibiting a hybrid morphology (labeled HY) is shown. This structure consists of a thin wetting layer of PLMA at the substrate with surface-perpendicular cylindrical (seen as round in plan-view) protrusions extending through the PS layer to the top surface. This is thought to be the three-dimensional equivalent of the two-dimensional hybrid (HY) structure seen in our calculations. (A more detailed discussion of this can be found in Reference 25.) The thinnest regions of film exhibit the PL morphology (Figure 2*c*).

One of the notable things about the behavior of PS-b-PLMA films is that an area of HL is not observed in the $L_0/2$ thickness regime. This is in contrast to e.g., PS-b-PBMA, which shows all of the same morphologies, FL, HY, PL, but with an additional region of HL between the FL and HY regimes. Although both materials exhibit symmetric wetting of the methacrylate block, the sequence of morphologies as film thickness decreases is different. When morphological data from all of the specimens were considered, they could be classed into two groups, distinguished by the existence of a stable HL phase. Entirely similar to SCF predictions for $R > 0$ in Figure 2*b*, methacrylates with more asymmetric surface/substrate interactions (e.g., ethyl, propyl, and butyl methacrylate) exhibited FL, HL, HY, and PL; those with more symmetric interactions (hexyl, octyl, and lauryl methacrylate) show only FL, HY, and PL (noted by the *right* gray arrow in Figure 2*b*).

Although an equally detailed experimental study of anti-symmetric film morphology does not currently exist, Morkved & Jaeger (37) have examined one system, namely bulk-lamellar PS-b-PMMA. They used TEM to probe films of this material deposited on electron-transparent silicon nitride windows known to be attractive to PMMA. The comparatively lower surface tension of PS favors this block at the free surface. Figure 2*a* shows a micrograph from their study. The section of film shown here decreases in thickness from the top of the image ($t \approx 3/2L_0$) to the bottom ($t \approx L_0/2$). Again, thickness-dependent morphological changes are apparent. The dark band of film with no lateral contrast at the top of the image is the AFL structure. There is a drop in film thickness (indicated with an arrow) of approximately $L_0/2$ to a shelf of film L_0 thick, which exhibits a morphology with a surface-perpendicular component. This is the three-dimensional analogue to the AHY discussed above. Finally, in the thinnest regions at the bottom of the image, contrast disappears, indicating a second surface-parallel form, now HL. This sequence mirrors a cascade of morphologies predicted for anti-symmetric systems, marked by the *left* gray arrow in Figure 2*b*. Note also, as discussed above, how this behavior is similar to that of the symmetric PS-b-PLMA films, only shifted to thicker films by approximately $L_0/2$. This supports the half-period principle of morphological behavior stated above.

SURFACE MORPHOLOGY AND CHEMICAL MOSAICS The conventional view of multicomponent polymer films posits that such systems express a chemically homogeneous surface consisting of the lowest surface tension (γ) constituent. For the most part, the behavior of BC films conforms to this postulate, as exemplified by, e.g., the FL and AFL morphologies. Likewise, systems with an inherent lateral structure, i.e., films of spheres and columns, most often exhibit homogeneous film surfaces even when surface homogeneity poses an entropic penalty, as in cases where the lower surface energy block is in the minority.

However, surface-perpendicular morphologies, such as the HY and PL structures, offer the possibility of film systems with heterogeneous surfaces—nanometer scale chemical mosaics—that could be harnessed as templates in a variety of applications. Our recent SCF analysis of these structures indicates that surface morphology is governed not only by differential surface energetics, but also by the interfacial (χ_{AB}) penalty exacted by so-called capping layers of the low-γ

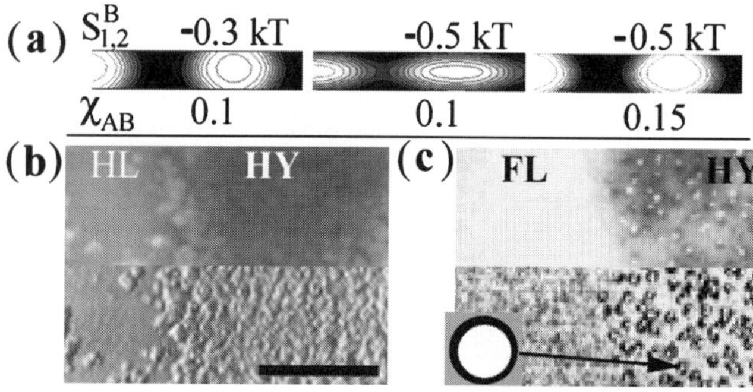

Figure 4 Surface heterogeneity and thin film morphology (49). (*a*) SCF calculations of PL morphology using various surface and interfacial energetics. Static parameters: $N = 200$, $R = 0.9$, $t = L_0/4$. *Left* and *right* bars, heterogeneous surfaces; *middle* bar, homogeneous surface. (*b*) AFM micrograph showing PS-b-PPMA ($\Delta\gamma = 7.4$ dynes/cm, $\chi_{AB} = 0.01$) thin film exhibiting HL and HY morphology, treated with cyanoethylacrylate (CEA) glue vapor. *Upper*: height; *lower*: phase. Polymerized CEA completely covers area with HY indicating a homogeneous PPMA capping layer. (*c*) AFM of PS-b-PLMA ($\Delta\gamma = 7.9$ dynes/cm, $\chi_{AB} = 0.083$) thin film also treated with CEA vapor. *Upper*: height; *lower*: phase. CEA nodules selectively decorate PLMA HY domains indicating a heterogeneous surface. Schematic in phase image shows hard CEA nodule (white) centered on softer PLMA domain. Scale bar indicates 250 nm for (*b*) and (*c*).

component (49). This interplay between surface energy and χ_{AB} is illustrated in Figure 4*a*, which shows the PL morphology calculated under different energetic conditions. Increasing the attraction of B-segments to the surfaces drives this system toward surface homogeneity (Figure 4*a*, *left* and *middle*). However, through a concurrent increase in χ_{AB} (Figure 4*a*, *left* and *middle*), surface heterogeneity is recaptured through the increased A/B interfacial penalty associated with capping.

Experimental observations of such distinctions are difficult because capping layers can be sufficiently thin [only ≈0.5 nm by some reports (50)] to elude reliable detection via X-ray photoelectron spectroscopy or other means. Recently, we adopted a surface-staining technique, developed by forensic scientists to enhance latent fingerprint traces (51, 52), that shows promise in illuminating surface heterogeneity. We treated laterally structured (HY forming) PS-b-PnMA films with cyanoethylacrylate (super glue) monomer vapor (49). The glue monomer preferentially polymerizes (anionically, via ambient H_2O initiation) over the more chemically similar methacrylate domains, forming hard nodules that can be imaged via AFM. These nodules formed a continuous coating over HY morphologies of some of the BC films, as shown for the example of PS-b-poly(propyl methacrylate), PS-b-PPMA, in Figure 4*b*, indicating a homogeneous methacrylate surface. For PS-b-PLMA films, however, the nodules were located specifically over methacrylate domains, as shown in Figure 4*c*, strongly indicating a surface mosaic. Interestingly, the surface energy difference ($\Delta\gamma$), i.e., the driving force toward a homogeneous surface, is larger for the PS-b-PLMA system than for PS-b-PPMA, as shown in Figure 4. However, PS-b-PLMA has a much higher value of χ_{AB} (see Figure 4) than PS-b-PPMA, making the penalty of a PLMA capping layer comparatively higher, thus resulting in surface heterogeneity for this system. Therefore, it seems that BC chemical mosaic surfaces are a possibility in systems with small $\Delta\gamma$ values and a large χ_{AB}. This suggests, for example, that the PS-b-PMMA system, which fulfills both of these criteria, may also exhibit a heterogeneous surface in its AHY morphology, as previous reports have implied (38, 53).

Developing Trends and Open Arenas

In this section, developing trends in BC thin film research are described, and generally, limited to basic research, i.e., work that more directly contributes to an understanding of the physics of substrate-supported BC film systems in areas not discussed above. Work that takes the field of BC films into new directions is emphasized. In addition, some areas in which little or no research has been completed are outlined.

OFF-SYMMETRIC COMPOSITIONS Most of the published research on BC films concerns bulk-lamellar systems. This is due, in part, to the relative simplicity of the system, in particular the dimensional reduction available to theoretical treatments. However, the inherent three-dimensional nature of the off-symmetric motifs—gyroid, cylinders and spheres—provides extra degrees of freedom for consideration and, further, make these forms interesting for certain technological applications, e.g., lithographic masks (54, 55). This is not to say that the extensive research effort invested into lamellar systems has no bearing on these systems. Rather, many of the tenets outlined above for compositionally symmetric BC films (e.g., a general surface-parallel domain orientation, the effect of the neutrality condition, etc) can and have been applied to systems of off-symmetric composition. However, some aspects of these thin film systems, in particular the consequences of their inherent interfacial curvature, probably cannot be fully understood without thorough experimentation and direct theoretical focus.

Of the other BC motifs, the cylindrical morphology has received the most attention. Basic experimental studies on confinement effects and domain orientation in substrate-supported cylinder-forming diblock films were first performed by Henkee et al (26) using TEM. Since then, studies have been published by Liu et al (56), Karim et al (34), and Harrison et al (43, 57, 58) among others (59–61). Of these, the work of Harrision et al (43, 57) is particularly noteworthy, in part because of the elegant imaging technique they used for studying these systems. A layer-by-layer scanning electron microscopy (SEM) method gave three-dimensional morphological information superior to the surface-localized imaging of AFM used in References 56, 59, 60, and TEM images (26), which become confusing if more than a monolayer of structure is present. By alternating a nonselective etching process with low-voltage SEM it is possible to track morphological behavior through the film from the free surface, achieving domain contrast through conventional OsO_4 staining. This method revealed detailed images of domain orientation, terrace edge defects, and novel morphologies, including hybrid forms. As in earlier studies (34), Harrison et al noted that for cylinder systems in which the minority component has a lower surface energy, the near-surface regime is composed of a planar wetting layer. This surface effect was also seen (see above) in freestanding films (22).

Theoretical descriptions of cylinder-forming films are sparse. This is unfortunate, because an adequate theoretical framework would help organize the various, sometimes conflicting, observations noted in the literature. Of the few efforts completed (19, 20), the work of Huinink et al (21), discussed above, stands apart for its ability to depict complex and observed phenomena in this system (22). However, as stated before, the symmetric boundary conditions used by these researchers, while an excellent foundation, are inadequate for describing substrate-supported films. An expanded theory might address some basic concerns regarding films of bulk-cylindrical diblocks. Liu and associates (56), for example, have noted that the persistence of surface-parallel layering in cylindrical systems is quite limited, compared with that in films of lamellar and spherical systems, which are known to maintain such layering for film thicknesses many times L_0.

Supported bulk-spherical BC film systems, despite their promise as nanoscopic lithographic masks, have been the subject of very few publications. To date, there does not seem to be a theoretical treatment of spherical thin film morphology. Experimental studies, for the most part, are geared toward simple observations of structure (26), imaging techniques (57), or device applications (55). Two recent efforts by Yokoyama and coworkers (62, 63), however, include detailed physical insight into the spherical phase in supported thin films. Using secondary mass ion spectrometry (SIMS) and AFM, they have probed both the through-film and lateral structure of this system under a variety of conditions, examining the effects of molecular weight, film thickness, and annealing time/temperature on film structure and surface topography (island and hole formation). An interested theorist would do well to address these findings.

Of the gyroid phase in thin films, almost nothing has been published. One experimental piece by Chan et al (64) deals with this film system, but the work is primarily directed toward technological applications. The gyroid morphology offers a unique symmetry and interconnected structure, and it would be interesting to know the conditions under which BCs exhibiting such a morphology maintain this structure as thin films. Given its (compositional) position between lamellae and cylinders in the classical BC phase diagram, the opportunity to observe surface-induced motif shifts from gyroid to these bordering forms seems likely.

LIQUID-CRYSTALLINE AND SEMI-CRYSTALLINE BC FILMS While we have hitherto considered amorphous materials, the inclusion of non-amorphous components into block copolymer molecules is not uncommon, and an understanding of their bulk behavior is fairly well established (see citations in References 65 and 66).

Research into liquid crystalline (LC) and semi-crystalline BC thin films, however, has only very recently begun.

One promise of BCs with an LC component (most systems still have one amorphous block) is the ability to alter and control morphology (of both LC and BC) through an interplay between the L_0-periodic BC microphase separation and the finer periodicity of LC ordering. Thin film geometries add surface energy considerations and confinement effects, now on two levels of periodicity, to this equation. The first observations of such complexity were by Wong and coworkers (65) who compared films of PS-b-LC (smectic) diblocks with symmetric and off-symmetric compositions. The symmetric material was found to exhibit a PL structure with smectic layering parallel to the surfaces in the LC domains. The asymmetric system was found to form FL with surface-perpendicular smectic layers. A later work by Sentenac et al (67) investigated the bulk and thin film behavior of a compositionally asymmetric PS-b-LC diblock with LC as the minority component. The films showed a novel morphology that exhibited both FL and a surface-parallel smectic phase. Finally, Hammond and coworkers (68) have recently studied thin films of volume-symmetric diblocks incorporating a side chain-LC block, in the $t < 2L_0$ thickness regime. They found a film structure exhibiting both a mixed-orientation morphology (an effect, it seems, of surface energy asymmetry) and an interesting surface topography that included terracing on both the L_0 and smectic length scales. Theoretical treatments (currently there are none) that explicitly include thin film effects on each level of order could greatly enhance our understanding of such systems.

Like LC-containing systems, BCs with a semi-crystalline component also offer an additional length scale of competing order manifest in the periodicity of the crystalline phase. In addition, the morphology of these systems is a strong function of the thermal history imparted through processing (69). The final microstructure these materials adopt depends upon whether microphase separation or crystallization occurs first. A few studies have considered this competition in BC thin films (66, 70–72). For example, in studying PS-b-poly(ethylene) (PS-b-PE), De Rosa et al (72) observed a variety of interesting and useful path-dependent structures, including surface-perpendicular crystalline PE cylinders and highly aligned PL. A similar effect was noted by these researchers (66) for triblock copolymers with PE end-blocks and an amorphous poly(ethylene)/poly(propylene) mid-block. This work and reorientation trends observed by Reiter et al (70) are described in more detail below.

BC BLENDS AND BC/HOMOPOLYMER BLENDS Blending block copolymers with other BCs or homopolymers can alter the morphology of the system in interesting ways. It is well known, for example, that addition of homopolymer A to an A-b-B diblock can cause a motif shift in bulk systems because the homopolymer associates with the chemically similar block (73). BC/BC blends have also shown intriguing microphase and dual-microphase behavior in bulk (50, 74). Furthermore, recent predictions by Leibler et al (75) propose that diblock/triblock blends could result in non-centro-symmetric morphologies that would have a variety of interesting physical and optical properties.

Thin film diblock/diblock blends were first explored by Mayes et al (76), who studied bimodal mixtures of long- and short-volume-symmetric PS-b-PMMA materials supported by silicon substrates. Using deuterated short chains and neutron reflectivity, these researchers found that for a range of compositions, the smaller diblocks localized at the PS/PMMA interfaces of larger FL structures. This is interesting because it opens the possibility of creating structures with localized functionality within the larger BC morphology. This study additionally demonstrated that the periodicity of such bimodal blends was the same as a unimodal diblock with an equivalent number-averaged molecular weight, as seen in bulk studies (74). Self-consistent field calculations have been used to model similar systems by Shi & Noolandi (77) and others (78), with similar results. Koneripalli et al (78) additionally examined supported diblock/diblock blend films of compositionally asymmetric PS-b-poly(vinyl pyridine) (PS-b-PVP) materials. These researchers found that if the final blend composition was volume-symmetric, the films formed the FL morphology even if the individual diblock constituents exhibited other motifs. Their study included concurrent SCF calculations that showed qualitative agreement with their experimental work.

A few studies of diblock/homopolymer blend thin film behavior are of note. Russell and coworkers (79) examined the PS-b-PMMA/PMMA and PS-b-PMMA/PS blend systems. Using deuterated homopolymers and neutron reflectometry, they probed the effect of homopolymer chain length on its ultimate distribution within the corresponding copolymer domains. When the homopolymer molecular weight was comparable to that of the corresponding block of the copolymer, the chains were accommodated in the center of the block domain (i.e., between the block "brushes"), leading to a period expansion. In contrast, shorter homopolymers were distributed more evenly throughout the block domain, resulting in little or no expansion. Qualitatively similar behavior was noted from SCF calculations of diblock/homopolymer blends by Shull and coworkers (80, 81). More recently, Green and coworkers (82, 83) reached similar conclusions employing AFM to study homopolymer N-dependence on the period (inferred through terrace heights) of PS-b-PMMA/PS and PS-b-PMMA/tetramethylbisphenol A polycarbonate (TMPC) blends, where TMPC exhibits favorable interactions with PS.

Each of the above blend studies focuses on films that are much thicker than L_0. An interesting route for future studies would be to consider film thicknesses that conflict with one or more of the natural length scales in the blend system. In the case of long/short diblock blends, for example, interesting behavior may develop in films when $t < L_0$ for the longer system but is still thick enough to accommodate the shorter component. Analogous work could be completed for diblock/homopolymer blends. Indeed, for very high–molecular weight homopolymers blended into diblocks, early studies made the surprising conclusion that macrophase separation is inhibited by confinement (84). This hypothesis has yet to be fully explored.

TRIBLOCK COPOLYMERS Adding a third block to the copolymer sequence brings added complexity to the thin film picture, especially in the case of A-B-C (versus A-B-A) forms. Indeed, the bulk state phase diagrams for amorphous A-B-C materials is only partially mapped to date, with reports of new morphologies appearing nearly every year (2, 85). While published studies of A-B-A triblock films (86–90) report qualitatively similar behavior to that described above for diblocks, the many specific morphological details that arise out of the inherent combinatorial complexity of A-B-C systems (91–94) make these reports hard to organize into a succinct narrative. This is especially true because an extensive theoretical framework for these systems does not exist. Only one study by Pickett & Balazs (92) considers the system from a theoretical standpoint. These researchers outline the energetics of lamellar domain orientation in A-B-C triblock films using SCF calculations. They find, for example, that if the film surfaces repulse blocks A and C, but are neutral or attractive to the central B block, the PL morphology is stabilized. These findings might be useful in explaining some recent observations, e.g., those of Stocker (91, 93) who reports PL in PS-b-PB-b-PMMA films.

A practical difficulty arises in imaging thin film surfaces of A-B-C triblock materials because the surface chemistry does not necessarily reflect the true system composition, which can make the species identification of morphological domains a challenge. To overcome this obstacle, Elbs and coworkers (94) describe a novel sample preparation technique, applied to PS-b-PVP-b-poly(tert butyl methacrylate) films, in which specimens were briefly exposed to solvent vapors carefully chosen to selectively swell a given block component. The surface reorganization that develops is apparently permanent if drying is accomplished quickly, thereby highlighting the locations of specific block domains under subsequent AFM analysis.

A volume-imaging AFM technique similar to the SEM technique of Harrison et al (57) was demonstrated on a PS-PB-PS film by Konrad and coworkers (89). Using alternating steps of AFM analysis and nonselective plasma etching, film morphology was tracked from the surface through the film, exploiting elasticity differences between PS and PB in phase imaging. A generic A-B-C system would most likely prove more problematic for this method. However, a combination of this technique and the solvent vapor treatment reported by Elbs et al could be useful for these more complex systems.

APPLICATIONS

The possible technological application of block copolymer thin films has been widely recognized. The length scale of microphase separation, 10–100 nm, is particularly interesting to technologists eager to develop a new generation of submicron scale electronic and optical devices. Such schemes would harness the spontaneous self-assembly of these materials to create nanoscopic device components or templates with which such components could be manufactured. Indeed, by tuning molecular parameters (N, composition, species), microphase domains with a variety of motifs, chemistries, and tailored size and periodicity might be created. Furthermore, recent developments in block copolymer synthesis, such as atom transfer radical polymerization (95), have greatly widened the spectrum of available copolymer materials or decreased their cost of manufacture.

Masks and Membranes

A prime example of a BC thin film application is their demonstrated utility as lithographic masks. In this scheme, ordered (and oriented) layers of microphase domains, deposited onto a suitable substrate, are selectively etched, via a plasma, ozone, or wet chemical means, to remove one of the block components. Holes in the resulting mask can then be used to transfer the BC motif pattern, via other etching or deposition steps, onto the substrate. This was demonstrated first by Mansky et al (96), employing a monolayer of hexagonally ordered diblock spheres. Indeed, the density of features that can be imparted to the substrate in this manner is staggering. Park et al (55), for example, reported 10^{11} etched-features/cm^2 in their refinement of this technique, also accomplished with thin films of sphere-forming diblocks. This system has since been combined with other microfabrication techniques to create equally dense arrays of metal dots (97) and GaAs nanocrystals (98). This same idea of selective etching of BC film domains has also shown potential for the formation of nanoporous membranes, as demonstrated recently by Liu and coworkers (99). In such an application cross-linking of the residual (unetched) matrix can make it stable enough for use, as shown by Nardin et al (90), who produced stable, freestanding, L_0-thick (10 nm) membranes with areas in excess of 1 mm^2 from A-B-A triblocks.

BC/Inorganic Composites

The idea of harnessing of BC morphology as a means of creating ultra-fine ordered dispersions of inorganic particles is well established (100). Several research initiatives have recently demonstrated that BCs can be used to create and/or distribute nanometer-sized, metallic (101, 102), magnetic (100, 103, 104), and ceramic (64, 105–109) structures. A plethora of devices, such as terabit-level memory storage media, become possible if the organization of the diblock domains that house these particles can be adequately controlled.

Three general strategies for patterning nanoscopic particulates in BC thin films have been explored: Surface decoration techniques involve the preferential segregation of deposited material to favored domains on a pre-oriented BC film surface (53, 110). In solution-blended composites, bare or surface-activated nano-particles are combined with BC in solution. During film processing, these particles sequester into preferred block domains (111, 112). A theoretical treatment of this phenomenon was published recently by Balazs and coworkers (113). Finally, so-called nanoreactors use inorganic precursors, such as metal salts, engineered to preferentially associate with one component of a BC. This can be accomplished by actually incorporating the precursor moieties as pendant groups on one of the block species in the system (100, 101), but BC/precursor blends have also been demonstrated to work (114–116). After films are cast and microphase separation occurs, the precursor molecules, now confined to one of the patterned domains, are reduced (or otherwise transformed) in situ to form particles. Later, the polymer may be removed, leaving a nano-patterned inorganic phase on the substrate (64, 115). Figure 5 illustrates a recent result from Leppert and coworkers, who employed sphere-forming PS-b-P4VP films to create periodic arrays of GaN nanoclusters (117).

Long-Range Order, Domain Orientation, and Designed Patterns

The above machinations for incorporating BC films into nanoscopic devices show promise. However, for many applications, technical challenges to these schemes are apparent. First and foremost is the issue of long-range order. Short-range ordering of microphase-separated BC films is excellent (as demonstrated in many published micrographs), but these films also typically exhibit the rich array of defects and grain structure seen in bulk systems. If BC films are to be used as templates in addressable memory applications, one of two conditions must be met: either long-range order needs to be greatly improved (indeed, perfected!) or more robust addressing schemes must be engineered. Furthermore, although the feature size and density offered by the first lithographic masks employing a monolayer of diblock spheres are impressive (55, 118), the aspect ratio of sharp features prepared

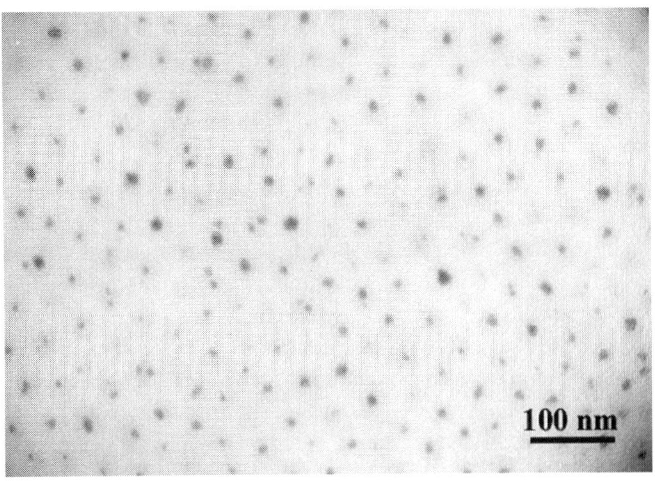

Figure 5 TEM image of film of PS-b-Poly(4-vinyl pyridine) doped with gallane-trimethylamine and processed to form a nanoparticle array. GaN nanoparticles are sequestered to spherical poly(4-vinyl pyridine) domains of the microphase-separated copolymer. Micrograph courtesy of Leppert (117).

by such ultra-thin masks is limited to about 1. If deeper features are to be created, the use of thicker films may be necessary, as might be accomplished, for example, with surface-perpendicular cylinder formations. Thus techniques to induce and control BC domain orientation in substrate-supported films remain important. Finally, the types of patterns natural BC films offer remain limited to the spatial motifs (spheres, columns etc) determined by thermodynamics. Widening the scope of self-assembly in film systems to include designed patterns would increase the usefulness of BC films. Several research efforts have explored strategies for surmounting one or more of these challenges, as discussed below.

ENGINEERING SURFACE EFFECTS The orientation of domains in block copolymer films is highly dependent upon the surface energy boundary conditions imposed on the system. In particular, surface fields that are neutral with respect to each of the blocks induce the formation of surface-perpendicular microphase domains (12). This effect has been exploited by Huang et al to impart a surface-perpendicular orientation in both lamellar and cylindrical PS-b-PMMA systems (28–30). Neutral surfaces were fabricated from PS/PMMA random copolymers (RCs) (27) of carefully tuned composition (30–32), localized at the film surfaces through the following means. At the substrate, RC chains of appropriate composition were end-grafted to the surface, forming a dense brush that could not diffuse into the cast diblock film. Additional RC chains of the same composition were end-functionalized with short poly(tetrafluoroethylene) (PTFE) segments. When incorporated into the diblock film, these RCs self-assembled at the free surface owing to their low surface energy chain ends, thereby creating a neutrality condition. Figure 6*a* shows a cylinder-forming DBC film specimen processed in this manner (33). Well-oriented cylindrical domains on a roughly hexagonal lattice are seen in this AFM micrograph. Demonstrations of the usefulness of these oriented films as lithographic masks and templates were recently published (33). Energetically neutral surfaces that induce domain orientations have also been reported by Peters et al (119), who chemically modified octadecyltrichlorosilane self-assembled monolayers with X-rays, thus demonstrating that the wetting properties of these layers can be tuned via the X-ray dosage.

In an effort to understand and control grain-boundary defect formation, movement, and elimination in diblock films with in-plane periodicity, Harrison et al (57, 58) studied such defects in specimens of cylinder- and sphere-forming systems, using the layer-by-layer SEM method described above. To examine the effect of surface interactions on defects and the degree of order, monolayer PS-b-PB films of surface-parallel cylinders (PB) were compared for two different substrate conditions: a bare silicon substrate and a substrate modified with an end-grafted PS brush (58). They found a lower defect density and greater persistence length for films with the brush-modified substrate, which they attributed to the reduced interaction strength between PB domains and the substrate. Strong substrate interactions can immobilize or pin defects, leaving them unable to participate in the processes by which these features collect and are eliminated.

(a)

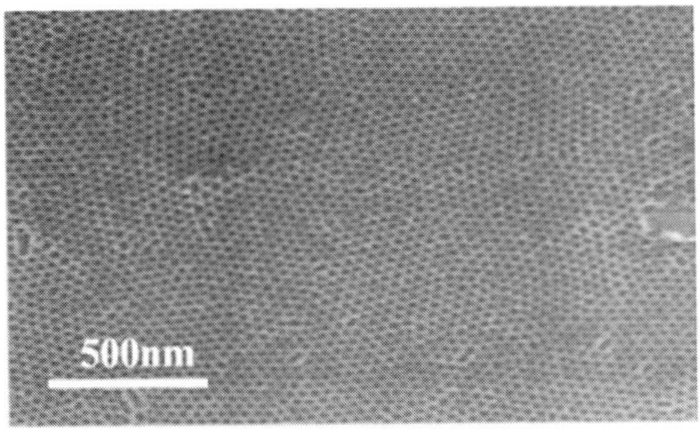

(b)

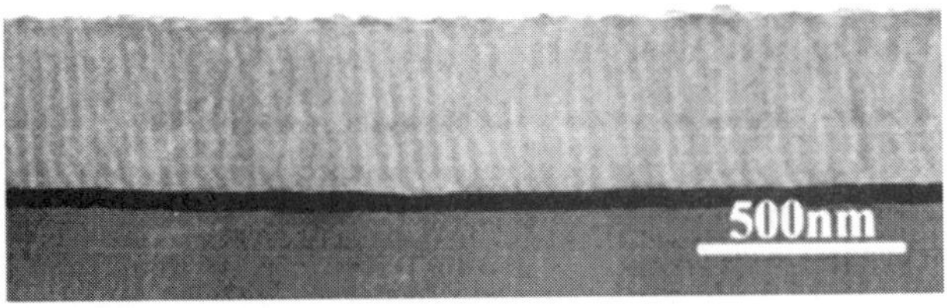

Figure 6 Oriented cylinders in diblock films. (*a*) Plan-view FESEM micrograph of PS-b-PMMA cylinders oriented perpendicular to the substrate plane via neutral surface energetics (28, 33). In this image, the PMMA phase has been preferentially etched away. Note the in-plane grain structure. (*b*) Cross-sectional TEM micrograph of PS-b-PMMA cylinders oriented by application of a through-film electric field (33). The film lies atop the dark band, an Au electrode layer, and shows surface-perpendicular cylinders that bridge the free and substrate surfaces. (reproduced with permission from 33, copyright 2000 Wiley & Sons, Inc).

APPLICATION OF EXTERNAL FIELDS Following bulk studies by Amundson and coworkers (120, 121), Morkved et al (122) examined the effect of an applied electric field on BC thin film morphology. Microfabricated electrodes were applied the E-field in the plane of the film during annealing. The resulting morphology exhibited in-plane cylinders oriented perpendicular to the electrode faces, i.e., aligned parallel to the E-field lines. Hence, as illustrated by Thurn-Albrecht et al (33, 123), application of the field perpendicular to the film plane induces cylinders into a surface-perpendicular morphology. A micrograph from their study is shown in Figure 6*b*. These researchers (123) have analyzed this situation theoretically as well and propose that the mechanism for alignment is found in the differential polarizability between the block components. Through these calculations, supported qualitatively by experiments by the same group (123), predictions regarding the critical field strength required to overcome the interfacial barriers to cylinder reorientation can be made. One advantage the E-field technique offers to microfabrication (in particular, as discussed above, for lithographic masks) is the thickness of films (up to 1 μm) that can be oriented (33). Indeed, very recently, researchers (124) have selectively etched E-field oriented films and deposited ferromagnetic cobalt material into the resulting cylindrical holes—a step toward block copolymer-based fabrication of terabit density memory storage media.

Given the success of E-field alignment and the general success of mechanical fields in inducing domain alignment (125) in bulk systems, it is surprising that the application of other fields to BC films has not been attempted. While magnetic fields would not be expected to affect pure BC materials, those doped with magnetic nanoparticles might respond amenably. Temperature gradients successfully used by Bodycomb and coworkers (126) to make millimeter thick single-crystal lamellar BC domains might also prove promising. It would be interesting to see the effect of in-plane temperature gradients on the long-range order of laterally organized BC film morphologies such as spheres and columns.

PATTERNED SUBSTRATES Substrates imparted with chemical and/or topographic patterns offer means of directing microphase separation in BC films. By using such substrates, local domain orientation, film surface morphology, and long-range order can be affected and sometimes controlled. In addition, substrate patterns can widen the scope of microphase domain structure beyond those motifs offered by nature.

Diblock films deposited onto chemically heterogeneous surfaces have recently received substantial attention. Recent years have seen a flood of publications reporting theoretical treatments of diblock films on chemically patterned substrates, led by the groundbreaking work of Balazs and coworkers (127–137). Generally, these studies discerned the effect of the substrate pattern periodicity on film morphology. In particular, when the period of the substrate pattern approaches molecular dimensions, these studies predict (almost universally) a surface-perpendicular orientation of the block domains.

This promise of morphological control dictates that experimental studies of diblock films on chemically patterned surfaces be pursued. Facilitated by the emergence of microcontact printing techniques (138), Heier et al (139–141) examined the effect of micron-scale substrate chemical heterogeneity on bulk-lamellar diblock films. This team demonstrated that contrasting wetting symmetries across patterned substrates induces a surface pattern of islands and holes that mirrors the substrate periodicity (140). They have used this effect to study the kinetics of island formation and coarsening (141). An X-ray lithography technique (134) recently allowed Yang and coworkers (142) to examine similar effects on the submicron scale.

Rockford et al (143) published the first observations of diblock films (compositionally symmetric PS-b-PMMA) on heterogeneous substrates with nanoscopic periodicity. The substrates they fabricated and employed had alternating stripes of gold and silicon, selectively attractive to PS and PMMA, respectively. They found that the PL morphology formed when the copolymer and pattern periods matched. Moreover, the only defects in this perpendicular morphology were due to flaws in the substrate pattern itself. When the substrate period was longer or shorter than L_0, diblock recognition of the underlying pattern was diminished. Similar observations were reported by Yang and coworkers (142) on surfaces produced by their lithographic technique.

Patterned substrate topography offers another means of manipulating BC film morphology. Li et al (144) examined symmetric BC films on micron-scale corrugated substrates. Interestingly, they found that islands were formed preferentially over the corrugation troughs, creating a surface that was anti-conformal with the substrate undulations. Moreover, this resulted in terrace-edge defects aligned perpendicular to the substrate corrugations. We have also studied the effect of surface topography on diblock films and demonstrated a lateral patterning technique based on the thickness dependence of film morphology and domain orientation (25, 48). This originally employed faceted silicon substrates (145) exhibiting sawtooth-profile corrugations of ~2-nm amplitude and periods in the 100-nm range. Thin films of PS-b-PBMA deposited on these substrates exhibited a flat free surface, in accordance with predictions by Turner & Joanny (146). This implies a periodic film thickness profile: thinner above corrugation peaks, thicker above troughs. Accordingly, if the average film thickness is chosen correctly, i.e., such that these thickness modulations occur about a critical thickness at which a morphological transition takes place, a lateral domain pattern develops that mirrors the substrate topography, as demonstrated in the micrograph in Figure 7*a*. Here the diblock film is, on average, about $L_0/3$ (~15 nm) thick. Near this film thickness, the HY morphology gives way to PL. The corrugated substrate (~210-nm period) laterally modulates the film thickness such that HY (dots) are formed over the troughs whereas the thinner morphology, PL, is formed on the peaks. As a result, a complex lateral patterning with a unique motif develops, as shown. A dual-scale (nanometer, micron) pattern can develop if micron-sized features are used, as in Figure 7*b*. In this case the film surface is nearly conformal over the 30-nm high, 1.5-μm wide substrate columns. However, the film thins slightly as it bends over the column edges. Choosing the film thickness to be just below $L_0/2$, where the HL to HY transition occurs, results in the preferential formation of HY (dots) on the edges of the substrate features. Thus through the use of substrate topography, diverse film patterns with a range of length scales can be created. This patterning technique has motivated a recent theoretical treatment by Podariu & Chakrabarti (147). These researchers compare HY formation on flat and corrugated substrates, demonstrating, as we previously observed (48) and discuss above, that these features preferentially form over topographic peaks and edges where the film is thinner.

CRYSTALLIZATION The path-dependent morphological behavior of semi-crystalline BCs can be exploited to create films with oriented domains and long-range order. Reiter and coworkers (70) demonstrated this behavior with lamellar PB-b-poly(ethylene oxide) (PB-b-PEO). Films of this material were spin cast onto substrates and annealed to induce formation of FL. Cooling leads to crystallization of the PEO within the layered microphase domains. Depending upon the temperature at which the PEO crystallized, various morphologies developed, including, for low temperatures, a highly oriented PL morphology. This reorientation was

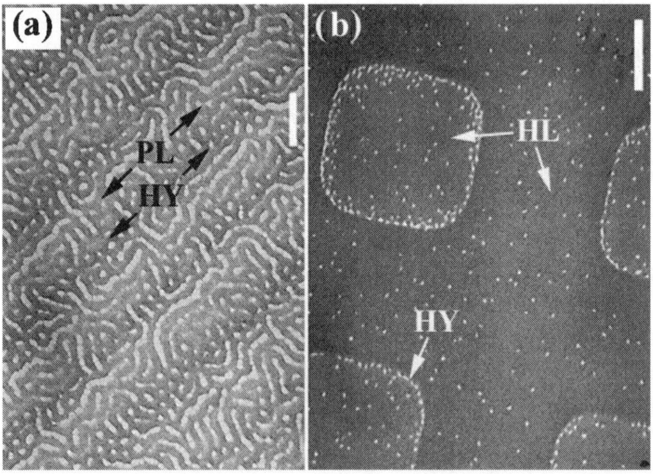

Figure 7 Patterning of DBC films via substrate topography (48). (*a*) TEM micrograph of PS-b-PBMA film peeled (25) from corrugated substrate with 210-nm period. PS domains appear dark. Thickness effects cause alternating HY and PL morphologies to form over corrugation troughs and peaks, respectively. Scale bar = 200 nm. (*b*) Dual-scale pattern of HY morphology. TEM micrograph of PS-b-PBMA film deposited on 30-nm high silicon posts with 1.5-mm diameter and 4-mm spacing. The HY morphology forms preferentially where the film is thinnest at the edges of the post. Areas without contrast are the HL morphology. Scale bar = 1 μm.

conjectured to accommodate the volume change that accompanies crystallization. De Rosa et al (66, 72) also examined crystallization-induced domain orientation. The key to their technique was the use of a crystallizable solvent (benzoic acid) into which the semi-crystalline PS-b-PE was dissolved. Directional solidification of the solvent between two glass plates led to well-oriented crystals of this small molecule that act as a template onto which the polymer crystallizes epitaxially (indeed, this might be considered a patterned surface technique!). Dependent upon BC composition, arrays of surface-perpendicular cylinders or PL were observed, both exhibiting good long-range order. The authors further noted the speed of this technique and its applicability to large-area films.

Optical Properties—Future Development of BC Films as Photonic Devices

It is well known that planar multi-layered dielectric systems exhibit useful optical properties, as evidenced by the widespread use of various band pass and notch optical filters. These "one-dimensional photonic crystals" selectively reflect a certain wavelength of light owing to their periodic structure and Fabry-Perot interference effects (148). The reflected wavelength (position of the optical band gap) is directly related to the layer periodicity and the dielectric contrast, i.e., the difference in the index of refraction between the component materials in the layered system. Indeed, if designed correctly, as demonstrated recently by Fink and coworkers (149), the band gaps of alternating multilayered systems shed their angular dependence. Such omnidirectional dielectric reflectors can serve as peerless wave guides capable of directing light though near hairpin turns without loss.

The self-assembled morphologies of BC systems make them intriguing candidates for photonic applications. This is especially true of BC film systems that spontaneously exhibit well-defined alternating layered structures. Fink et al (150) have succinctly outlined the materials parameters involved in transforming BC systems into effective photonic materials. Of these, two are most important: First, common BC systems exhibit L_0 values in the 10–100-nm range, whereas photonically active systems require periodicities of hundreds of nanometers. However, ultra-high molecular weight BCs are currently synthesizable, if not yet routine, and, as we have seen (79), increasing the period of FL layers is possible through the addition of homopolymer. In thick film systems, Urbas and coworkers (151, 152) recently proved the efficacy of this concept. Through the addition of two homopolymers (one of each block species), they increased the L_0 of a lamellar-forming diblock into the photonic regime, also demonstrating that the position of the band gap could be tuned through the amount of added homopolymer. These authors found, however, that defects in the lamellar structure were problematic. It appears that the increased order offered by approaches described above would be of benefit. The second barrier to block copolymer photonic activity is dielectric contrast. Most polymer systems have nearly identical indices of refraction. Nevertheless, this can be altered, as discussed above, by the selective doping of microphase domains with high index particles (111, 149, 150).

The real promise BC thin films offer to photonics applications might be found in their in-plane ordering. Here, laterally distributed domains would exhibit an in-plane two-dimensional optical band structure that could be harnessed in a variety of ways (148, 150). For such applications, control of domain order, orientation, and pattern are required. The means for such control are available, as described above. The question is whether they can be applied to higher-L_0 systems, which are known to pose a significant kinetic challenge. Another challenge will be in characterizing in-plane photonic film structures as they are being developed. While various spectroscopic techniques provide an ensemble picture of photonic activity, essential information about these systems will be found in the optical activity of small groups of microphase domains and single defects. The size of these structures places them beyond the reach of conventional far-field optical microscopies. Fortunately, the recent emergence of near-field scanning optical microscopy (NSOM) offers the opportunity to image structures below the diffraction limit with light (see 153 for a review of this technique). Indeed, NSOM has been established as a tool for probing the local optical structure of inorganic two-dimensional photonic crystals (154, 155), and initial examinations of photonic BC films are currently underway (156).

SUMMARY AND CONCLUSIONS

The behavior of BC films is rooted in a balance between surface energetics, which find increased importance in film systems, and confinement effects, i.e., conflicts between the equilibrium period of microphase separation, L_0, and the film thickness. In even the simplest situations, compositionally symmetric diblocks under symmetric surface energetics, these governing principles are the source of a rich array of morphological behavior. Indeed, the prevalence of surface-parallel domain layering, the thickness dependence of domain orientation, and the reorientation effect of surface neutrality can be witnessed under symmetric boundary conditions. However, some phenomena, in particular those associated with substrate-supported films, require a consideration of surface energy asymmetry, i.e., different segmental affinities for the substrate and free surfaces. When these effects are included, a more comprehensive picture of diblock film behavior develops. Hybrid morphologies, which include both surface parallel and surface perpendicular components, seem to be a direct consequence of boundary asymmetry. Moreover, through this more complete picture, an organizing principle of morphology, based around $L_0/2$-thick units, can be outlined for lamellar film systems that unites the symmetric and anti-symmetric wetting regimes and structures observed over a range of thickness. Such a principle may lend insight when examining more complex systems such as asymmetric BC compositions, chemistries including crystalline and liquid crystalline components, and A-B-C triblock copolymers, discussed above.

Recent research efforts have suggested the utility of thin BC films as components in a new generation of optical and electronic devices. Lithographic and microfabrication techniques using BC films offer unprecedented feature dimensions and densities. Moreover, elegant methods of harnessing their microphase separation to produce fine, ordered dispersions of inorganic material are in development. However, for many potential applications, e.g., addressable memory devices, limitations arise from the imperfect long-range order and restricted choice of motifs. Headway has been made through exploitation of surface and confinement effects, application of external fields, and use of chemically and topographically patterned substrates. These approaches for morphological control evolved directly from research on the fundamental physics of block copolymer films. Thus understanding the physics of block copolymer film structure remains a worthwhile pursuit.

ACKNOWLEDGMENTS

This work was supported primarily by the MRSEC program of the National Science Foundation under Award DMR-98-08941. MJF thanks the Research Associateship Program of the National Research Council. The authors thank Dr. Valerie Leppert for providing the micrograph in Figure 5.

Visit the Annual Reviews home page at www.AnnualReviews.org

LITERATURE CITED

1. Bates FS, Fredrickson GH. 1990. *Annu. Rev. Phys. Chem.* 41:525
2. Thomas EL, Lescanec RL. 1994. *Philos. Trans. R. Soc. London Ser. A* 348:149
3. Matsen MW. 1998. *Curr. Opin. Colloid Interface Sci.* 3:40
4. Gido SP, Gunther J, Thomas EL, Hoffman D. 1993. *Macromolecules* 26:4506
5. Gido SP, Thomas EL. 1994. *Macromolecules* 27:849
6. Russell TP, Coulon G, Deline VR, Miller DC. 1989. *Macromolecules* 22:4600
7. Kämpf G, Hoffmann M, Krömer H. 1970. *Ber. Bunsenges. Phys. Chem.* 71:851
8. Fredrickson GH. 1987. *Macromolecules* 20:2535
9. Coulon G, Collin B, Ausserre D, Chatenay D, Russell TP. 1990. *J. Phys.* 51:2801
10. Ausserre D, Chatenay D, Coulon G, Collin B. 1990. *J. Phys.* 51:2571
11. Turner MS. 1992. *Phys. Rev. Lett.* 69:1788
12. Walton DG, Kellogg GJ, Mayes AM, Lambooy P, Russell TP. 1994. *Macromolecules* 27:6225
13. Pickett GT, Balazs AC. 1997. *Macromolecules* 30:3097
14. Matsen MW. 1997. *J. Chem. Phys.* 106:7781
15. Tang WH, Witten TA. 1998. *Macromolecules* 31:3130
16. Geisinger T, Muller M, Binder K. 1999. *J. Chem. Phys.* 111:5241
17. Geisinger T, Muller M, Binder K. 1999. *J. Chem. Phys.* 111:5251
18. Kikuchi M, Binder K. 1994. *J. Chem. Phys.* 101:3367
19. Suh KY, Kim YS, Lee HH. 1998. *J. Chem. Phys.* 108:1253
20. Turner MS, Rubinstein M, Marques CM. 1994. *Macromolecules* 27:4986
21. Huinink HP, Brokken-Zijp JCM, van Dijk MA, Sevink GJA. 2000. *J. Chem. Phys.* 112:2452
22. Radzilowski LH, Carvalho BL, Thomas EL. 1996. *J. Polym. Sci. B* 34:3081
23. Lambooy P, Russell TP, Kellogg GJ, Mayes AM, Gallagher PD, Satija SK. 1994. *Phys. Rev. Lett.* 76:2899
24. Koneripalli N, Singh N, Levicky R, Bates FS, Gallagher PD, Satija SK. 1995. *Macromolecules* 28:2897
25. Fasolka MJ, Banerjee P, Mayes AM, Pickett G, Balazs AC. 2000. *Macromolecules* 33:5702
26. Henkee CS, Thomas EL, Fetters LJ. 1988. *J. Mater. Sci.* 23:1685
27. Kellogg GJ, Walton DG, Mayes AM, Lambooy P, Russell TP, et al. 1996. *Phys. Rev. Lett.* 76:2503
28. Huang E, Rockford L, Russell TP, Hawker CJ. 1998. *Nature* 395:757
29. Huang E, Russell TP, Harrison C, Chaikin PM, Register RA, et al. 1998. *Macromolecules* 31:7641
30. Huang E, Pruzinsky S, Russell TP, Mays J, Hawker CJ. 1999. *Macromolecules* 32:5299
31. Mansky P, Liu Y, Huang E, Russell TP, Hawker C. 1997. *Science* 275:1458
32. Mansky P, Russell TP, Hawker CJ, Mays J, Cook DC, Satija SK. 1997. *Phys. Rev. Lett.* 79:237
33. Thurn-Albrecht T, Steiner R, DeRouchey J, Stafford CM, Huang E, et al. 2000. *Adv. Mater.* 12:787
34. Karim A, Singh N, Sikka M, Bates FS, Dozier WD, Felcher GP. 1994. *J. Chem. Phys.* 100:1620
35. Huang E, Mansky P, Russell TP, Harrison C, Chaikin PM, et al. 2000. *Macromolecules* 33:80
36. Tang WH. 2000. *Macromolecules* 33:1370
37. Morkved TL, Jaeger HM. 1997. *Europhys. Lett.* 40:643

38. Russell TP, Menelle A, Anastasiadis SH, Satija SK, Majkrzak CF. 1991. *Macromolecules* 24:6263
39. Scheutjens JMHM, Fleer GJ. 1979. *J. Phys. Chem.* 83:1619
40. Kikuchi M, Binder K. 1993. *Europhys. Lett.* 21:427
41. Pickett GT, Witten TA, Nagel SR. 1993. *Macromolecules* 26:3194
42. Anastasiadis SH, Russell TP, Satija SK, Majkrzak CF. 1989. *Phys. Rev. Lett.* 62:1852
43. Harrison C, Park M, Chaikin P, Register RA, Adamson DH, Yao N. 1998. *Macromolecules* 31:2185
44. Russell TP, Menelle A, Anastasiadis SH, Satija SK, Majkrzak CF. 1992. *Makromol. Chem.-Macromol. Symp.* 62:157
45. Anastasiadis SH, Russell TP, Satija SK, Majkrzak CF. 1990. *J. Chem. Phys.* 92:5677
46. Menelle A, Russell TP, Anastasiadis SH, Satija SK, Majkrzak CF. 1992. *Phys. Rev. Lett.* 68:67
47. Carvalho BL, Thomas EL. 1994. *Phys. Rev. Lett.* 73:3321
48. Fasolka MJ, Harris DJ, Mayes AM, Yoon M, Mochrie SGJ. 1997. *Phys. Rev. Lett.* 79:3018
49. Fasolka MJ. 2000. PhD thesis. *The morphology and lateral patterning of diblock copolymer thin films.* MIT, Cambridge, MA. 174 pp.
50. Hasegawa H, Hashimoto T. 1992. *Polymer* 33:475
51. Kendall FG, Rehn BW. 1983. *J. Forensic Sci.* 28:777
52. Kobus HJ, Warrener RN, Stoilovic M. 1983. *Forensic Sci. Int.* 23:233
53. Zehner RW, Lopes WA, Morkved TL, Jaeger H, Sita LR. 1998. *Langmuir* 14:241
54. Mansky P, Chaikin PM, Thomas EL. 1995. *J. Mater. Sci.* 30:1987
55. Park M, Harrison C, Chaikin PM, Register RA, Adamson DH. 1997. *Science* 276:1401
56. Liu Y, Zhao W, Zheng X, King A, Singh A, et al. 1994. *Macromolecules* 27:4000
57. Harrison C, Park M, Chaikin PM, Register RA, Adamson DH, Yao N. 1998. *Polymer* 39:2733
58. Harrison C, Chaikin PM, Huse DA, Register RA, Adamson DH, et al. 2000. *Macromolecules* 33:857
59. Nick L, Lippitz A, Unger W, Kindermann A, Fuhrmann J. 1995. *Langmuir* 11:1912
60. Vandijk MA, Vandenborg R. 1995. *Macromolecules* 28:6773
61. Hahm J, Sibener SJ. 2000. *Langmuir* 16:4766
62. Yokoyama H, Mates TE, Kramer EJ. 2000. *Macromolecules* 33:1888
63. Yokoyama H, Kramer EJ, Rafailovich MH, Sokolov J, Schwarz SA. 1998. *Macromolecules* 31:8826
64. Chan VZH, Hoffman J, Lee VY, Iatrou H, Avgeropoulos A, et al. 1999. *Science* 286:1716
65. Wong GCL, Commandeur J, Fischer H, de Jeu WH. 1996. *Phys. Rev. Lett.* 77:5221
66. De Rosa C, Park C, Lotz B, Wittmann JC, Fetters LJ, Thomas EL. 2000. *Macromolecules* 33:4871
67. Sentenac D, Demirel AL, Lub J, de Jeu WH. 1999. *Macromolecules* 32:3235
68. Wu JS, Fasolka MJ, Hammond PT. 2000. *Macromolecules* 33:1108
69. Cohen RE, Cheng PL, Douzinas K, Kofinas P, Berney CV. 1990. *Macromolecules* 23:324
70. Reiter G, Castelein G, Hoerner P, Riess G, Blumen A, Sommer JU. 1999. *Phys. Rev. Lett.* 83:3844
71. Lammertink RGH, Hempenius MA, Vancso GJ. 2000. *Langmuir* 16:6245
72. De Rosa C, Park C, Thomas EL, Lotz B. 2000. *Nature* 405:433
73. Winey KI, Thomas EL, Fetters LJ. 1991. *Macromolecules* 24:6182
74. Hashimoto T, Yamasaki K, Koizumi S, Hasegawa H. 1993. *Macromolecules* 26:2895
75. Leibler L, Gay C, Erukhimovich I. 1999. *Europhys. Lett.* 46:549
76. Mayes AM, Russell TP, Deline VR, Satija

SK, Majkrzak CF. 1994. *Macromolecules* 27:7447
77. Shi AC, Noolandi J. 1994. *Macromolecules* 27:2936
78. Koneripalli N, Levicky R, Bates FS, Matsen MW, Satija SK, et al. 1998. *Macromolecules* 31:3498
79. Mayes AM, Russell TP, Satija SK, Majkrzak CF. 1992. *Macromolecules* 25: 6523
80. Shull KR, Winey KI. 1992. *Macromolecules* 25:2637
81. Shull KR, Mayes AM, Russell TP. 1993. *Macromolecules* 26:3929
82. Orso KA, Green PF. 1999. *Macromolecules* 32:1087
83. Smith MD, Green PF, Saunders R. 1999. *Macromolecules* 32:8392
84. Pan T, Huang KL, Balazs AC, Kunz MS, Mayes AM, Russell TP. 1993. *Macromolecules* 26:2860
85. Brinkmann S, Stadler R, Thomas EL. 1998. *Macromolecules* 31:6566
86. Vandenberg R, deGroot H, Vandijk MA, Denley DR. 1994. *Polymer* 35:5778
87. Kim G, Libera M. 1998. *Macromolecules* 31:2569
88. O'Connor SM, Gehrke SH, Retzinger GS. 1999. *Langmuir* 15:2580
89. Konrad M, Knoll A, Krausch G, Magerle R. 2000. *Macromolecules* 33:5518
90. Nardin C, Winterhalter M, Meier W. 2000. *Langmuir* 16:7708
91. Stocker W, Beckmann J, Stadler R, Rabe JP. 1996. *Macromolecules* 29:7502
92. Pickett GT, Balazs AC. 1998. *Macromol. Theory Simul.* 7:249
93. Stocker W. 1998. *Macromolecules* 31: 5536
94. Elbs H, Fukunaga K, Stadler R, Sauer G, Magerle R, Krausch G. 1999. *Macromolecules* 32:1204
95. Matyjaszewski K. 1999. *Macromol. Symp.* 143:257
96. Mansky P, Harrison CK, Chaikin PM, Register RA, Yao N. 1996. *Appl. Phys. Lett.* 68:2586
97. Park M, Adamson DH, Chaikin PM, Register RA. 1999. In *ACS PMSE Proc.* 81:12–13
98. Li RR, Dapkus PD, Thompson ME, Jeong WG, Harrison C, et al. 2000. *Appl. Phys. Lett.* 76:1689
99. Liu GJ, Ding JF, Hashimoto T, Kimishima K, Winnik FM, Nigam S. 1999. *Chem. Mater.* 11:2233
100. Ciebien JF, Clay RT, Sohn BH, Cohen RE. 1998. *New J. Chem.* 22:685
101. Chan YNC, Craig GSW, Schrock RR, Cohen RE. 1992. *Chem. Mater.* 4:885
102. Ciebien JF, Cohen RE, Duran A. 1998. *Supramol. Sci.* 5:31
103. Sohn BH, Cohen RE. 1997. *Chem. Mater.* 9:264
104. Sohn BH, Cohen RE, Papaefthymiou GC. 1998. *J. Magn. Magn. Mater.* 182:216
105. Sankaran V, Cummins CC, Schrock RR, Cohen RE, Silbey RJ. 1990. *J. Am. Chem. Soc.* 112:6858
106. Cummins CC, Schrock RR, Cohen RE. 1992. *Chem. Mater.* 4:27
107. Kane RS, Cohen RE, Silbey R. 1999. *Chem. Mater.* 11:90
108. Kane RS, Cohen RE, Silbey R. 1999. *Langmuir* 15:39
109. Lee T, Yao N, Aksay IA. 1997. *Langmuir* 13:3866
110. Morkved TL, Wiltzius P, Jaeger HM, Grier DG, Witten TA. 1994. *Appl. Phys. Lett.* 64:422
111. Mattoussi H, Radzilowski LH, Dabbousi BO, Fogg DE, Schrock RR, et al. 1999. *J. Appl. Phys.* 86:4390
112. Limary R, Swinnea S, Green PF. 2000. *Macromolecules* 33:5227
113. Ginzburg VV, Gibbons C, Qiu F, Peng GW, Balazs AC. 2000. *Macromolecules* 33:6140
114. Spatz JP, Roescher A, Moller M. 1996. *Adv. Mater.* 8:337
115. Spatz JP, Mossmer S, Hartmann C, Moller M, Herzog T, et al. 2000. *Langmuir* 16:407
116. Eibeck P, Spatz JP, Mossmer S, Moller M,

Herzog T, Ziemann P. 1999. *Nanostruct. Mater.* 12:383
117. Leppert VJ, Risbud SH, Stender M, Power PP, Banerjee P, et al. 2000. *J. Am. Chem. Soc.* In review
118. Harrison C, Park M, Chaikin PM, Register RA, Adamson DH. 1998. *J. Vac. Sci. Technol. B* 16:544
119. Peters RD, Yang XM, Kim TK, Sohn BH, Nealey PF. 2000. *Langmuir* 16:4625
120. Amundson K, Helfand E, Davis DD, Quan X, Patel SS, Smith SD. 1991. *Macromolecules* 24:6546
121. Amundson K, Helfand E, Quan XN, Hudson SD, Smith SD. 1994. *Macromolecules* 27:6559
122. Morkved TL, Lu M, Urbas AM, Ehrichs EE, Jaeger HM, et al. 1996. *Science* 273:931
123. Thurn-Albrecht T, DeRouchey J, Russell TP, Jaeger HM. 2000. *Macromolecules* 33:3250
124. Thurn-Albrecht T, Schotter J, Kästle GA, Emley N, Tuominen MT, et al. 2000. *Science* 290:2126
125. Hamley IW. 1998. *The Physics of Block Copolymers*. Oxford, UK: Oxford Univ. Press
126. Bodycomb J, Funaki Y, Kimishima K, Hashimoto T. 1999. *Macromolecules* 32:2075
127. Huang KL, Balazs AC. 1991. *Phys. Rev. Lett.* 66:620
128. Balazs AC, Gempe MC, Zhou ZX. 1991. *Macromolecules* 24:4918
129. Balazs AC, Huang KL, McElwain P, Brady JE. 1991. *Macromolecules* 24:714
130. Chakrabarti A, Chen H. 1998. *J. Polym. Sci. B* 36:3127
131. Petera D, Muthukumar M. 1998. *J. Chem. Phys.* 109:5101
132. Wang Q, Nath SK, Graham MD, Nealey PF, de Pablo JJ. 2000. *J. Chem. Phys.* 112:9996
133. Wang Q, Yan QL, Nealey PF, de Pablo JJ. 2000. *Macromolecules* 33:4512
134. Yang XM, Peters RD, Kim TK, Nealey PF. 1999. *J. Vac. Sci. Technol. B* 17:3203
135. Chen H, Chakrabarti A. 1998. *J. Chem. Phys.* 108:6897
136. Pereira GG, Williams DRM. 1998. *Phys. Rev. Lett.* 80:2849
137. Pereira GG, Williams DRM. 1999. *Langmuir* 15:2125
138. Kumar A, Whitesides GM. 1993. *Appl. Phys. Lett.* 63:2002
139. Heier J, Kramer EJ, Walheim S, Krausch G. 1997. *Macromolecules* 30:6610
140. Heier J, Genzer J, Kramer EJ, Bates FS, Walheim S, Krausch G. 1999. *J. Chem. Phys.* 111:11101
141. Heier J, Kramer EJ, Groenewold J, Fredrickson GH. 2000. *Macromolecules* 33:6060
142. Yang XM, Peters RD, Nealey PF. 2000. *Macromolecules* 33:9575
143. Rockford L, Liu Y, Mansky P, Russell TP, Yoon M, Mochrie SGJ. 1999. *Phys. Rev. Lett.* 82:2602
144. Li Z, Qu S, Rafailovich MH, Sokolov J, Tolan M, Turner MS, et al. 1997. *Macromolecules* 30:8410
145. Mochrie SGJ, Song S, Yoon MR, Abernathy DL, Stephenson GB. 1996. *Physica B* 221:105
146. Turner MS, Joanny JF. 1992. *Macromolecules* 25:6681
147. Podariu I, Chakrabarti A. 2000. *J. Chem. Phys.* 113:6423
148. Jannopoulos JD, Mead R, Winn JN. 1995. *Photonic Crystals, Molding the Flow of Light*. Princeton, NJ: Princeton Univ. Press
149. Fink Y, Winn JN, Fan SH, Chen CP, Michel J, et al. 1998. *Science* 282:1679
150. Fink Y, Urbas AM, Bawendi MG, Joannopoulos JD, Thomas EL. 1999. *J. Lightwave Technol.* 17:1963
151. Urbas A, Fink Y, Thomas EL. 1999. *Macromolecules* 32:4748
152. Urbas A, Sharp R, Fink Y, Thomas EL, Xenidou M, Fetters LJ. 2000. *Adv. Mater.* 12:812
153. Dunn RC. 1999. *Chem. Rev.* 99:2891

154. McDaniel EB, Hsu JWP, Goldner LS, Tonucci RJ, Shirley EL, Bryant GW. 1997. *Phys. Rev. B* 55:10878
155. Bryant GW, Shirley EL, Goldner LS, McDaniel EB, Hsu JWP, Tonucci RJ. 1998. *Phys. Rev. B* 58:2131
156. Fasolka MJ, Hwang J, Goldner LS, Urbas A, DeRege P, Thomas EL. 2001. *Near-field optical images of photonic block copolymer morphology*. Presented at Gordon Res. Conf.: Polymers. Ventura, CA

Annu. Rev. Mater. Res. 2001. 31:357–71

Mechanisms Involved in Osteoblast Response to Implant Surface Morphology

Barbara D Boyan,[1] Christoph H Lohmann,[1,2] David D Dean,[1] Victor L Sylvia,[1] David L Cochran,[1] and Zvi Schwartz[1,3]

[1]*University of Texas Health Science Center at San Antonio, San Antonio, Texas 78229-3900*

[2]*Georg-August Universität, Göttingen, Germany*

[3]*Hebrew University Hadassah Faculty of Dental Medicine, Jerusalem, Israel*

e-mail: BoyanB@uthscsa.edu; LohmannCH@t-online.de; DeanD@uthscsa.edu; Sylvia@uthscsa.edu; Cochran@uthscsa.edu; Zvis1@cc.huji.ac.il

Key Words surface roughness, protein kinase C, PGE_2, titanium

■ **Abstract** Osteoblasts respond to surface topography with altered morphology, proliferation, and differentiation. The effects observed vary with cell culture model and the topographical features of the surface. In general, increased surface roughness is associated with decreased proliferation and increased differentiation. Cell responses to hormones, growth factors, and cytokines are altered as well, as is autocrine production of these factors. The cells interact with the surface via integrin receptors, and their expression is also surface roughness-dependent. Integrin binding to cell attachment proteins activates signal transduction cascades, including those mediated by protein kinase C and phospholipase A_2. These signaling pathways are also used by regulatory factors, which results in synergistic responses. Prostaglandins are important mediators of the surface effects, and both constitutive and inducible cyclooxygenase are involved.

INTRODUCTION

It is now well understood that surface morphology modulates the response of cells to an implant material (see 1, 2 for reviews). Surface morphology affects attachment, proliferation, extracellular matrix synthesis, and growth factor and cytokine production, as well as the response of cells to systemic and local regulatory agents. These observations suggest that specific interactions of bone cells with their substrate will result in altered phenotypic expression. The purpose of this review is to examine how cells that are in the osteoblast lineage interact with the surface and to assess what is known about the underlying mechanisms involved.

OSTEOBLAST RESPONSE TO SURFACE MORPHOLOGY IN VIVO

There is some controversy as to whether osteoblasts actually interact directly with implant surfaces in vivo. Histologic analysis of the bone/implant interface indicates that fibrous tissue is present (3–5). The thickness of this fibrous layer varies depending on the material and the degree of micromotion at the interface (6). Thus there is more connective tissue when stainless steel implants are used and less connective tissue when titanium (Ti) implants are used. Similarly, there is less connective tissue when implants are stabilized and more connective tissue when they are not. There are some instances when there appears to be direct bonding of the newly formed bone with the implant surface, particularly when the surface is coated with hydroxyapatite (7, 8, 8a); however, once remodeling begins, the mineral interface is replaced with connective tissue.

It is unclear if the connective tissue interface arises because fibroblast attachment to the surface is favored over osteoblast attachment or if characteristics of the surface modulate the commitment and differentiation of mesenchymal precursor cells along a fibroblastic or osteoblastic lineage. Another possibility is that mineralization may be prevented or inhibited at the bone implant surface.

In reality, the cells that colonize the surface never actually encounter the metal implant. However, the chemical composition and physical aspects of the material dictate the composition and conformation of serum components that adsorb to the surface (9–11), and this may have profound effects on subsequent cell recruitment and attachment (12, 13). If a surface has a rough morphology, surface area is increased, and surface energy may also be affected, with the net result of a significantly different adsorption layer than might be attained on a smooth surface.

In vivo, the situation is further complicated by the vascularity of the implanted tissue and resultant oxygen tension, the presence of infection, and the size of the defect. Time-course studies show that the defect site is first filled with a clot and/or hematoma (14). The first cells to actually come in contact with the implant are neutrophils, followed by macrophages. The osteoblast progenitors may migrate toward the implant from the endosteal surface of the surrounding bone bed, forming bone by creeping substitution. In addition, bone may form within the site by differentiation of osteoprogenitor cells that have migrated through the vasculature or marrow. Regardless of mechanism, however, these cells are attachment dependent, so their ability to form bone at the interface requires that they have a substrate upon which to move. How they interact with that substrate may be equally important as how they interact with the implant surface.

The question then arises as to whether it really matters what the physical characteristics of the actual implant are because most of the osteoblasts will not see the implant directly. There is abundant evidence showing that surface morphology does play an important role. Surfaces that have been modified with beads and fibers can result in a mechanical interlock that improves the pullout strength of the implant. However, micro-roughness also affects the interaction between the implant and the surrounding bone. Studies in miniature pigs (17) have shown that implant removal torque value is correlated with surface morphology, and Ti implants, in general, that have surfaces with greater micro-roughness exhibit greater pullout strength. A long-term evaluation has also provided positive results in humans (18).

In the case of micro-roughness, mechanical interlock is not a factor. Therefore, other aspects of cell response must be considered. Events at the surface clearly affect the surrounding bone. Studies examining local effects of various implants in rats following tibial marrow ablation show that endosteal bone formation in the treated limb varies with the type of material used (19–21). Moreover, these studies show that no implant is without effects. Regardless of the type of material used, synthesis of the extracellular matrix by osteoblasts and calcification of the matrix are altered compared with normal bone healing.

Implants also exert a systemic effect that is dependent on the material used. Ablation of tibial marrow activates endosteal bone formation in other bones, indicating a systemic effect of the injury (22). All implant materials tested thus far alter this systemic response to some degree. Materials such as Ti that support osteointegration have the least effect, whereas materials such as stainless steel, which result in a greater degree of fibrous tissue interface, have the greatest effect (19, 20, 23). This suggests that the implant may act as a bioreactor, providing a surface that serves as a physicochemical catalyst. It also suggests that cells may interact either directly with the surface, releasing bioactive factors that act both locally and systemically, or indirectly with the surface by responding to factors released by cells at the bone/implant interface.

That cells distant from the implant surface can be affected by events on the surface is clear, based on the in vivo studies described above. The anatomy of peri-implant bone is such that factors released by surface-associated cells, whether or not they are in the osteoblast lineage, can have direct access to cells within the surrounding bone bed through canaliculi that communicate with the bone/implant interface (Figure 1, see color insert). The presence of canaliculi extending from the implant surface through the surrounding bone has been verified by transmission electron microscopy (24). A number of studies indicate that corrosion products and ions from the implant interact with bone cells through this mechanism (25–28). It is reasonable to assume that cells within the peri-implant bone may also contribute to the net tissue response to an implant surface as well.

OSTEOBLAST RESPONSE TO IMPLANT SURFACE MORPHOLOGY IN VITRO

Cell Models and the Osteoblast Lineage

Osteoblasts are cells that form bone. These cells have a set of distinguishing characteristics that include the ability to synthesize osteoid or bone matrix and to mineralize the osteoid, which results in calcified bone. A number of markers have been identified that characterize typical osteoblasts in culture. These cells respond to 1α,25-$(OH)_2D_3$, a metabolite of vitamin D_3 that causes an increase in alkaline phosphatase activity and production of osteocalcin. In addition, osteoblasts have receptors for parathyroid hormone (PTH), which also regulates the differentiation of the cells.

Alkaline phosphatase is an enzyme that is associated with calcification, although its exact role is not known. Almost all cells have alkaline phosphatase in their plasma membranes, although cells that calcify their extracellular matrix, including odontoblasts (tooth forming cells), cementoblasts (cementum forming cells), and growth plate chondrocytes (cartilage cells at the ends of long bones), tend to have higher levels of this enzyme. Moreover, activity of the enzyme increases when osteoblasts produce osteoid, owing in large part to the production of extracellular membrane organelles called matrix vesicles. This is an early marker of osteoblast differentiation because activity is greatest just before mineralization actually begins. The effect of alkaline phosphatase is to increase local levels of inorganic phosphate, one of the components of apatite, the mineral phase of bone. Because von Kossa stains actually indicate levels of free phosphate ions, and not calcification per se, many investigators have incorrectly assumed that a positive reaction shows the presence of actively mineralizing osteoblasts, when they may actually be seeing evidence of high levels of phosphatase activity such as that caused by alkaline phosphatase.

Osteocalcin is a small molecular weight protein produced by osteoblasts as a late marker of differentiation. Other cells also make osteocalcin, particularly odontoblasts, but in most cell culture systems used to examine cells in the osteoblastic lineage, it is reasonable to assume that any osteocalcin being produced is produced by bone-forming cells. For this reason, it is a particularly effective marker of a well-differentiated osteoblast.

Because osteoblasts are of mesenchymal cell origin, they also produce type I collagen. Measurement of this collagen type is a good indicator of the synthetic capacity of the cells, but it is not a good indicator of their osteoblastic phenotype because many mesenchymal cells, most notably fibroblasts, also synthesize collagen type I. Chondrocytes, which are also mesenchymal cells, produce type II collagen in vivo, but in vitro these cells may also produce predominantly type I collagen.

This discussion of phenotypic markers is of particular importance because the many models used to examine osteoblast response to implant surfaces use cells at vastly different states of phenotypic maturation in the osteoblast lineage and with considerably different levels of homogeneity (Table 1). In vivo, the cells that migrate to the implant site are primarily mesenchymal cells that have not yet committed to a specific lineage cascade. That these are multi-potent cells is demonstrated by the fact that the progeny of these cells may form a variety of tissue types adjacent to the implant. If oxygen tension is low, cartilage formation occurs. If certain signals are present, adipose tissue may result. When micromotion is present, fibrous connective tissue is the consequence. The goal, then, is to provide an environment that either solicits osteoprogenitor cells only, or one that promotes

TABLE 1 Common cell culture models used to examine the response of cells in the osteoblast lineage to surface morphology

Cell model	Source	Phenotype
MG63	Human osteosarcoma	Immature osteoblast
FRC	Fetal rat calvaria	Multi-potent osteoprogenitor cells and osteoblasts
MC-3T3-E1	Mouse non-transformed cell line	Osteoblast-like
OCT-1	Transgenic mouse Large T-antigen Osteocalcin promoter	Mature osteoblast-like cells
MLO-Y4	Transgenic mouse Large T antigen Osteocalcin promoter	Osteocyte-like
NHOst	Commercial preparations of human bone	Osteoblast-like
ROS 17/2.8	Rat osteosarcoma	Osteoblast-like
2T9	Transgenic mouse Large T-antigen BMP-2 promoter	Osteoprogenitor cells

differentiation of the mesenchymal cells along the osteoblastic lineage. Whether a chosen marker is relevant depends in large part on where in the lineage the test population is found.

Many of the studies concerning cell response to implant surfaces have used MG63 cells. The MG63 cell line was originally derived from a human osteosarcoma (29, 30). The cells exhibit characteristics of relatively immature osteoblasts. They synthesize osteoid and exhibit increases in alkaline phosphatase and osteocalcin when treated with 1α,25-$(OH)_2D_3$ (30–32); however, to date there are no reports that these cells can actually mineralize their osteoid, even when they are provided with a source of inorganic phosphate. These cells provide an excellent model for examining the effects of surface morphology because they behave distinctly differently on smooth and rough surfaces, and these responses correlate with changes in cell morphology that reflect maturation states in the osteoblastic lineage. When MG63 cells are cultured on smooth surfaces, whether tissue culture plastic or smooth Ti, they assume a flattened morphology and exhibit relatively small changes in alkaline phosphatase or osteocalcin in response to 1α,25-$(OH)_2D_3$ (33). Moreover, 1α,25-$(OH)_2D_3$ has no effect on PGE_2 release or transforming growth factor beta-1 (TGF-β1) levels (33). However, when these cells are cultured on Ti surfaces with average roughness values (Ra) of 4–7 μm, they assume a more cuboidal morphology with cytoplasmic extensions. Basal levels of alkaline phosphatase activity and osteocalcin production are greater, and there is a synergistic increase when the cells are treated with 1α,25-$(OH)_2D_3$. Similarly, basal levels of PGE_2 and TGF-β1 are higher, and 1α,25-$(OH)_2D_3$ causes an increase in both.

There is always concern that because MG63 cells are transformed, they may not respond to surface morphologies in a manner typical of normal osteoblasts. For this reason, many groups have opted to use fetal rat calvarial cells (FRC) as their model system. This model also requires caveats, however. Fetal cells are often very different from adult cells. In addition, the method used to release cells from the calvaria can have a substantial impact on the FRC population that is obtained. If the bone chips are pre-digested with collagenase, the cells that migrate from the bone chips will be enriched for osteoprogenitor cells. If the bone chips are placed into culture without collagenase pre-treatment, other cell types may also be present, if not dominant, in the culture. Certainly, the cultures will be more heterogeneous. Finally, if a sequential digestion method is used, there will be a greater proportion of committed osteoblasts. It is important to remember that fetal rat calvarial cells will almost never see a dental implant in vivo, so the relevancy of the results to the adult human remains an issue.

To overcome this last consideration, laboratories are beginning to use normal human osteoblasts obtained from adult bone tissue. These cells are subject to normal human variability, making it imperative to use multiple batches of cells before accepting a result as typical. If researchers harvest their cells from tissue themselves, they must take into consideration the variability, as well as the issues described for FRC cells above.

A number of other osteoblast cell lines are commonly used in implant surface research in part because of the vast literature using these cells to study cell regulation. MC-3T3-E1 cells are a mouse osteoblast-like cell line that exhibits characteristics of mature secretory osteoblasts. These cells synthesize osteoid, respond to 1α,25-$(OH)_2D_3$ and PTH, and under the right experimental conditions will calcify the osteoid (34–36). ROS 17/2.8 cells are a rat osteosarcoma cell line that exhibits similar properties (37, 38). As noted above for MG63 cells and FRC cells, MC-3T3-E1 cells and ROS 17/2.8 cells are typically used to study the initial events involved in the interaction of an osteoblast with the surface of a material: attachment, proliferation, matrix synthesis, and matrix calcification.

Only a relatively small portion of new peri-implant bone, if any, occurs on the surface of the implant. Much of the bone growth actually involves outgrowth of bone toward the implant surface from the endosteal surface of the surrounding bone bed. In addition, once bone is formed, a subset of osteoblasts undergoes programmed cell death, apoptosis, but essentially nothing is known about the effects of surface morphology on this process. Another subset continues to differentiate within the lineage, becoming surrounded by the mineralized osteoid. These terminally differentiated osteoblasts, now called osteocytes, are responsible for maintaining bone. They respond to mechanical signals within the tissue, as well as to hormones and local regulatory factors. It is now understood that they communicate with the surface of bone through a complex series of canaliculi by cytoplasmic

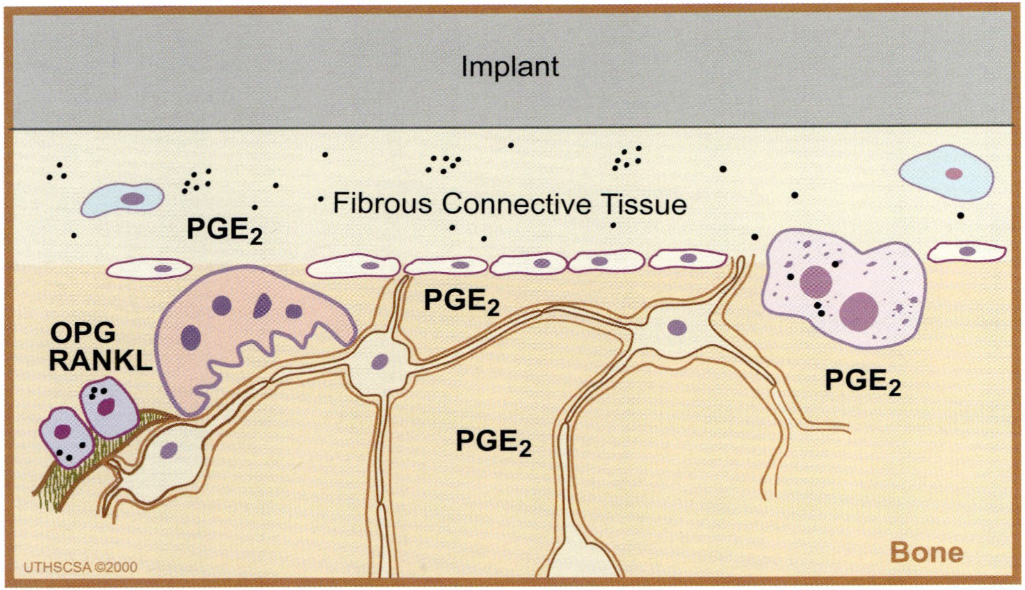

Figure 1 Peri-implant tissue at the bone/material interface. In vivo, materials such as Ti are in contact with multiple cell types, both directly and indirectly. Following the initial bone-healing response, a dynamic equilibrium is established. Bone lining cells and/or fibroblastic cells may provide the interface. A fibrous connective tissue is present, although its thickness may vary considerably depending on the success of osteointegration. Osteocytes are in direct communication with this interface via osteocyte canaliculi, as well as with osteoblasts at bone-remodeling sites. Osteoblasts modulate bone resorption through production of osteoprotegerin (OPG) and RANK ligand (RANK), as well as by production of cytokines such as prostaglandin E_2 (PGE_2). Ti and Ti6Al4V are well tolerated biologically; even so, ions released from implant surfaces may elicit an inflammatory response, resulting in recruitment of monocytes and macrophages to the surface. The complexity of the in vivo implant/bone interface may differ considerably from in vitro observations using cell culture models.

extensions that can form gap junctions with neighboring osteocytes as well as with osteoblasts.

In vivo, mature osteoblasts and osteocytes do not contact the implant surface directly. However, these same canaliculi serve as a mechanism for interaction with the bone/implant interface. We have used two cell culture models to begin to understand how mature osteoblasts and osteocytes are affected by implant surface roughness. Both OCT-1 cells and MLO-Y4 cells were obtained from a transgenic mouse that had been transfected with the SV-40 large T-antigen driven by the osteocalcin promoter (39, 40). The OCT-1 cells exhibit lower levels of alkaline phosphatase than would be typical of a less mature osteoblast, but they produce relatively high levels of osteocalcin. The MLO-Y4 cells are morphologically similar to osteocytes in vivo in that they have a small cell body with long cytoplasmic extensions. They have low levels of alkaline phosphatase that are not responsive to 1α,25-$(OH)_2D_3$, and they produce high levels of osteocalcin. When these cells are grown on Ti surfaces of increasing roughness, they behave very differently from either MG63 cells or FRC cells and differently from each other (41).

Effects of Surface Roughness on Osteoblast Response to Regulatory Agents

The studies reviewed above show that surface morphology has a direct effect on phenotypic expression of cells in the osteoblast lineage. As the surface becomes rougher, MG63 cells and FRC cells exhibit a more differentiated phenotype based on decreased proliferation and increased activity of alkaline phosphatase and osteocalcin production. More differentiated osteoblasts, the OCT-1 cells, respond with decreased cell number and alkaline phosphatase, but increased osteocalcin, PGE_2, and TGF-β1 production, whereas the terminally differentiated MLO-Y4 cells exhibit a decrease in cell number, no change in alkaline phosphatase, and increased production of osteocalcin, PGE_2, and TGF-β1.

The response of osteoblasts to systemic factors that regulate their proliferation and differentiation is also affected by surface roughness. As surface roughness increases, the effects of the vitamin D metabolite 1α,25-$(OH)_2D_3$ on proliferation of MG63 cells (33) and FRC cells (41) decrease, but the effects on phenotypic expression become more pronounced. This is commensurate with the hypothesis that surface roughness enhances the expression of a more mature phenotype and the cells then respond to 1α,25-$(OH)_2D_3$ in this context. This hypothesis is supported by the fact that 1α,25-$(OH)_2D_3$ does not affect proliferation of OCT-1 cells or MLO-Y4 cells, nor does it affect activity of the early differentiation marker alkaline phosphatase, but it does increase osteocalcin levels (41).

Surface roughness also affects the response of cells to local regulatory factors such as BMP-2. When Ong et al (42) cultured an osteoblast progenitor cell line (2T9 cells) on rough and smooth Ti surfaces, BMP-2 treatment resulted in prolonged alkaline phosphatase–specific activity and more rapid osteocalcin production in cells grown on surfaces that had been ground with 600 grit silicon carbide paper. The BMP-2-treated cells also exhibited an enhanced response to 1α,25-$(OH)_2D_3$ when cultured on the rougher surface (43). The actual difference in surface roughness between the smooth and rough surfaces used by Ong et al (44) was much smaller than the differences between the surface roughnesses used in the studies described above, indicating the sensitivity of osteoblastic cells to surface morphology.

In addition to modulating how cells respond to exogenous factors, surface roughness affects production of local regulatory factors by the cells. As roughness increases, MG63 cells release increased levels of prostaglandin, including PGE_1 and PGE_2, into their conditioned medium (45). Similarly, FRC cells, OCT-1 cells, and MLO-Y4 cells all exhibit this surface-dependent increase in PGE_2 production (41). Moreover, the amount of PGE_2 released is modulated by 1α,25-$(OH)_2D_3$ in a roughness-dependent manner (33). On smooth surfaces, 1α,25-$(OH)_2D_3$ has no effect on PGE_2 production, but on rougher surfaces the effect of the vitamin D metabolite is synergistic with surface roughness. Again, this supports the hypothesis that the effects of the hormone are secondary to the effects of surface morphology on phenotypic expression, since previous studies indicated that basal levels of PGE_1 and PGE_2, as well as 1α,25-$(OH)_2D_3$ regulation of their production, depend on the maturation state of the osteoblast (31, 46). Surface roughness also elicits marked increases in PGE_2 production by OCT-1 and MLO-Y4 cells, but 1α,25-$(OH)_2D_3$ does not regulate production of this prostanoid by these cells (41).

The effects of surface morphology on local factor production are relatively complex. As roughness increases, MG63 cells release an increasing amount of TGF-β1 into their conditioned medium, essentially all of which is latent in form (45, 47). There is also a corresponding increase in the amount of TGF-β1 stored in the extracellular matrix with increasing roughness. This release of TGF-β1 into the media is regulated by 1α,25-$(OH)_2D_3$ in a surface-dependent manner. The vitamin D metabolite has no effect on TGF-β1 levels in cultures grown on smooth surfaces, but it causes an increase in growth factor production on rough surfaces (33). Moreover, the effect of 1α,25-$(OH)_2D_3$ on TGF-β1 is dependent on the state of maturation of the cells within the osteoblastic lineage. Whereas the vitamin D metabolite causes an increase in TGF-β1 levels in MG63 cells and FRC cells grown on rough surfaces, it has no effect on OCT-1 cells and actually decreases the amount of TGF-β1 in the conditioned media of MLO-Y4 osteocytic cells (41).

These studies have not differentiated between latent and active growth factor, either with respect to surface roughness or as a function of 1α,25-$(OH)_2D_3$ treatment. Nonetheless, they clearly demonstrate that a factor as important to osteoblastic differentiation as TGF-β1 is affected by the substrate upon which the cells are grown. In bone, TGF-β1 is associated with the matrix through an interaction with its latency-associated peptide and latent TGF-β1-binding protein-1 (LTBP-1) (48), and this latent complex interacts with fibrillin as part of the extracellular matrix (49). Because extracellular matrix production is increased on rougher surfaces (50), it is not surprising that the amount of TGF-β1 in the extracellular matrix is also increased. This may have important consequences downstream. During bone remodeling, osteoclasts resorb bone mineral by decreasing local pH, which results in activation of latent TGF-β1 (51, 52); subsequent release of proteolytic enzymes also contributes to the activation process (53). The increase in local concentration of active TGF-β1 causes an inhibition of bone resorption through action on osteoclasts and contributes to bone formation by action on osteoblasts (54, 55).

MECHANISMS MEDIATING OSTEOBLAST RESPONSE TO SURFACE MORPHOLOGY

The mechanisms by which osteoblastic cells are regulated by the morphology of their substrate are not yet clear, but a picture is beginning to emerge that can help us to understand the complexities of the processes involved. Osteoblasts are attachment-dependent cells, which means that they must form specific cell surface interactions in order to express their differentiated phenotype.

Attachment to the Surface

Because osteoblasts are attachment-dependent cells, they must first attach to a surface before they can produce and mineralize their extracellular matrix. In the context of cell culture, attachment is a term that indicates the number of cells actually adhering to a surface with respect to the number of cells that had an opportunity to attach. This is somewhat different from adhesion, which is the relative strength of the attachment, or how difficult it is to dislodge the cell from the surface. Once cells attach to a surface, they may spread to a greater or lesser extent. A well-spread cell is assumed to be well adapted to culture, but this assessment is based on conventional tissue culture observations using a plastic or glass surface that has been treated by glow-discharge to make it suitable for cell attachment and spreading.

A number of studies have shown that cells attach well to Ti and Ti alloy surfaces. The amount of attachment appears to be greater on Ti surfaces than on tissue culture polystyrene or on cobalt chrome surfaces (56). Adhesion of the cells is also affected by the morphology of the surface; as surface roughness increases, adhesion of osteoblasts to the surface increases. Our group has empirically confirmed this observation. Cells cultured on smooth Ti can be easily released from the substratum with a brief exposure to trypsin, whereas those cells cultured on rougher Ti surfaces require a minimum of two incubations with the enzyme, and even then cells remain adherent to the surface (57).

Not only do the cells adhere to the surface more tightly, but also they assume a distinctly different morphology. Whereas cells on smooth Ti and Ti6Al4V surfaces are well spread in a tissue-culture sense, those cells cultured on rough Ti and Ti6Al4V surfaces are less well spread and appear to attach to the surface through cytoplasmic extensions (58, 59). This rearrangement of the cytoskeletal elements of the cell has been noted by others (56, 60) and can have major consequences to the cell, as information about the surface is transduced to the nucleus.

Roughness is only one parameter of surface morphology. Topography also plays a role. Brunette noted that parallel grooves of a polished surface needed to be close enough together for a single cell to recognize that it is on a rough surface (61, 62). In our own studies, we have observed that when MG63 cells are grown on Ti with a mixed topography (e.g. areas of roughness together with smooth surfaces), the culture displays characteristics typical of the dominant morphology (50). Anselme et al (60) found that human osteoblasts oriented in a parallel order on polished Ti6Al4V, but on sandblasted surfaces they assumed a stellate shape. Adhesion was decreased on the less organized surfaces, as was proliferation. The extracellular matrix, as well as the orientation of cytoskeletal proteins, was also affected by the substrate.

Role of Integrins and Signaling Pathways

All of the studies described above indicate that the manner in which the cell interacts with the surface is fundamental to how the cell will express its phenotype, including its response to regulatory factors. Cells bind to surfaces indirectly by binding to attachment proteins that adhere to the surface (63). These proteins include fibronectin, vitronectin, and osteopontin, although other proteins and serum components may contribute to the organic film that forms on the implant surface immediately upon its exposure to tissue-culture medium or to biological fluids (64). Adsorption of mineral ions, particularly Ca^{2+} and inorganic phosphate (65), also plays an important role, not only in modulating adsorption of protein but also in modulating cell response. There is some evidence that cells may bind directly to the material as well (66).

Integrins have been implicated in the mechanisms by which the cells bind to their respective attachment proteins. The integrins are a class of molecules that extend across the plasma membrane of cells, linking the cytoskeletal elements to binding sites on the attachment proteins adsorbed on the cell surface. The characteristics of this class of proteins have been reviewed in a number of papers (67–69). Integrin expression is specific to a cell type and may change as the cell undergoes differentiation or as a function of the type of surface to which the cell is exposed. The nature of the integrins expressed by the cells determines the kinds and numbers of focal contacts the cell can make with its substratum (Table 2). When osteoblasts are cultured on rough Ti surfaces, the focal contacts appear to be distributed peripherally, and cytoskeletal organization is enhanced in comparison with cells cultured on smooth polystyrene surfaces (56). Because the cytoskeleton extends from the focal contact to the nuclear matrix, the morphology of the surface can have a direct effect on new gene expression and subsequent phenotypic expression, of the cell.

Integrins also play a critical role in how a cell responds to a surface because of their signaling properties. Activation of an integrin receptor through binding to an attachment protein initiates a series of reactions, termed signal transduction cascades. At least one consequence of this is activation of protein kinase C (PKC),

TABLE 2 Integrin subunit expression in focal adhesions by human osteoblasts on tissue culture polystyrene, and on polished (smooth) and roughened (rough) Ti6Al4V surfaces (from 74)

Integrin subunit	Polystyrene	Smooth	Rough
α_2	✓	✓	✓
α_3	✓	✓	—
α_4	✓	✓	✓
α_5	✓	—	—
α_v	✓	✓	✓
β_1	✓	✓	✓
β_3	✓	✓	✓

which results in a series of phosphorylations. This kinase activity may then up- or down-regulate cellular processes, thus culminating in phosphorylation of MAP (mitogen-activated protein) kinase. Once MAP kinase is phosphorylated, it can enter the nucleus and phosphorylate transcription factors that lead to changes in gene expression.

In our laboratory, we have shown that inhibition of PKC activity with chelerythrine has no effect on proliferation of confluent cultures of MG63 cells, regardless of surface roughness (70). However, it blocks the surface-dependent increase in alkaline phosphatase activity, suggesting that differentiation of the cells is affected. Even though the response of these cells to 1α,25-$(OH)_2D_3$ is increased on rough Ti, the stimulatory effect of this vitamin D_3 metabolite on alkaline phosphatase is not via PKC. In contrast, 1α,25-$(OH)_2D_3$-dependent alkaline phosphatase is affected by inhibition of the protein kinase A (PKA) signaling pathway regardless of the surface morphology, whereas the effect of surface roughness on alkaline phosphatase is independent of PKA (71).

Part of the activation of PKC is likely through activation of phospholipase C and subsequent increases in diacylglycerol. In addition, activation of phospholipase A_2 results in increases in arachidonic acid production. Studies in our laboratory show that arachidonic acid itself can cause an increase in PKC, as well as serve as a substrate for cyclooxygenase activity, which results in an increase in production of prostaglandin E1 (PGE_1) and PGE_2. Release of both prostaglandins is increased on rougher Ti surfaces (45).

Role of Prostaglandins

Prostaglandins are necessary for normal osteoblast differentiation, although their effects are concentration dependent. At low concentrations, PGE_2 stimulates alkaline phosphatase activity and osteocalcin production. However, at high concentrations, PGE_2 can inhibit osteoblast function and in vivo can activate bone

TABLE 3 Both Cox-1 and Cox-2 mediate the effects of surface roughness on MG63 osteoblast-like cells[a]

	Inhibition of surface roughness effect		
	Indomethacin	**Resveratrol**	**NS-398**
PGE_2 production	↓↓↓	↓↓	↓↓
Alkaline phosphatase	↓↓↓	—	—
Osteocalcin production	↓↓↓	↓	↓↓
TGF- 1 production	↓↓↓	↓↓	↓
Proliferation	—	—	—

[a]MG63 cells were cultured on Ti disks with smooth ($R_a < 0.2$ m) or rough (R_a 4–7 m) surfaces. At confluence, the cultures were treated with the general cyclooxygenase (Cox) inhibitor indomethacin, or with either the Cox-1 inhibitor resveratrol or the Cox-2 inhibitor NS-398. The cultures were examined at 24 h. Total inhibition of the surface roughness effect is indicated by ↓↓↓; partial inhibition is indicated by ↓↓ or ↓; no effect is shown by —. Data are derived from Boyan et al (73).

resorption. Our studies have shown that prostaglandins mediate the initial response of MG63 cells to Ti (47, 72). When MG63 cells are incubated in culture media containing indomethacin, a general cyclooxygenase (Cox) inhibitor, the effects of surface roughness on proliferation and differentiation of the cells are blocked. However, indomethacin has no effect on these parameters in confluent cultures, regardless of surface morphology. In contrast, indomethacin blocks the effects of 1α,25-$(OH)_2D_3$ on confluent cultures, even on the roughest Ti surfaces (72). The inhibitory effect of indomethacin on the response of MG63 cells to surface roughness is time dependent, becoming less pronounced as the cells approach confluence (47).

Clinically, non-steroidal anti-inflammatory agents (NSAIDs) such as indomethacin are used to manage post-surgical pain, in part because they decrease the production of high levels of prostaglandins. Newer NSAIDs discriminate between the two isoforms of cyclooxygenase. Cyclooxygenase-1 (Cox-1) is a constitutive enzyme, meaning that it is always present in cells at a constant level of activity, whereas Cox-2 is inducible and has been implicated in the inflammatory process leading to pain. We have examined the role of each of these cyclooxygenase isoforms in the response of MG63 cells to surface roughness (73). In these studies, Cox-1 was selectively inhibited with resveratrol, and Cox-2 was inhibited with NS-398 (Table 3). The results indicate that both forms of the enzyme are involved, suggesting that the differential modulation of prostaglandin production and cell response to prostaglandins may be far more complex than originally thought.

SUMMARY

The studies described above indicate that osteoblasts are sensitive to surface roughness. The cell model selected and the morphological properties of the surface greatly affect the type of responses that occur and have led to much of the confusion in the literature to date. Despite these caveats, it is clear that surface roughness must be within the scale of the cell to be perceived by the cell, and within this limit, rougher surfaces support the expression of a more differentiated osteoblastic phenotype based on increased alkaline phosphatase activity and osteocalcin production. Osteoblasts interact with the surface indirectly through integrin-mediated mechanisms, involving phospholipase A_2 and protein kinase C. These responses are also affected by the presence of non-proteinaceous constituents on the surface, such as mineral ions and glycosaminoglycans, and the presence of regulatory factors, including growth factors, cytokines, and hormones. The synergistic response of osteoblasts to surface roughness and regulatory factors may be due in part to differential activation of various signal transduction pathways, leading to new gene transcription. Prostaglandins are important mediators of the effects of surface roughness on osteoblasts and both cyclooygenase-1 and cyclooxygenase-2 are involved. Understanding how cells interact with their surface provides clues to optimal surface design, and at the same time, provides a method for identifying pharmacologic methods for controlling tissue response during and after healing next to an implant.

ACKNOWLEDGMENTS

The authors acknowledge the assistance of their staff and students in the research described in this paper. Work done in the authors' laboratories was funded by the National Institutes of Health, the ITI Foundation (Waldenburg, Switzerland), and the Center for the Enhancement of the Biology/Biomaterials Interface at the University of Texas Health Science Center at San Antonio (UTHSCSA). We also acknowledge the contributions of our colleague, Dr. Lynda Bonewald (UTHSCSA), and particularly her generous gifts of osteoblast-like and osteocyte-like cell lines.

LITERATURE CITED

1. Boyan BD, Hummert TW, Dean DD, Schwartz Z. 1996. *Biomaterials* 17:137–46
2. Schwartz Z, Lohmann CH, Oefinger J, Bonewald LF, Dean DD, Boyan BD. 1999. *Adv. Dent. Res.* 13:38–48
3. Albrektsson T, Branemark PI, Hansson HA, Kasemo B, Larsson K, et al. 1983. *Ann. Biomed. Eng.* 11:1–27
4. Linder L, Albrektsson T, Branemark PI, Hansson HA, Ivarsson B, et al. 1983. *Acta Orthop. Scand.* 54:45–52
5. Linder L, Carlsson A, Marsal L, Bjursten LM, Branemark PI. 1988. *J. Bone Joint Surg.* 70B:550–55
6. Albrektsson T, Hansson HA. 1986. *Biomaterials* 7:201–5
7. Haddad RJ Jr, Cook SD, Thomas KA. 1987. *J. Bone Joint Surg.* 69A:1459–66
8. Thomas KA, Kay JF, Cook SD, Jarcho M. 1987. *J. Biomed. Mater. Res.* 21:1395–414

8a. Listgarten MA, Buser D, Steinemann SG, Donath K, Lang NP, Weber HP. 1992. *J. Dent. Res.* 71:364–71

9. Meyer AE, Baier RE, Natiella JR,

Meenaghan MA. 1988. *J. Oral Implantol.* 14:363–79
10. Pankowsky DA, Ziats NP, Topham NS, Ratnoff OS, Anderson JM. 1990. *J. Vasc. Surg.* 11:599–606
11. Pearson BS, Klebe RJ, Boyan BD, Moskowicz D. 1988. *J. Dent. Res.* 67:515–17
12. Howlett CR, Evans MD, Walsh WR, Johnson G, Steele JG. 1994. *Biomaterials* 15:213–22
13. Terranova VP, Aumailley M, Sultan LH, Martin GR, Kleinman HK. 1986. *J. Cell Physiol.* 127:473–79
14. Schmitz JP, Schwartz Z, Hollinger JO, Boyan BD. 1990. *Acta Anat.* 138:185–92
15. Deleted in proof
16. Deleted in proof
17. Buser D, Schenk R, Steinemann S, Fiorellini J, Fox C, Stich H. 1991. *J. Biomed. Mater. Res.* 25:889–902
18. Buser D, Mericske-Stern R, Bernard JP, Behneke A, Behneke N, et al. 1997. *Clin. Oral Implants. Res.* 8:161–72
19. Schwartz Z, Amir D, Boyan BD, Cochavy D, Muller-Mai C, et al. 1991. *Calcif. Tissue Int.* 49:359–64
20. Sela J, Shani J, Kohavi D, Soskolne WA, Itzhak K, et al. 1995. *Biomaterials* 16:1373–80
21. Braun G, Kohavi D, Amir D, Luna MH, Caloss R, et al. 1995. *Clin. Oral Implants Res.* 6:1–13
22. Sela J, Gross UM, Kohavi D, Shani J, Dean DD, et al. 2000. *Crit. Rev. Oral Biol. Med.* 11:423–36
23. Schwartz Z, Swain LD, Marshall TS, Sela J, Gross U, et al. 1992. *Calcif. Tissue Int.* 51:429–37
24. Steflik DE, Corpe RS, Lake FT, Young TR, Sisk AL, et al. 1998. *J. Biomed. Mater. Res.* 39:611–20
25. Yao J, Cs-Szabo G, Jacobs JJ, Kuettner KE, Glant TT. 1997. *J. Bone Joint Surg.* 79:107–12
26. Gross UM. 1988. *J. Dent. Edu.* 52:798–803
27. Jacobs J, Silverton C, Hallab N, Skipor A, Patterson L, et al. 1999. *Clin. Orthop. Rel. Res.* 358:173–80
28. Urban R, Jacobs J, Gilbert J, Galante J. 1994. *J. Bone Joint Surg.* 76:1345–59
29. Franceschi RT, James WM, Zerlauth G. 1985. *J. Cell Physiol.* 123:401–9
30. Franceschi RT, Romano PR, Park KY. 1988. *J. Biol. Chem.* 263:18938–45
31. Bonewald LF, Kester MB, Schwartz Z, Swain LD, Khare AG, et al. 1992. *J. Biol. Chem.* 267:8943–49
32. Boyan BD, Schwartz Z, Bonewald LF, Swain LD. 1989. *J. Biol. Chem.* 264:11879–86
33. Boyan BD, Batzer R, Kieswetter K, Liu Y, Cochran DL, et al. 1998. *J. Biomed. Mater. Res.* 39:77–85
34. Kodama H, Amagai Y, Sudo H, Kasai S, Yamamoto S. 1981. *Jpn. J. Oral Biol.* 23:899–901
35. Kurihara N, Ishizuka S, Kiyoki M, Haketa Y, Ikeda K, Kumegawa M. 1986. *Endocrinology* 118:940–47
36. Sudo H, Kodama H, Amagai Y, Yamamoto S, Kasai S. 1983. *J. Cell Biol.* 96:191–98
37. Chen TL, Bates RL, Dudley A, Hammonds RG Jr, Amento EP. 1991. *J. Bone Miner. Res.* 6:1387–93
38. Majeska RJ, Rodan GA. 1982. *J. Biol. Chem.* 257:3362–65
39. Chen D, Chen H, Feng JQ, Windle JJ, Koop BA, et al. 1995. *Mol. Cell. Differ.* 3:193–212
40. Kato Y, Windle JJ, Koop BA, Mundy GR, Bonewald LF. 1997. *J. Bone Miner. Res.* 12:2014–23
41. Lohmann CH, Bonewald LF, Sisk MA, Sylvia VL, Cochran DL, et al. 2000. *J. Bone Miner. Res.* 15:1169–80
42. Ong JL, Carnes DL, Cardenas HL, Cavin R. 1997. *Implant. Dent.* 6:19–24
43. Ong JL, Cardenas HL, Cavin R, Carnes DL Jr. 1997. *Int. J. Oral Maxillofac. Implant.* 12:649–54
44. Ong JL, Prince CW, Raikar GN, Lucas LC. 1996. *Implant. Dent.* 5:83–88
45. Kieswetter K, Schwartz Z, Hummert TW,

Cochran DL, Simpson J, et al. 1996. *J. Biomed. Mater. Res.* 32:55–63
46. Schwartz Z, Dennis R, Bonewald LF, Swain LD, Gomez R, Boyan BD. 1992. *Bone* 13:51–58
47. Sisk M, Lohmann CH, Cochran DL, Sylvia VL, Simpson JP, et al. 2001. *Clin. Oral Implant. Res.* 12:52–61
48. Dallas SL, Park-Snyder S, Miyazono K, Twardzik D, Mundy GR, Bonewald LF. 1994. *J. Biol. Chem.* 269:6815–22
49. Dallas SL, Miyazono K, Skerry TM, Mundy GR, Bonewald LF. 1995. *J. Cell Biol.* 131:539–49
50. Martin JY, Schwartz Z, Hummert TW, Schraub DM, Simpson J, et al. 1995. *J. Biomed. Mater. Res.* 29:389–401
51. Oursler MJ. 1994. *J. Bone Miner. Res.* 9:443–52
52. Oreffo RO, Mundy GR, Seyedin SM, Bonewald LF. 1989. *Biochem. Biophys. Res. Commun.* 158:817–23
53. Lyons RM, Keski-Oja J, Moses HL. 1988. *J. Cell Biol.* 106:1659–65
54. Bonewald LF, Mundy GR. 1990. *Clin. Orthop. Rel. Res.* 250:261–76
55. Bonewald LF. 1996. In *Principles of Bone Biology*, ed. JP Bilezikian, LG Raisz, GA Rodan, pp. 647–59. San Diego: Academic
56. Sinha RK, Morris F, Shah SA, Tuan RS. 1994. *Clin. Orthop. Rel. Res.* 305:258–72
57. Schwartz Z, Martin JY, Dean DD, Simpson J, Cochran DL, Boyan BD. 1996. *J. Biomed. Mater. Res.* 30:145–55
58. Lincks J, Boyan BD, Blanchard CR, Lohmann CH, Liu Y, et al. 1998. *Biomaterials* 19:2219–32
59. Boyan BD, Lincks J, Lohmann CH, Sylvia VL, Cochran DL, et al. 1999. *J. Orthop. Res.* 17:446–57
60. Anselme K, Bigerelle M, Noel B, Dufresne E, Judas D, et al. 2000. *J. Biomed. Mater. Res.* 49:155–66
61. Brunette DM. 1986. *Exp. Cell Res.* 167:203–17
62. Brunette DM, Ratkay J, Chehroudi B. 1991. In *Bone-Biomaterials Interface*, ed. JE Davies, pp. 170–79. Toronto: Univ. Toronto Press
63. Degasne I, Basle MF, Demais V, Hure G, Lesourd M, et al. 1999. *Calcif. Tissue Int.* 64:499–507
64. Stanford CM, Solursh M, Keller JC. 1999. *J. Biomed. Mater. Res.* 47:345–52
65. Ong J, Chittur K, Lucas L. 1994. *J. Biomed. Mater. Res.* 27:1337–46
66. Gronowicz G, McCarthy MB. 1996. *J. Orthop. Res.* 14:878–87
67. Craig W, Cheng S, Mullen D, Blevitt J, Pierschbacher M. 1995. *Biopolymers* 37:157–75
68. LeBaron R, Athanasiou K. 2000. *Tissue Eng.* 6:85–103
69. Anselme K. 2000. *Biomaterials* 21:667–81
70. Lohmann CH, Sagun R Jr, Sylvia VL, Cochran DL, Dean DD, et al. 1999. *J. Biomed. Mater. Res.* 47:139–51
71. Boyan BD, Sylvia VL, Liu Y, Sagun R, Cochran DL, et al. 1999. *Biomaterials* 20:2305–10
72. Batzer R, Liu Y, Cochran DL, Szmuckler-Moncler S, Dean DD, et al. 1998. *J. Biomed. Mater. Res.* 41:489–96
73. Boyan BD, Lohmann CH, Sisk M, Liu Y, Sylvia VL, et al. 2001. *J. Biomed. Mater. Res.* In press
74. Sinha RK, Tuan RS. 1996. *Bone* 18:451–57

Annu. Rev. Mater. Res. 2001. 31:373–85

The Role of Materials Research in Ceramics and Archaeology[1]

Pamela Vandiver
Smithsonian Center for Materials Research and Education, Smithsonian Institution, Washington, DC 20560; e-mail: vandiverp@scmre.si.edu

Key Words archaeological science, pyrotechnology, glass and glazes, plasters, craft history, ethnoarchaeology, archeomaterials, history of technology

■ **Abstract** Materials research has been applied successfully to the study of archaeological ceramics for the last fifty years. To learn about our history and the human condition is not just to analyze and preserve the objects but also to investigate and understand the knowledge and skills used to produce and use them. Many researchers have probed the limits and methods of such studies, always mindful that a glimpse at ancient reality lies in the details of time and place, context of finds, and experimentally produced data, usually compared with standards that were collected in an equivalent ethnographic setting or that were fabricated in a laboratory in order to elucidate the critical questions in a technology that could be understood in no other way. The basis of most studies of ancient technology has been established as microstructure; composition and firing; methods and sequence of manufacture; differentiation of use; use-wear and post-depositional processes; technological variability that can be interpreted as a pattern of stasis or innovation, which can be related to cultural continuity or change; and interpretation that can involve technology, subsistence trade, organization, and symbolic group- and self-definition.

INTRODUCTION

In his preface to the translation of Pushkin's *Eugene Onegin*, Vladimir Nabokov observed (1), "In art as in science there is no delight without the detail, and it is on the details that I have tried to fix the reader's attention. Let me repeat that unless these are thoroughly understood and remembered, all 'general ideas' (so easily acquired, so profitably resold) must necessarily remain but worn passports allowing their bearers short cuts from one area of ignorance to another." Using modern materials analysis to characterize ancient ceramics as they exemplify our cultural heritage is much the same problem. Just as an author deconstructs the details of a situation or character he imagines as meaningful and real enough to reassemble and record, we seek to analyze and interpret the ancient record starting with the study of "things" (in this case, ceramic things) and the context of their deposition in the archaeological record. We try to establish the details of date, place, context, materials, technology, significance, and cultural meaning—the same expectations we have of an informative newspaper article. The ceramic art object or ceramic archaeological artifact or site with ceramic architecture actually represents specific knowledge, meaning, and culture. To learn about our history and the human condition is not just to analyze and preserve the objects but also to investigate and understand the knowledge and skills used to produce and use them (2). In the quest to learn from objects, materials science has special philosophical and practical roles. The structure-property-processing-performance paradigm of materials science and engineering provides the template and the methodology, and materials research laboratories provide some of the cutting-edge tools to do the job.

RESULTS OF PAST STUDIES

Two previous articles have appeared in the *Annual Review of Materials Science* (4, 5) and one in the *Annual Review of Anthropology* (6) that have aimed to use analyses of materials, their processing and transformation into useful goods, as a way of understanding cultural phenomena. In 1986 Smith (4) stated that recent developments in materials science had allowed the visualization of material structure beyond human vision such that direct investigation of objects would lead to a critical understanding of the role of materials development in human history. His experiences with the translation of Medieval and Renaissance period European texts had shown that scholars often had misunderstood and improperly recorded the practices of craftsmen. Smith emphasized that modern researchers, in common with ancient craftsmen, learn by "doing things," and both have a special knowledge of the direct interaction with materials that reveals links among composition, properties, and processing. Such experiences, he maintained, have an aesthetic quality that arouses curiosity and allows the identification of xenomorphs or misfits, especially at boundaries where instability, interaction, and transformation occur. Isolation of such phenomena he considered as prime locations of concentrated investigation and as sources of possible discovery. However, such invention is impossible to document in the archaeological record, and only objects that represent technological development, or technological innovation and change (as when something new is brought to the market and becomes commonplace) are excavated. Smith viewed materials history as the replacement of one material or process by another for utilitarian purposes, but he believed that the subtle advances in materials technology that occurred before advances in mechanical technology resulted from the symbolic use of materials to satisfy philosophical questions and to demonstrate and control seemingly magical phenomena. His thoughts have provided many questions for those who study material culture and cultural heritage.

Smith's call for the study of archaeology from a materials point of view had been foreshadowed in 1985 by Kramer (6) who reviewed and promoted the study of modern potters in traditional settings before their wares and livelihoods were replaced by alternative materials and technologies. Kramer, like Smith, called for generation of a body of empirical data, but by using comparative ethnographic studies of traditional potters and their wares to evaluate the archaeological interpretations of ceramic evidence. The conclusions often enrich or contradict archaeological models that attempt to explain production, distribution and exchange, consumption, reuse, and discard. One purpose of ethnoarchaeology is to evaluate factors that promote or discourage preservation of clay-based products in the archaeological record, as, for instance, when a traditional village is abandoned and its degradation can be observed. Another is to is to evaluate factors that favor stylistic conservatism or innovation and change, as the meaning of change in the archaeological record is one of the critical questions. For instance, when does a change in pottery style equal a change in people? A third is to provide information on organization of ceramic production, and the shift from home-based seasonal production to workshop to full-time factory production with its attendant implication of an increase social stratification, class or caste differentiation in which artisans, craftsmen, and workers are low in status and often denied freedom to migrate, effectively being slaves. Archaeological studies of pottery produce information primarily about production, consumption, and exchange, whereas one would like to learn more about cultural organization, operation, and evolution, and especially such timely topics as gender and social identity. One area of study that is proving fruitful in this quest is in instances where historical documentation, observation of pottery communities, and archeological investigation overlap and intersect.

In 1989, in addition to traditional historical documentation, Kingery (5) used materials analysis of the products of three industries—the seventeenth century development of European porcelain at Meissen, the late nineteenth century Edison carbon microphone and electric light, and the early twentieth century Nernst zirconia-yttria lighting filament—in order to research the early development of modern materials technology and especially modern ceramics with its significant impact on the electrical and electronics communication and information industries. He placed the development of these socially desired technologies within a model of technological innovation and engineering science that is now familiar to us all in the twenty-first century. Essentially, a research laboratory conducts materials science—discovery and scientific explanation—in order to enable new materials, devices, and products that modify our lives and evolve into new socio-technical systems, such as transportation, communication, or education. Kingery was effectively using the methods prescribed by Smith and Kramer to investigate the recent historical past. In order to study modern technology, he found useful three ways of viewing, classifying, or investigating a technology. These perspectives, borrowed from studies of the history of technology, are product-centered, industry-centered, and task-focused.

The above three articles from the Annual Reviews publications represent the kinds of problems and viewpoints that were central to defining and developing the field of ceramics and archaeology in the1980s. They are focused both methodologically and heuristically. If we review articles in *Science* in the 1960s, we find a different set of questions being pursued, that is, questions that used materials analysis of archaeological artifacts or art objects specifically chosen to explore the outer bounds of the field and to provide an overview and philosophical direction. Wertime (7) attempted an industry-wide review of the development of metallurgy during the fifth to third millennia B.C. in Southwest Asia. Smith (8, 9) emphasized materials, not matter, as the basic unit of human experience, and the processing of ceramics and metals as a way of altering properties for aesthetic and utilitarian, social, and symbolic functions. Smith emphasized as fundamental the dependence of properties upon structure, and thus enunciated what has become the structure-processing-properties-performance paradigm of materials science. Wertime (10) then, even more globally, defined his field of study as pyrotechnology, the use of fire to fabricate plasters, ceramics, metals, glazes, and glasses, and he catalogued analytical methods, primarily chemical, then useful in the study of these materials. Then in 1978, Kranzberg & Smith (11) synthesized these ideas in order to make a national case for the support and funding of materials science and engineering as a separate discipline. Wertime & Wertime (12) in 1979 organized a conference to explore case studies of early pyrotechnologies and to document the evolution of the first fire-using industries. A second conference, called Archaeological Ceramics, was held a year later to focus on ancient ceramics, generously defined as non-metals, so that many technologies heretofore considered unrelated to clay-based ceramics were reported (13). The result was that the field of archaeological ceramics was enlarged to more closely approximate the field defined by the major professional organization, the American Ceramic Society, as pigments, cement, concrete and plasters, whitewares, and glasses among others. These developments cleared the way for new questions of cross-cultural comparisons of technologies, cross-craft interactions among craftsmen, and a questioning of the way we define a craft activity or specialization or industry in an ancient culture (14). The term archeomaterials was coined (15). The effect on archaeological field research has been to increase the significance of technology and materials identification and to promote the search for direct evidence of craft activities, especially workshops, furnaces, industrial debris, manufacturing elements and tools, caches of raw materials, and objects in successive processing stages in order to learn what people did, how they did it, and what they had to know to practice the technology.

METHODS, TOOLS, AND TECHNIQUES

Archaeological excavations conducted in the last century produced artifacts that were displayed in museums and organized and described by material and by technology, with stone and bone being the oldest and coming from the lowest strata, and with bronze and iron, respectively, being more recent. The presence of ceramics spanned the later stone strata and was nearly ubiquitous in higher strata up to the present, and thus was left out in the naming of periods as Old and New Stone Ages, Bronze Age, and Iron Age. In the Old Stone Age, or Paleolithic period, stone artifacts were mainly produced by percussion and flaking, where as in the New Stone Age, or Neolithic period, tools were often smoothed by grinding, but this generality has exceptions; for instance, grindstones for pigments were made and used in the Paleolithic period. With continued excavation, observations amassed a body of data that separated cultures by economic or subsistence system and social organization into the hunter-gatherer with the longest history, the farmer from the Neolithic period, and the urban-state dweller, the rural farmer, and cultural outsider or outlaw with the shortest history. The understanding of the details of this development is one of the crowning achievements of archaeology. However, the old museum-based periodization has been retained throughout the elaboration of our socio-economic, politico-military, and artistic magico-religious philosophical understanding of the significance of the ancient record.

Identification of the technologies specific to each period or economic system have led to four kinds of studies. In one case, some sought to microscopically study a technology and reproduce it, often examining the replicated artifact or processing stage using the same analytical methods as for the original artifact, Semenov's (16) investigation of stone tool surfaces in the 1920s being the first. In another case, researchers sought to produce compendia that enumerated various technologies, often retrospectively from the point of view of the modern practice and organization of a technology; one of the best and most commonly used is by Hodges (17). These compendia derive from the study of the history of western technology, and, although meant to apply to the archaeological record, often are based on such Eurocentric nineteenth century texts as Diderot's *Encyclopedia of Arts and Industries*. Thus these general guides help to establish our expectations of what will be found during an investigation but must be used with a measure of skepticism to avoid producing a retrospective technology. Some compendia are specific to particular culture areas and include technical analyses, such as the study of Egyptian materials and technologies by Lucas (18) and recent update of the text (19). In a third case, reconstructing stages in the production, use, and discard of a particular type or style of artifact through careful observation of the group of artifacts and their archaeological contexts in order to establish the nature and diversity of a technology was the methodology first set forth by Leroi-Gourhan in the 1940s (20). He called "chaîne operatoire," or the chain or sequence of operations, as the critical step of this systems approach to establishing the life history of an artifact. He then relied philosophically on the French school of structuralism for the intellectual tools to generalize to the technology and the socio-technical system into which it is embedded. In the fourth case, in order to be sure that we are seeing sharply through the changes or veil of age, a recent trend, especially for very old artifacts, has been to examine processes of deposition, degradation, corrosion, and weathering in order to separate "ecofact" from artifact, that is natural from human processing; for example to distinguish an ancient ceramic object made in a forest fire from one fired by people. The term technofact has been used to describe tools and incomplete objects in various stages of production. The significance here for methodology is that we are using or making and then testing artificial standards, the processing, structure, and properties of which are known, in order to compare with aged objects that we seek to know. Each of the four approaches outlined above has been a powerful technique in the study of pyrotechnology and archaeomaterials in general and archaeological ceramics, in particular.

Prior to the development of direct dating by radiocarbon and luminescence techniques, relative chronologies were established by the technique of serial dating. Essentially, the relative placement of diagnostic pottery shards in superimposed archaeological strata allowed comparison of relative age in different excavation squares or trenches at one site, as well as comparison of age between sites. The stylistic grouping and counting of wares by strata led to plots of abundance through time, called battleship curves, that provided a measure of material culture change from one period to another. To this refined connoisseur-ship of broken, specially shaped or decorated pottery that could only be learned through apprenticeship and practice came other strategies to aid in the determination of date. One approach, exemplified in the work of Caley (21, 22), was to analyze the chemical composition of metals, especially coins, and glasses in order to correlate changes in composition with a time period. Another approach was to attempt to make connections between one archaeological site and another of similar date by the recognition of imported trade goods and their chemical and technical identification to particular source areas or a general provenance. Shepard (23) and others tested ceramics with unusual decorations or anomalous fabrics that were probably intrusive into an archaeological site by using the geological methods of petrographic analysis to identify minerals and mineral assemblages, to quantify their relative amounts, and to relate them to possible source regions with similar suites of minerals and particular types of associated geologies. Often it was only possible to relate pottery shard imports to regions where that particular type was more abundant. Thus many studies have sought to establish the nature of the exchange by mapping abundances of artifacts at central and peripheral areas. Dating has now become a specialized craft (24), and the early chemical and structural studies have developed into more specialized techniques that have been successful in establishing aspects of regional economy, exchange systems, long-distance trade, and the source or provenience of artifacts (25).

The analytical methods to deconstruct and interpret ancient ceramic technology and to determine raw material sources and the provenance of ceramics are well established. Until we generate new questions, the tools and techniques remain static. McGovern (26) reviewed the status of scientific studies in archaeology, as well as some of the contemporary methods, for the study of archaeological materials. Much of the accumulated methodology is available in handbooks; three that are comprehensive and classics are by Shepard (23), Rye (27), and Rice (28). Although Shepard uses the type-style classification of pottery in her own studies, she states more technology can be learned from the misfits—over- or under-fired shards, misshapen ones with fabrication errors entrained, or ones with evidence of use or reuse. She presents photographs of various pottery surfaces and calls for careful microscopic examination of surface textures. Rye uses radiography to image the inner structure of ceramics and compare it with surface traces in order to elucidate methods of manufacture and the sequence of manufacturing steps. Rice provides a thorough review of previous studies and approaches to ceramic problems, but her encyclopedic text often does not provide sufficient specific technical data to permit the researcher to solve problems, such as how to calculate a coefficient of thermal expansion or determine a ceramic firing temperature, but she provides references.

Many other texts use a case study approach, as do conference and journal articles. Kingery & Vandiver (29, 29a) determine composition, microstructure, and firing temperature to interpret the appearance and production of several individual ceramic objects significant in the history of ceramic technology. A landmark ceramic exhibition at the British Museum is detailed in a catalog *Pottery in the Making*, edited by Freestone & Gaimster (30), that contains significant review articles on ceramics in various culture areas and periods from a technological and materials analysis perspective. This exhibition parallels an earlier one, entitled Masterpieces of Glass, that remains an authoritative source on the fabrication of ancient glass objects (31). Two ethnographies with significant technical content are the classics by Wulff (32) and Rye & Evans (33). Wulff places pottery within the larger context of various cultural uses of clays in Iran, and Rye & Evans investigate the glazed pottery of Pakistan. Many case studies are contained in conference proceedings. Especially significant and relevant are those that expanded the ecological and environmental context of ceramic production and use and that have been edited by Matson (34) and van der Leeuw (35). Regular conferences with case studies in archaeological ceramics are the Archaeological Chemistry conferences of the American Chemical Society, the Archaeometry conferences sponsored by the Laboratory for Archaeometry and the History of Art at Oxford, Ceramics and Civilization sponsored by the American Ceramic Society, Materials Issues in Art and Archaeology sponsored by the Materials Research Society, the International Symposium on Ancient Ceramics sponsored by the Shanghai Institute for Ceramics, as well as parts of conferences of the Archaeological Institute of America, the Society of American Archaeology, the Society of Archaeological Sciences, and others. For conservation the prime conferences are those sponsored by the International Institute of Conservation, the American Institute of Conservation, and the International Conservation Committee for Museums.

The most recent developments in ceramic analysis have been studies to identify organic residues concentrated from various parts of cooking and storage jar walls in order to reconstruct subsistence patterns. McGovern et al (36) have reconstructed a history of beer and wine, and Evershed (37) has determined the contents of medieval British stews. Organic chemistry, particularly biogeochemistry, combined with inorganic materials analysis, is promising to contribute to better understanding through environmental reconstruction of sites. Two particularly noteworthy topics are the analyses of ancient soils for the traces of activities trapped within, as well as a detailed depositional history that provides additional context, and the survey and reconstruction of the ancient landscape to interrelate human and natural processes. For instance, does environmental degradation begin with sedentary Neolithic populations? What is the carrying capacity of various ecological niches for various activities and population densities? Such studies should generate new knowledge about the mixed use of site catchement areas and areas of interaction on the peripheries of settlements. In the first case, microexcavation of blocks lifted from the site and transported to the laboratory provides the samples to be studied and tested. In the second case, techniques of remote sensing, geological coring, and prospecting are used.

Others (38, 39) have discussed ways to better remove cultural materials from the field and to conserve them through a process of assessment, documentation, stabilization, and, if necessary, intervention to preserve original material, appearance, and sometimes use, through reversible treatments. Evaluation and improvement of treatments may well become a significant endeavor for materials analysts because many objects are composite structures, such as low-fired alkaline or lead-glazed ceramics, or high-fired, high-calcium-glazed ceramics in which the body and glaze do not fit well (40). They are recognized by glazes that are crazed or flaking or by a porous body that retains salts. Current conservation treatments favor the preservation of either the body, by removing salts and delaminating the glaze by delayed crazing and slow crack growth, or the glaze, by not immersing the ceramic in water. In addition, many ceramics are a mixture of crystals and glass or an inhomogeneous glass that weathers preferentially at the least stable part of the microstructure and composition. Many low-fired earthenwares consist of clay barely dehydroxylated and held together with only a small amount of glass at the sintered contact joints. Many of these wares are difficult to conserve and lift in the field, and many do not make it to the laboratory, including entire ware types. Microstructural changes in porous materials are often small but cumulative, difficult to document, and could benefit from characterization and analysis of materials and degradation processes (41).

The process of archaeological excavation is innately destructive and descriptive and can only be conducted once. In recognition of the limitations of this methodology, the task of archaeology has been enlarged in the last few years to include not only preservation of artifacts and the reporting of the results of excavation and analyses, but also public education and outreach, site planning, and oftentimes site preservation, and even onsite museum planning and training. In addition, as archaeological sites become more scarce and precious, the need to prevent looting and theft of cultural heritage and the illicit plundering of sites, as well as their unnecessary development, requires vigorous cultural resource management. The task of ceramic technical studies will expand from reconstructing and explaining the nature of the objects and their context, along with an understanding of the knowledge, skills, and decisions used in production, use, exchange, and discard, to one of defining and preserving material culture as a significant way learning about past meaning and values. One change for ceramic studies is that many analyses will need to be non-destructive, and many museum laboratories already are developing such strategies and equipment, especially in France and Japan. The specific significance of research projects and tests will need to be made more explicit in terms of, not only scholarly value, but also public value. Sampling strategies will be articulated more carefully so as to be both representative and hierarchical, that is, proceeding from non-invasive methods for large numbers of artifacts to more destructive sampling of fewer numbers of artifacts for more cost- and time-intensive tests. Materials analysts and archaeologists are each becoming more interdisciplinary and entering one another's spheres of expertise. Interdisciplinary graduate programs in both fields are readily available at many universities. Finally, tasks following excavation may expand to the preservation of a technology through its onsite or factory replication, as well as to interpretive public outreach through local ceramic workshops.

TYPES OF PROBLEMS IN ARCHAEOLOGICAL CERAMICS

Trace element analysis is the primary means by which studies of trade, provenance, and sourcing are conducted, not only for ceramics, but for obsidian and other materials as well. Neff (25) and Bishop & Lange (42) have reviewed the current status and development of the field. Petrographic analysis, bulk chemical analysis, and microstructural studies often complement and refine the conclusions. Onsite resource survey for clay and other raw material sources is essential to transforming a study of provenance into one of sourcing. Anticipation of the geological variability is also crucial to success. Problem definition is the most significant aspect of any archaeological materials research. Once an initial study is carried out that "gets the picture right," other sources of information and studies can be brought to bear to amplify and enlarge our interpretation of the ancient world. For instance, Sayre's (43) classic study that established compositional categories of ancient glasses has been buttressed, not only by other compositional studies, but also by the study of texts and replicate processing experiments by Oppenheim (44).

Technological studies primarily have sought to establish pattern by structure. That structure can be macrostructure or microstructure. Smith (45) in his book, *History of Metallography*, detailed the success of microscopy in understanding the composition, processing, and properties of both modern and traditional metals. Lang & Middleton (46) show how macrostructure can be imaged using traditional X-radiography and xeroradiography to detect fabrication methods and sequences, defects in production, and conservation treatments that help authenticate museum objects, but also lead to an understanding of production technology. Microstructural imaging by scanning electron microscopy has found a special function in archaeological ceramic studies, which is determination of firing temperature. Kingery & Frierman (47) and Maniatis & Tite (48) have pioneered its use by refiring fragments of a ceramic to known increments of temperature and then comparing the degree of sintering of these standards to the unknown fragment that has not been refired. The firing temperature is established just below the point at which the microstructure is observed to change, especially by having the walls of pores smooth, sintered joints increased in size, and small pores round. This method is often supplemented by refiring tests that observe changes in refired color and linear firing shrinkage with increasing temperature, as well as changes in phase assemblage, as determined by X-ray diffraction or differential thermal analysis.

One of the reasons that studies of metals, both modern and ancient, progressed more rapidly in the twentieth century than studies of ceramics (and the development of new ceramic materials) is simply because clay particles are so much smaller than the grain size in most metals processed by traditional means that probes of much finer magnification are required to see the submicron and micron particle sizes. In 1992, Kingery (49), in a study of the history of sintering, alluded to the fact, as did Smith (50), that materials interactions can be viewed at interfaces with a variety of high magnification probes. This ability to study interfaces, postulate reactions, and understand fracture surfaces has been critical in the study of both clay and non-clay-based ceramics.

The study of ancient ceramics has expanded from pottery studies to numerous related materials; for instance, the Paleolithic period processing and use of pigments and ceramics (51, 52) and the Neolithic period plasters (53, 54)—studies that help us understand the place and use of materials technologies in various cultures. Also studied have been the technologies, such as faience (55), glass, and metals (56), that developed for urban elites as correlates of the hierarchical distribution of power and prestige, the accumulation of wealth, and the growth of bureaucracy. The special place of slags as a bridge between glass and metal processing also has received treatment (57). Many of the early luminaries of the field hoped for a unified history of technology in the days when the history of technology was envisioned as a progression of the newest, biggest, and best of enlightenment developments, often primarily mechanical. What has resulted is a diverse and rich picture of ancient technology that can support a complex range of interpretations from reconstruction of craft, or cross-craft interactions, cross-cultural comparisons of economy, technology, and organization, as well as regional economy and patterns of long-distance exchange and socio-technical systems.

The yield of information from a study depends on the sampling strategy, and different types of problems can be solved by different sets of samples; for instance, the intensively analyzed single sample that can prove the presence of a technology in a given time and place and can lead to an understanding of the practice of a technology, but cannot demonstrate the absence or range of a technology or technological practice. Selected from a single site or site locus, a group of samples can be chosen to demonstrate the range of technological variables—color and texture, oxidation and reduction, low- and high-firing temperature range. Technical analysis can show the range of technological variables necessary to produce a particular visual effect or a particular property, such as strength or toughness, impermeability, or translucency. From a group of sites or a single site through a depth of time, we can survey the range of variables that indicate conservatism or change in the practice of a technology. Comparison with geological or replicated standards can prove illuminating, as can a variety of other sampling strategies. For instance, at one site in the absence of pottery or pottery profiles, the small sherds stratigraphically excavated are being analyzed to develop a technological typology.

It is my opinion that without a physical basis in data generated by analysis of artifacts and objects, the reconstruction of history becomes clouded as it moves away from direct interpretation of artifacts to the interpretation of what others have written. Smith was impressed by the dichotomy between those who wrote about technology and those who practiced various crafts in Europe during the Medieval and Renaissance periods. However, that same dichotomy exists today when some write about technology and its meaning without identifying the materials or processes or "getting the technology right." Oftentimes the basis is dealing with only the written historical references that are peripheral to the technologies. Many contemporary references elucidate the technology and provide a resource to initiate problem formulation (58–68).

CONCLUSION

It is hoped that this review has convinced the reader that we can reach back through time (without a machine for time travel) to develop a critical understanding of how ancient people used craft and technology to solve problems of survival and organization and to make symbols or representations of what was important in their world, especially for its maintenance, longevity, and beautification. This process should be known not just as the study of material culture and cultural heritage, but also as an important field of materials research.

LITERATURE CITED

1. Nabokov V. 1990. *Pushkin's Eugene Onegin*, p. 8. Princeton, NJ: Princeton Univ. Press
2. Vandiver PB. 2001. *Mater. Res. Sci. Bull.* 26(1):13–14
3. Vandiver PB. 1988. In *Materials Issues in Art and Archaelogy*, ed. EV Sayre, PB Vandiver, C Stevenson, 123:89–102. Warrendale, PA: Mater. Res. Soc. Symp. Proc.
4. Smith CS. 1986. *Annu. Rev. Mater. Sci.* 16:1–11
5. Kingery WD. 1989. *Annu. Rev. Mater. Sci.* 19:1–20
6. Kramer C. 1985. *Annu. Rev. Anthropol.* 14:77–102
7. Wertime TA. 1964. *Science* 146:1257–67
8. Smith CS. 1965. *Science* 148:908–17
9. Smith CS. 1968. *Science* 162:637–44
10. Wertime TA. 1973. *Science* 182:875–87
11. Kranzberg M, Smith CS. 1978. In *Materials Science and Engineering: Its Evolution, Practice and Prospects*, pp. 1–39. Lausanne: Elsevier Sequoia
12. Wertime TA, Wertime SF, eds. 1982. *Early Pyrotechnology: The Evolution of the First*

Fire-Using Industries. Washington, DC: Smithsonian Inst. Press
13. Olin JS, Franklin AD, eds. 1982. *Archaeological Ceramics*. Washington, DC: Smithsonian Inst. Press
14. Kingery WD, ed. 19984–1997. *Ceramics and Civilization*. Westerville, OH: Am. Ceram. Sci. Vol. 1–8
15. *Archeomaterials*. 1987–1993. Vol. 1–7
16. Semenov SA. 1964. *Prehistoric Technology: An Experimental Study of the Oldest Tools and Artefacts from Traces of Manufacture and Ware*. London: Cory, Adams & Mackay
17. Hodges HWM. 1964. *Artifacts: An Introduction to Primitive Technology*. New York: Praeger
18. Lucas A. 1962. *Ancient Egyptian Materials and Industries*. 4th Ed. revised by JR Harris. London: Arnold
19. Nicholson PT, Shaw I, eds. 2000. *Ancient Egyptian Materials and Technology*. Cambridge, UK: Cambridge Univ. Press
20. Leroi-Gourhan L. 1943–1945. *Evolution et Techniques*. Paris: Michel. Vol. 1–2
21. Caley ER. 1964. *Analyses of Ancient Metals*. New York: Macmillan.
22. Caley ER. 1962. *Analyses of Ancient Glasses, 1790–1957, a Comprehensive and Critical Survey*. Corning, NY: Corning Museum of Glass
23. Shepard AO. 1956. *Ceramics for the Archaeologist*. Washington, DC: Carnegie Inst. Pub.
24. See issues of *Radiocarbon*, 1959–2000, and Feathers J. 2001. *Ancient TL*. 18(2): 51–61
25. Neff H. 1992. *Chemical Characterization of Ceramic Pastes in Archaeology*. Madison, WI: Prehistory Press
26. McGovern PE, ed. 1995. *Am. J. Archaeol.* 99:79–142
27. Rye OS. 1981. *Pottery Technology: Principles and Reconstruction*. Seattle, WA: Taraxacum
28. Rice PM. 1987. *Pottery Analysis: A Sourcebook*. Chicago, IL: Univ. Chicago Press
29. Kingery WD, Vandiver PB. 1986. *Ceramic Masterpieces*. New York: Free Press
29a. Kingery WD, Vandiver PB. 2000. *Ceramic Masterpieces*. Westerville, OH: Am. Ceram. Soc. 2nd Ed.
30. Freestone I, Gaimster D, eds. 1997. *Pottery in the Making: Ceramic Traditions*. London: British Museum Press; Washington, DC: Smithsonian Inst. Press
31. Harden DB, et al. 1968. *Masterpieces of Glass*. London: British Museum Press
32. Wulff H. 1966. *The Traditional Crafts of Persia*. Cambridge, MA: MIT Press
33. Rye OS, Evans C. 1976. *Traditional Pottery Techniques of Pakistan: Field and Laboratory Studies*. Washington, DC: Smithsonian Inst. Press. Vol. 21
34. Matson FR, ed. 1965. *Ceramics and Man*. New York: Wenner-Gren Found. Anthrop. Vol. 41
35. van der Leeuw SE, ed. 1984. *The Many Dimensions of Pottery*. Amsterdam: Univ. Amsterdam
36. McGovern PE, Biers WR. 1990. *Organic Contents of Ancient Vessels*. Philadelphia, PA: Univ. Pennsylvania Press
37. Evershed RP, Charters S, Quye A. 1995. In *Materials Issues in Art and Archaeology IV*, ed. PB Vandiver, 352:85–96. Warrendale, PA. Mater. Res. Soc. Symp. Proc.
38. Cronyn JM. 1990. *The Elements of Archaeological Conservation*. New York: Routledge
39. Buys S, Oakley V. 1998. *The Conservation and Restoration of Ceramics*. Oxford, UK: Butterworth-Heinemann
40. Vandiver PB. 1992. In *Corrosion of Glass Ceramics and Ceramic Superconductors*, ed. DE Clarke, BK Voitos, pp. 393–430. Park Ridge, NJ: Noyes
41. Scherer GW, Flatt R, Wheeler G. 2001. *MRS Bull.* 26(1):44–50
42. Bishop RC, Lange FW, eds. 1991. *The Ceramic Legacy of Anna O. Shepard*. Denver, CO: Niwot

43. Sayre EV, Smith RW. 1961. *Science* 133(3467):1824–26
44. Oppenheim AL. 1970. *Glass and Glass Making in Ancient Mesopotamia*. Corning NY: Corning Museum of Glass
45. Smith CS. 1960. *A History of Metallography*. Chicago, IL: Univ. Chicago Press
46. Lang J, Middleton A. 1997. *Radiography of Cultural Materials*. Oxford/Boston, MA: Butterworth-Heinemann
47. Kingery WD, Frierman JD. 1974. *Proc. Prehist.* 40:204–5
48. Maniatis Y, Tite MS. 1981. *J. Archaeol. Sci.* 8:59–76
49. Kingery WD. 1992. *Solid State Phenom. B* 25/26:1–10
50. Smith CS. 1981. *Search for Structure: Selected Essays on Science, Art and History*. Cambridge, MA: MIT Press.
51. Vandiver PB. 2001. *Paleolithic Pigment Processing: A Soft Stone Technology*. In press
52. Vandiver PB, Soffer O, Klima B, Svoboda J. 1989. *Science* 246:1002–8
53. Gourdin WH, Kingery WD. 1975. *J. Field Archaeol.* 2:133–50
54. Kingery WD, Vandiver PB, Prickett M. 1988. *J. Field Archaeol.* 15:219–44
55. Kaczmarczyk A, Hedges REM. 1983. *Ancient Egyptian Faience*. Warminster, UK: Aris & Phillips
56. Moorey PRS. 1985. *Materials and Manufacture in Ancient Mesopotamia*. Oxford, UK: BAR Int. Series 237. Reprinted, revised 1994. Ancient *Mesopotamian Materials and Industries: the Archaeological Evidence*. Oxford/NewYork: Clarendon
57. Bachmann HG. 1982. *The Identification of Slags from Archaeological Sites*. London: Inst. Archaeology Occas. Pub. No. 6
58. Parmelee CW. 1993. *Ceramic Glazes*. Marietta, OH: Ceramic Book Lit. Service
59. Scholes SR. 1965. *Modern Glass Practice*. 7th Ed. revised by CH Greene. 1975. Boston: Cahners
60. Singer F. 1963. *Industrial Ceramics*. London: Chapman & Hall
61. Norton FH. 1956. *Elements of Ceramics*. Boston, MA: Addison-Wesley
62. Norton FH. 1968. *Refractories*. New York: McGraw-Hill
63. Norton FH. 1970. *Fine Ceramics*. New York: McGraw-Hill
64. Norton FH. 1956. *Ceramics for the Artist Potter*. Boston: Addison-Wesley
65. Rhodes D. 1973. *Clay and Glazes for the Potter*. New York: Chilton
66. Rhodes D. 1959. *Stoneware and Porcelain*. New York: Chilton
67. Rhodes D. 1968. *Kilns*. New York: Chilton
68. Henkes VE, Onoda GY, Carty WM. 1996. *Science of Whitewares*. Westerville, OH: Am. Ceram. Soc.

Annu. Rev. Mater. Res. 2001. 31:387–404

SYNTHETIC CELLS–SELF-ASSEMBLING POLYMER MEMBRANES AND BIOADHESIVE COLLOIDS

Daniel A Hammer and Dennis E Discher
School of Engineering and Applied Science, University of Pennsylvania, Philadelphia, PA 19104; e-mail: hammer@seas.upenn.edu

Key Words polymer vesicles, biocolloids, artificial cells

■ **Abstract** We summarize developments in the construction of synthetic cells made from polymers, with a particular focus on mimicking the structure and behavior of blood cells. Two basic themes emerge—the use of block copolymers to make polmer vesicles and the functionalization of colloidal or polymeric microspheres with cell-like adhesive properties. Both platforms provide a means for building the complex hierarchy that is characteristic of biological cells, while also incorporating novel and perhaps superior properties of material strength, specific targeting, and controlled release.

INTRODUCTION

Biological cells are complex, hierarchical materials. As such, they perform a number of functions that include self-propulsion, adhesion and arrest, evolution and differentiation, replication and division, as well as nutrient transport and secretion. Any recent text on cell and molecular biology vividly demonstrates such tasks and also emphasizes how much has been learned over the last decade about the structural and molecular basis of cellular function. From the materials point of view, physicochemical principles such as self-assembly in aqueous solution pervade modern cell biology. Recognition of such principles clearly suggests synthetic routes to cell mimicry and function. This brief review specifically focuses on two recent innovations in these directions that make selective use of modern materials and methods to engineer synthetic cells.

The field of artificial cells is not new. The first efforts were perhaps those of Chang (6) who microencapsulated the isolated contents of the simple red cell. The method used centered on inverse emulsification of the cellular material into an organic such as chloroform:cyclohexane, followed by stabilization of the dispersed droplets through addition of surfactant plus an interfacially accumulating polymer. The interfacial stabilization proved sufficient to allow separation of the cell-size droplets by simple centrifugation. The membrane that resulted was found to be semipermeable so that encapsulated cell enzymes such as carbonic anhydrase were found to act with reasonable kinetics on a permeating substrate. A reported drawback to this classic approach was the unavoidable denaturation of protein at the interface with the continuous organic phase. As a consequence, it seems that the in vivo activity and survival of the microcapsules were limited by denatured surface proteins.

The early results of Chang and coworkers point to the fact that Nature tends to use a more molecularly structured approach to equilibrium self-assembly in aqueous media. Recognition of this feature presents challenges in chemical synthesis, some of which have been met in the last few years with the generation of, for example, block copolymers that self-assemble in water (21). Moreover, the ability to engineer enhanced activity into mimicking structures holds the promise to go beyond cells and make super-cellular materials for biological or medical applications. As Chang and coworkers recognized long ago, a number of opportunities exist in nutrient transport and drug delivery, as well as in other fields. As a specific example, the search for blood substitutes is strongly motivated by frequent blood shortages, as well as blood-borne contagions such as HIV and hepatitis C. Recognition of application, as well as opportunities in design and characterization of soft nano-structured materials, seems extremely likely to spur synthetic and physical chemists, as well as materials scientists, to produce novel molecules and materials for de novo cellular engineering.

In the spirit of an annual review, we briefly summarize recent work in the area of artificial blood cells, both red cells and white cells. Synthetic polymers constitute the basic material approach with added functionality achieved via attached biomolecules. Two specific accomplishments are reviewed and placed in context: the construction of encapsulating vesicles from amphiphilic block copolymers that self-assemble in water and the construction of polymeric microspheres that mimic the adhesion of white cells or leukocytes.

BIO-INSPIRED POLYMER MEMBRANES

Biomembrane Schematic: A Glycocalyx, Fluid Bilayer, and Solid Network

The central structural feature of a prototypical biomembrane is a lamellar bilayer of self-assembled lipids (Figure 1, see color insert). Proteins invariably associate with the membrane, however, and these impart additional, critical bio-functionality. The proteins are either transmembraneous (with a portion inserted through part or all of the membrane) or peripheral (with only a small intercalation into the membrane and located on either the cytoplasmic or extracellular side of the membrane). On the outer face of the membrane is a "forest" of proteins that is extensively glycosylated and therefore high in oxygen content. In addition, on the cytoplasmic face of most biomembranes is a proteinaceous cytoskeleton that is anchored to the membrane via transmembrane proteins. From detailed studies of the simplest of cells, the red cell, properties of the membrane ranging from stability to compatability emerge as a composite of both the bilayer membrane and the underlying cytoskeleton, including how precisely the layers are connected. Whereas the lipid bilayer constrains the overall size, i.e. area of the cell and establishes a semipermeable selective barrier (e.g. 27), the glycocalyx has both active and passive roles in cellular interactions, such as signaling, adhesion, and suppression of interactions. The network, in comparison, stabilizes the membrane against shear (reviewed recently in 31).

Self-Assembling Polymer Membranes

There is a well-recognized similarity between the overall structure of a biomembrane, in particular the lipid bilayer, and the lamellar phase often found in block copolymer melts and dispersions. This has led to various efforts to make self-assembled membranes from block copolymers, as summarized below.

POLYMER VESICLES FORMED IN SOLVENT-WATER MIXTURES When diluted into solvents such as *N,N*-dimethylformamide (DMF) and tetrahydrofuran (THF), many synthetic polymers have been found to form vesicle-like, sub-micron diameter structures. A recent example is a large polyethyleneoxide-polystyrene diblock copolymer of Yu & Eisenberg (40), EOx-PSy ($x = 15$, $y = 240$). This diblock forms vesicles when water is slowly added to organic solvent, and the precipitated structures appear stable after several days of dialysis into water. The spherical and deflated tubular vesicles that form have a wall thickness of 20–25 nm after being dried and metallized for transmission electron microscopy (TEM). This thickness is a small fraction of $PS_{240's}$ contour length of over 60 nm and suggests relatively unextended, perhaps entangled, core chains. Because of the very small hydrophilic PEO block in this particular synthetic, aggregates are said not to form directly in aqueous solution. A similar copolymer with $x = 45$ (or about 7% hydrophilic) generated an occasional vesicle at 1.5 wt% polymer in solution, but with $x = 80$ (or about 12% hydrophilic), no vesicles were generated. Replacing neutral PEO with polyacrylic acid (PAA) and using even longer PS chains (up to 500 units) also leads to vesicles in a solvent-selective manner, as reviewed extensively in Yu et al (41). Consistently, the resulting membranes ranged from 20 to 40 nm in thickness. In these systems it appears likely that the organic solvent is imbibed by the hydrophobic core, fluidizing it, lowering the glass transition of PS, and enabling sufficient topological rearrangement of a polymer lamellae to close upon itself and form at least sub-micron diameter spherical vesicles.

A structurally distinct synthetic that also makes vesicles is a pentablock of polyethyleneoxide-polymethylphenylsilane (PMPS-PEO)2-PMPS of 27 kDa (22). This copolymer is about 50% hydrophilic with a polydispersity of 1.6 (as the weight-to-number average molecular weight). This polydispersity is at the upper end of most reported vesicle-forming polymers described to date. The two PEO blocks were designed to cap, both inside and outside the vesicles, the stiff hydrophobic PMPS segments. The latter span the 8–12-nm-thick hydrophobic core revealed in negative-staining TEM. The finding that addition of bulk PMPS to water generated poor dispersions of a few small vesicles suggests that this design is almost lipid-like in its overall amphiphilicity.

Another recent and novel example of vesicle formation in solvent started with polyisoprene-*block*-poly(2-cinnamoylethyl methacrylate) (PI88-*b*-CEMA230) (11). Interestingly, cross-linking of the PCEMA shell and conversion of the PI chains to poly(2,3-dihydroxyl-2-methyl-butane) generated water-soluble and rhodamine-loadable nanospheres. Given the 230 monomers of the hydrophobic chain, a shell thickness for these structures of about 20 nm suggests an unusual, perhaps entangled, inner membrane structure. An even more recent example of a polymerizable polymer membrane started with poly(2-methyloxazoline)-poly(dimethylsiloxane)-poly(2-methyloxazoline) (PMOXA-PDMS-PMOXA) as a triblock of 9 kDa that was 40% hydrophilic (28). Vesicles were precipitated from ethanol with addition of water, and the two functional ends of each constituent chain were activated by ultraviolet light to polymerize the membrane. Solidity of the final membrane was primarily suggested by a collapsed, hollow sphere morphology observable under scanning electron microscopy.

POLYMER VESICLES THAT ASSEMBLE IN WATER A subset of polymer vesicles have now also been shown to assemble directly under solvent-free aqueous conditions just as lipid vesicles do. Among these super-amphiphiles that, as polymers in the conventional sense, are significantly larger than lipids is the chiral construct PS40-(isocyano-L-alanine-L-alanine)$_x$ (9). For $x = 10$, but not $x = 20$ or 30, small collapsed vesicles with diameters ranging from tens of nanometers to several hundred nanometers, and a bilayer thickness of 16 nm were observed by TEM to exist under mildly acidic conditions. Interestingly, bilayer filaments and superhelical rods were found to coexist under the same solution conditions, perhaps reflective of repulsive destabilization of the more planar membrane structures. However, comparatively neutral vesicles made in a range of aqueous media with polyethyleneoxide-polyethylethylene, specifically EO_{40}-EE_{37}, also coexist, although dominantly, with spherical and worm-like micelles (Figure 2*A*) (12). Once again, the membrane of these vesicles was found to be hyperthick: cryo-TEM resolved a hydrophobic core thickness of 8 nm. These PEO-PEE polymersomes are reviewed further below in relation to membrane properties. Very similar initial results with a polyethyleneoxide-polybutadiene (PEO-PBD) diblock have also been recently described (24a). Importantly, this non-hydrogenated precursor to PEE can be covalently cross-linked, thereby allowing a membrane elastic network to be established as highlighted in Figure 1.

Finally, a relatively conventional pluronic consisting mostly of polypropyleneoxide (EO_5-PO_{68}-EO_5; 10% PEO by weight) has also been reported to yield small vesicles observable by cryo-TEM (32). However, a mean vesicle half-life of only a few hours more than likely reflects the high oxygen content of the weakly hydrophobic polypropyleneoxide block. The relatively thin polymer membranes of 3–5 nm, compared with the 25-nm contour length of PO68, probably reflects a tendency for interfacial localization of the mid-block oxygen. By factoring this oxygen into the hydrophilic fraction, what unifies all of the above vesicle-forming

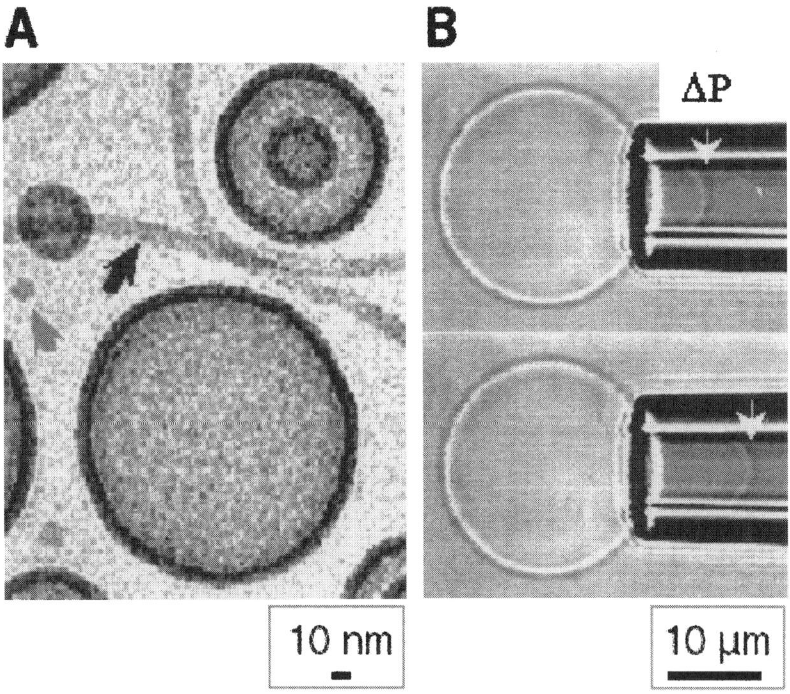

Figure 2 Polymer vesicle images (12). (*A*) Cryo-transmission electron micrograph (CTEM) of vesicles made of PEO-PEE. (*B*) Optical image of a giant polymersome. The PEO-PEE vesicle is suspended in aqueous solution (250-mM sucrose) and pulled into a micropipette by an increasing negative pressure, ΔP.

super-amphiphiles with natural lipids is a common hydrophilic weight or volume fraction in the range of 20 to 40%.

Micromaterial Characterizations

INTERFACIAL ELASTICITY AND PERMEATION TO RUPTURE: PROPERTIES OF PEO-PEE VESICLES Because the stability of bulk copolymer mesophases is well-appreciated as depending on the product of the Flory interaction energy, χ, and chain length, N, one would expect that these same parameters underlie at least some measurable properties of polymer vesicles. For instance, as a relative measure of the dominant χ, the interfacial tension, γ_{HP}, at the membrane surface defined by the HP-block intersection ought to prove physically reasonable. In addition, given a similar χ as a lipid membrane, the relatively larger N of a polymer membrane ought to yield a more cohesive structure.

Membrane properties such as interfacial elasticity and membrane cohesiveness are directly accessible through measurements on giant, cell-sized PEO-PEE vesicles (diameters >5 μm). Figure 2*B* shows a many-micron-sized vesicle (of radius R_s) being aspirated into a micropipette (of radius R_p): a negative pressure ΔP increases the membrane area by ΔA relative to the original vesicle surface area, A_0. A proportionality between the pressure-imposed surface tension, $\tau = 1/2\ \Delta P\ R_p/(1 - R_p/R_s)$, and relative area dilation, $\alpha = \Delta A/A_0$, yields an area elastic modulus, $K_a = \tau/\alpha$ with units of interfacial tension (Figure 3*A*). A value of $K_a \approx 120$ mN/m reported for PEO-PEE vesicles, together with the common view that $\gamma_{HP} = {}^1\!/\!_4\ K_a \approx 30$ mN/m, proves consistent with an H-P: oil-water interface. Importantly, the enhanced mean thickness, *d*, would appear to have no influence on this interfacial tension. This is consistent with a broad range in K_a for lipid membranes (from 100–1000 mN/m) (25) despite a very narrow range for d_{lipid}. In contrast, the bending elastic constant, K_b, that was reported to be ~35 k_BT (12) appears extremely well approximated by an elastic plate model prediction of ${}^1\!/\!_{48}\ K_a\ d^2$, which is derivable for a bilayer of two uncoupled monolayers.

In spite of lipid-like elasticity, the stably sustainable critical strain, α_c, of the PEO-PEE vesicles (0.19 ± 0.02) considerably exceeds the range of 0.03–0.06 typical of natural lipid membranes (Figure 3*B*). Similar to bending and also membrane permeability (12), this cohesive stability (corresponding to ~0.5 to 1 k_BT/nm^2) appears likely to have its origins in membrane thickness. Because K_a provides a measure of γ_{HP}, and γ_{HP} is related to χ, comparisons of α_c for different effective *N* ought to be made with membranes having similar K_a. The unsaturated lipid DAPC mixed with cholesterol (1:1) has a $K_a = 102$ mN/m, or $\gamma_{HP} \approx 25$ mN/m, (25), and it has an $\alpha_c = 0.043$. If a critical yield stress of $\sigma_c = K_a\ \alpha_c/d \approx 1$ MPa is postulated as common to both this particular but very representative lipid membrane and the PEO-PEE polymersomes, then one would naively expect both critical yield tensions and strains to increase linearly with membrane thickness. However, an estimate based linearly on *d* alone is simply too low; instead, a more hydrophilic pore would introduce the outer corona thickness, δ. The hydrated PEO chain is believed to extend $\delta \approx 4$ nm above each interface, whereas a lipid bilayer has a total thickness of 4–5 nm. A ratio of $(d + \delta)$ for the two systems allows rationalization of the rupture tension measured for the first polymer membranes.

Encapsulants and Complex Media Stability—Biomedical Directions

Complex solutions can modulate, and in particular undermine, the stability of polymer vesicles through mechanisms that, for example, reduce or change molecular packing. Indeed, difficulties in the encapsulation of globular proteins can arise with the surfactant-like character of denatured proteins, although the hydrated PEO brush of the present copolymers would tend to oppose interfacial access. The work of Chang (6) mentioned in the INTRODUCTION highlights the relevance of interfacial denaturation phenomena. It is nonetheless found with polymersomes that encapsulation can be as straightforward as adding solid pieces of bulk diblock to an aqueous solution of desired encapsulant and waiting 24 h

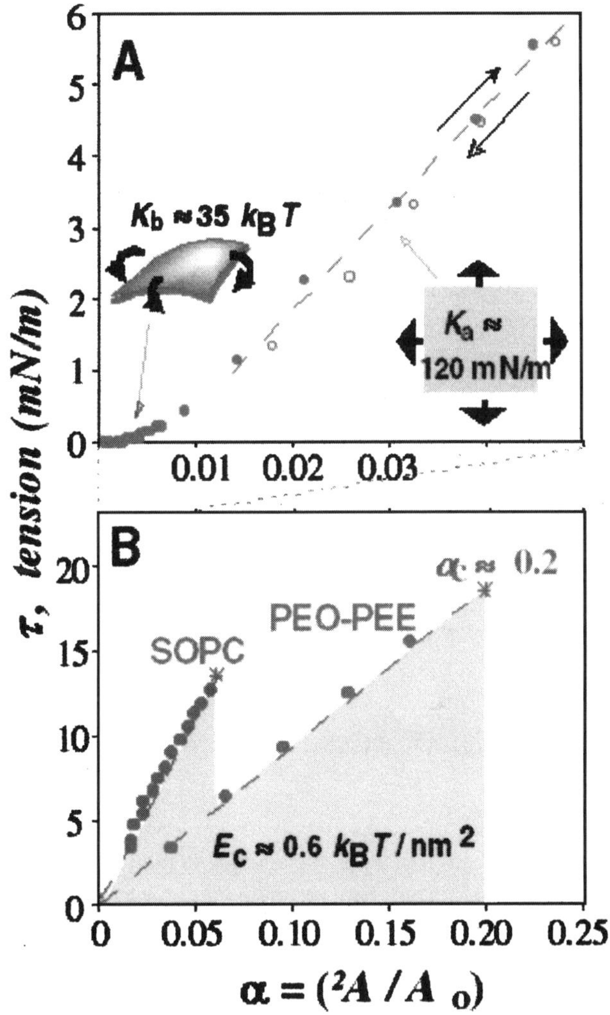

Figure 3 Micromechanical characterization of PEO-PEE vesicles by aspiration. (*A*) Relationship between the ΔP-induced tension, τ, and membrane expansion α, reveals the two invariant modes of deformation of a soft fluid membrane. These modes are illustrated in insets together with the corresponding elastic bending and area expansion moduli. (*B*) Membrane toughness was determined by aspiration to the point of rupture (asterisk). The figure demonstrates the difference in both critical area strain, α_c, and the cohesive energy density, E_c, between a common lipid vesicle, SOPC, and a PEO-PEE vesicle.

(24a). To develop a sense of encapsulation efficiency, three encapsulants were examined: myoglobin, hemoglobin, and albumin. All three proteins could be entrapped during vesicle formation, although the efficiency of encapsulation, as well as its limitations, has as of yet to be fully understood. Nonetheless, it is clear that encapsulation of macromolecules can be done with polymersomes.

More complex than a solution of a single protein, blood plasma has a number of surface-active agents, including fatty acids and carrier proteins. In addition, such colloidal entities generate osmotic depletion forces by becoming excluded from narrow gaps between juxtaposed membranes; surfactant bridging between adjacent vesicles can then act to more rigidly couple vesicles together. Stability in plasma thus has at least two aspects: membrane patency and colloidal stability. These polymer vesicles are found to be stable for at least five days in plasma when suspended at room temperature and kept well-mixed in a quasi-physiological manner (24a). Furthermore, colloidal stability is demonstrated out to almost 60 h in quiescent plasma; aggregation of settled-out vesicles occurs at longer times. However, polymersomes in the formed aggregates do not appear to coarsen into either giant vesicles or bulk lamellar phases, despite the close proximity and elevated density in the aggregates. This indicates that polymer membrane fusion is not readily achieved—a finding most likely attributable to repulsion by the PEO brush. Nonetheless, the dense aggregates are not exceedingly fragile assemblies either because they could not be sheared apart by rapid dragging through solution and generated surface shear stresses as high as $\sim$1 dyne/cm^2. Regardless, the likelihood of forming such aggregates in vivo seems remote owing to incessant fluid motion.

Interactions with Biological Cells: Phagocyte Challenge In Vitro

With polymersomes at least partially stable in blood plasma, a logical next step was to assess the interaction of a polymersome with the cellular components of blood, particularly granulocytes, which are the predominant circulating phagocytes. The basic in vitro experiment is illustrated in the video excerpts of Figure 4, which show either a yeast particle or a polymersome being brought into contact with blood cells. When placed in contact with a red cell, neither a polymersome nor a yeast particle exhibits any adhesion or other clear cellular response, despite the presence of 20% plasma. In comparison, 1–2 min of contact between a yeast particle and a white cell (known as a neutrophil or differentiated leukocyte) leads to strong adhesion. This generally develops further as the white cell spreads and actively engulfs the yeast particle (Figure 4*A*, *left*). This type of control experiment was generally complete within 3 to 4 min ($n = 3$ particles) (Figure 4*B*) and was found to occur even faster at 37°C. In contrast, over the same length of time and much longer, PEO-PEE polymersomes appeared inert to these phagocytic cells ($n = 5$ vesicles). Several such experiments are summarized in the timelines of

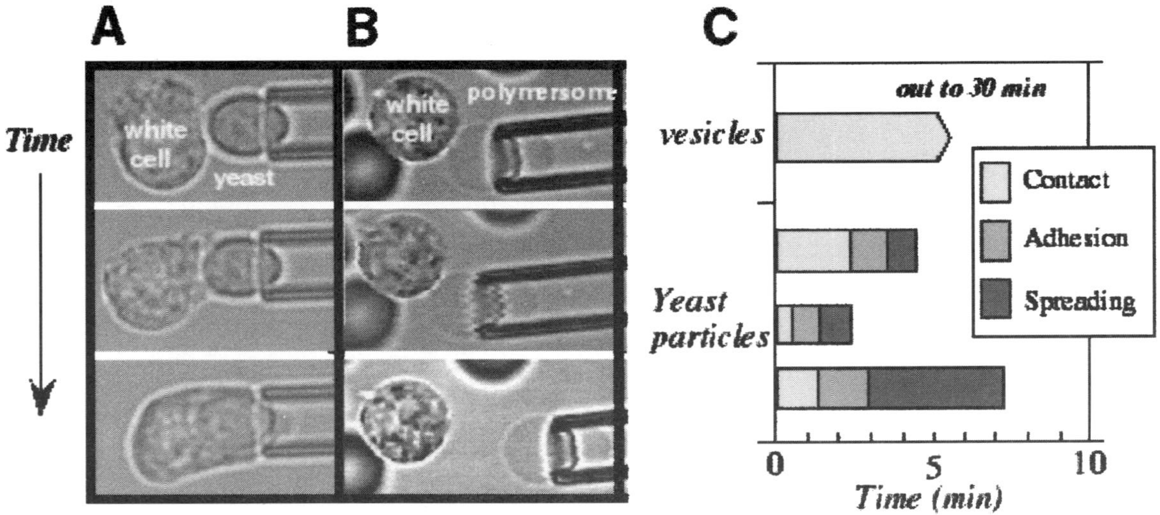

Figure 4 Phagocyte challenge in vitro. Yeast particles (*A*) or PEO-PEE polymersomes (*B*) were mixed in 20% citrated plasma and manipulated into contact with freshly isolated and washed white blood cells or red blood cells. (*C*) Within about 5 min, white cells adhered to and engulfed the yeast particles, consistent with earlier studies (42). Polymersomes appeared inert in contact with a white cell for as long as 30 min. Interactions with red cells were non-existent as well. The micropipette inner diameter is 5 μm. Experiments shown were done at room temperature and lasted 30–60 min in total duration.

Figure 4*C*. Results similar to those for PEO-PEE were also found with EO_{26}-BD_{46} (BD = butadiene), at least out to 5–10 min, despite this second polymer's shorter PEO chain of about only 1 kDa.

Similar in vitro tests of polymersomes with macrophages (another differentiated type of leukocyte) as well as more quiescent cells, including endothelial cells and myoblasts, further indicate the inertness of the polymersome surface. The dense PEO brush of the polymersomes seems to act like the glycocalyx on cellular exofaces, preventing the deposition of phagocytic ligands such as plasma C3b on the vesicle surface and/or repelling phagocyte adhesion. The in vitro protein-rejecting properties of PEO-coatings, including nanoparticles (30), are well-appreciated (23). Ultimately, the lack of polymersome recognition by phagocytes would seem important to prolonged circulation upon intravital injection. Emerging studies involving intravenous injections of ~100-nm vesicles into rats appear to substantiate this conjecture.

Cross-Linked Polymersome Membranes

Most recently, the double bonds embedded in the hydrophobic core of PEO-PBD have allowed us to fully cross-link the membrane of giant unilamellar vesicles (BM Discher H Bermudez, Y-Y Won, DA Hammer, FS Bates & DE Discher, submitted). This has been attempted for decades with lipids but without success (H Ringsdorff, personal communication). We speculate that this is because polymerization strains the nanostructure (39) and lipids cannot handle the strain, whereas polymersomes can. This problem is exacerbated by the low number of available sites for cross-linking that have been tried in lipid systems. It is clear that the cross-linked polymer membranes are imparted with a shear rigidity that makes them respond in deformation similar in quality, most notably, to red cell membranes (see Figure 1). Hence, the development of hemo–polymersomes that have material properties of red cells and encapsulate oxygen-binding globins is rapidly emerging.

BIOADHESIVE COLLOIDS

Leukocyte Adhesion

A second approach to using synthetic materials to recreate the dynamic function of biological cells has been the development of chemically derivatized particles that recreate the dynamics of leukocyte adhesion. In physiology, cells in the circulation must traffic from blood into tissues. To do this, cells must adhere to blood vessel walls via specific receptor-ligand binding. Many cells exhibit this type of adhesive trafficking, such as leukocytes (as part of their physiological function of entering tissue to eliminate invaders), stem cells (which leave blood vessels and enter bone marrow to proliferate and differentiate to give rise to blood cells), and tumor cells (during metastasis, in which they enter the circulation to spread throughout the body). Therefore, adhesion of blood borne cells to blood vessel walls is a ubiquitous physiological process.

By far, the best studied of these processes is leukocyte trafficking, and although our knowledge of the surface molecules involved in this type of adhesion is not complete, there has been some definitive elucidation of which adhesion molecule pairs participate in leukocyte trafficking. Interestingly, the trafficking of white cells out of the circulation occurs in two steps (for a more complete but recent review, see 14). The first step, called rolling, is a dynamic friction in which the white cell is slowed down but not stopped by interactions with the endothelial cells that line the blood vessel wall. This step is mediated by very specialized molecules, including selectins and their ligands, as well as the integrin heterodimer $\alpha_4\beta_7$. Rolling is followed by firm adhesion, which in leukocytes is mediated by integrin heterodimers in the β_2 family. β_2 integrins must undergo a complex structural change to enter a state where it can bind strongly with its ligand (typically a member of the Ig superfamily), which then leads to firm adhesion. This structural change is induced by signal transduction pathways within cells, which are initiated through either the occupancy of selectin molecules during rolling or the binding of chemokine ligands to chemokine receptors on cells (14).

The recreation of the rolling step of the adhesion cascade using synthetic chemistry has immense value in both understanding how adhesion works and also designing cellular mimetics that can target and exploit the same pathways that cells use to traffic throughout the body. It is not clear which leukocyte cell surface receptors actually mediate rolling adhesion on endothelium. The uncertainty comes from the fact that selectins are selective lectins; (*a*) lectins are proteins that bind to carbohydrates, and (*b*) the carbohydrate decorations on a single protein scaffold can vary widely. Each selectin appears to respond to different carbohydrate decorations, which in turn are presented on different protein backbones on cells. One common feature of all selectin ligands is the requirement for a sialylated-fucosylated carbohydrate ligand, such as sialyl-Lewis^x. These carbohydrates are often linked to hydroxyl groups on threonines or serines (thus described as *O*-linked glycosylation). The carbohydrate decorations may act in concert with other structural features of the selectin ligand. For example, P-selectin glycoprotein ligand-1 (PSGL-1), which is found on leukocytes and which supports rolling on P-selectin, has a threonine at position 16 that is glycosylated, but it has been postulated that tyrosine sulfation is also required for full PSGL-1 activity (26). Therefore, sialyl-Lewis^x, or a closely related carbohydrate, can be viewed as the minimum functional binding element of the selectins (17). The current view of receptor-ligand biochemistry that supports leukocyte rolling adhesion is shown in Figure 5 (see color insert).

The location in the vasculature where leukocytes bind is controlled by the expression of ligands (selectins or selectin ligands) on the upper surface of endothelial cells that line blood vessels. Endothelial cells can express both P- and E-selectin, in response to cues it receives from the environment. The control of leukocyte recruitment in inflammation (when there is an infection in tissue) is through the generation of tissue factors (cytokines), released from survelliance cells, that cause endothelial cells to express E- and P-selectin, thus recruiting white cells.

After rolling, leukocytes bind firmly through firm adhesion mediated by the β_2 integrins binding to molecules in the Ig superfamily (intercellular adhesion molecule-1, ICAM-1).

Leukocyte Mimics

The fact that all three selectins share a common carbohydrate epitope, sialyl-Lewis^x as a ligand (17), led to the first cell-free colloidal mimetics of leukocyte adhesion. The strategy behind construction of colloidal mimetics to leukocyte adhesion is shown in Figure 6. Biotinylated forms of sialyl-Lewis^x were bound to streptavidin-coated polystyrene microspheres, and their adhesion was measured on E-selectin-coated surfaces under shear flow in a parallel plate flow chamber (3, 4). Slow rolling, defined as rolling of several to tens of microns per second at shear stresses up to 2 dynes/cm^2, as seen for leukocytes on endothelium or purified E-selectin surfaces (24), was observed. Cellular rolling is also distinguished by a variance in the rolling velocity (fluctuation in instantaneous velocity about its average value); this dynamic feature of the rolling was also observed with the cell-free system (3, 4). The rolling was specific because antibodies against sLe^x or E-selectin fully inhibited the rolling. The rolling of sLe^x-coated spheres on E-selectin, and the close similarity in rolling dynamics between the reconstituted microspheres and neutrophils indicated that receptor-ligand physical chemistry was responsible for the control of neutrophil adhesion and suggested that features such as cell deformability, roughness, and intracellular signaling were not determinants of neutrophil rolling dynamics. These results seemed to explain the ability of glutaraldhyde-fixed neutrophils to roll E-selectin-coated surfaces (16).

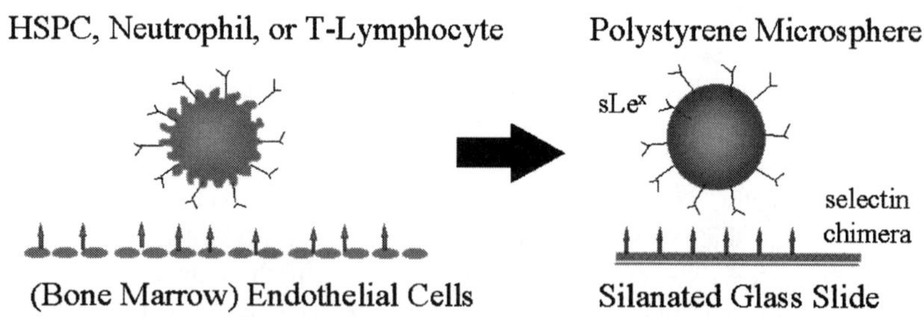

Figure 6 In cell-free systems, the adhesion of leukocytes (lymphocytes or neutrophils) and hematopoietic stem and progenitor cells to endothelial cells can be recreated by attaching small molecule ligands (parts of selectin adhesion ligands) to microspheres. The endothelium can be recreated by attaching endothelial surface molecules (such as E- and P-selectin) to glass slides. The dynamics of adhesion displayed by these microspheres interacting with selectin-coated surfaces, as measured in a flow chamber, closely approximates that of leukocytes adhering to endothelium under flow.

Subsequently, rolling was also observed using sLe^x-coated beads on both P- and L-selectin surfaces (31, 19 respectively). The result for P-selectin was informative because it had been previously thought that PSGL-1, the ligand for P-selectin, had to be both glycosylated and tyrosine-sulfated for rolling activity (26). However, sLe^x-coated beads showed robust rolling on P-selectin (31). The rolling dynamics of sLe^x-beads on P-selectin could be compared to that engineered with PSGL-1-coated beads. The ability of PSGL-1-coated spheres to adhere to P-selectin-coated surfaces under flow was first shown by Goetz and coworkers (18). In our experiments, a chimeric form of PSGL-1, containing the last N-terminal 19 amino acids of PSGL-1 linked to the Fc portion of a human IgG$_1$, was attached to protein A-coated microspheres as the same density at which PSGL-1 is found on human neutrophils. The rolling of neutrophils, sLe^x-coated spheres, and PSGL-1-coated spheres could be compared (Figure 7*B*). Neutrophils rolled the slowest, PSGL-1-coated spheres rolled only slightly faster, and sLe^x-coated spheres rolled the fastest but still displayed robust rolling. These results indicate again that roughness, deformabilty, and signaling have a limited effect on the dynamics of cell rolling (because PSGL-1-coated beads closely approximate the rolling of neutrophils) and that tyrosine sulfation enhances but is not necessary for robust rolling on P-selectin (sLe^x-beads do display adequate rolling on P-selectin).

This latter point remains a controversy in the selectin field because mutation of all the tyrosine to phenyalanines in PSGL-1 expressed in a cellular system seems to abolish rolling on P-selectin (26). Further experiments with purified amino acids from the terminal 19 amino acids of PSGL-1 indicate that peptides that are glycosylated but not sulfated do support rolling on P-selectin (31), consistent with the previously-cited result that sLe^x-coated beads support rolling on P-selectin. One self-consistent explanation for this apparent discrepancy is that PSGL-1 in which tyrosine is mutated to phenyalanine may be improperly processed and glycosylated in cellular systems, which further argues for testing the chemical dependence of rolling using systems that involve molecular synthesis and reconstitution for recreation of proper activity.

On L-selectin, sLe^x-beads could be used to recreate the rolling that had been observed in cellular systems using L-selectin and some of its carbohydrate-bearing physiological ligands. L-selectin-mediated rolling is much faster than that seen with the other selectins, presumably because of the faster off-rates of L-selectin with the carbohydrate (15). Rolling velocities observed with beads were in the high tens to hundreds of microns per seconds, and thus could be viewed as a fast skipping of the beads over the surface. The rolling velocity decreased with increasing L-selectin surface density (Figure 8*B*, see color insert).

An interesting dynamic feature of L-selectin-mediated rolling is the shear threshold effect. The shear threshold effect describes the increase then decrease in adhesion as the shear rate increases, which is seen with L-selectin-mediated cell adhesion. It was first identified with neutrophils (which bear L-selectin) on peripheral-node addressin (PNAd) surfaces (PNAd is an sLe^x-bearing ligand found on endothelium in immune organs such as lymph nodes) (15). The maximum in

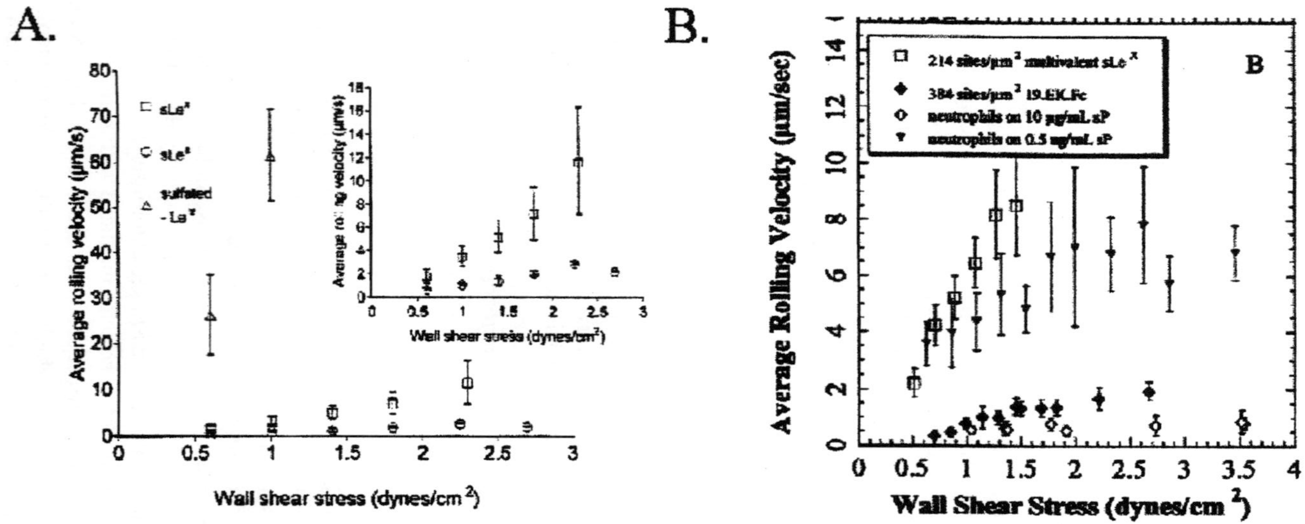

Figure 7 Data on the adhesion of carbohydrate and ligand-coated microspheres to selectin-coated glass slides using the cell free system. (*A*) Data on the rolling of beads coated with sLe^x, sialyl-Lewis^a (an isomer of sLe^x), and HSO_3-Le^x on E-selectin-coated surfaces under flow show that sLe^x or sLe^a-coated spheres roll at velocities of several microns per second over wall shear stresses up to 2.5 dynes/cm^2. HSO_3-Le^x-coated beads show weak adhesion on E-selectin, and sLe^a is a slightly better ligand than sLe^x for E-selectin. Thus the cell free system can distinguish fine differences in carbohydrate chemistry (3). (*B*) Cell-free rolling of beads on P-selectin. The adhesion of beads coated with sLe^x or a chimera of PSGL-1 (19.EK.Fc, which has the last 19 amino acids of PSGL-1 tied to the Fc portion of human IgG_1 via an enterokinase susceptible domain) were compared with neutrophils for adhesion under flow to P-selectin physiosorbed to glass slides. PSGL-1 coated beads (closed diamonds) roll almost as slowly as neutrophils (open diamonds). SLe^x-coated spheres roll significantly, but much more quickly than neutrophils (open squares), suggesting that PSGL-1 has features beyond sialyl-Lewis^x, which facilitate rolling (31).

adhesion, as measured by adhesion flux–number of adherent particles bound per unit time per area—is typically between 0.7 and 1 dyne/cm^2 (see Figure 8*A*) (15). It has now been shown that the shear threshold effect could be recreated in a cell-free system using sLe^x-containing beads on L-selectin-coated surfaces (19) (Figure 8*C,D*). The existence of the maximum seems to depend on the density of L-selectin. On high densities of L-selectin (as found with L-selectin densities created by physisorption with 5 μg/ml L-selectin), adhesion of sLe^x-coated beads steadily decreased with increasing shear stress. However, on lower densities of L-selectin, adhesion clearly goes through a maximum with shear stress (19). The effect of ligand density in controlling the shear threshold had not been observed before these experiments. Although a number of hypotheses for the control of the shear threshold effect have been postulated (cell roughness and micron-scale deformability, for example), it is clear from these cell-free adhesion experiments that the origin of the shear threshold must be in the physical chemisty of L-selectin binding its ligand. Hints of the shear threshold phenomena have also been observed in cellular aggregation experiments (33) and in antibody-antigen adhesion (35); it appears to be a fundamental phenomenon that governs molecular association for tethered receptors and ligands, and cell-free systems can help with its further elucidation.

Other Adhesive Systems

Leukocyte rolling, or dynamic friction, is a unique dynamic state of adhesion. As mentioned before, relatively few molecules support this type of adhesion. One can question what the functional properties of adhesion molecules need to be to generate this dynamic state of adhesion, and whether the same dynamics can be achieved with other molecules. This question has been extensively explored with antigen-antibody pairs whose affinities are chosen to match those of selectin-carbohydrate interactions. Some transient adhesion has been observed with low-affinity antigen-antibody interactions, but in general it is short-lived and either leads to no adhesion or firm adhesion (thus the transient or rolling adhesion is meta-stable). For example, Chen and coworkers showed that neutrophils adhere transiently to an antibody against sLe^x immobilized in a flow chamber, but the rolling was not sustained, was observed only with a period of pre-incubation, and was not observed at shear stresses above 1 dyne/cm^2 (7). Swift showed that antibody-mediated binding of rat basophilic leukemia (RBL) cells to antigen-coated surfaces was transient, but always ended in firm adhesion, for an antibody whose K_d for its antigen was within an order of magnitude of that for selectin-carbohydrate interactions (35, 37). Other very low–affinity pairs gave rise to either no adhesion or firm adhesion, but no state in between (36, 38). A possible explanation for the failure of these molecular systems to match rolling response is that adhesion dynamics depends on functional parameters that describe their binding other than affinity. Using computer simulations, it has been shown that the dynamics of adhesion might depend most closely on the mechanico-chemical properties of adhesion molecules or the dynamics of their failure in response to force (5, 10, 20), as first postulated by Bell (1). In addition to the kinetics of dissociation, a bond's sensitivity to force has been predicted to govern the adhesive response of cells. Thus molecular interactions that have the same affinity as selectin-carbohydrate interactions would not necessarily give rise to the same dynamics of adhesion because the mechanico-chemical properties are not set by the affinity and are not necessarily the same.

Targeting Applications

Recently, the technology to make functional leukocyte mimetics has recently been extended with the clear goal of targeting for biomedical applications. Goetz' laboratory has shown that nanospheres coated with antibodies against P- and E-selectin can bind to these substrates under flow (2). Gimbrone's laboratory showed that vesicles functionalized with an antibody against E-selectin could be targeted to an inflammatory site in mice (34). It has been shown that small microspheres coated with PSGL-1 roll in vessels induced to express P- or E-selectin (or both) in mice (29).

A further development in this area is to combine the well-established technologies for making biodegradable microspheres from polymers, such as polylactic-glycolic acid (PLGA), with cell-free rolling and adhesion (8). Goetz and coworkers showed that biodegradable microspheres physisorbed with antibodies to E- and P-selectin showed limited, weak adhesion under a limited set of shear stresses to E- and and P-selectin surfaces. Recently, the Hammer laboratory has shown that sLe^x can be coupled to streptavidin-derivatized PLGA microspheres and that these spheres can roll on P- selectin surfaces (Figure 9, see color insert). It is well known that such particles can be used for long-time–scale release of biological agents (8), and combined with functionalization that mimics the adhesive behavior of leukocytes, they may provide a useful tool for drug delivery, especially in inflammatory applications.

IMMEDIATE POSSIBILITIES

Clear avenues of future effort in synthetic cells lie in the engineering of polymersomes with specifically designed adhesive functionality. Surface groups on the polymer have already been derivativized with specific ligands such as biotin, allowing the hierarchical design of a wide array of adhesive functionality. Leuko-polymersomes would provide the dual control of adhesion, through the selection of the type and density of leukocyte adhesion ligand, and synthetic cell rheology, through the selection of polymer type, length, and enacapsulant.

ACKNOWLEDGMENTS

The authors are grateful for the support of the Penn MRSEC (DMR-9632598) and NASA Materials Science program. Synthesis of block copolymers by Frank

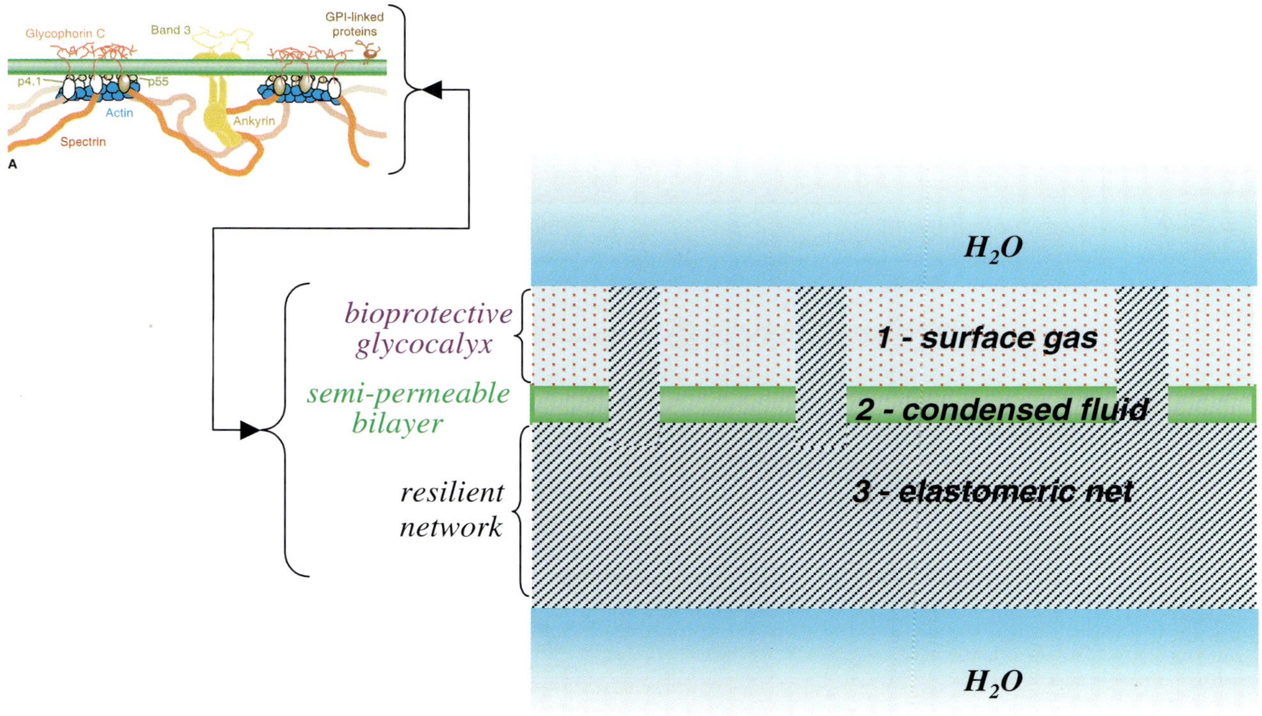

Figure 1 Typical biomembrane (13) and its coarse-grained, trilaminate representation.

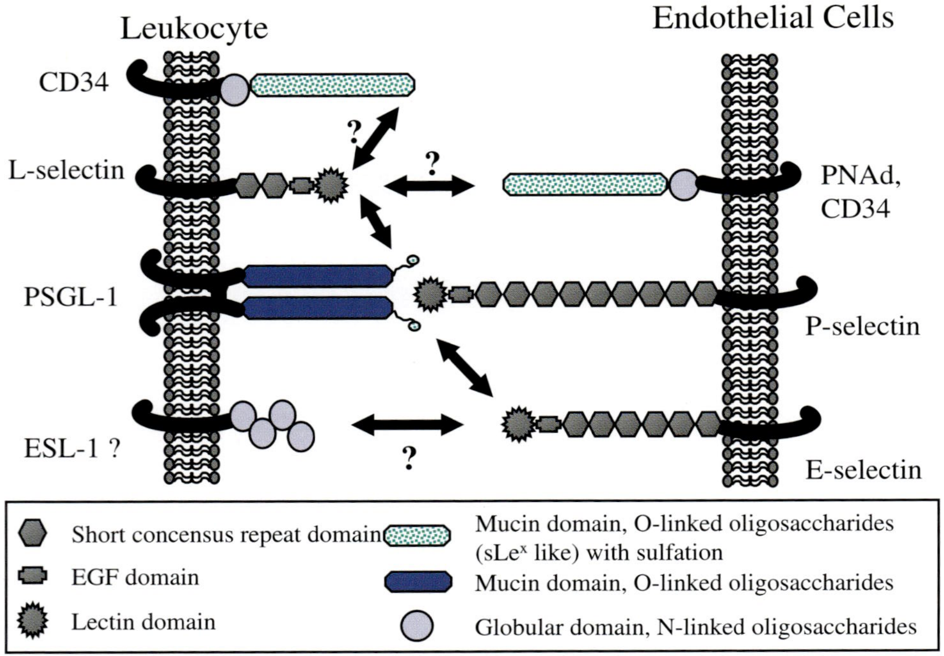

Figure 5 Adhesion molecules in the immune system used during leukocyte adhesion to endothelium. Selectin family molecules exist on leukocytes (L-selectin) and on endothelium (P- and E-selectin). Each binds to carbohydrate-bearing ligands, principally with O-linked glycosylation domains that bear the sialyated-fucosylated ligands sialyl-Lewis^x (sLe^x). L-selectin binds to peripheral node addressin and CD34; P-selectin binds to leukocyte P-selectin glycoprotein ligand-1 (PSGL-1); and E-selectin binds E-selectin ligand-1 (ESL-1). Other important structural domains include the EGF domain and consensus-binding repeat motifs on selectins and globular and mucin domains on selectin ligands.

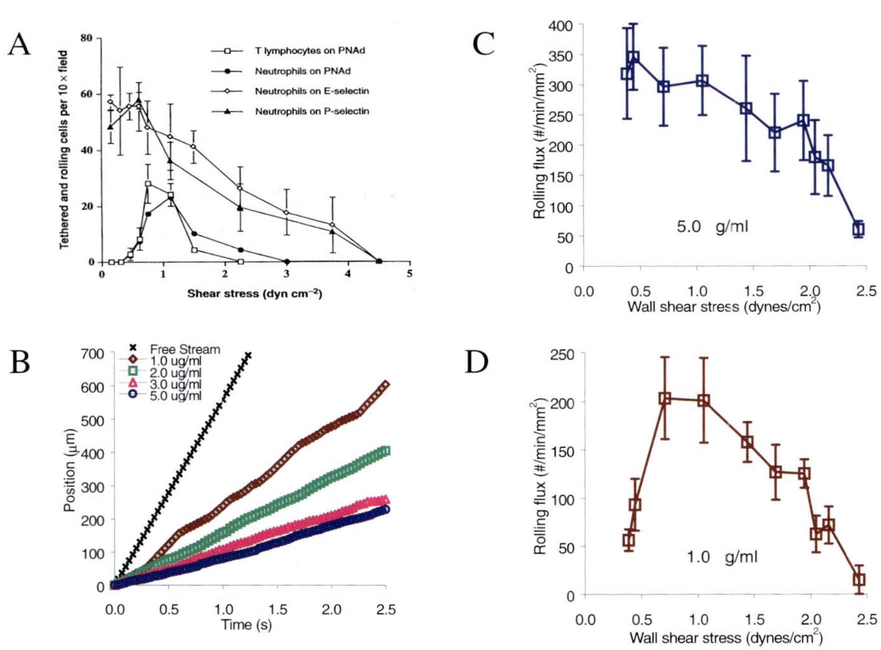

Figure 8 The shear threshold effect. (*A*) Neutrophils and lymophocytes were perfused over PNAd addressin, E-selectin, or P-selectin-coated surfaces under flow. On PNAd surfaces (where adhesion was mediated by leukocyte L-selectin binding to PNAd), there was a clear maximum in the adhesion with the shear stress. This maximum, not seen during adhesion via E- and P-selectin, is the shear threshold effect [reprinted by permission from *Nature* © 1996 Macmillan Mag. Ltd. (15)]. (*B*) sLe^x-beads roll over L-selectin-coated surfaces under flow. Increasing the density of L-selectin decreases the rolling velocity below that of a free stream particle (L-selectin concentrations are given as solution concentrations) (19). (*C,D*) Adhesion flux of sLe^x-coated beads on L-selectin shows a recreation of the shear threshold effect with a cell-free system at low, but not high, densities of L-selectin (19).

A. B.

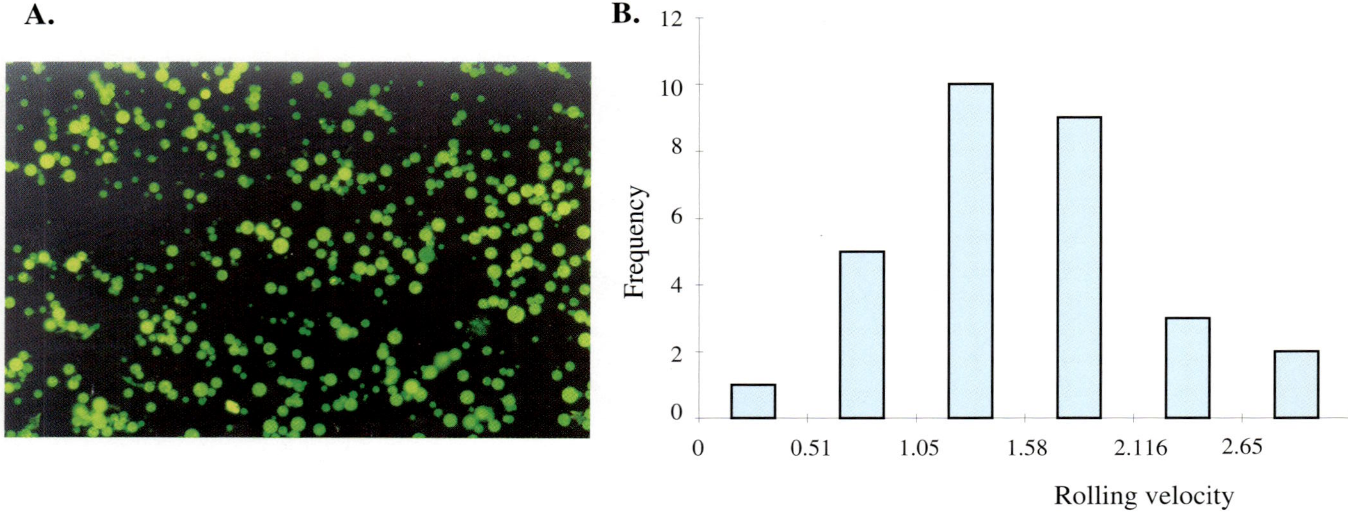

Figure 9 Cellular mimetics can be modified for potential drug delivery. (*A*) Micron-sized microspheres made from polylactic-glycolic acid polymers can encapsulate carboxyfluorescein, as viewed with a fluorescence microscope. (*B*) PLGA microspheres, covalently linked via carbodiimide with neutravidin, then decorated with biotin-SLe^x, roll on P-selectin-coated surfaces under flow, with rolling velocities in several microns/second. Heterogeneity in rolling velocity is the result of both heterogeneity in bead size (between 1 and 7 μm) and differences in sLe^x density. The shear stress was 1.235 dynes/cm^2, and the average amount of sLe^x/bead was 1.2×10^{-6} μg.

Bates' group at the University of Minnesota is very gratefully acknowledged. The authors acknowledge the scientific contributions of B Discher, Y-Y Won, H Bermudez, C-M Lee, S Rodgers, D Brunk, D Tees, A Greenberg, M Santore, and L Eniola.

Visit the Annual Reviews home page at www.AnnualReviews.org

LITERATURE CITED

1. Bell GI. 1978. *Science* 200:618–27
2. Blackwell JE, Dickerson JB, Berg EL, Goetz DJ. 2001. *Ann. Biomed. Eng.* Submitted
3. Brunk DK, Goetz DJ, Hammer DA. 1996. *Biophys. J.* 71:2902–7
4. Brunk DK, Hammer DA. 1997. *Biophys. J.* 72:2820–33
5. Chang K-C, Tees DFJ, Hammer DA. 2000. *Proc. Natl. Acad. Sci. USA* 97:11262–67
6. Chang TMS. 1964. *Science* 146:524–25
7. Chen SQ, Alon R, Fuhlbrigge RC, Springer TA. 1997. *Proc. Natl. Acad. Sci. USA* 94:3172–77
8. Cohen S, Yoshioka T, Lucarelli M, Hwang LH, Langer R. 1991. *Pharm. Res.* 8:713–20
9. Cornelissen JJLM, Fischer M, Sommerdijk NAJM, Nolte RJM. 1998. *Science* 280:1427–30
10. Dembo M, Torney DC, Saxman K, Hammer DA. 1988. *Proc. R. Soc. London Ser. B* 234:55–83
11. Ding J, Liu G. 1998. *J. Phys. Chem. B* 102:6107–13
12. Discher BM, Won Y-Y, Ege DS, Lee JC, Bates FS, et al. 1999. *Science* 1284:1143–46
13. Discher DE. 2000. *Curr. Opin. Hematol.* 7:117–22
14. Ebnet K, Vestweber D. 1999. *Histochem. Cell Biol.* 112:1–23
15. Finger EB, Puri KD, Alon R, Lawrence MB, von Andrian UH, Springer TA. 1996. *Nature* 379:266–69
16. Finger EB, Bruehl RE, Bainton DF, Springer TA. 1996. *J. Immunol.* 157:5085–96
17. Foxall C, Watson SR, Dowbenko D, Fennie C, Lasky LA, et al. 1992. *J. Cell Biol.* 117:895–902
18. Goetz DJ, Greif DM, Ding H, Camphausen RT, Howes S, et al. 1997. *J. Cell Biol.* 137:509–19
19. Greenberg AW, Brunk DK, Hammer DA. 2000. *Biophys. J.* 79:2391–402
20. Hammer DA, Apte SM. 1992. *Biophys. J.* 63:35–57
21. Hillmyer MA, Bates FS. 1996. *Macromolecules* 29:6994–7002
22. Holder SJ, Hiorns RC, Sommerdijk NAJM, Williams SJ, Jones RG, Nolte RJM. 1998. *Chem. Commun.* 14:1445–46
23. Lasic DD, Papahadjopoulos D, eds. 1998. *Medical Applications of Liposomes*. Amsterdam: Elsevier. 794 pp.
24. Lawrence MB, Springer TA. 1991. *Cell* 65:859–73

24a. Lee JC-M, Bermudez H, Discher BM, Sheehan M, Won Y-Y, et al. 2001. *Biotech. Bioeng.* 73:135–45

25. Lipowsky R, Sackmann E, eds. 1995. In *From Cells to Vesicles*, pp. 1–5A. Amsterdam: Elsevier Sci. Vol. 1A
26. McEver RP, Cummings RD. 1997. *J. Clin. Invest.* 100:S97–103
27. Mohandas N, Evans EA. 1994. *Annu. Rev. Biophys. Biomol. Struct.* 23:787–818
28. Nardin C, Hirt T, Leukel J, Meier W. 2000. *Langmuir* 16:1035–41
29. Norman KE, Katopodis AG, Thoma G, Kolbinger F, Hicks AE, et al. 2000. *Blood* 96:3585–91

30. Peracchia MT, Harnisch S, Pinto-Alphandary H, Gulik A, Dedieu JC, et al. 1999. *Biomaterials* 20:1269–75
31. Rodgers SD, Raymond T, Camphausen RT, Hammer DA. 2000. *Biophys. J.* 79:694–706
32. Schillen K, Bryskhe K, Mel'nikova YS. 1999. *Macromolecules* 32:6885–88
33. Simon SI, Neelamegham S, Taylor A, Smith CW. 1998. *Cell Adhesion Commun.* 6:263–76
34. Spagg DD, Alford DR, Greferath R, Larsen CE, Lee KD, et al. 1997. *Proc. Natl. Acad. Sci. USA* 94:8795–800
35. Swift DG. 1998. *The extent and rate of antigen-antibody mediated adhesion of rat basophilic leukemia cells to surfaces in a linear shear field*. PhD thesis. Univ. Pennsylvania. 340 pp.
36. Swift DG, Posner RG, Hammer DA. 1998. *Biophys. J.* 75:2597–611
37. Swift DG, Posner R, Hammer DA. 2001. *Biophys. J.* In revision
38. Tempelman LA, Hammer DA. 1994. *Biophys. J.* 66:1231–43
39. Won YY, Davis HT, Bates FS. 1999. *Science* 283:960–63
40. Yu K, Eisenberg A. 1998. *Macromolecules* 31:3509–18
41. Yu YS, Zhang LF, Eisenberg A. 1998. *Macromolecules* 31:1144–54
42. Evans E. 1989. *Cell Motil. Cytoskelet.* 14(4):544–51

Subject Index

D

Cumulative Indexes

CONTRIBUTING AUTHORS, VOLUMES 27–31

CHAPTER TITLES, VOLUMES 27–31

Prefatory

Experimental and Theoretical Methods

Preparation, Processing, and Structural Change

Properties and Phenomena

Special Materials

Skutterudites: A Phonon-Glass-Electron-Crystal Approach to Advanced Thermoelectric Materials Research — GS Nolas, DT Morelli, TM Tritt — 29:89–116

Nanocrystalline Diamond Films — DM Gruen — 29:211–59

Structure